T0134100

ARTIFICIAL INTELLIGENCE APPLICATIONS IN ELECTRICAL TRANSMISSION AND DISTRIBUTION SYSTEMS PROTECTION

ARTIFICIAL INTELLIGENCE APPLICATIONS IN ELECTRICAL TRANSMISSION AND DISTRIBUTION SYSTEMS PROTECTION

Edited by
Almoataz Y. Abdelaziz,
Shady Hossam Eldeen Abdel Aleem,
and Anamika Yadav

CRC Press is an imprint of the
Taylor & Francis Group, an **informa** business

First edition published 2022
by CRC Press
6000 Broken Sound Parkway NW, Suite 300, Boca Raton, FL 33487-2742

and by CRC Press
2 Park Square, Milton Park, Abingdon, Oxon, OX14 4RN

CRC Press is an imprint of Taylor & Francis Group, LLC

Library of Congress Cataloging-in-Publication Data
A catalog record has been requested for this book

ISBN: 978-0-367-55234-3 (hbk)
ISBN: 978-0-367-55238-1 (pbk)
ISBN: 978-0-367-55237-4 (ebk)

DOI: 10.1201/9780367552374

Typeset in Times
by MPS Limited, Dehradun

Contents

Preface

Power networks are currently more sophisticated due to growing interconnected systems, flexible alternating current transmission system (FACTS) technology, high voltage direct current (HVDC) transmission lines, distributed generations (DGs), and micro-grids. Therefore, the protection of high voltage AC (HVAC) power transmission and distribution lines is vital to provide these systems' reliability and security. Electromechanical, electronic, digital, numerical, and today's smart relays represent a trend in the field of protection schemes of the power system network. Since the performance of some schemes of the conventional relays has some shortcomings, the new methods such as the pattern-recognition-based methods can improve the schemes' performance. Simultaneously, using meters, PMUs, intelligent electronic devices (IEDs), communication systems, and automation systems make available information about the whole power system. A protective relay to protect the power system networks from abnormal conditions is necessary with an accurate operation and smart functions. The main focus of this book is to consider the application of artificial intelligence (AI) methods in the field of protection of different types and topologies of transmission and distribution lines to cover various aspects of transmission and distribution lines protection from academia and industry perspectives.

AI can successfully help in solving real-world problems in transmission and distribution systems. AI-based schemes are fast, adaptive, and robust and are applicable without any knowledge of the system parameters. Fault detection, fault detection during power swing, high impedance fault detection, fault classification, fault location, and others are examples of AI applications in power systems. These schemes can be used as auxiliary functions in conventional-based methods for the protection purpose. Compared to conventional methods, AI offers new advantages to deal with uncertainties and is a useful tool for solving complex and non-linear problems. Besides, AI-based solutions are good alternatives to make a decision when some data of the system is missing, or some parameters of the system are not available.

In this book, the editors and the authors tried to identify the main challenges of the existing and future generation of transmission and distribution systems while offering new solutions based on AI methods. This book is mainly focused on AI applications in electrical HVAC and HVDC transmission and distribution systems protection, but each of this book's chapters begins with the fundamental structure of the problem required for a basic understanding of the methods described. This book is sorted out and organized into 18 chapters.

Chapter 1: AI encompasses metaheuristic algorithms. The researchers have applied these algorithms to find optimal solutions to many engineering optimization problems. Typically, optimization problems are related to minimizing or maximizing objective functions subject to various constraints. Several metaheuristic algorithms have been proposed and used by researchers in the last three decades to solve engineering optimization problems. A bibliographical assessment and customary background of research in electrical transmission and distribution systems protection is provided. More prominence is accorded to contemporary techniques. An attempt has been made to give the reader an insight into different meta-heuristic algorithms employed for various protection functions in transmission and distribution systems, which can assist power engineers and researchers to establish the suitability of the method for their respective problems.

Chapter 2: In this chapter, a general overview of the distributed generation and its advantages over the conventional power system is given. Further, several technical and operating issues that arise due to the incorporation of DG are also explained. The problems of DG interconnection and its probable solutions have been summarized. Furthermore, applying the novel concept of AI has been presented to sort out the protection issues of DG interconnection. Also, a case study is given, and its results are discussed. It has been thoroughly explained how one of the protection issues (islanding detection) of DG interconnection can be resolved. This chapter will help the readers carry out further research in the area of islanding detection and restoration of protection coordination in the presence of DG by utilizing various AI-based techniques.

Chapter 3: In this chapter, an Ensemble Tree Classifier (ETC) model-based fault classification scheme has been proposed to classify all types of shunt faults, such as commonly occurred shunt faults (CSFs), cross-country faults (CCFs), and evolving faults (EVFs). In this proposed ETC-model, standard deviations of Discrete Wavelet Transform (DWT) coefficients of full-cycle data of locally measured three-phase current signals are used as an input to erudite different decision tree modules such as Bagged Decision Tree (BGDT), AdaBoost Decision Tree (ABDT), and RUSBoost Decision Tree (RBDT). A practical 400 kV power transmission network of Chhattisgarh state has been modeled and simulated in MATLAB®/Simulink software to employ the proposed fault classification scheme. Complete data sets have been presented at a wide range of fault scenarios, thereby applying an illustrative signal processing technique such as DWT. Further, the performance assessment has been carried out by comparing different performance metrics such as accuracy, sensitivity (or) true positive rate (TPR), and specificity (or) true negative rate (TNR), etc. The simulation results revealed the proposed ETC-model adaptability for classifying all typical and atypical shunt faults that need to be cleared as early as. The performance metrics attained by the proposed ETC-model for classification of CSFs/CCFs/EVFs is appreciable, and it gives a research insight to adapt the same in practical power transmission network to improve reliability.

Chapter 4: Detection and classification of faults on the transmission lines are affected remarkably by many factors that contaminate the phase voltages and line currents signals measured at the relay point and the errors caused by the circuit elements. The protection system should respond to the faults efficiently despite these factors and should be able to discriminate between permanent faults and the other abnormalities of the power system. That's why there is a persistent need to use powerful techniques to ensure the distance relay's accurate operation. In this chapter, different techniques are presented to help develop the distance protection system and provide accurate detection and classification of faults within a very short period, improving the efficiency of the protection system in protecting the transmission lines.

Chapter 5: This chapter aims to visualize the most recent developments for fault location schemes for transmission and distribution systems using AI tools. Today, AI represents an ideal alternative for performing those complex and unsolved problems that can not be processed using conventional tools. First, traditional methodologies of fault location are shortly reviewed. Then, contributions with AI for fault location purposes are addressed. Finally, smart grid requirements and their contributions for fault location tasks are visualized.

Chapter 6: This chapter presents an integrated approach for fault detection, classification, and location in medium voltage underground cables. The integrated approach used a three-phase current measured at a single terminal of the system, which was modeled using autoregressive modeling (AR) signal modeling. All three stages of the protection scheme used AR coefficients extracted from the three-phase current. The fault detection approach detected the fault using the critical pole calculated from the AR coefficients. The fault classification approach used the extreme learning machine (ELM) binary classifiers and ground index to recognize the fault type. The fault location approach used ELM in regression mode to find the fault distance. The training process of fault classification and location ELMs helped in fixing the structure of ELMs for the integrated approach. Trained ELMs are tested to evaluate the performance metrics of the integrated approach. Test results confirm the robustness of the proposed integrated approach to protecting the medium voltage underground cable. The comparative evaluation shows the proposed integrated approach outperforms the other schemes considered. The proposed integrated approach also provides reliable data to the power system operator during fault to restore the supply with less outage time.

Chapter 7: This chapter presents the deep learning neural network technique to detect and classify high impedance faults (HIFs). Relays are based on a novel low-frequency diagnostic vector, third and fifth harmonic function. The currents and voltages signals are fed to the deep learning neural network for fault discrimination, with a deep neural network trained to satisfy the third and fifth harmonic image of the voltage and current, differential operator, initial condition boundary conditions. This chapter aims to design a robust deep learning neural network-based smart relay, predicting and classifying the high impedance of low-current defects in radial electrical delivery systems. Varietics of faults and system conditions were simulated to determine the reliability and sensitivity of the proposed procedure. The achieved results show the validity of the developed approach.

Chapter 8: In this chapter, applying different AI techniques such as artificial neural network, fuzzy logic, support vector machine, and a combination of AI-based techniques with other transient tools like Wavelet to protect a multi-terminal transmission line is discussed. These techniques primarily develop a statistical model and train it with a database representing a wide variation of the fault and other abnormalities. These techniques' primary task for three-terminal line protection includes fault detection, faulty section identification/classification, and fault location.

Chapter 9: This chapter thoroughly addresses the additional critical protection issues observed in the compensated transmission network and describes multiple data mining-based protection methodologies. The described procedures can detect the normal shunt and transform fault events irrespective of varying network topologies and fault conditions. Recently, wide-area measurement devices are inevitably implemented over the grid, which generates a large measurement dataset. Therefore, compared with mathematical network modeling based protection methods, advanced data mining based protection methodologies are more pertinent and competent in providing adaptive protection to the compensated transmission network. The faster response, lesser complexity, and better competency of handling large measurement data set make the data-mining-based protection schemes more superior and adaptive than network-modeling-based methods.

Chapter 10: Different AI-based algorithms have been recommended in the literature in the last three decades to protect FACTS compensated transmission line. AI-based algorithms are suitable for the non-linear and complex problem. A variety of machine learning and AI techniques have been presented for fault analysis in FACTS compensated transmission lines. In this chapter, a comprehensive review of different AI-based algorithms and feature extraction techniques is given to protect the transmission line incorporating FACTS controller. This exhaustive review of literature in the FACTS-based transmission line will help the researchers begin their work and choose a suitable method for the respective power system network. Lastly, the new status with a recommendation towards the future direction of research for FACTS compensated transmission line has also been presented.

Chapter 11: This chapter introduces the principles of phasor measurement units (PMUs) and wide-area monitoring systems (WAMS) used to protect the transmission networks. The PMUs-based WAMS perform three jobs; data collection and acquisition, data transportation and monitoring, control, and protection. PMUs perform the first job, and their communication infrastructure provides the second job. The last job is provided by the energy management system that includes control and protection actions. The conventional mathematical methods and advanced AI used to allocate the PMUs and WAMS optimally are presented, and a comparison is conducted. A case study of finding the minimum number of PMUs with full system monitoring using an advanced AI technique is introduced. Finally, the modern applications of PMUs and WAMS in protecting the power systems are covered in this chapter.

Chapter 12: This chapter introduced a novel expert system (ES) for optimal coordination of directional overcurrent relays (DOCRs) in meshed networks, in which the coordination knowledge is represented in a rule style. The introduced ES has inherent immunity from trapping in a local minimum. Moreover, it has got rid of miscoordination, which may arise due to rounding off the time delay setting values. Hence, it is suitable for coordinating electromechanical and digital DOCR, off-line and on-line, respectively. Extensive simulation of the introduced ES proved its efficiency in setting the DOCR optimally.

Chapter 13: The network of the electrical system is one of the most complex structures that has been developed by human civilization. The complexity of the operation of power systems is rising as the electric system is rapidly evolving. Maintaining power system stability and preserve network reliability is the main task for protective relays. The main objective of protective relays is to isolate and disconnect only fault sections rapidly and keep healthy operation sections. In power system protection, DOCRs are widely applied in distribution and sub-transmission systems. The problem of coordination is considered a non-linear problem, and this problem contains many constraints. In this chapter, the concern of coordination is expressed using various optimization techniques. The water cycle algorithm (WCA) and its modified-WCA (MWCA) are suggested to find a coordination solution. Moth-flame optimization (MFO) and improved-MFO (IMFO) are also presented to find a solution for coordination in the conventional and non-conventional characteristic relay curve. The suggested techniques' robustness and effectiveness are assessed using different networks and compared with other and recent optimization algorithms. The results prove the superiority of the proposed methods in solving DOCRs coordination problems without any miscoordination between relay pairs. Also, the reduction in the total operating time of primary relays reaches more than 59% with the usage of the non-standard characteristic relay curve.

Chapter 14: Although DC microgrid offers numerous operational and potential benefits compared to AC microgrid, it cannot be protected by AC existing protective approaches and infrastructures. Implementing a DC microgrid cannot be realized without designing an effective protection system, which plays an integral part in the DC microgrid system design. To face associated challenges with DC microgrid protection, first, DC fault current in two main stages of capacitor discharge stage and freewheeling diodes stage is investigated. Second, technical issues that should be considered in designing a DC microgrid are described. Third, the grounding structure can also affect the safety of the microgrid. It significantly impacts the protection schemes' ability to detect the grid's fault and survivability under a faulty situation. Hence, five grounding structure, including IT, TN-C, TN-S, T-N-C-S, and TT, and different grounding strategies (ungrounded, solidly grounded, thyristor grounded, and diode ground), are discussed and compared. The intrinsic difference between AC and DC fault current necessitates the utilization of protective devices (PDs) that comply with DC fault current requirements. The structure and operation principles of five classes of PDs are described and compared from different aspects, including cost, losses, response time, and size. Two main categories of fault detection and isolation methods, including unit and non-unit protection with their subcategory studies, are discussed in detail. Although each of the investigated methods has specific advantages and disadvantages, most of them are evaluated directly on grounded radial or ring configurations under low fault impedance. Hence, identifying faults on the meshed type configurations with high impedance fault need more sophisticated schemes. As a result, to achieve such a sophisticated protection scheme, it should provide definite protection. AI-based protection schemes can satisfy all DC microgrids' protection requirements. Feature extraction techniques are the first candidates to enhance the performance of AI-based protection. A protection system based on the AI will be realized by extracting desired features and then training the ANN with efficient chosen parameters.

Chapter 15: Shunt active power filter (SAPF) is a potent power quality conditioner due to its multi-functionality nature and reduced inverter rating requirement. In this chapter, a soft computing-based DC-link voltage control technique for SAPF has been proposed to improve harmonic compensation performance. Soft computing, i.e., fuzzy logic-based controller, has been designed in such a way to decline the harmonics due to load perturbation and mismatch of system parameters. Furthermore, a discontinuous space vector pulse width modulation (DSVPWM) has been utilized for switching pattern generation. A control approach based on the hybrid synchronous reference frame (HSRF) method has been employed to assist in reference generation. The simulation and experimental results show the potency and reliability of the SAPF with the proposed control approach.

Chapter 16: Among the protection engineers, HVDC grid protection is one of the most challenging issues allocating a large proportion of new researches to itself. HVDC grid protection corresponds to the conventional AC grid and accomplishes the same objectives. But there are considerable differences in practice, such as the high rate of rising of current due to the lack of line inductance and the absence of neutral zero crossing. Such differences lead to a specific breaker technology, which makes the HVDC grid protection scheme more complicated. Therefore, establishing a fast, intelligent protection scheme to clear fault current within 3–5 ms is necessary, one-tenth of the allowed clearing time for the AC protection system. In addition, a wide variety of contingencies would affect the detection process arising in several factors like grid topology, severe transient discharge, noise, the effect of output DC filter, grounding topology, etc. Thus, an AI-based protection scheme for HVDC applications is inevitable to meet the requirement and limitations. In this chapter, HVDC protection challenges are introduced. Afterward, the DC fault phenomenon is investigated to evaluate the fault current contribution and its major influencing parameters. Ultimately, a comparative study of the AI-based fault detection, classification, and location schemes in HVDC applications is presented, considering inputs, extracted features, sampling frequency, test systems, and result's accuracy. This chapter may be viewed as a guideline for future researchers to be acquainted with different techniques.

Chapter 17: With the widening application of HVDC systems, related issues such as their protection requirements have attracted much attention. Indeed, the various technologies employed in HVDC systems have their particular complexities and protection challenges. Therefore, it is essential to design appropriate fault detection, classification, and location schemes with specific consideration for each technology. The pattern-recognition and machine-learning techniques can play a highly influential role in developing such plans for HVDC systems and overcome the complexities and challenges by relying on their inherent capabilities. In this chapter, after briefly stating the advantages and applications of HVDC systems, current-source converter-based HVDC (CSC-HVDC) and voltage-source converter-based HVDC (VSC-HVDC) systems are introduced, and their specific protection challenges are described. Next, the performed studies and achieved advances in fault detection, classification, and location in CSC-HVDC and VSC-HVDC systems are reviewed with specific focus given to the applied pattern-recognition and machine-learning procedures. The related intelligent schemes are discussed considering two main aspects, including the extracted/selected input features and the employed learning algorithms/models.

Chapter 18: Comprehensive exploration of MT-HVDC systems is generally initiated because of its applications of bulk power transfer capability, the non-synchronized interconnection of AC grids, adequate controllability, and high efficiency. Control and protection of DC grids are the emerging explorations. Because of the abrupt building up of DC fault current, fault diagnosis is the bottom line of protection of MT-HVDC systems. Much research is conducted to develop new techniques and modify existing techniques based on speed and accuracy so that the harmful effects of the sudden rise of DC fault current could be limited. Machine learning is found to be highly applicable in fault classification and location. In this chapter, support vector machines-based learning approaches for fault classification and location are developed and implemented for four-terminal HVDC systems. Normal and faulty test system states are classified, and fault location is identified based on the frequency domain-based training and testing data employed for support vector machines.

Finally, this book aims to introduce good practice with new research outcomes and ideas that present different AI applications in electrical HVAC and HVDC transmission and distribution systems protection from various perspectives. It is a useful tool for the network planners, designers, operators, and practicing engineers of modern power systems concerned with this topic.

Almoataz Y. Abdelaziz,
Cairo, Egypt;
Shady Hossam Eldeen Abdel Aleem,
Cairo Egypt;
Anamika Yadav,
Raipur, CG, India
July 20, 2021

Editors

Dr. Almoataz Y. Abdelaziz (SM'15) received B.Sc. and M.Sc. degrees in electrical engineering from Ain Shams University, Cairo, Egypt, in 1985 and 1990, respectively, and a Ph.D. degree in electrical engineering according to the channel system between Ain Shams University, Egypt, and Brunel University, U.K., in 1996. He has been a professor of electrical power engineering with Ain Shams University since 2007. He was the vice dean for Education and Students Affairs in Faculty of Engineering and Technology, Future University in Egypt from 2018–2019. He has authored or coauthored more than 450 refereed journal and conference papers, 30 book chapters, and 6 edited books with Elsevier, Springer, and CRC Press. In addition, he has supervised 80 master's and 35 Ph.D. theses. His research areas include the applications of artificial intelligence and evolutionary and heuristic optimization techniques to power system planning, operation, and control. Dr. Abdelaziz is a senior member in IEEE and member in the Egyptian Sub-Committees of IEC and CIGRE. He has been awarded many prizes for distinct researches and for international publishing from Ain Shams University and Future University in Egypt. He is the chairman of the IEEE Education Society chapter in Egypt. He is a senior editor of *Ain Shams Engineering Journal*, an editor of *Electric Power Components and Systems* journal, an editorial board member, an editor, an associate editor, and an editorial advisory board member for many international journals.

Emails: almoatazabdelaziz@hotmail.com, ayabdelaziz63@gmail.com

Dr. Shady Hossam Eldeen Abdel Aleem received B.Sc., M.Sc., and Ph.D. degrees in electrical power and machines from the Faculty of Engineering, Helwan University, Egypt, in 2002, and the Faculty of Engineering, Cairo University, Egypt, in 2010 and 2013, respectively. From September 2018 to September 2019, he has been an associate professor at the 15th May Higher Institute of Engineering and the quality assurance unit director. Since September 2019, he has been an adjunct associate professor in the Arab Academy for Science, Technology & Maritime Transport, College of Engineering and Technology, Smart Village Campus for teaching power quality energy efficiency, wind energy, and energy conversion courses. Also, he is a consultant of power quality studies in ETA Electric Company, Egypt. His research interests include harmonic problems in power systems, power quality, renewable energy, smart grid, energy efficiency, optimization, green energy, and economics. Dr. Shady is the author or co-author of many refereed journals and conference papers. He has published 100 plus journal and conference papers, 18 plus book chapters, and 7 edited books with the Institution of Engineering and Technology (IET) (2), Elsevier (3), Springer (1), and InTech (1). He was awarded the State Encouragement Award in Engineering Sciences in 2017 from

Egypt. He was also awarded the medal of distinction from the first class of the Egyptian State Award in 2020 from Egypt. Dr. Shady is a member of the Institute of Electrical and Electronics Engineers (IEEE). Dr. Shady is also a member of the Institution of Engineering and Technology (IET). He is an editor/associate editor for the *International Journal of Renewable Energy Technology, Vehicle Dynamics, IET Journal of Engineering, Technology and Economics of Smart Grids and Sustainable Energy*, and *International Journal of Electrical Engineering Education*.

Emails: engyshady@ieee.org; engyshady@gmail.com; engy-shady@hotmail.com

Dr. Anamika Yadav (SM'2014) received a bachelor of engineering degree in electrical engineering from RGPV, Bhopal, in 2002; a master of technology degree in integrated power system from V.N.I.T., Nagpur, India, in 2006; and a Ph.D. degree from CSVTU, Bhilai, India, in 2010 at National Institute of Technology (NIT) Raipur as research center. She was an assistant engineer with the Chhattisgarh State Electricity Board, Raipur, from 2004 to 2009. She served as an assistant professor in the Department of Electrical Engineering, NIT Raipur, during 2009–2018. Presently she is an associate professor with the Department of Electrical Engineering, NIT Raipur, Raipur, since 2018. Additionally, she is also serving as associate dean (Research and Consultancy) at NIT Raipur since 2018. She has supervised 7 PhD thesis and 22 M.Tech thesis. Her research interests include digital protection and automation, smart grid technologies and applications, distributed generation, micro-grid, and application of artificial intelligence, power system protection, FACTS technologies, etc. She received the Institution of Engineers India Young Engineers Award in the Electrical Engineering Division during 2015 to 2016, VIFFA young faculty award 2015, VIFRA young scientist award 2015, and Chhattisgarh young scientist award in 2016 given by Chief Minister of CG. She is also a senior member of IEEE from 2014. She has more than 120 publications in International journals and conferences and around 1,800 citations as per Google scholar.

Emails: ayadav.ele@nitrr.ac.in, anamikajugnu4@gmail.com

Contributors

Almoataz Y. Abdelaziz
Faculty of Engineering and Technology
Future University in Egypt
Cairo, Egypt

Abdelazeem A. Abdelsalam
Electrical Engineering Department
Suez Canal University
Ismailia, Egypt

Dalia Allam
Department of Electrical Engineering
Faculty of Engineering
Fayoum University
Fayoum, Egypt

Valabhoju Ashok
Department of Electrical Engineering
National Institute of Technology
Raipur, C.G., India

M. H. Awadalla
Department of Electrical and Computer Engineering
Sultan Qaboos University, SQU
Oman

Prashant Bedekar
Department of Electrical Engineering
Government College of Engineering
Amravati, Maharashtra

Bhavesh Kumar R. Bhalja
Department of Electrical Engineering
Indian Institute of Technology Roorkee
Uttarakhand, India

M. M. Eissa
Department of Electrical and Power Engineering
Faculty of Engineering at Helwan
Helwan University

Nagy I. Elkalashy
Electrical Engineering Department
Faculty of Engineering
Menoufia University
Shebin El-Kom, Egypt

Mahmoud A. Elsadd
Electrical Engineering Department
Faculty of Engineering
Menoufia University
Shebin El-Kom, Egypt

Mohammad Farshad
Department of Electrical Engineering
Faculty of Basic Sciences and Engineering
Gonbad Kavous University
Gonbad Kavous, Iran

Ammar Adnan Hajjar
Department of Electrical Engineering
Mechanical and Electrical College
University of Tishreen
Latakia, Syria

Karim M. Hassanin
Electrical Engineering Department
Suez Canal University
Ismailia, Egypt

Amir Imani
Faculty of Electrical and Computer Engineering
Semnan University
Semnan, Iran

Francisco Jurado
Department of Electrical Engineering
University of Jaén
Jaén, Spain

Vijay Kale
Department of Electrical Engineering
Visvesvaraya National Institute of Technology
Nagpur, Maharashtra, India

Salah Kamel
Department of Electrical Engineering
Faculty of Engineering
Aswan University
Aswan, Egypt

M. Karthikeyan
Department of Electrical and Electronics Engineering
Velammal Engineering College
Chennai, Tamil Nadu, India

Tamer A. Kawady
Electrical Engineering Department
Faculty of Engineering
Menoufia University
Shebin El-Kom, Egypt

Ahmed Korashy
Department of Electrical Engineering
Faculty of Engineering
Aswan University
Aswan, Egypt

Bhupendra Kumar
Department of Electrical Engineering
Vishwavidyalaya Engineering College
Lakhanpur, Chhattisgarh, India

Yogesh M. Makwana
Department of Electrical Engineering
Government Engineering College Dahod
Gujarat, India

Zahra Moravej
Faculty of Electrical and Computer Engineering
Semnan University
Semnan, Iran

Raheel Muzzammel
Department of Electrical Engineering
University of Lahore
Lahore, Punjab, Pakistan

Loai Nasrat
Department of Electrical Engineering
Faculty of Engineering
Aswan University
Aswan, Egypt

Mohammad Pazoki
School of Engineering
Damghan University
Damghan, Iran

Pravat Kumar Ray
Department of Electrical Engineering
National Institute of Technology
Rourkela, India

Ali Raza
Department of Electrical Engineering
University of Lahore
Lahore, Punjab, Pakistan

R. Rengaraj
Department of Electrical and Electronics Engineering
Sri Sivasubramaniya Nadar College of Engineering College
Kalavakkam, Kancheepuram District, Tamil Nadu, India

R. K. Saket
Department of Electrical Engineering
Indian Institute of Technology, BHU
Varanasi, India

Morteza Shamsoddini
Department of Electrical Engineering
Amirkabir University of Technology
Tehran, Iran

A. M. Sharaf
Energy Systems Inc
Fredericton, NB, Canada

S. K. Singh
Department of Electrical Engineering
Shri Ramswroop Memorial University
Lucknow, India

Sushree Diptimayee Swain
Department of Electrical Engineering
O.P. Jindal University
Raigarh, Chhattisgarh, India

Behrooz Vahidi
Department of Electrical Engineering
Amirkabir University of Technology
Tehran, Iran

D. N. Vishwakarma
Department of Electrical Engineering
Indian Institute of Technology, BHU
Varanasi, India

Anamika Yadav
Department of Electrical Engineering
National Institute of Technology
Raipur, Chhattisgarh, India

1 Application of Metaheuristic Algorithms in Various Aspects of Electrical Transmission and Systems Protection

Vijay Kale[1], *Anamika Yadav*[2], *and Prashant Bedekar*[3]

[1]Department of Electrical Engineering, Visvesvaraya National Institute of Technology, Maharashtra, India

[2]Department of Electrical Engineering, National Institute of Technology, Chhattisgarh, India

[3]Department of Electrical Engineering, Government College of Engineering, Maharashtra, India

1.1 INTRODUCTION

A problem of finding the best possible solution from the set of feasible solutions is an optimization problem. Metaheuristic algorithms, which form a significant part of Artificial Intelligence (AI), have been used by the researchers to find optimal solutions to various problems in the engineering domain. During the activities like designing, constructing, or maintaining an engineering system, the engineers are required to take many scientific and management decisions. These decisions should result in maximizing the desired benefits. Optimization, in its broadest sense, can be applied to solve any engineering problem. Optimization in electrical engineering areas is crucial to support the complex task of efficiently generating, transmitting, distributing and utilizing electricity. The electrical transmission and distribution systems protection is one such area of power system where optimization is needed. Coordinating overcurrent relays in distribution system optimally, placing PMUs at substations optimally, estimating fault sections in distribution systems, and finding fault location on transmission lines are the optimization problems considered in this chapter.

DOI: 10.1201/9780367552374-1

Some of the applications in Electrical Engineering suitable for applying optimization techniques include:

1. Design of transformers [1], motors [2], and generators [3]
2. Design of control systems [4]
3. Unit commitment for thermal power plants [5]
4. Load forecasting [6]
5. Transmission expansion planning [7]
6. Reactive power planning [8]
7. Relay coordination [9–18]
8. PMU placement [19–25]
9. Estimation of fault section on distribution network [26–36]
10. Estimation of fault location on transmission lines [37–45]

Researchers have applied various conventional as well as AI-based techniques as a means of effectively dealing with these optimization problems. This chapter discusses various research problems [(7) to (10) in the list above] in the area of electrical transmission and distribution systems protection. Each problem will be defined briefly, highlighting its significance. The problem definition will be followed by the mathematical formulation of the problem, its illustration, and various researchers' methodologies to solve it. There are many metaheuristics algorithms proposed and used by the researchers in the last three decades to solve the variety of optimization problems. Some popular algorithms like the Genetic Algorithm, Differential Evolution technique, Simulated Annealing method, Particle Swarm Optimization method, Firefly algorithm, Cuckoo search technique, and Grey Wolf optimizer will be discussed in brief.

The domain knowledge of an engineer goes to formulate the engineering problem as an optimization problem. The optimization problem is represented mathematically in a certain way, as given below.

1.2 MATHEMATICAL REPRESENTATION OF OPTIMIZATION PROBLEM

Typically, an optimization problem is related to minimizing or maximizing objective functions subject to variety of constraints. The generic mathematical form of optimization problem is given below.

$$\text{Optimize} \quad z = f_i(x), \quad (i = 1, 2, \ldots \ldots M) \tag{1.1}$$

$$\text{subjectto} \quad g_j(x) = 0 \quad (j = 1, 2 \ldots \ldots ..J) \tag{1.2}$$

$$h_k(x) \leq 0 \quad (k = 1, 2 \ldots \ldots K) \tag{1.3}$$

where $f_i(x)$, $g_k(x)$, *and* $h_j(x)$ are the functions of decision vector:

$$\boldsymbol{x} = (x_1, x_2, \ldots \ldots x_n);$$

The elements of design vector $\boldsymbol{x}$ are termed as design variables or decision variables. Space spanned by design variables is called search space. The function $f_i(x)$ is an objective function to be optimized over the n-variable vector, $\boldsymbol{x}$. It is also called as cost function or loss function, or error function. These terms are synonymous. However, the terms *objective function* or *cost function* is used more in the optimization problem. Similarly, the terms *loss function* or *error function* is used in the estimation problem. The functions $g_i(x)$ represent equality constraints, while functions $h_j(x)$ represents inequality constraints.

Single objective optimization problems have only one objective function, while most of the real-world optimization problems are multi-objective function problems. Un-constrained optimization problems have no constraints. Optimization problems may have equality and inequality constraints. Linear programming problem (LPP) has all the objective functions and constraints that are linear functions of the decision variables. Nonlinear programming problem (NLPP) has one or more objective functions and constraints that are non-linear functions of the decision variables. It may be noted that the word "programming" means planning or optimization, and it is not related to computer programming.

The following sections will describe some of the metaheuristic algorithms briefly, followed by their applications to optimization problems in the domain of electrical transmission and distribution systems protection.

1.3 METAHEURISTIC ALGORITHMS

Metaheuristic algorithms are considered as a part of artificial intelligence. There are a number of ways to classify these algorithms. One of the ways to classify metaheuristic algorithms is given below.

1. *Population-based algorithms:* These methods repetitively modify a group of individual solutions. The population advances toward an optimal solution over successive iterations. GA is an example of population-based algorithms.
2. *Trajectory-based algorithms:* These techniques use a single point (solution) that traces a trajectory in the search space. These methods have non-zero probability to reach the global optimum. A Simulated Annealing algorithm is a trajectory-based method.

Conventional gradient-based methods, and other deterministic search methods, start with a guess and tries to find the optimum solution iteratively. For unimodal objective functions, methods converge to the global optimum. However, for multimodal objective functions, the search may get trapped at the local optimum. To overcome this problem, researchers proposed metaheuristic algorithms with components like intensification and diversification. Diversification makes it possible to explore the search space globally so that solution does not get trapped in local optima. Intensification is related to focusing on a search in a local region to obtain the current worthy solution. When used in combination with the selection of best solutions, these two components ensure that the solutions will meet to optimality.

Metaheuristic algorithms founded on some "intelligent" biological behavior of animals are called as "Nature Inspired Metaheuristic Algorithms." These methods search large spaces for candidate solutions. They begin from a set of points and approach towards a better solution guided by heuristic and empirical rules. Genetic algorithm [46], the nature inspired, population-based metaheuristic algorithm was the first to be developed. It set the stage for more metaheuristic algorithms to be developed. Some of them are described briefly below.

John Holland developed Genetic Algorithm (GA). It is a perception of biological evolution founded by Charles Darwin's theory of natural section. GA is an iterative optimization technique that recurrently applies operators such as selection, crossover, and mutation to a group or population of solutions until some convergence criterion has been satisfied.

Researchers R. Storn and K. Price modified GA and called it the Differential Evolution (DE) algorithm. In this algorithm, solutions are treated as real numbers obliviating the need for encoding and decoding required in GA. Also, mutation plays a prominent role in DE as compared to GA.

Simulated Annealing (SA) algorithm mimics the annealing process in material processing. Annealing involves heating a metal or glass material and cooling it in a controlled manner to reduce its structure defects. The algorithm starts with the initial solution and initial temperature. During each iteration, the next solution is obtained by the Random Walk technique, and the temperature is gradually reduced. Random Walk refers to the random process, which comprises a series of consecutive random steps whose step size can be fixed or varying. Next, solution is accepted if it is better. If it is a worse solution, it is accepted with some probability. This avoids the possibility of a solution being trapped in local minima.

Particle Swarm Optimization (PSO) technique is based on the group behavior of fish or birds. PSO searches the solution space of an objective function by adjusting particles' trajectories as the piecewise paths. The algorithm starts with initializing the locations (solutions) and velocities of all particles and choosing the particle with the best fitness value as the initial global best. The velocity and location of each particle are updated during every iteration to find its current best. At the end of each iteration, it gives the current global best. The procedure is reiterated until the stopping criterion is attained.

Firefly algorithm is a kind of swarm intelligence algorithm motivated by the flashing behavior of fireflies. Xin-She Yang proposed it at the University of Cambridge. During each iteration, each firefly moves towards a more attractive firefly. These updates of the locations of fireflies (solutions) come to an end when the termination criterion is achieved.

The cuckoo search algorithm is based on the obligate brood parasitism of some cuckoo species, whereby they exploit a suitable host to increase their population. A fraction of worse nests (solutions) are abandoned, and new nests are generated using the Levy Flight technique, and the current best nest is found during each iteration.

Levy Flight refers to the random process, which consists of taking a series of consecutive random steps whose step length obeys the levy distribution. This iterative procedure continues until the stopping condition becomes true.

The Grey Wolf Optimizer (GWO) algorithm proposed by Mirjalili et al. in the year 2014 imitates the management hierarchy and shooting mechanism of gray wolves. For simulating the leadership hierarchy, four types of grey wolves are employed, such as alpha, beta, delta, and omega. Furthermore, other important steps of searching for prey, encircling prey, attacking prey, and hunting are employed to perform optimization.

The optimization process in most stochastic algorithms is repeated until some stopping criterion is attained. Possible stopping criteria are:

- When the objective function value becomes smaller than the threshold set ε.
- When the change in the objective function value is smaller than Δ ε.
- When the maximum number of iterations has reached
- When the maximum number of iterations has reached without improvement of the objective function f(x).

The following points may be noted regarding metaheuristic algorithms.

- Researchers keep on proposing the modifications to popular metaheuristic algorithms. Therefore, many variants of the same algorithm may exist. For example, the Genetic Algorithm is not a single algorithm but a family of algorithms.
- The parameters of most of the optimization algorithms are set or specified first. For example, the choice of factors of GA such as population size, selection criterion, crossover probability, and mutation rate need careful consideration.
- It is generally accepted that universally best optimization algorithm does not exist. In other words, there is no single algorithm that performs equally well for all problems.
- Metaheuristics algorithm may not be the first choice for a given problem. They can be used when conventional method fails. Many researchers have combined the use of conventional and metaheuristic algorithms to apply in the field of transmission and distribution system protection. The choice of algorithm for a given problem depends on many factors including type of problem, consideration for advantages and limitations of algorithm and expertise of the researcher.

Some of the applications of these metaheuristic algorithms in the area of electrical transmission and distribution systems protection are discussed in the following sections.

1.4 OPTIMAL RELAY COORDINATION

The most commonly used primary protection for distribution systems is overcurrent relaying. It is a backup protection for transmission systems. Trippings of the relays

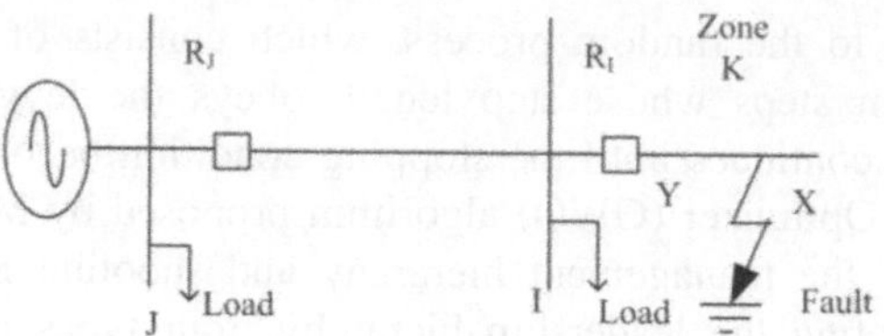

FIGURE 1.1 Radial distribution system.

are frequently due to improper settings as compared to faults. Relays mal-operate due to improper coordination between primary and backup relay pairs. Therefore, in order to maintain the selectivity, coordination between primary and backup pairs of relays is required.

With reference to a textbook system [47] as shown in Figure 1.1, the relay coordination problem may be stated as: Given the magnitudes of all the loads and the fault currents at all the buses, how to set the Over-Current Relays (OCRs) at buses "I" and "J" so that the entire feeder gets over-current protection arranged as primary and backup protection. The relay R_I provides the primary protection for faults in zone k. Relay R_J gives backup or secondary protection to faults in zone k. There are two settings for the OCR viz. Time Multiplier Setting (TMS) and Plug Setting (PS).

Generally, load flow and short circuit studies have to be carried out by writing programs or using simulation software before coordinating over-current relays. The typical steps in coordinating OCRs are as follows:

1. Select the ratios of the CTs. The rating of the relay current coil decides the secondary current of the CT. Similarly, the maximum load current carried by the CT decides the its primary current.
2. The TMS of tail end relay is assigned a value of 0.1 (or minimum setting value available), and its PS is set at 1.
3. Adjust pickup of the backup relay in such a way that it should pick up for a value less than for a short circuit at the far end of the adjoining line section (X) under minimum generation conditions.
4. In order to obtain backup for a fault beyond the breaker in the adjoining line (Y), adjust the time delay (TMS) of the backup relay. Near end fault (maximum fault current) is considered. This is because if the selectivity is obtained under such conditions, it is sure to be obtained for lower currents.

However, systems may not always be radial and may have loops; therefore, either break points have to be solved or optimization procedures need to be used. The previous manual procedures can be applied only to smaller systems and they do not give optimal results.

1.4.1 Formulation of Relay Coordination Problem

The coordination of relays has been stated as a minimization problem [9]. In an optimal relay coordination problem, the summation of time of operation of all the primary relays is minimized

$$\text{Minimize: } z = \sum_{i=1}^{n} \sum_{k=1}^{l} w_{i,k} t_{i,k} \tag{1.4}$$

where, $w_{i,k}$ represents the probability of a given fault that may occur at k[th] location of the protective zones and $t_{i,k}$ is the operating time of ith relay for the fault at k[th] location. All of the weights in (1.4) can be considered as 1. This is on the basis of the assumption of equal probability of fault occurrence at all locations of protective zones. If we consider only one fault location for each line, (1.4) is reduced to equation (1.5).

$$\text{Minimize: } \quad z = \sum_{i=1}^{n} t_i \tag{1.5}$$

Here "z" is the objective function to be minimized, "*i*" denotes the relay number, and "*n*" is the total number of relays. The time taken by the primary relay to operate is t_i. Other forms of the objective function have also been proposed in the literature. The objective function is subjected to the following constraints.

a. *Coordination Constraint:* Coordination is done by the condition,

$$t_j - t_i \geq CT \tag{1.6}$$

where, t_j is the time of operation of the backup relay and t_i is time of operation of the primary relay. CTI is Coordination Time Interval, which generally falls in the range 0.1–0.5s. This time is dependent on how fast the circuit breaker operates.

b. Relay Operating Time Bounds
Relays require certain minimum time to operate. Also, they must not be too slow. This gives rise to the constraint,

$$t_{i\,\min} \leq t_i \leq t_{i\,\max} \tag{1.7}$$

where, $t_{i,min}$ and $t_{i,max}$ are the minimum and maximum operating time of the ith relay, respectively. These are defined in accordance with protection scheme requirements.

c. Constraints on Time Multiplier Setting (TMS)
The TMS is constrained to lie within the settings provided by the relay manufacturer.

$$TMS_i^{\min} \leq TMS_i \leq TMS_i^{\max}. \tag{1.8}$$

where TMS_i^{max} and TMS_i^{min} are the maximum and minimum values of TMS of the ith relay, respectively. Generally, the range of TMS is between 0.1–1.1 in steps of 0.1.

d. Constraints on Plug Setting
The Plug Setting of the relay should be kept at a value more than normal load current but less compared to the minimum fault current

$$PS_i^{min} \leq PS_i \leq PS_i^{max}. \quad (1.9)$$

where PS_i^{min} and PS_i^{max} are the least and highest values of PS_i of the relay R_i. In most research papers the range of PS is chosen between 0.5 and 2.5 in steps of 0.5.

e. Constraint Due to Operating Characteristics of Relays

There are various operating characteristics of OCR. Inverse Definite Minimum Time (IDMT) relay has characteristics given by the following equation

$$t_i = \frac{0.14 * (TMS)}{(I_{relay}/PS)^{(0.02)} - 1} \quad (1.10)$$

where I_{relay}= I/CT ratio and t_i denotes time of operation of the relay.

In general, the relay coordination problem is nonlinear with TMS and PS as design variables. Objective function of (5) is modified by substituting the value of t_i from (1.10). The relay coordination problem as NLPP can be given as:

$$\text{Minimize} z = \sum_{i=1}^{n} \frac{0.14 * (TMS_i)}{(I_{relay}/PS_i)^{(0.02)} - 1} \quad (1.11)$$

subject to the constraints given by (1.6), (1.7), (1.8), and (1.9).

However, with TMS as design variable and PS as constant, the relay operating time varies linearly with the TMS, and the problem becomes linear programming problem (LPP).

In the linear case, as PS is constant, (1.10) becomes:

$$t_i = a_i(TMS) \quad (1.12)$$

where a_i is now a constant value and given by the following equation.

$$a_i = \frac{0.14}{PSM^{(0.02)} - 1} \quad (1.13)$$

Therefore, relay coordination problem as LPP can be written as follows

$$\text{Minimize:} \quad z = \sum_{i=1}^{N} a_i(TMS)_i \quad (1.14)$$

subject to the constraints given by (1.6), (1.7), (1.8), and (1.9).

1.4.2 Illustrative Example

Figure 1.2 shows a single end fed system with parallel feeders with four relays. This system is considered to illustrate the formulation of the relay coordination problem as an NLP problem.

Here, the fault on any of the parallel lines is fed not only from the faulted line but also from the healthy line. Relays 1 and 4 can be non-directional but relays 2 and 3 must be directional to implement the zones of protection, as shown in Figure 1.2. Two fault points, A and B, are considered. Relay 1 is providing backup protection to relay 3 for fault B. Similarly, relay 4 is providing backup protection to relay 2 for fault A. Let the currents seen by relays 1 and 2 for fault A be 10 A and 3.3 A, respectively, based on the CT ratios and fault current. Similarly, let the currents seen by the relays 3 and 4 for fault B be 3.3 A and 10 A, respectively, based on the CT ratios and fault current.The minimum allowable operating time for each relay is considered as 0.1 sec, and the CTI is taken as 0.3 sec.

The objective function of relay coordination problem as NLPP (11) is reproduced here for convenience.

$$\text{Minimize } z = \sum_{i=1}^{n} \frac{0.14 * (TMS_i)}{(I_{relay}/PS_i)^{(0.02)} - 1}$$

If TMS settings of relays 1 to 4 are represented by x_1 to x_4 and PS settings of relays are represented by x_5 to x_8, then the above objective function can be written as follows:

$$\begin{aligned}\text{Minimize } z = {} & (0.14x_1/((10/x_5)^{0.02} - 1)) + (0.14x_2/((3.33/x_6)^{0.02} - 1)) \\ & + (0.14x_3/((3.33/x_7)^{0.02} - 1)) + (0.14x_4/((10/x_8)^{0.02} - 1))\end{aligned} \quad (1.15)$$

Subject to coordination constraint:

$$\begin{aligned}&\text{Minimize} z = (0.14x_4/((3.33/x_8)^{0.02} - 1)) - (0.14x_2/((3.33/x_6)^{0.02} - 1)) > =0.3; \\ &(0.14x_1/((3.33/x_5)^{0.02} - 1)) - (0.14x_3/((3.33/x_7)^{0.02} - 1)) > =0.3;\end{aligned} \quad (1.16)$$

Other constraints can be written as per equations (1.7) to (1.9).

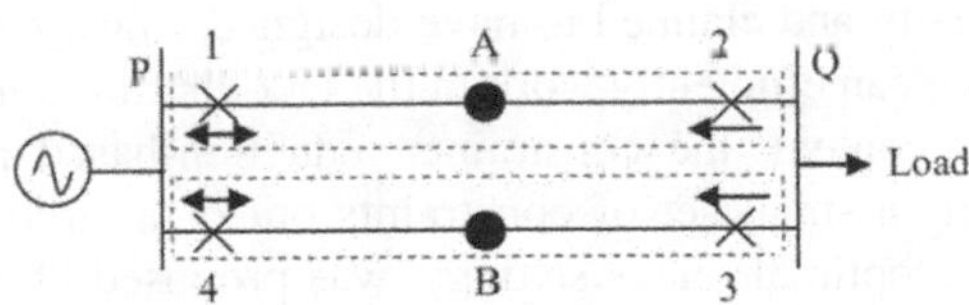

FIGURE 1.2 Parallel feeder system.

1.4.3 State of Research in Optimal Relay Coordination

Researchers have been attracted to the topic of optimal relay coordination (ORC) for decades. Genetic algorithm initiated the era of application of metaheuristic algorithms to ORC [10]. Researchers observed that population size and the number of generations affect the relay settings. Many journal papers used GA and other metaheuristic algorithms to solve the ORC problem.

The combination of Genetic Algorithm and Non-Linear Programming approach was used to address the relay coordination problems [11]. Here, the initial values of Time Multiplier Setting and Plug Setting were determined using Genetic Algorithm. The optimum values of the relay settings are then found by applying Sequential Quadratic Programming method, yielding better results.

Seeker optimization technique was used to solve the problem of coordination of directional over-current relays formulated as a mixed-integer nonlinear programming problem [12]. The seeker technique, based on the act of human searching, determines adaptively the search direction, and step length. It is maintained that the concept of seeker mutation enables the algorithm to escape from local optima.

A new approach based on the Adaptive Differential Evolution (ADE) algorithm was proposed for solving the ORC problem by selecting discrete values of the decision variables [13]. In an ordinary Differential Evolution (DE) algorithm, the scaling factors being constant, the chance of the solutions to obtain stuck to local optima is very high. Also, the parameter tuning process is a time-consuming trial and error method. ADE avoids such troublesome tuning and also avoids convergence to the local optima by making the control parameters of the conventional DE algorithm time-varying and adaptive. A biogeography-based optimization algorithm was used to solve the optimal coordination of directional over-current relays [14].

The integration of Distributed Generation (DG) into the distribution system has technical, economic, and environmental benefits, and therefore it is becoming commonplace. A distribution system with DGs is a multi-source system in which DGs can be connected or disconnected to meet the high or low load requirements. Also, during fault conditions, DGs can get disconnected due to their low fault ride-through capabilities. These changes in connections, change the number of DGs and the system's operational status, which in turn change the fault currents seen by relays and the ranges of maximum and minimum fault currents for which the relay settings work coordinately. This paper realizes that an increase in the proliferation of DGs makes the relay coordination problem a highly constrained optimization problem, and changes in the number of DGs make the problem more complex, which can decelerate the overall performance of relays. The researchers attempted to reduce the complexity and claimed to have designed a fast, robust, and adaptive protection scheme that can efficiently work in the DG-distribution system's variable environment. In this context, the constraints reduction-based relay coordination method, which selects a small set of constraints out of a large set of constraints while determining the optimal relays settings, was proposed [15].

Wind power has come to be the mainstream of renewable energy systems in several countries among renewable energy resources. It is regarded as a reliable and financially reasonable source of electricity. Wind power plants have been vastly employed as the means of power generation in the smart grids as a distributed generation system. Overcurrent prevention is one of the most crucial areas of protection in wind farms. The protection scheme has been proposed based on IEC 60255-151:2009 standard to achieve optimized OCR settings based on the genetic algorithm [16].

Power demand has been growing exponentially, and there has been significant capacity addition to the generation centrally as well as distribution generation locally in the distribution system. The increased generation has increased the fault current levels, resulting in the necessity of upgrades and replacement of equipment in substations. Fault Current Limiter (FCL) is offered as a solution to be used in the power system to postpone the costly upgrade and replacement of substation equipment. FCL is a device used to limit the prospective fault current when a fault occurs in a transmission or distribution network without complete disconnection. Superconducting FCL (SFCL) is one of the promising designs of FCL. However, the operation of over current relays (OCRs) is affected by the use of SFCL in the system causing trip delays. As a substitute for OCRs' periodic resetting, it is suggested to add a voltage component to OCR [17]. The suggested OCR is operated by not only the amplitude of the current but also the voltage across the SFCL, avoiding trip delays.

The use of SFCL reduces the fault current to be less than the circuit breaker's rated interruption capability. However, circuit breaker application in a system requires evaluation of fault current and the transient recovery voltage (TRV) characteristics. The impact of the operation of various designs of SFCL on TRV capability of the circuit breaker is highlighted in [18].

1.5 OPTIMAL PMU PLACEMENT

Wide-area monitoring, protection, and control (WAMPAC) systems use system-wide information to limit the propagation of large disturbances. Phasor Measurement Units (PMUs) are the crucial elements and enablers of WAMPAC providing synchronized phasor measurements.

In electric power systems, synchronized phasor measurements are the measurement techniques which are of choice. They provide positive sequence voltage and current measurements synchronized to within a microsecond. GPS (Global Positioning System) and the techniques of sampled data processing, which have been developed for applications to computer relaying, have made this possible. In addition, PMUs also measure rate of change of frequency and local frequency. Measurement of harmonics, zero and negative sequence quantities, and current and voltage of individual phase can be done by customizing the PMUs.

Many problems of power system protection, such as series compensated lines protection, multi-terminal lines protection, and proper setting of "out-of-step relays," can be solved with the help of PMUs. In many situations, the reliable measurement of a remote voltage or current on the same reference as local variables has made a substantial improvement in protection functions possible.

The cost of PMU and associated communication facilities is high. Each line is incident on two and only two buses. If one of these two buses is observed by placing a PMU, then the other automatically becomes observed. So, it is uneconomical to have all buses having PMU installed. Hence, finding optimal number of PMUs and their optimum placement is a matter of concern for power system engineer. The problem of determining optimum (minimum) number of PMUs and their proper placement to make an entire system observable has been a topic of research, for a long time.

1.5.1 Formulation of PMU Placement Problem

A PMU can measure the voltage and the current phasors for the lines which are incident on the bus at which PMU is placed. The objective of the problem of PMU placement is making the system observable with a minimum number of PMUs placed.

For the n-bus system, the problem of optimum PMU placement can be stated as:

$$\min p = \sum_{i=1}^{n} w_i x_i \tag{1.17}$$

under the constraints:

$$\mathbf{Ax} \geq v \tag{1.18}$$

where

p: objective function to be minimized

x: vector of binary decision variable, the entries of which are:

$$x_i = 1 \text{ if PMU to be installed at } i^{th} \text{ bus}$$
$$= 0 \text{ otherwise}$$

w_i: PMU cost (to be installed at i^{th} bus). All the weights are set to unity to obtain the problem of minimizing number of PMUs.

Matrix **A** is called the connectivity matrix (which is binary in nature) for the system and its entries are:

$$A_{i,j} \text{ is 1 if } i = j \text{ or } i \text{ and } j \text{ are adjacent nodes}$$
$$= 0 \text{ for other cases}$$

v is a vector whose length is n. The elements of vector v are set to 2, which confirm that each bus is being observed by a minimum of two PMUs. This ensures that the bus remains observable even if one PMU or one line is out of order. If outage of line or outage of PMU is not considered, then the elements of vector v are set to 1.

Once we get the optimum number of PMUs (i.e. optimum p), part two is solved, which determines the locations of PMUs that are optimum. Optimum location of

PMUs ensures maximum redundancy in bus observation. This problem is formulated as:

$$Maximize\ G = sum(\mathbf{Ax}) - n \quad (1.19)$$

subject to the constraints:

$$\boldsymbol{Ax} \geq v \quad (1.20)$$

and

$$\sum_{i=1}^{n} x_1 = p \quad (1.21)$$

where

p, **x**, **A**, n, and **v** have the same meaning as that in the first part of the problem and sum(**A x**) represents sum of elements of vector (**A x**), and G is redundancy in observed buses.

1.5.2 Illustrative Example

For formulating the PMU placement problem, we consider a system consisting of five buses. Figure 1.3 shows the system under consideration. The number of variables will be five, because there are five buses (n = 5). These variables are considered as x_1 to x_5. Each variable can take either value zero or value one. Value zero indicates that PMU is not installed at that bus, whereas one indicates that PMU is to be installed at that bus. Each line is incident on two and only two buses. If one of these two buses is observed, then the other is automatically observed.

Table 1.1 shows the observed buses due to placement of PMUs at different buses.

With this, part one of the problem in which we find minimum PMUs is stated as

$$\min p = x_1 + x_2 + x_3 + x_4 + x_5 \quad (1.22)$$

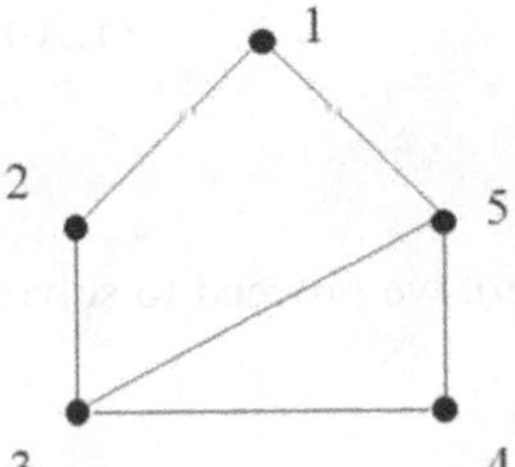

FIGURE 1.3 A five-bus system.

TABLE 1.1
Observed Buses Due to Placement of PMU

Bus at which PMU is placed	Buses observed
1	1, 2, 5
2	1, 2, 3
3	2, 3, 4, 5
4	3, 4, 5
5	1, 3, 4, 5

with the constraint as

$$\mathbf{Ax} \geq v \tag{1.23}$$

where matrix **A** is called the connectivity matrix, which is binary in nature. Using Table 1.1, the connectivity matrix can be defined, as shown (for the system shown in Figure 1.3):

$$\mathbf{A} = \begin{bmatrix} 1 & 1 & 0 & 0 & 1 \\ 1 & 1 & 1 & 0 & 0 \\ 0 & 1 & 1 & 1 & 1 \\ 0 & 0 & 1 & 1 & 1 \\ 1 & 0 & 1 & 1 & 1 \end{bmatrix} \tag{1.24}$$

x is a vector containing variables which are unknown:

$$= \begin{bmatrix} x_1 \\ x_2 \\ x_3 \\ x_4 \\ x_5 \end{bmatrix} \tag{1.25}$$

$\boldsymbol{v}$ is a vector whose length is n and is given as below:

$$= \begin{bmatrix} 2 \\ 2 \\ 2 \\ 2 \\ 2 \end{bmatrix} \tag{1.26}$$

Once we get the optimum number of PMUs (i.e. optimum p), we proceed to solve part two, as given below.

$$Maximize\ G = sum(\mathbf{Ax}) - n \tag{1.27}$$

under the constraints:

$$Ax \geq \mathbf{v} \tag{1.28}$$

and

$$x_1 + x_2 + x_3 + x_4 + x_5 = p \tag{1.29}$$

This part determines the locations for PMUs that will be the optimum locations.

1.5.3 State of Research in Problem of PMU Placement

An optimal PMU placement (OPP) problem is considered as a problem involving optimization of two objectives (which are conflicting) simultaneously. One objective is to minimize the number of PMUs and the other is to maximize the redundancy in measurements [19]. These are conflicting objectives, as the improvement of one leads to deterioration of the other. Instead of a unique optimal solution, there exists a set of best trade-offs between competing objectives, the so-called Pareto-optimal solutions. A specially tailored Nondominated Sorting Genetic Algorithm (NSGA) for a PMU placement problem was proposed to find these Pareto-optimal solutions from which the most desirable one is chosen.

The number of PMUs can be reduced further if the power system consists of the zero injection buses. Zero injection buses are the buses with no generation or load connected. Two formulations of the OPP problem are presented in [20] and solved using a metaheuristic technique, called Chemical Reaction Optimization (CRO). CRO loosely mimics the interactions between molecules in a chemical reaction process. In the formulation of first, zero injections were not considered, whereas zero injections were taken into account in the second.

Generally, the existing OPP objective functions determine the minimum number and optimal location of a set of PMUs, assuming that the entire power system remains a single observable island or network. However, some severe faults may expose parts of the network to angle, frequency, or voltage instability. In that case, maintaining system integrity and operating the system as entirely interconnected is very difficult and may cause the propagation of local weaknesses to other parts of the system. As a solution, controlled islanding (CI) is employed by system operators, in which the interconnected power system is separated into several planned islands before catastrophic events. After system splitting, a wide-area blackout can be avoided because the local instability is isolated and prevented from further spreading. With this in view, the PMU placement problem is addressed [21], considering controlled islanding so that the power network remains observable under controlled islanding conditions as well as normal operating conditions.

The PMU cost includes hardware cost, design, planning cost, cost of commissioning, and limitations of channel. It requires a communication network that is robust, so as to maintain hierarchy between Phasor Data Concentrator (PDCs),

PMUs, and control center. The solution obtained without considering these factors may give an uneconomic planning. So, realistic cost and the associated factors must be analyzed before the installing the PMUs. The approach [22] tries to find out a solution to OPP problem considering these cost factors, which helps in the initial planning and design stage of PMU deployment. To determine the solution of OPP problem, the Symbiotic Organism Search optimization method has been used, which considers limitations of channel and redundancy maximization.

Problem of State Estimation of power system can be addressed with optimal PMU placement. Improved PSO algorithm has been suggested for this purpose [23]. Another metaheuristic algorithm called the Sine Cosine Algorithm, was applied to find the optimal number and placement of PMUs to obtain full observability of distribution system [24]. In OPP problems, it is assumed that the communication network exists with good coverage and with sufficient bandwidth. However, this assumption may be invalid for a practical scenario. A CL (Communication link) placement problem is recently included in the OPP problem. This ensures full observability in a power system [25]. It is contended that PMUs alone will not make a system observable. To transmit PMU data to PDC, an appropriate communication network is a must. Thus, an optimal PMU-CL placement (OPLP) problem is an extension of a conventional OPP problem that investigates the optimal placement of CLs and PMUs simultaneously.

1.6 ESTIMATION OF FAULT SECTION ON DISTRIBUTION NETWORK

When a fault occurs in a distribution network, it should be quickly repaired so as to cause minimal disruption in service. To carry out quick repairs, identifying the faulty section in a complex distribution network is necessary. Identification of the faulty section requires data of observed statuses of protection devices such as relays and circuit breakers (CBs). Often this task becomes difficult not only because of maloperation of primary relays and CBs but also due to multiple faults occurring in the system at just about the same time.

1.6.1 Formulation of Fault Section Estimation Problem as an Optimization Problem

One can approach the problem of FSE by combining Hebb's learning rule with GA [26]. The block diagram as shown in Figure 1.4 indicates the Hebb rule-GA method. This approach consists of two parts. In the first part, the weight matrix is obtained from the training data set using the Hebb rule. The training data set contains the statuses of relays and CBs as input patterns (P) and corresponding faulted line sections as targets or desired outputs. The faulted section or target (T) is decided by the operation of related relays and CBs in real time (I). Thus, we write the equations for each target using the weights obtained using Hebb rule. In the second part of the approach, the objective function is to minimize the squared error is formed. The error term in the objective function is formed by subtracting the

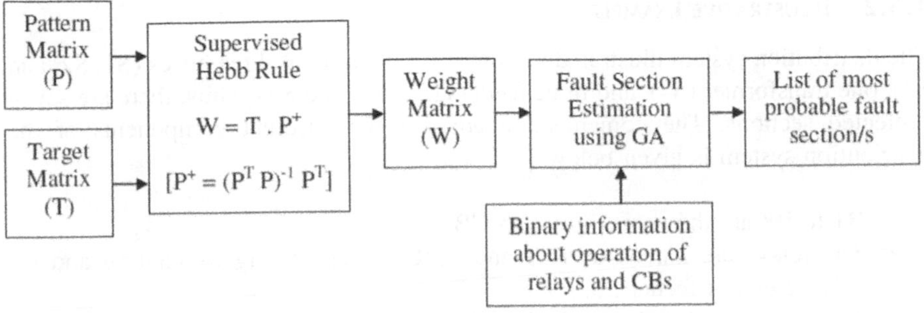

FIGURE 1.4 FSE problem solving method.

chromosome (possible solution) of GA method from the value of the target obtained by status information of relays and CBs. When the objective function is minimized using GA, the most probable fault section is obtained.

The information in the form of "0" and "1" corresponding to statuses of relays and CBs and other related information is elaborated as follows:

- Operation of relay is indicated by "1" and "0" designates that relay has not operated.
- Operation of CB is indicated by "1" and "0" designates that CB is closed.
- Faulted section is indicated by "1" and "0" designates un-faulted section.
- The number of inputs, I, are decided by the "m" number of relays and CBs.
- The number of targets, T, is decided by the "n" number of sections.
- Finally, the Hebb rule and the input target pattern generated are used to obtain the weight matrix.

The expression for target functions (faulty section/s) T_i is written as:

$$T_i = \sum_{j=1}^{m} (W_{ij} I_j) \tag{1.30}$$

where $i = 1, 2, \ldots, n$.

$I_j = j^{th}$ test input

The fault section estimation, an unconstrained optimization problem is given as minimizing the sum of squared errors

$$f = \sum_{i=1}^{n} [(T_i - S_i)^2] \tag{1.31}$$

where "$\boldsymbol{n}$" is the number of sections.

The individual solution of the population selected gives the value of S_i and by using (1.30) the values of T_i can be obtained. When the function is minimized or global minimum value of squared error is obtained using GA, we get the value of S_i. When this value of S_i obtained by function minimization process is greater than or equal to 0.5, the ith section is considered faulty.

1.6.2 Illustrative Example

The distribution system illustrated in Figure 1.5 consists of three buses (S1, S2, and S3), one transformer (S4), and three feeders (S5, S6, and S7). Thus, there are seven protected sections. The nomenclature adapted for different components of the distribution system is given below:

- B1 to B8 are the names given to CBs.
- The relays are named as P5S and P5R, corresponding to sending and receiving end of feeder S5.
- The relays are labeled as P6S and P6R, corresponding to sending and receiving end of feeder S6.
- The relay K56 is a backup relay for P5S and P6S.
- The P6S is a backup relay for P5R and P5S is a backup relay for P6R.

The faulted line section will be determined using the above approach for the following distribution system.

The information regarding primary backup relay pairs is given in Table 1.2. There are 18 input training patterns, and each pattern consists of 22 inputs as there are 14 relays and 8 CBs in the study system. The input training patterns are shown in Table 1.3.

Table 1.4 shows the target (desired output) matrix corresponding to input training patterns of the matrix (Table 1.3). As numbers of sections are 7, the targets are also 7. The ith row of the target matrix corresponds to the corresponding i^{th} row of the input pattern matrix of Table 1.3.

The target functions given below in equations (1.32)–(1.38) are written using the weight matrix and (1.30). It may be noted that the numerical values in the target functions are the entries from the weight matrix (not shown here). Also, the 22 inputs are labeled using variable p(i) for statuses of relays and CBs for convenience.

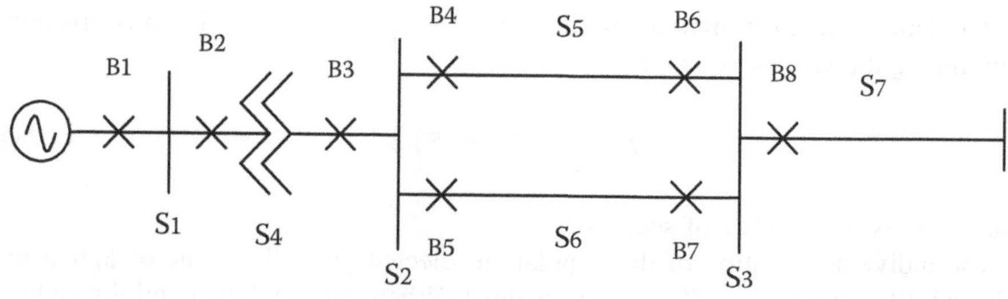

FIGURE 1.5 Distribution system for illustration.

TABLE 1.2
Information Related to Relays

Section	Description	Primary Relay	Backup Relay
S_1	Busbar	P1	K1
S_2	Busbar	P2	K2
S_3	Busbar	P3	K3
S_4	Transformer	P4	K4
S_5	Feeder	P5S & P5R	K56 & P6S
S_6	Feeder	P6S & P6R	K56 & P5S
S_7	Feeder	P7	P5S & P6S

$$\begin{aligned} T(1) = {} & 0.2973 * (p(1) + p(10)) + 0.0595 * (p(2) + p(11)) \\ & + 0.0162 * (p(3) + p(12)) - 0.0865 * (p(4) - p(7) + p(13)) \\ & - 0.0486 * p(5) - 0.027 * p(6) + 0.0432 * p(8) + 0.5946 * p(15) \\ & + 0.1081 * (p(16) - p(19)) - 0.0216 * (p(17) + p(21)) \\ & + 0.0703 * p(18) + 0.0054 * p(20) \end{aligned} \tag{1.32}$$

$$\begin{aligned} T(2) = {} & 0.0595 * (p(1) + p(10)) + 0.3919 * (p(2) + p(11)) \\ & - 0.0568 * (p(3) + p(12)) + 0.1027 * (p(4) - p(7) + p(13)) \\ & - 0.4297 * p(5) + 0.0946 * p(6) - 0.1514 * p(8) + 0.1189 * p(15) \\ & - 0.1784 * (p(16) - p(19)) + 0.0757 * (p(17) + p(21)) \\ & + 0.3541 * p(18) - 0.0189 * p(20) \end{aligned} \tag{1.33}$$

$$\begin{aligned} T(3) = {} & 0.0162 * (p(1) + p(10)) - 0.0568 * (p(2) + p(11)) \\ & + 0.4027 * (p(3) + p(12)) - 0.0811 * (p(4) - p(7) + p(13)) \\ & - 0.0081 * p(5) - 0.3378 * p(6) - 0.2595 * p(8) + 0.0324 * p(15) \\ & - 0.0486 * (p(16) - p(19)) + 0.1297 * (p(17) + p(21)) \\ & - 0.1216 * p(18) + 0.4676 * p(20) \end{aligned} \tag{1.34}$$

TABLE 1.3
FSE Training Pattern

SN	Primary Relays									Backup Relays					Circuit Breakers							
	P1	P2	P3	P4	P5S	P5R	P6S	P6R	P7	K1	K2	K3	K4	K56	B1	B2	B3	B4	B5	B6	B7	B8
1	1	0	0	0	0	0	0	0	0	0	0	0	0	0	1	1	0	0		0	0	0
2	0	1	0	0	0	0	0	0	0	0	0	0	0	0	0	0	1	1	1	0	0	0
3	0	0	1	0	0	0	0	0	0	0	0	0	0	0	0	0	0	0	0	1	1	0
4	0	0	0	1	0	0	0	0	0	0	0	0	0	0	0	1	1	0	0	0	0	0
5	0	0	0	0	1	1	0	0	0	0	0	0	0	0	0	0	0	1	0	1	0	0
6	0	0	0	0	0	0	1	1	0	0	0	0	0	0	0	0	0	0	1	0	1	0
7	0	0	0	0	0	0	0	0	1	0	0	0	0	0	0	0	0	0	0	0	0	1
8	0	0	0	0	1	0	1	0	0	0	0	0	0	0	0	0	0	1	1	0	0	0
9	0	0	0	0	0	0	0	0	0	1	0	0	0	0	1	1	0	0	0	0	0	0
10	0	0	0	0	0	0	0	0	0	0	1	0	0	0	0	0	1	1	1	0	0	0
11	0	0	0	0	0	0	0	0	0	0	0	1	0	0	0	0	0	0	0	1	1	0
12	0	0	0	0	0	0	0	0	0	0	0	0	1	0	0	1	1	0	0	0	0	0
13	0	0	0	0	1	0	1	0	0	0	0	0	0	0	0	0	0	1	1	0	0	0
14	0	0	0	0	1	0	1	0	1	0	0	0	0	0	0	0	0	1	1	0	0	0
15	0	0	0	0	0	0	0	1	0	0	0	0	0	1	0	0	1	0	0	0	1	0
16	0	0	0	0	1	0	1	0	0	0	0	0	0	1	0	0	1	1	0	0	0	0
17	0	0	0	0	1	1	1	1	0	0	0	0	0	0	0	0	1	1	1	1	0	0
18	0	0	1	0	0	0	1	0	0	0	0	0	0	0	0	0	0	0	1	1	0	0

TABLE 1.4
Target Patterns Corresponding Input Patterns of Table 1.3

SN	T1	T2	T3	T4	T5	T6	T7
1	1						
2		1					
3			1				
4				1			
5					1		
6						1	
7							1
8					1		
9	1						
10		1					
11			1				
12				1			
13							1
14							1
15						1	
16						1	
17					1	1	
18			1				

$$\begin{aligned}T(4) = {} & -0.2432 * (p(1) + p(10)) - 0.1486 * (p(2) + p(11)) \\ & - 0.0405 * (p(3) + p(12)) + 0.2162 * (p(4) - p(7) + p(13)) \\ & + 0.1216 * p(5) + 0.0676 * p(6) - 0.1081 * p(8) - 0.4865 * p(15) \\ & + 0.7297 * p(16) + 0.0541 * (p(17) + p(21)) - 0.1757 * p(18) \\ & + 0.2703 * p(19) - 0.0135 * p(20)\end{aligned} \tag{1.35}$$

$$\begin{aligned}T(5) = {} & -0.1595 * (p(1) + p(10)) - 0.1919 * (p(2) + p(11)) \\ & - 0.0432 * (p(3) + p(12)) - 0.2027 * p(4) + 0.3297 * p(5) \\ & + 0.4045 * p(6) - 0.2973 * p(7) + 0.0514 * p(8) \\ & - 0.5 * (p(9) - p(14) - p(22)) - 0.3189 * p(15) + 0.4781 * p(16) \\ & - 0.2757 * (p(17) + p(21)) - 0.0541 * p(18) + 0.5216 * p(19) \\ & + 0.3189 * p(20)\end{aligned} \tag{1.36}$$

$$\begin{aligned} T(6) = {} & 0.1784 * (p(1) + p(10)) + 0.1757 * (p(2) + p(11)) \\ & - 0.1703 * (p(3) + p(12)) + 0.3081 * (p(4) + p(13)) - 0.2892 * p(5) \\ & + 0.2838 * p(6) + 0.6919 * p(7) + 0.5459 * p(8) + 0.3568 * p(15) \\ & - 0.5351 * p(16) + 0.227 * (p(17) + p(21)) + 0.0622 * p(18) \\ & - 0.4649 * p(19) - 0.0568 * p(20) \end{aligned} \tag{1.37}$$

$$\begin{aligned} T(7) = {} & -0.0216 * (p(1) + p(10)) - 0.2743 * (p(2) + p(11)) \\ & + 0.0797 * (p(3) + p(12)) - 0.2919 * (p(4) + p(13)) + 0.6608 * p(5) \\ & - 0.4662 * p(6) - 0.2081 * p(7) + 0.0459 * p(8) \\ & + 0.5 * (p(9) - p(14) + p(22)) - 0.0432 * p(15) \\ & + 0.0649 * (p(16) - p(19)) + 0.227 * (p(17) + p(21)) + 0.1122 * p(18) \\ & - 0.3068 * p(20) \end{aligned} \tag{1.38}$$

where p(i) indicates the ith input of the pattern under consideration (i = 1, …, 22).

The equations (1.32)–(1.38) show that the targets are linear algebraic functions of statuses of the relays and circuit breakers. Using (1.31), the objective function is formed and solved by GA.

1.6.3 State of Research in Fault Section Estimation

Distribution systems may consist of overhead lines with cables inserted at a certain distance from two substations. Such a scenario was considered in [27] and a technique based on processing of transient signals was proposed to find the faulted sections. Researchers also have been using various AI approaches for solving FSE problems. One of the AI approaches to estimate the possible fault section is an expert system, as suggested by [28]. Based on the knowledge about protective systems and information of the operated relays and circuit breakers, the expert system makes inferences whenever a fault occurs. By analyzing the information, the expert system is not only able to detect the probable section of fault but also is able to pinpoint the scenario such as a sequence of relay operations and the tripping of CBs, leading to present fault situation.

Another AI tool viz. ANN, finds its use for online FSE [29]. The proposed ANN system learns fault scenarios. ANN system makes a reasonable generalization of learned scenarios by automatically adding the sample data into its training set. Fuzzy Petri Net technique was proposed in [30] for FSE in a distribution system with distributed generation. Binary status data from protection devices and fault indicators as well as voltage and current analog data has been used for the purpose.

GA was applied to solve the problems of FSE [31]. The GA method was proposed to identify the faulty sub-networks when formulating problem of FSE as a 0-1 integer programming problem. A refined GA method was applied [32], where a systematic and sound mathematical model is applied to FSE. In this method, the structure and function of a power system's protection system are encoded as a probability casualty matrix. The probabilistic causal relationship among section fault, protective relay trips, and circuit breaker action are formulated into a matrix. The two main issues related to relay viz. uncertainties of relay and CB failure were claimed to have been addressed using this method as it takes into account the reliability indices of protective relays and CBs.

Yet another technique, the set covering theory and Tabu Search (TS) technique was proposed [33] for fault section estimation in power systems. It is claimed that in a large-scale power system, this method has the potential for faster on-line fault diagnosis with flexibility. Researchers of paper [34] compared the Evolutionary Programming (EP) and GA for FSE and reported different parameters that affect the EP convergence.

For fault diagnosis in the distribution system, a technique using the fuzzy rule matrix transformation was proposed [35]. The inference procedure consists of fuzzy Cause-Effect Networks representing the fault section's causalities and the actions of protective devices. Possible fault sections can be detected after performing simple matrix operations. The proposed method has the ability to handle uncertainty and has no problem with convergence during the diagnosis procedure.

A combination of the Unconstrained Binary Programming (UBP) model and an Adaptive Genetic Algorithm (AGA) method have been adapted by [36] to solve the FSE problem in power systems. The Parsimonious Covering Theory (a formal model of diagnostic reasoning) for associating the alarms of the protective relay functions and the expected states of the protective relay functions are used to express the UBP model. Parameters such as maximum number of iterations, crossover probability, and mutation probability are adjusted adaptively during each iteration of AGA.

1.7 ESTIMATION OF FAULT LOCATION ON TRANSMISSION LINES

Power systems are growing larger and are becoming increasingly complex in nature. This is a reflection of the increasing dependence of society on continuous and reliable availability of electrical energy. The faults on the power system cannot be averted. However, it is possible to alleviate the effects of the faults. This is the objective of a protection system, which is accomplished by quickly segregating the faulted component from the rest of the system. Thus, faults on the transmission lines result in short-term to long-term power outages for consumers and may lead to significant losses to both the utility and the consumer.

The permanent faults typically result in mechanical damages. Repairs need to be carried out before reinstating the line for use. The problem is difficult due to the existence of various variable factors like dc offset, fault resistance, mutual coupling, presence of FACTS devices, etc. The restoration can be expedited if the fault location is known with high accuracy. Also, the accuracy of fault location estimation is vital, whenever repairs are be carried out in difficult terrain and bad weather. Therefore, fault location estimation has been the research topic of choice for decades.

1.7.1 Formulation of Fault Location Estimation Problem as an Optimization Problem

The transmission line can be represented by series R-L model or a single PI section. The instantaneous values of the voltage signal at the relay location are available through potential transformers (PTs). Also, voltage at relay location is related through a differential equation. This differential equation for voltage can be solved using a meta-heuristic algorithm, such as the Genetic Algorithm. The initial population for GA is formed with random assumed values of series parameters of transmission lines. These calculated instantaneous voltage values are then compared with instantaneous values of the voltage signal measured using PT.

The fitness function is obtained with the help of the sum of square of errors. If the error, the difference between measured voltage and calculated voltage, is represented by er, then the error at the ith instant of time can be written as

$$er_i = v_i - v_{i,calulated} \tag{1.39}$$

In the above equation, v_i is the value of voltage sample at the ith instant of time, and $v_{i,calculated}$ is the value of calculated voltage at the ith instant of time using the GA algorithm. The value $v_{i,calculated}$ is obtained using (1.42), in case of lumped series representation. Similarly, the value $v_{i,calculated}$ is obtained using (1.45) in case of single pi section representation of line, as explained in the next section.

The sum of square of errors can be written as

$$\Sigma\,(error)^2 = (er_2)^2 + (er_3)^2 + \ldots + (er_{n-1})^2 \tag{1.40}$$

It should be noted that the errors at t_1 and t_n cannot be calculated since the derivatives of current and voltage cannot be computed at the very first and the very last (n^{th}) sampling instant. Minimization of the sum of squared errors is the objective. Since, as per convention, the fitness function should have a higher value for a better solution, the fitness function can be coded as negative of $\Sigma(error)^2$, or can be coded as sum of squares of errors subtracted from a large positive number (N). The fitness function becomes:

$$F(x) = N - \Sigma(error)^2 \tag{1.41}$$

This fitness function should be maximized. Thus, it is an unconstrained optimization problem. The problem can be addressed using the Genetic Algorithm to obtain the global optimum [37].

1.7.2 Illustrative Example

The lumped-series transmission-line model is shown in Figure 1.6. Resistance R and inductance L are the series parameters of the line from location where relay is placed to the point where the fault has occurred.

From the basic principle of circuit theory, *v(t)* and *i(t)* are related by:

$$v_k = Ri_k + L\frac{di_k}{dt} \tag{1.42}$$

Let v_k and i_k be the voltage and current samples at time t_k and let Δt be the sampling time interval. The derivative of the current at the kth time interval with respect to time (di_k/dt) can be approximately determined using (1.43) given below.

$$\frac{di_k}{dt} = \frac{i_{k+1} - i_{k-1}}{2(\Delta t)} \tag{1.43}$$

A transmission line represented by a pi section is shown in Figure 1.7. The fault resistance is assumed to be sufficiently small. Also, the line capacitance at the far end and in-feed from the remote end is neglected.

From the principle of circuit theory, *v(t)* and *i(t)* are related by:

$$v = R(i - i_c) + L\frac{d(i - i_c)}{dt} \tag{1.44}$$

In the above equation, if we substitute $C\frac{dv}{dt}$ for i_c, we get the following equation:

$$v = Ri + L\frac{di}{dt} - RC\frac{dv}{dt} - LC\frac{d^2v}{dt^2} \tag{1.45}$$

Equation (1.43) gives the way to compute the derivative term involving current. The voltage derivatives can be computed using the following equations:

$$\frac{dv_k}{dt} = \frac{v_{k+1} - v_{k-1}}{2(\Delta t)} \tag{1.46}$$

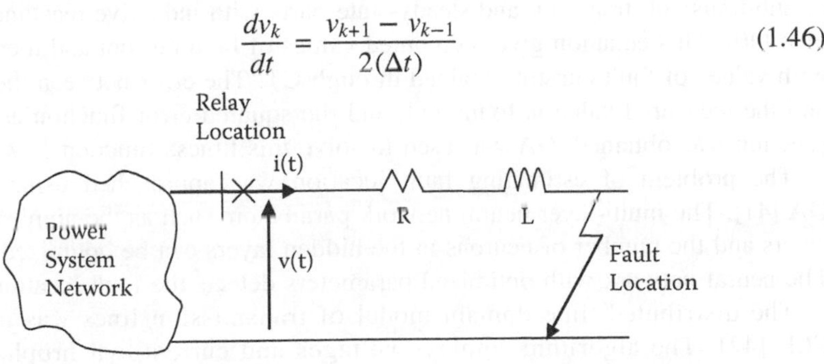

FIGURE 1.6 Lumped-series model of transmission line.

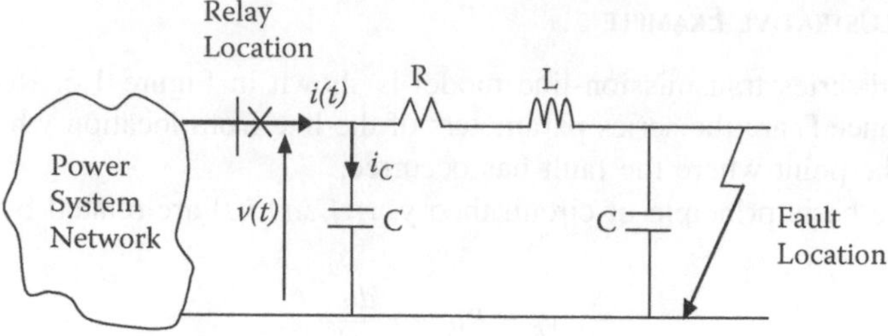

FIGURE 1.7 Transmission line pi model.

$$\frac{d^2v_k}{dt^2} = \frac{v_{k+1} - 2v_k + v_{k-1}}{(\Delta t)^2} \tag{1.47}$$

The series parameters of the line are used to find the fault location. This can be achieved by converting the problem of determining R and L into a nonlinear optimization problem based on a suitable fitness function, as explained in section 1.7.1.

1.7.3 State of Research in Fault Location Estimation

Approaches utilizing fuzzy logic and neural network have been applied extensively for fault location estimation (FLE). Positive sequence current and voltage signals were used as inputs to the fuzzy logic-based system with fault location as output. Mamdani-type fuzzy inference system was employed and the fault location error was validated using the chi square $\chi 2$ test [38]. FLE in series compensated transmission lines is quite difficult because a non-linear current dependent circuit appears between the substation and fault point. A fault location estimation scheme using an artificial neural network was presented [39] for multi-location faults (faults which occur at different locations at the same time in different phases), transforming faults as well as for commonly occurring shunt faults in the Thyristor Controlled Series Capacitor (TCSC) compensated transmission line.

Fault current in time domain was expressed in the form of an equation consisting of sub-transient, transient, and steady-state parts with inductive reactance as a part of it [40]. This equation gives computed values of fault current and then compared with values of fault current obtained through CT. The error between the computed and the measured value is found out, and the squared error function as the fitness function was obtained. GA was used to solve this fitness function.

The problem of estimating fault location was approached using ANN and GA [41]. The multi-layer neural network parameters such as the number of hidden layers and the number of neurons in the hidden layers can be optimized using GA. The neural network with optimized parameters detects the fault location.

The distributed time domain model of transmission lines was utilized for FLE [42]. The algorithm employs voltages and current synchrophasors from both ends of the transmission line. Since the fault point's voltage should be the same regardless of the data used for calculating it, the two derived voltages should be equal at all sampling instants. However, initially, they will not be the same, resulting in the error. The objective function, obtained by summing the squared errors at all instants of time, was minimized using GA. At the actual location of a fault, the value of the objective function should be at its minimum; therefore, this value should be calculated for all possible locations of the fault along the transmission line. The supposed location of the fault with the minimum value of the objective function, among other locations, is selected as the actual fault location.

The unsynchronized measurement from the three bus terminals of the teed feeder system is used by the algorithm proposed in the paper [43]. Negative sequence quantities from the bus computes the voltage at the fault occurring in the line section between a bus and the tee point. Further, the negative sequence tee-point quantities calculate the fault point voltage. These differences between the two voltages result in formation of objective functions. In order to minimize the objective function, a modified version of genetic algorithm is applied.

For hybrid cable-overhead multiple branch distribution lines with distributed generators, a combined fault location method of Binary Particle Swarm Optimization (BPSO) and a Genetic Algorithm have been proposed [44]. For a transmission line, a distributed parameter model was used. After the occurrence of fault, the fourth cycle sample data was extracted to calculate positive sequence voltage and current signals at two ends of the line. This is done to avoid the influence of higher harmonics and decaying DC components. The fitness function to be minimized is formed on the basis of the fact that the voltage amplitudes at the fault point calculated by electrical quantities at two ends of the faulted line section should be equal. On the fault location, the effects of a few parameters such as different fault section lines, fault distances, fault inception angles, DG capacity, and fault resistance were studied/obtained. The Hybrid Strategy Genetic Algorithm was applied to locate the fault on a distribution network with a distributed generation [45].

1.8 CONCLUSION

A comprehensive overview of metaheuristic algorithms' applications in various aspects of electrical transmission and distribution systems protection has been presented in this chapter. The chapter discusses applications: optimal relay coordination, optimal PMU placement, estimation of fault section in the distribution network, and fault location estimation on transmission lines. A useful insight is provided by the literature survey to the researchers, scientists working in the area of the application of the metaheuristic algorithms to some aspects protection of electrical transmission and distribution systems.

REFERENCES

[1] Bin Xia, Gil-Gyun Jeong, and Chang-Seop Koh, "Co-Kriging Assisted PSO Algorithm and its Application to Optimal Transposition Design of Power Transformer Windings for the Reduction of Circulating Current Loss", *IEEE Transactions on Magnetics*, Vol. 52, No. 3, pp. 1–4, March 2016.

[2] Vandana Rallabandi, Jie Wu, Ping Zhou, David G. Dorrell, and Dan M. Ionel, "Optimal Design of a Switched Reluctance Motor with magnetically Disconnected Rotor Modules Using a Design of Experiments Differential Evolution FEA-Based Method", *IEEE Transactions on Magnetics*, Vol. 54, No. 11, pp. 1–5, November 2018.

[3] Hye-Ung Shin, and Kyo-Beum Lee, "Optimal Design of a Switched Reluctance Generator for Small Wind Power System using a Genetic Algorithm", 9th International Conference on Power Electronics-ECCE Asia Korea, pp. 2209–2214, June 2015.

[4] Salah Soued, Mohamed A. Ebrahim, Haitham S. Ramadan, and Mohamed Becherif, "Optimal Blade Pitch Control for Enhancing the Dynamic Performance of Wind Power Plants via Metaheuristic Optimizers", *IET Electric Power Applications*, Vol. 11, No. 8, pp. 1432–1440, 2017.

[5] Jian Zhao, Shixin Liu, Mengchu Zhou, Xiwang Guo, and Liang Qi, "An Improved Binary Cuckoo Search Algorithm for Solving Unit Commitment Problems: Methodological Description", *IEEE Access*, Vol. 6, pp. 43535–43545, 2018.

[6] Gwo-Ching Liao,and Ta-Peng Tsao, "Application of a Fuzzy Neural Network Combined with a Chaos Genetic Algorithm and Simulated Annealing to Short-Term Load Forecasting", *IEEE Transactions on Evolutionary Computation*, Vol. 10, No. 3, pp. 330–340, 2006.

[7] Ibrahim Alhamrouni, Azhar Khairuddin, Ali Khorasani Ferdavani, and Mohamed Salem, "Transmission Expansion Planning using AC-Based Differential Evolution Algorithm", *IET Generation, Transmission & Distribution*, Vol. 8, No. 10, pp. 1637–1644, 2014.

[8] Abdullah M. Shaheen a, Shimaa R. Spea, Sobhy M. Farrag, and Mohammed A. Abido, "A review of Meta-Heuristic Algorithms for Reactive Power Planning Problem", *Elsevier Ain Shams Engineering Journal*, Vol. 9, No. 2, pp. 215–231, 2018.

[9] Alberto J. Urdaneta, Nadira Ramon, and Luis G. Parez Jimenez, "Optimal Coordination of Directional Relays in Interconnected Power System", *IEEE Transactions on Power Delivery*, Vol. 3, No. 3, pp. 903–911, July;1988.

[10] C. W. So, K. K. Li, K. T. Lai, and K. Y. Fung, "Application of Genetic Algorithm for Overcurrent Relay Coordination", 4th International Conference on Advances in Power System Control, Operation and Management, APSCOM-97, Hong Kong, pp. 283–287, November;1997.

[11] Prashant Prabhakar Bedekar and Sudhir Ramkrishna Bhide, "Optimum Coordination of Directional Overcurrent Relays Using the Hybrid GA-NLP Approach", *IEEE Transactions on Power Delivery*, Vol. 26, No. 1, pp. 109–119, January;2011.

[12] Turaj Amraee, "Coordination of Directional Over-current Relay using Seeker Algorithm", *IEEE Transactions on Power Delivery*," Vol. 27, No. 3, pp. 1415–1422, 2012.

[13] Joymala Moirangthem, K. R. Krishnanand, Subhransu Sekhar Dash, and Ramas Ramaswami, "Adaptive Differential Evolution Algorithm for Solving Non-Linear Coordination Problem of Directional Overcurrent Relays", *IET Generation, Transmission & Distribution*, Vol. 7, No. 4, pp. 329–336, April-2013.

[14] Fadhel A. Albasri, Ali R. Alroomi, and Jawad H. Talaq, "Optimal Coordination of Directional Overcurrent Relays using Biogeography-Based Optimization Algorithms", *IEEE Transactions on Power Delivery*, Vol. 30, No. 4, pp. 1810–1820, August; 2015.

[15] Ekta Purwar, D. N. Vishwakarma, and S. P. Singh, "A Novel Constraints Reduction-Based Optimal Relay Coordination Method considering Variable Operational Status of Distribution System with DGs", *IEEE Transactions on Smart Grid*, Vol. 10, No. 1, pp. 889–898, 2019.

[16] Nima Rezaei, M. N. Uddin, I. K. Amin, M. L. Othman, and M. Marsadek, "Genetic Algorithm-based Optimization of Overcurrent Relay Coordination for Improved Protection of DFIG Operated Wind Farms", *IEEE Transactions on Industry Applications*, Vol. 55, No. 6, pp. 5727–5736, 2019.

[17] Sung-Hun Lim, and Seung-Taek Lim, "Analysis on Coordination of Over-Current Relay using Voltage Component in a Power Distribution System with a SFCL", *IEEE Transactions on Applied Superconductivity*, Vol. 29, No. 5, pp. 1–5, August; 2019.

[18] H. J. Bahirat, S. A. Khaparde, S. Kodle, and V. Dabeer, "Impact on Superconducting Fault Current Limiters on Circuit Breaker Capability", National Power Systems Conference (NPSC), pp. 1–6, 2016.

[19] Borka Milosevic, and Miroslav Begovic, "Nondominated Sorting Genetic Algorithm for Optimal Phasor Measurement Placement", *IEEE Transactions on Power Systems*, Vol. 18, No. 1, pp. 69–75, February 2003.

[20] Jin Xu, Miles H. F. Wen, Victor O. K. Li, and Ka-Cheong Leung, "Optimal PMU Placement for Wide-Area Monitoring using Chemical Reaction Optimization", Proc. IEEE PES Innovative Smart Grid Technologies Conference (ISGT), Washington, DC, USA, pp. 1–6, February 2013.

[21] Lei Huang, Yuanzhang Sun, Jian Xu, Wenzhong Gao, Jun Zhang, and Ziping Wu, "Optimal PMU Placement considering Controlled Islanding of Power System", *IEEE Transactions on Power Systems*, Vol. 29, No. 2, pp. 742–755, March 2014.

[22] Soumesh Chatterjee, Biman Kr. Saha Roy, and Pronob K Ghosh, "Optimal Placement of PMU considering Practical Costs in Wide Area Network", 14th IEEE India Council International Conference (INDICON), pp. 1–6, 2017.

[23] Alaa Abdelwahab Saleh, Ahmed S. Adail, Amir A. Wadoud, "Optimal Phasor Measurement Units Placement for Full Observability of Power System Using Improved Particle Swarm Optimization", *IET Generation, Transmission & Distribution*, Vol. 11, No. 7, pp. 1794–1800, 2017.

[24] Mosbah Laouamer, Abdellah Kouzou, R.D. Mohammedi, A. Tlemçani, "Optimal PMU Placement in Power Grid using Sine Cosine Algorithm", International Conference on Applied Smart Systems, pp. 1–5, 2018.

[25] Xingzheng Zhu, Miles H. F. Wen, Victor O. K. Li, and Ka-Cheong Leung, "Optimal PMU-Communication Link Placement for Smart Grid Wide-Area Measurement Systems", *IEEE Transactions on Smart Grid*, Vol. 10, No. 4, pp. 4446–4456, July 2019.

[26] Prashant P. Bedekar, Sudhir R. Bhide, and Vijay S. Kale, "Fault Section Estimation in Power System using Hebb's Rule and Continuous Genetic Algorithm", *Electrical Power and Energy Systems*, Vol. 33, No. 3, pp. 457–465, 2011.

[27] Aleksandr Kulikov, Anton Loskutov, Pavel Pelevin, "The Method of Faulted Section Estimation for Combined Overhead and Cable Power Lines Using Double-Ended Measurements", IEEE International Ural Conference on Electrical Power Engineering, pp. 70–75, 2020.

[28] Chihiro Fukui, and Junzo Kawakami, "An Expert System for Fault Section Estimation using Information from Protective Relays and Circuit Breakers", *IEEE Transactions on Power Delivery*, Vol. PWRD-1, No. 4, pp. 83–90, 1986.
[29] Hong-Tzer Yang, Wen-Yeau Chang, and Ching-Lien Huang, "A New Neural Networks Approach to Online Fault Section Estimation using Information of Protective Relays and Circuit Breakers", *IEEE Transactions on Power Delivery*, Vol. 9, No. 1, pp. 220–230, 1994.
[30] Iman Kiaei, and Saeed Lotfifard, "Fault Section Identification in Smart Distribution Systems Using Multi-Source Data Based on Fuzzy Petri Nets", *IEEE Transactions on Smart Grid*, Vol. 11, No. 1, January 2020.
[31] Fushuan Wen, and Zhenxiang Han, "Fault Section Estimation in Power System Using a Genetic Algorithm", *Electric Power System Research*, Vol. 34, pp. 165–172, 1995.
[32] F.S. Wen, and C.S. Chang, "Probabilistic Approach for Fault-Section Estimation in Power Systems Based on A Refined Genetic Algorithm", *IET Proceedings on Generation, Transmission and Distribution*, Vol. 144, No. 2, pp. 160–168, 1997.
[33] Fushuan Wen, and C.S. Chang, "A Tabu Search Approach to Fault-Section Estimation in Power Systems", *Electric Power System Research*, Vol. 40, pp. 63–73, 1997.
[34] L. L. Lai, A. G. Sichinie, and B. J. Gwyn, "Comparison between Evolutionary Programming and a Genetic Algorithm for Fault-Section Estimation", *IET Proceedings - Generation, Transmission and Distribution*, Vol. 145, No. 5, pp. 616–620, 1998.
[35] Wen-Hui Chen, "Fault Section Estimation using Fuzzy Matrix-based Reasoning Methods", *IEEE Transactions on Power Delivery*, Vol. 26, No. 1, pp. 205–213, 2011.
[36] Esaú Figueroa Escoto, and Fábio Bertequini Leão, "Fault Section Estimation in Power Systems using an Adaptive Genetic Algorithm", IEEE Power and Energy Society General Meeting (PESGM), pp. 1–5, 2016.
[37] P.P. Bedekar, and S.R. Bhide, "Genetic Algorithm based Fault Locator for Transmission Lines", Power Research - A Journal of CPRI, Vol. 14, No. 1, pp. 19–26, February 2019.
[38] Anamika Yadav, and Aleena Swetapadma, "Enhancing the Performance of Transmission LineDirectional Relaying, Fault Classification and Fault Location Schemes using Fuzzy Inference System", *IET Generation, Transmission & Distribution*, Vol. 9, No. 6, pp. 580–591, 2015.
[39] Aleena Swetapadma, and Anamika Yadav, "Improved Fault Location Algorithm for Multilocation Faults, Transforming Faults and Shunt Faults in Thyristor-Controlled Series Capacitor Compensated Transmission Line", *IET Generation, Transmission & Distribution*, Vol. 9, No. 13, pp. 1597–1607, 2015.
[40] K. M. EL-Naggar, "A Genetic Based Fault Location Algorithm for Transmission Lines", 16th International Conference and Exhibition on Electricity Distribution, Vol. 3, pp. 1–5, 2001.
[41] R.K. Aggarwal, S.L. Blond, P. Beaumont, G. Baber, and F. Kawano "High Frequency Fault Location Method for Transmission Lines based On Artificial Neural Network and Genetic Algorithm using Current Signals only", 11th IET International Conference on Developments in Power Systems Protection, pp. 1–6, 2012.
[42] Masoud Ghiafeh Davoudi Javad, and Ebadollah Kamyab, "Time Domain Fault Location on Transmission Lines using Genetic Algorithm", 11th International Conference on Environment and Electrical Engineering, pp. 1–6, 2012.

[43] Shoaib Hussain, A. H. Osman, and S. Miura "Genetic Algorithm based Fault Locator Scheme for Three-Terminal Transmission Lines using Asynchronous Measurements", 6th International Conference on Modeling, Simulation, and Applied Optimization, pp. 1–6, 2015.

[44] Tao Jin, and Hongnan Li, "Fault Location Method for Distribution Lines with Distributed Generators Based on a Novel Hybrid BPSOGA", *IET Generation, Transmission & Distribution*, Vol. 10, No. 10, pp. 2454–2463, 2016.

[45] Zhang Yong, Wan Shanming, and Zhou Jianghua, "Fault Location Method for Distributed Power Distribution Network based on Hybrid Strategy Genetic Algorithm", Asia Energy and Electrical Engineering Symposium, pp. 572–578, 2020.

[46] Xin She Yang, "Nature Inspired Metaheuristic Algorithm", Second Edition, Luniver Press, 2010.

[47] Y. G. Paithankar, and S. R. Bhide, *"Fundamentals of Power System Protection"*, Second Edition, Prentice Hall of India Private Limited, 2013.

2 AI-Based Scheme for the Protection of Power Systems Networks Due to Incorporation of Distributed Generations

Bhavesh Kumar R. Bhalja[1] and *Yogesh M. Makwana*[2]

[1]Department of Electrical Engineering, Indian Institute of Technology Roorkee, Uttarakhand, India

[2]Electrical Engineering Department, Government Engineering College Dahod, Gujarat, India

2.1 INTRODUCTION TO DISTRIBUTED GENERATION (DG)

The conventional practice of a power system is to generate the electrical energy at the generating station(s), which would be available to the load-center/consumer through transmission and distribution network. In this practice, the generating station is located far away from the load-center or consumer. The generated power has to transfer over a long transmission and distribution lines or feeders. During this power flow, lots of technical and non-technical issues need to look after by the electrical power supplier. Further, the availability of the power to the load depends upon the normal and abnormal operating conditions of the power system network. In addition, the conventional technology utilizes the non-renewable sources for the generation of electrical energy which may raise several environment issues. In this case, the per unit charge of energy generation may become higher for the consumer. In order to minimize such higher charges and to improve the reliability of the power supply to the end user, a novel concept of distributed generation (DG) is currently receiving attention worldwide.

2.1.1 What Is Distributed Genfrating (DG)?

Due to the advancement in the power electronic technology, other allied engineering technologies, and implementation of liberalization policies, the trend of electrical power generation has shifted from central power generation to the distributed generation (DG). Further, the demand and share of DGs in the power system is increasing rapidly due to

DOI: 10.1201/9780367552374-2

continuous increase in demand for reliable electric supply, scarcity of the fossil fuels, unavailability of land for developing huge transmission and distribution networks, and the technical advancement in utilizing renewable energy sources. Hence, it has become more reasonable to develop a small power plant of the order of a few kWs to several MWs at the location of the load demand. The existing power system model has to be changed from the conventional large generator systems to a small low voltage microgrid system consisting of many small rating generators. This changes the energy flow structure to bidirectional flow instead of a centralized flow, as shown in Figure 2.1. The low rating distributed plants are utilizing distributed energy resources with an inverter or rotating machine-based generators. Such plants are known as DG plants. The DG plant caters the electrical energy demand of the local area. The consumers are operating such plants in parallel with the existing utility grid supply by interconnecting the DG plant(s) with the utility grid. Therefore, the surplus power is distributed over an existing utility grid network [7]–[9].

2.1.2 Advantages of DG over Conventional Power Generation

The DG offers various benefits over conventional power generation, which are listed below:

- It has a low capital cost due to its small size.
- Generally, it is located at the consumer end.

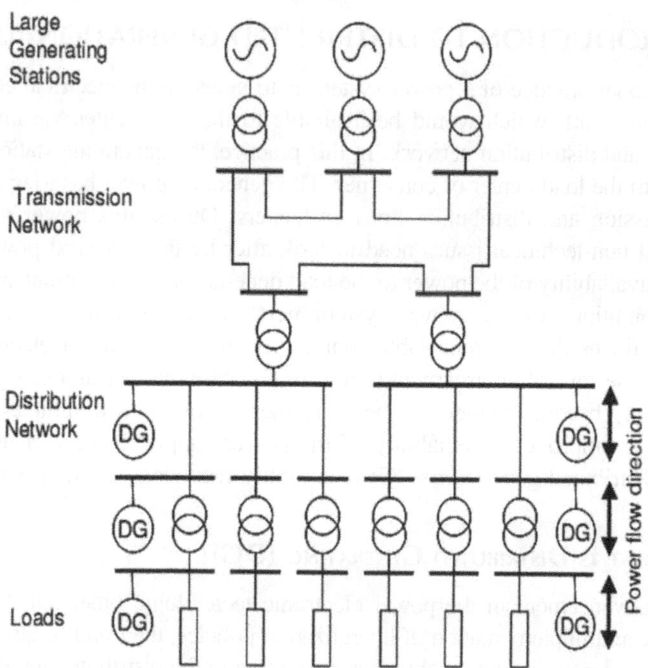

FIGURE 2.1 Decentralized power network.

- It delivers power at a higher efficiency as compared to a large system.
- It requires less space and infrastructure as compared to a large system.
- It provides power to the local area and hence, reduces pressure on transmission and distribution network.
- It can be made up of renewable and non-renewable types of energy sources. Therefore, the pollutant emission production will be very less.
- It increases system reliability by providing backup or stand by power.

Mostly, the DG plant is of small size that includes both conventional and non-conventional type of energy sources. Presently, various kinds of power generation technologies are in the development stage or in use which include: photovoltaics, wind turbines, batteries, microturbines, small steam turbines, small combustion turbines, fuel cells, small-scale hydroelectric power, etc. Based on these resources, the developed DG plants are as follows:

- Small hydro plant.
- Microturbines plants that run on high-energy fossil fuel such as oil, natural gas, propane, and diesel.
- Solar-energy-based plant.
- Wind-energy-based plant or Windfarm.
- Biomass-based plant.
- Fuel-cell-based plant.

The above-mentioned plants are utilizing electrical power generation technology based on either the rotating machine or the inverter. These plants can also be categorized with respect to their capacity/size. Table 2.1 shows the capacity wise DG classification of the plants [7]. It is to be noted from Table 2.1 that the DG capacity is ranging from 1 W to 300 MW. Accordingly, the size of the DG plant is also ranging from micro plant to large plant. Further, in order to supply excess power to the utility grid network, the DGs are connected to the main grid supply. Different energy resources are utilized to produce electrical energy. Table 2.2 shows the maximum capacity of DG plants with renewable and non-renewable energy resources [7].

According to the load demand of the area and availability of the energy source, the capacity and type of the plant is decided. It is to be noted from Table 2.2 that the PV array plant offers the minimum capacity of power generation whereas, the maximum power can be generated with combined cycle plant. Along with the mitigation of local energy demand, the use of DG directly or indirectly offers several other advantages which are listed below:

- Its setup time is short.
- Its installation and running cost are less.
- It helps in reduction of the greenhouse effect.
- It has lower transmission losses.
- Its voltage profile is almost constant.
- It improves the reliability of supply.

TABLE 2.1
Type of plants

Type of plant	Capacity of Plant
Micro	1 W to 5 kW
Small	5 kW to 5 MW
Medium	5 MW to 50 MW
Large	50 MW to 300 MW

TABLE 2.2
Maximum capacity of DG plants

Source of energy-based power generation	Capacity of plant (MW)
Non-renewable energy-based plant	
• Combined Cycle Turbine	35 to 400
• IC Engines	0.005 to 10
• Micro Turbine	0.035 to 1
Renewable energy-based plant	
• Micro Hydro Plant	0.025 to 1
• Small Hydro Plant	1 to 100
• Wind Turbine Power Plant	200×10^{-6} to 3
• PV array (Solar Power Plant)	20×10^{-6} to 0.100
• Solar Thermal Power Plant	1 to 80
• Biomass Power Generation	0.100 to 20
• Fuel Cell	0.200 to 5
• Geothermal Power Generation	5 to 100
• Ocean Energy	0.100 to 1
• Battery Storage	0.500 to 5

2.1.3 Applications of DG

Generally, the DG is interconnected with the secondary distribution voltage level. According to the type and duration of load demand, the applications of DG are given below:

1. *Stand by DG:* In this type of application, DGs are used as a stand by supply for the load like a hospital, process industries, etc., during the main grid outage.
2. *Standalone DG:* In this case, DGs are disconnected from the utility grid and they are used as the main source of electrical power generation. It provides power to the non-electrified or remote area.

3. *Peak load sharing DG:* In this case, DGs are used under peak load demand to assist the industrial customer with reduced per unit energy cost.
4. *Combined Heat and Power (CHP) DG:* In this type of application, DGs are providing CHP with higher efficiency. The heat produced in this situation will be reused for other applications.
5. *Base load DG:* In this type of application, DGs are used as a base load plant and support the main utility grid.

2.2 IMPACT OF INTEGRATION OF DISTRIBUTED GENERATION ON THE POWER SYSTEM

Integration of DGs on the power system faces a series of challenges. The conventional electrical transmission and distribution networks are developed in such a way that large power generating stations push electrical energy to small consumers who are often very far away from the generating stations. A similar example can be seen in the water distribution of a large city, where big pipes are carrying a huge quantity of water and they become smaller and smaller as the water reaches the consumer's tap. The existing power system network is not intended for the incorporation of additional DGs into the network. DG power networks are often microgrids that have to be connected to the local low-voltage distribution systems. This is not the case with the existing large power system networks. Addition of DGs to the low voltage distribution systems can lead to decrease in system stability, reduced power quality to the consumers, and difficulty in designing a protection system. Several studies have been conducted pertaining to the analysis of technical obstacles while integrating the distribution generators to the existing low voltage power system network. The probable issues raise during DG integration into the existing network would be operating (economic) issues, technical issues, and protection/safety issues [10], [11].

2.3 PROBLEMS DURING DG INTERCONNECTION

2.3.1 Operating (Economic) Issues

The major issues concerned with the application of DGs in a distribution network are categorized as per the following:

1. Energy efficiency
2. Economics of the energy generation and utilization
3. Reliability of the system
4. Environmental issues
5. Interconnection with an existing utility grid network

The utilization of different types of DGs (inverter and synchronous generator) and the integration of these DGs with the existing network, causes deviations in the signal of voltages. Hence, it can be said that among the above-mentioned issues, an interconnection of DG with the existing utility grid network is a great challenge for the power system engineers. Moreover, along with certain technical and non-technical problems, it can raise an issue of development of protocol and licensing of the integration of DG with the DN.

2.3.2 Technical Issues

The DG plant is mostly located at the consumer end to supply the energy demand of the load. Similarly, the utility is also supplying power through the grid network to the consumers. In case of peak load demand, the load sharing is carried out between the DG and the utility. However, during off peak period, DG supplies an excess power over the grid network as it is interconnected with the utility grid network. Besides, the various advantage of DG interconnection, it creates several other problems such as degradation of power quality and loss of utility control over the existing grid network. Along with these, there are several other issues which need to be considered during the interconnection of DGs with the existing utility grid network. These are listed below.

- Sudden change of state of operation from ON-grid to OFF-grid mode
- Voltage flicker
- Insertion of harmonic contents
- Synchronization/re-synchronization of microgrid

2.3.3 Protection/Safety Issues

Generally, interconnection of DG with the utility is performed by interconnecting switch/breaker which in turn changes the mode of operation i.e. ON-grid or OFF-grid. Hence, the flow of current also changes. During ON-grid operation, as the flow of current is bi-directional i.e. into the power network, it creates several protection/safety issues [10], [11]. These issues are listed below.

- The safety of the utility personals working with the utility grid network.
- Safety of the customers.
- Protection of existing distribution network and consumer's property.
- Protection against the abnormal operation of DG.
- Protection coordination of existing distribution network.

In addition, during integrated mode operation, the load is supplied by the combined power supply from the utility and the DG. In this case, the sudden disconnection of utility makes the DG feed the load. Therefore, the sudden change in power demand on the DG fluctuates the voltage signal of the DG and load. At the same time, the DG must be capable enough to mitigate the sudden rise in load demand. The sudden deviation in the voltage magnitude and frequency from its nominal operating range damages the DG or equipment of the DN/end user.

Furthermore, it is well known that the conventional protection schemes used for DN are designed for the unidirectional power flow. Hence, all existing protection schemes provide satisfactory results when the direction of current flow is unidirectional. However, an integration of DG into DN changes the direction of current flow in the network. This makes the protective devices to mal-operate or in some cases, it may remain un-operated during various abnormal/fault conditions. In case of a fault situation, the protective device located at the fault location must be able to withstand the additional fault current supplied by the DG. Hence, it can be said that the integration of DG into the existing radial DN increases the short circuit level of the network. This leads to the replacement of existing protective devices with a higher short circuit current withstand capacity. This shows that the major protection problems caused by the integration of DG with an existing DN are (i) Formation of an electrical island (sudden disconnection of DG) and (ii) Loss of existing protection coordination among protective devices. These two problems are very critical for protection engineers. It would be a duty of protection engineer to identify the said two situations and proposed a probable solution. Due to popularity of artificial intelligence (AI), researchers have proposed different solutions of the above-mentioned two problems based on AI techniques. The same is discussed in the following sections. Before understanding the solution of the said two issues, it would be better to comprehend the issues of islanding detection and loss of protection coordination due to DG interconnection thoroughly.

2.4 ISLANDING (FORMATION OF ELECTRICAL ISLAND)

The condition in which the part of an electrical power network consisting of one or more numbers of DGs and load disconnected from the rest of the network is called islanding [12]. Figure 2.2 shows the radial network containing a DG.

It is seen from Figure 2.2 that the DG is interconnected at the secondary distribution. Further, as per Figure 2.2, a Circuit Breaker (CB) is used to connect or disconnect the DG with the utility. An opening of the CB disconnects the part of the electrical network and creates an electrical island. The islanding is either intentional or unintentional. An intentional islanding is a planned event under the supervision of experts. In this case, the equipment and devices are designed to cope with this event. Furthermore, during the intentional islanding, the voltage and frequency of the island network are monitored and provisions are made to keep them constant or within the prescribed limit. Conversely, in case of an unintentional islanding, the small part of the distribution network containing DGs and loads disconnects suddenly from the rest of the DN. In this situation, DGs are still energizing a small network through a Point of Common Coupling (PCC). During such condition, the voltage and frequency of the island network exceed its nominal values. This deviation in the value of voltage magnitude and frequency is large enough to damage the equipment at the consumer end. Moreover, managing a network following an islanding is a very crucial task as the sudden disconnection creates high-frequency oscillations in the network, which may lead to maloperation of another relay in the network. In addition, other major problems such as (i) power quality issue, (ii) personal safety, and (iii) out of synchronism reclose may arise due to the unintentional islanding situation.

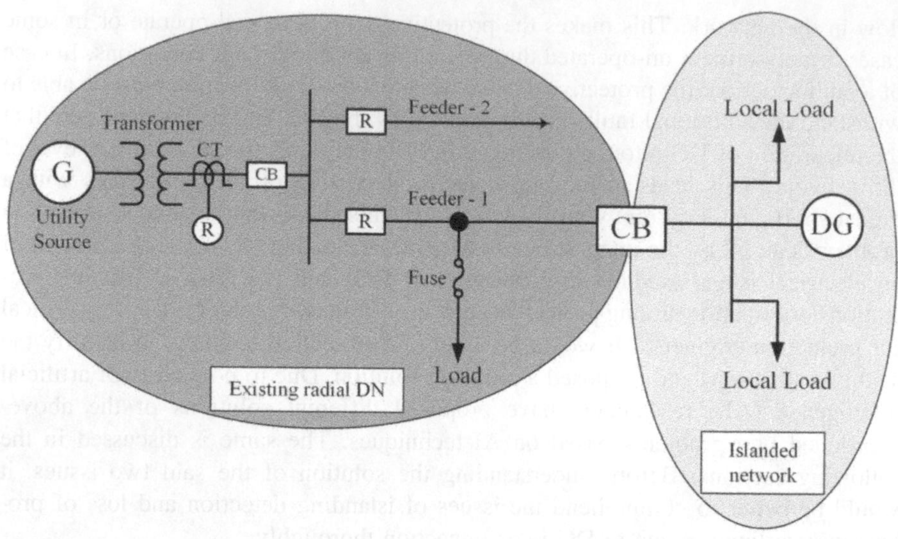

FIGURE 2.2 Formation of electrical island.

2.4.1 Power Quality Issue

The power system operator is responsible for the quality of power provided to the consumers. However, during an islanding situation, the voltage and frequency of the power provided to the customers vary significantly and, sometimes, it may exceed the statutory limits. The power quality issue deals with Over Voltage (OV), Under Voltage (UV), voltage imbalance, waveform distortion (harmonic), voltage fluctuations, Over Frequency (OF), and Under Frequency (UF). Further, during this situation, the utility has no control over it. Hence, it is recommended to look after the power quality issue during islanding.

2.4.2 Personnel Safety

The power system is designed to work as a passive network with "top-down" unidirectional power flow. When the DGs are interconnected with the distribution network, the network power flow becomes bidirectional. In the eventuality of fault in the upstream network, the faulty part will be disconnected from the network by opening a breaker. However, the DG still remains energized and continues to supply the power to the disconnected network. Meanwhile, personnel sent out for maintenance work may get an electric shock during maintenance work due to power supplied by the DG in an island network. This poses a safety hazard to utility maintenance personnel and the general public.

2.4.3 Out of Synchronism Reclose

An auto recloser is commonly used in DN to restore service after a fault. It has been reported in the literature that the recloser effectively improves the reliability of supply during temporary faults. During the attempt of reclosing, DG may not be in synchronism, even though a recloser is trying to reconnect the DG with the utility grid. This may cause overvoltage, overcurrent and severe torque transients which subsequently put rotating machines and other equipment at risk.

2.5 ISLANDING DETECTION

In order to avoid damage at the consumer end as well as to the DN, and also to provide personal safety, the islanding situation must be detected as quickly as possible. According to IEEE standard 1547-2003, the DG in an islanded network must be disconnected within 2 s from the inception of the event [5], [6]. To fulfil the requirement of the said standard, an accurate and fast islanding detection method is required. Hence, various researchers have proposed different methods of islanding detection. These methods are operated remotely or locally. According to its mode of operation and acquisition of signal(s), these methods are categorized as (1) Remote method and (2) Local method. The remote method is a communication-assisted method, whereas the local method is implemented at the location of the DG. Further, in the remote method, the islanding detection is carried out at the remote location and a trip signal is issued to disconnect the target DG. Conversely, in the case of the local method, the islanding detection is performed at the location of DG. In this method, different electrical parameters and its rate of change are monitored locally. The significant deflection or deviation in these parameters are used to detect an islanding situation. On detection of an islanding situation, a trip signal will be executed to disconnect the target DG. The local method is further categorized as (i) Active method, (ii) Passive method, and (iii) Hybrid method [12].

2.5.1 Remote Method

This method is a communication-based method. Figure 2.3 shows the single line diagram of a network containing a typical communication-based islanding detection scheme.

In this method, the status of recloser or CB that can cause an island is continuously monitored. During the opening of recloser or CB due to any fault/abnormal condition in the network, a trip signal is initiated to disconnect the DG. This method does not depend upon the type and size of DG. Further, it gives satisfactory results during islanding detection with minimum or almost zero non-detection zones. However, the main drawback of this method is that it is very expensive and, it requires an almost structural level modification in the existing network. Furthermore, it increases a level of complexity due to the placement of signal transmitters and receivers at all possible switching locations from where the island can be formed. Moreover, in case of loss of communication between transmitter and receiver, the Loss of Mains (LOM) protection is affected largely [13].

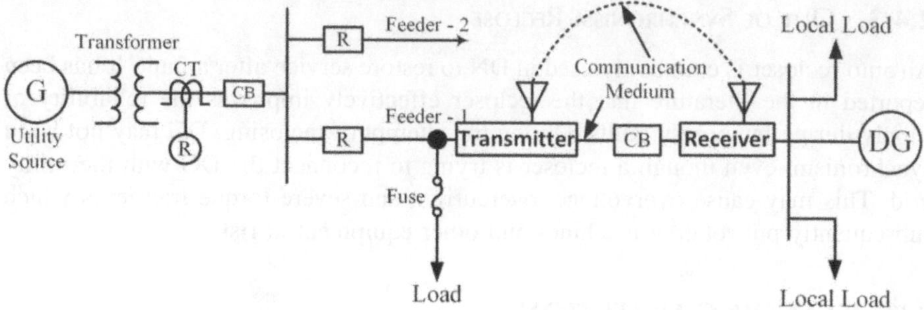

FIGURE 2.3 A typical network containing communication-assisted islanding detection.

2.5.2 Active Islanding Detection Method

Active method is one of the local methods. Figure 2.4 shows the typical active method of islanding detection. As shown in Figure 2.4, an external turbulence is continually injected into the DG. The islanding situation is detected by observing the response of the system followed by the external signal injection. The system response involves the measurement of various electrical parameters such as voltage, current, frequency, active power, reactive power, and harmonics at the DG location. During normal operating condition, the change in the measured parameters due to external turbulence is not noticeable. However, in the case of islanding, the external turbulent changes one of the measured parameters to a larger magnitude which in turn detects the event as an islanding situation. The foremost advantage of this method is the minimum value of NDZ. However, the major disadvantages of this method are the degradation of power quality and higher detection time due to external signal injection [12], [14], [15]. Many researchers have proposed various active islanding detection schemes.

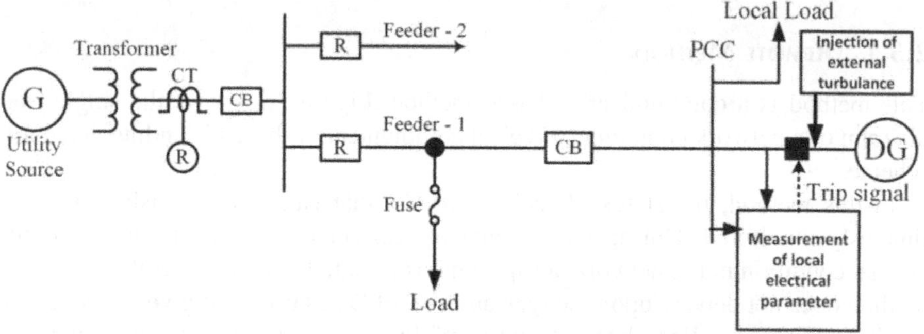

FIGURE 2.4 A typical network containing active islanding detection method.

2.5.3 Passive Islanding Detection Method

Figure 2.5 shows the single line diagram of a typical network, which is used for illustration of the passive islanding detection method. As shown in Figure 2.5,

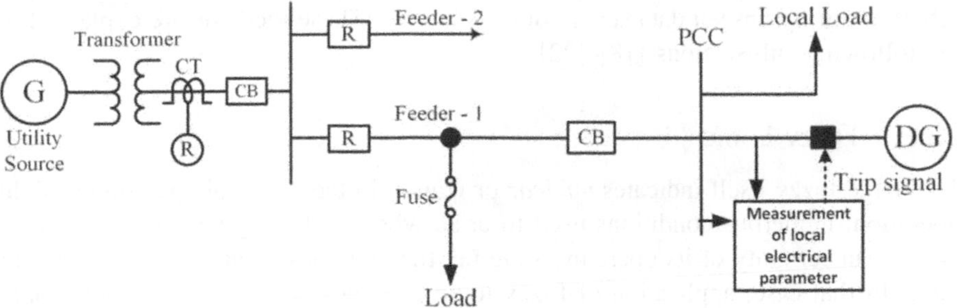

FIGURE 2.5 A typical network containing passive islanding detection method.

variations in the magnitude, phase, and frequency of various electrical parameters such as voltage, current, impedance, active power, reactive power, harmonic distortion, and phase angle are monitored at the terminal of the target DG. Moreover, the variation in the rate of change of these parameters with respect to the time as well as with respect to other parameters is also examined for the detection of islanding situation. In addition, different approaches using various transforms, classifiers, and artificial intelligence on the acquired signals or on its features are also utilized for islanding detection. The passive method is easy to implement and less expensive as compared to the communication-based and active method. However, this method suffers from the higher value of NDZ and nuisance tripping [12], [16].

2.5.4 Hybrid Method of Islanding Detection

The hybrid method is a combination of both active and passive methods of islanding detection. In this method, an active technique is applied only if islanding detection is suspected by a passive technique. In this case, the measurement of electrical parameters is carried out at the DG locations only [17].

2.6 APPLICATION OF ARTIFICIAL INTELLIGENCE FOR ISLANDING DETECTION

Recently, with the advancement in the field of artificial intelligence (fuzzy logic, pattern recognition (PR), data mining and decision tree (DT)), several researchers have proposed different methods of islanding detection based on these techniques. In these schemes, magnitude, frequency, or other extracted features of the acquired signals have been utilized for the detection of islanding situation. These schemes require a large number of data sets related to islanding and non-islanding events. Initially, these data sets are used to train the AI-based model for effective differentiation between islanding and non-islanding event. In a real-time operation, the generated event is passing through the trained modal for classification as an islanding or non-islanding event. If the event is detected as an islanding event, the relay initiates a trip signal for disconnection of the target DG. However, practical implementation of these schemes is very complex. Further, the performance of such schemes during unseen data set is not satisfactory. These methods are explained in the following sub-sections [18]–[22].

2.6.1 Fuzzy Logic

The name fuzzy itself indicates *unclear* or *vague*. In the practical operating or field operation, numerous conditions used to arise where a dealing person faces a problem of uncertainty of its operating state i.e. the state of operation is either true or false. In that case, application of fuzzy logic provides better solution with proper intellectual [19]. By this way, the person would be able to determine the probable solution for uncertainties. In comparison with Boolean logic system (truth value system), the fuzzy logic does not provide absolute true (1) and false (0) state of operation. However, the fuzzy logic also offers an intermediate state of operation, which could be said as state of partially true and partially false. Figure 2.6 gives an idea about the states of Boolean and fuzzy logic. As shown in Figure 2.6(a), there is glass of water. The state of water could be either hot or cold. However, as per fuzzy logic shown in Figure 2.6(b), the glass of water could be any one of the hot, medium hot/cold, and cold.

The architecture of fuzzy logic contains several parts such as (i) rule base, (ii) fuzzification, (iii) inference engine, and (iv) defuzzification. In rule base, the expert develops certain rules based on linguistic information for smooth decision making with the help of IF and THEN conditions. However, in recent development, by the use of effective method of fuzzy logic design, the number of fuzzy rules could be minimalized. The fuzzification converts input into fuzzy sets. These inputs are called Crisps. The Crisps are the sensors measured output for control system processing. These sensors could be temperature, voltage, currents, pressure, speed, etc. Based on the input, it is a duty of Inference Engine to decide which fuzzy rule (s) is applicable to fire. With respect to the combined fire rules, the further control action would be taken. The defuzzification is the process of conversion of fuzzy sets obtained from inference into the output crisp value. For this process, different methods have been developed. Among these available methods, the best suited method with its expert system is used to minimize the error.

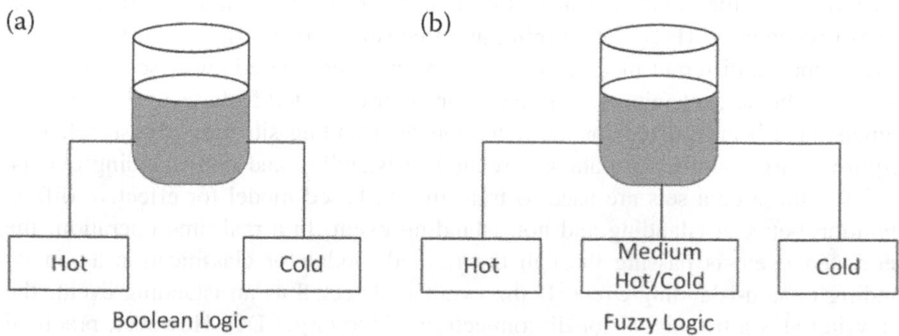

FIGURE 2.6 A typical network containing passive islanding detection method.

Looking at the various advantages such as easiness of operation, use of simple mathematical concepts, human reasoning, decision making and minimum memory requirement of fuzzy logic, various real life applications have been developed based on fuzzy logic. The same concept has been applied in the area of power system protection. Various researchers have proposed different techniques of the detection of islanding situation. In case of fuzzy-logic-based islanding detection, two major criteria namely (1) the selection of appropriate feature signal or extracted information of a signal and (2) classification of different states such as islanding and non-islanding, are required to be accomplished [19]. To select an appropriate signal or feature of a signal, a data mining process has to be carried out. Whereas, for effective classification, fuzzy classifier is initialized by the decision tree. In some techniques of islanding detection, the combination of fuzzy membership functions (MFs) and the rule base have been used to develop the fuzzy rule base. This method has been developed for online islanding detection and it is able to detect islanding situation with uncertainties of noise. In other methods, the band pass filter has been applied in place of discrete wavelet transform (DWT).

2.6.2 Artificial Neural Network (ANN)

An artificial neural network (ANN) is a method of supervised learning with the numerous elements called neurons. This neuron has an ability to take an appropriate decision and feed this information to other neurons. The neuron structure replicates the human brain with a neuro structure. Each neuron in ANN is interconnected with other neuron(s) in a specific layer(s). With these multiple layers of neurons, the ANN can solve any problem or question in a practical way [20]. To make them learn about the situation for decision, the ANN need enough number of event or situation training samples with computing power. As shown in Figure 2.7, there are several layers, namely (i) input layer, (ii) hidden layer, and (iii) output layer. Here, the hidden layer could be one or more than one. In this structure, an input layer accepts the inputs and output layer gives predictions on the given input. Based on the number of hidden layers, the ANN is categorized as: (1) Feed-forward neural networks, (2) Recurrent neural networks, (3) Convolutional neural networks, (4) Deconvolutional neural networks, and (5) Modular neural networks.

In ANN or supervised learning, there is a need to generate large number of data sets. For each event or input, the generated data would be given an appropriate label. Here, in supervised learning, training provides the input and tells the network about the output. Further, the neurons of ANN would require to train from these input data sets, hence, these data sets are called as training data sets. Based on these training data sets, the ANN would predict the probable output. The ANN gives several advantages such as parallel processing, ability to learn/model nonlinear or complex relationships between input and output, capability of fault tolerance, and ability to learn hidden relationship between data. With these advantages, ANN found real-life applications in various fields. Looking at its advantages, various researchers have proposed different ANN-based methods of islanding detection. In these methods, the classification between non-islanding and islanding has been achieved based on the supervised learning of the neurons of ANN. Therefore, large

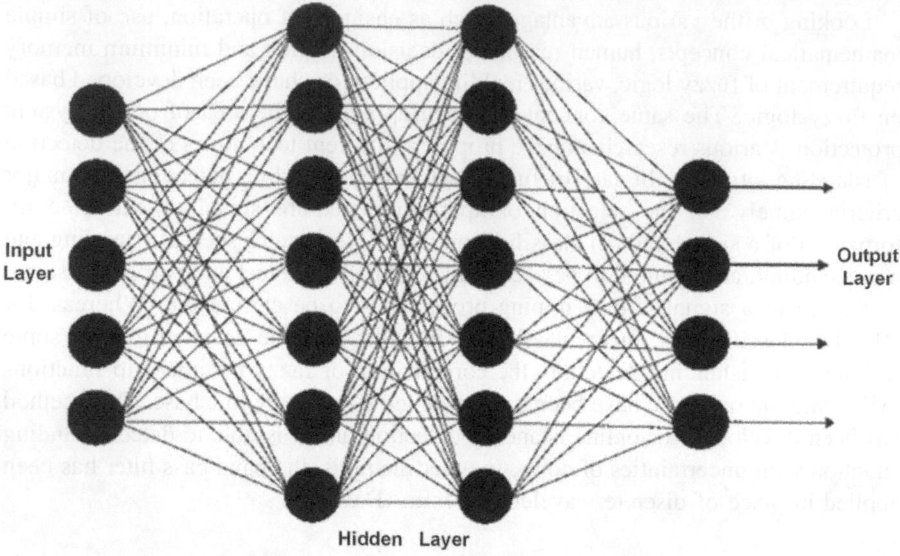

FIGURE 2.7 Layers of ANN.

number of data sets of various events have been generated on the software simulation package or real-time hardware (prototype). The data sets have been further analysed for an appropriate signal or various information have been extracted from these signals with the help of various transforms and mathematical formulations. The signals of various operating conditions, which are given as input to the ANN, are required to label with an appropriate target value. The target value indicates the type of event i.e. non-islanding or islanding. The neurons of the ANN will be trained based on these input training data sets. Here, in case of various methods of islanding detection based on ANN, the signals of current, voltage, power, power factor, rate of change of frequency, rate of change of real power, rate of change of reactive power, impedance, etc. are utilized. The extracted information from these acquired signals are utilized by various researchers to propose an accurate islanding detection scheme. In these methods, different types of ANN with multi-levels of hidden layers of the ANN have been proposed. The trained neurons of the ANN would be able to predict or discriminate between islanding and non-islanding situation effectively [20], [21].

2.6.3 Machine Learning Classifier

Supervised learning is also one of the methods of machine learning. In this method, the patterns of the input data would be recognized by the specific algorithm of machine learning. Classification problems can be solved with the help of various algorithms such as linear classifiers, support vector machines, decision trees, k-nearest neighbor, random forest, and relevance vector machine. In this classifier, the selected model will find the features within the data that would correlate to either category (target) and create the mapping function. The major applications of the machine learning methods are in the field of classification analysis, regression analysis, clustering analysis, and density estimation [22]. In classification analysis, an input pattern is labeled with an appropriate target value. This target value could be either binary ("1" or "0") or multiple for different case patterns. In order to achieve effective classification, the classifier model is required to train with the help of available patterns. These patterns are in terms of extracted feature(s) of acquired signal(s). Each pattern of an event has its own signature of a specific category. Further, the extracted feature of the acquired signal is given as an input to the classifier. The classifier model finds the most "like pattern" and classify the input pattern in that category. This technique has been adopted by various researchers to classify two different categories such as non-islanding and islanding. The patterns of all non-islanding situations have been marked with a specific target value, whereas all the islanding situation patterns have been marked with different target values. Large number of data sets for the said islanding and non-islanding situation are generated to train the classifier model. The data sets which are used to train the classifier model is known as training data set. In a later stage, the testing of the trained model is required to be carried out with help of testing data set. The testing of already trained model gives an accuracy of the classifier model in terms of discrimination of different set of events [22]–[27].

Among the above-mentioned AI techniques (2.6.1, 2.6.2, and 2.6.3) of islanding detection, a technique based on machine learning is explained in the next section as a case study.

2.7 CASE STUDY OF CLASSIFIER (MACHINE LEARNING)-BASED ISLANDING DETECTION

In this case study, a relevance vector machine (RVM) is used as a classifier model. In order to achieve accurate discrimination between islanding and non-islanding event, various extracted features of the acquired voltage and current signals have been tested. In this testing, variation in the percentage of training data set and width parameter of the classifier have been considered. With the rigorous analysis, it has been found that the negative sequence component of current (I_2) has performed better in terms of discrimination efficiency with a lower percentage of training data sets as compared to other feature signals. Moreover, to develop a large number of different patterns of islanding and non-islanding events, various islanding and non-islanding events with variable operating parameters have been generated. In this work, IEEE 34 bus network has been modelled in real-time digital simulator (RTDS) environment. This model is designed in such a way that it generates a large number of islanding situations with a variable percentage of power mismatch between load and generation. At the same time, it is also capable to generate different non-islanding events such as fault, sudden load change, switching of a capacitor bank, starting of an induction motor, etc. The acquired signal of currents during these events is then utilized for the training of the RVM classifier.

2.7.1 Relevance Vector Machine

The RVM works on Bayesian platform for implementing regression and classification with corresponding sparsity properties. The total number of relevance vectors (RVs) required by RVM are much less in comparison to the total number of support vectors (SVs) required by a support vector machine (SVM). This is a major advantage of RVM over SVM [27], [28]. The input feature vector of the proposed model with its target class label is given by (2.1):

$$\{x_i, t_i\}_{i=1}^{N} \tag{2.1}$$

where $\{t_i\}$ is the target class label and $\{x_i\}$ is the input feature vector. In general, for a machine learning, the output of the proposed model, given as a function $y(x)$, has to be predicted for an input feature vector $\{x_i\}$ as given in (2.2):

$$y(x; w) = \sum_{i=1}^{M} w_i \bullet \phi_i(x) + w_0 \tag{2.2}$$

where w_0 is a bias parameter, $\phi_i(x) = (\phi_1(x), \ \phi_2(x), \ \phi_3(x), \ \ldots\ldots, \phi_M(x))^T$ are basic functions, and $w = (w_1, \ w_2, \ w_3, \ \ldots\ldots, w_M)^T$ are weights. In the proposed RVM classifier model, by utilizing I_2, the training set has been formulated. This data set targets are given as 1 and 0 for non-islanding and islanding event, respectively. The (2.2) is the prediction after training and can be written as given in equation (2.3):

$$y(x) = \sum_{i=1}^{M} w_i \bullet K(x, x_i) + w_0 \tag{2.3}$$

where $K(\cdot, \cdot)$ is a kernel function. After applying the logistic sigmoid function, the RVM classifier model can be written as shown in equation (2.4):

$$p(d = 0|x) = \frac{1}{1 + e^{-y(x)}} \tag{2.4}$$

where p is the probability and $d \in \{0, 1\}$ is the class label. Equation (2.5) shows the probability after applying the Bernoulli distribution for $p(t|x)$:

$$p(t|w) = \prod_{i=1}^{N} \sigma\{y(x_i; w)\}^{t_i}[1 - \sigma\{y(x_i; w)\}]^{1-t_i} \tag{2.5}$$

where t is a scalar label vector given by $t = [t_1, \ldots, t_N]^{\mathrm{T}}$ with target $t_i \in \{0, 1\}$ and the sigmoid function is given by, $\sigma(y) = 1/(1 + e^{-y(x)})$. It is to be noted that for obtaining the marginal likelihood in classification, the weights cannot be integrated analytically and hence, an approximated procedure is followed. Equation (2.6) gives the posterior distribution over the weight by utilizing Bayes' rule.

$$posterior\ \ (p(w|t,\alpha,\sigma^2)) = \frac{likelyhood \times prior}{evidence} = \frac{p(t|w,\sigma^2)p(w|\alpha)}{P(t|\alpha,\sigma^2)} \tag{2.6}$$

where α is a set of hyperparameters. At the most probable parameters W_{MP}, as given by equation (2.7), the posterior $p(w|t,\alpha,\sigma^2) \propto p(t|w,\sigma^2)p(w|\alpha)$ will have a strong peak:

$$W_{MP} = \Sigma\phi^T Bt \tag{2.7}$$

where $\Sigma = (\phi^T B\phi + A)^{-1}$, Σ is the posterior covariance matrix for a Gaussian approximation to the posterior over weights centered at W_{MP}. The value of $A = \text{diag}(\alpha_1, \alpha_2 \ldots\ldots \alpha_N)$, $B = \text{diag}(\beta_1, \beta_2 \ldots\ldots \beta_N)$ where, $\beta_N = \sigma\{y(x_n)\}[1 - \sigma\{y(x_n)\}]$. The above process is repeated until a convergence criterion is met. The hyperparameters, resulted by the above method, is used to estimate the target value for input feature vector. At the Gaussian distribution with a sharp peak, the variance is very small and given by equation (2.8):

$$p(W_{MP}|\alpha,\sigma^2) \approx \frac{1}{\sigma_{W_{MN}}} s \tag{2.8}$$

where $\sigma_{W_{MN}}$ denotes the standard deviation of the posterior distribution. Further, the equation for $p(t|\alpha,\sigma^2)$ is given as equation (2.9):

$$p(t|\alpha,\sigma^2) \approx p(t|W_{MP},\sigma^2) \times p(W_{MP}|\alpha)\sigma_{W_{MP}} \tag{2.9}$$

The best-fit likelihood for a given hyperparameter σ^2 is given as the first right-hand side term of the above equation. If the value of α is very high then$p(W_{MP}|\alpha) = 1/\sigma_w$ and in such a case, the second term of the right-hand side of equation (2.9) will be $\sigma_{W_{MN}}/\sigma_w$, which is also defined as Occam factor. The complexity of the model can be calculated by multiplying the Occam factor with the accuracy of the model for given data [27]–[29].

2.7.2 Simulation and Test Cases

By modeling a standard IEEE 34 bus system in the RTDS environment, the authenticity of the proposed technique has been tested. The developed models of various existing equipment such as distribution load, distributed line, spot load, voltage regulator, and distribution transformer have been reconfigured with 50-Hz operating frequency. The considered single line diagram of IEEE 34 bus system, modeled in the RTDS environment, is shown in Figure 2.8. In this model, two DGs with their interconnecting circuit breakers (CBs), DG1 and DG2, are interconnected

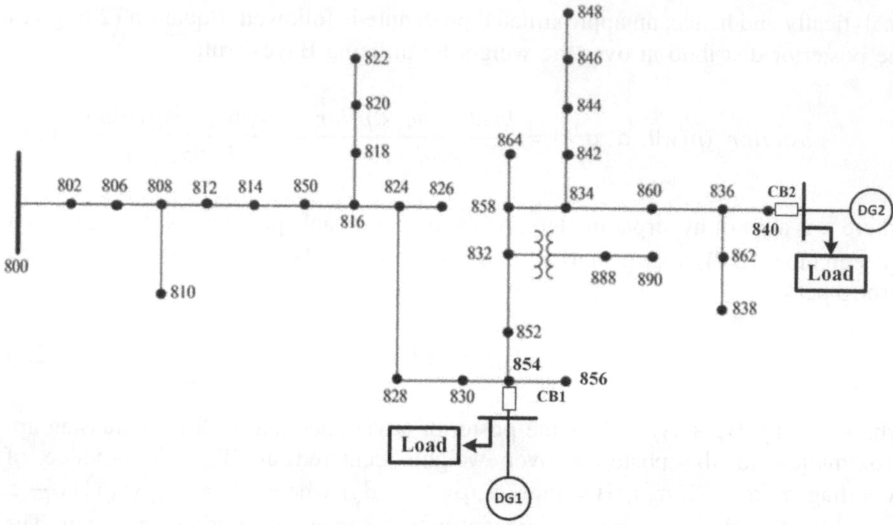

FIGURE 2.8 A standard IEEE 34 node network.

with bus numbers 854 and 840, respectively. The CB1 and CB2 are used to integrate the DG1 and DG2 with the IEEE 34 node network, respectively. The DG1 and DG2 are supplying the generated power to the local loads. At the same time, the excess power is supplied to the network. Here, an islanding situation is formed by sudden disconnection of DG from the utility supply or sudden loss of power in utility supply. In both the cases, it is required to disconnect/shutdown the DG from the load. Therefore, a dedicated relaying algorithm is required, which detects the sudden loss of utility supply at the DG side. As per IEEE standard 1547-2003, the DG must be disconnected from the load within maximum of 2 s from the inception of islanding situation. On the developed model, large number of islanding situations with variation in percentage value of real and reactive power mismatch between generation and load have been simulated to generate the test case for islanding situations. At the same time, to generate non-islanding test cases, the simulation has been performed for various conditions such as sudden variation in load, different types of fault in the network with varying fault parameters (fault resistance and fault location), sudden switching of large-sized capacitor bank and starting of a large-sized induction motor.

2.7.3 Feature Vector Formation

In order to achieve higher accuracy in classification between islanding and non-islanding situation, an input signal to the classifier must be unique in terms of pattern for both islanding and non-islanding situations. Therefore, the signal acquired from the terminal of the target DG (for passive islanding detection) or extracted feature of the acquired signal(s) has to utilized as an input signal for the classifier model. In the presented case, the acquired signals of three phase currents and voltages, their extracted features, and their combinations have been utilized for the formation of feature vector. The feature vector comprises of the input signal with the respective target value (either 0 or 1) in the last column. In the proposed technique, the signals of currents and voltages are acquired with a sampling frequency of 4 kHz. Therefore, in one cycle, 80 samples of current and voltages are acquired. In the formation of feature vector, in one row 80 columns for samples of input signal and in the last column the target value is inserted. In the presented work, 140 cases of non-islanding situation and 53 cases of islanding events have been generated. Table 2.3 shows the formation of feature vector with the feature input signal data and the respective target value.

2.7.4 Training of RVM Classifier

In the training of the RVM classifier, the feature vector for the selected training data set along with their class labels as per (2.1) are given as an input to the RVM classifier. Here, in the presented case, the feature vector is formed from 40% of the total data of the input signal. On the other hand, this feature vector has only two class labels i.e. "1" indicates non-islanding whereas "0" refers to islanding event. Afterwards, the said training data set has been initialized with weight as given by (2.2). Once the training data set is obtained, the next step is to determine the optimal parameter settings of RVM. In this process, the type of kernel function and the width parameter (w) need to decide. Therefore, proper kernel function (in this case Gaussian function) is selected for the prediction of input training data set as per

TABLE 2.3
Formation of feature vector for training data sets

Various simulation cases	Samples	Target
Non-islanding Case 1	$1_{row}\times80_{column}$	1
Non-islanding Case 2	$2_{row}\times80_{column}$	1
‖	‖	1
‖	‖	‖
‖	‖	‖
‖	‖	‖
‖	‖	‖
Non-islanding Case 140	$140_{row}\times80_{column}$	1
Islanding Case 1	$1_{row}\times80_{column}$	0
Islanding Case 2	$2_{row}\times80_{column}$	0
‖	‖	‖
‖	‖	‖
‖	‖	‖
Islanding Case.53	$53_{row}\times80_{column}$	0
Total training cases193	$193_{row}\times80_{column}$	1 or 0

equation (2.3). Thereafter, a separable Gaussian prior, with a distinct hyperparameter for each weight is calculated. Then, the most suitable marginal likelihood for the given hyperparameter is determined. The hyperparameters are used in estimating the target value for the input feature vector. The posterior distribution over the weight from Bayes rule and the posterior distribution standard deviation are calculated. The RVM training by such a method will give the training vector associated with the remaining non-zero weights that are called as *relevance vectors*. Finally, testing of the given data set (different from the data set used in training process) is carried out directly on the already trained model of RVM classifier.

2.7.5 Result and Discussion

The performance of the proposed classifier for various non-islanding and islanding cases has been validated. By utilizing an already well-trained RVM classifier (by considering a width of 0.003), the proposed classifier has been tested. Table 2.4 shows the simulation results for the incorrectly and correctly classified cases in addition with discrimination accuracy. Further, various non-islanding events and the total numbers of cases are also given in Table 2.4. It is observed from Table 2.4 that the proposed classifier based on RVM is capable of identifying non-islanding situations with 98.69% accuracy. This indicates the stability of the proposed scheme against undesired tripping. Similarly, islanding situations have been identified with 98.36% accuracy. This indicates the superiority of the proposed RVM classifier in accurately detecting the islanding situations. Discrimination between non-islanding events at different loading conditions and islanding situations at different percentage of power mismatches have been analyzed for the proposed classifier.

2.8 PROTECTION MISCOORDINATION DUE TO DG INTERCONNECTION

2.8.1 Issue of Protection Miscoordination

An interconnection of DG with an existing distribution network disturbs the existing protection coordination among various protective devices of the network. The said issue can be easily understood from Figure 2.9. As observed from Figure 2.9, various protective devices such as Overcurrent Relay (OCR), recloser, and fuse are coordinated based on unidirectional power flow in radial distribution network [1], [2]. When the DG is interconnected with radial DN, the DG current flows into the network towards the load as well as the source [3], [4]. Hence, the lateral end protective devices see more current than the source end device. In this condition, existing protection coordination may be lost. The said situation of loss of protection coordination is explained below.

Let I_S be the utility source current, I_R be the current seen by recloser, I_{Fuse} be the current flowing through the fuse and, I_{DG} be the current supplied by the DG.

When no DG is connected, the current flowing through the recloser and fuse are same. Further, the current supplied by the DG is zero. This situation is described by (2.10):

TABLE 2.4
Proposed classifier discrimination accuracy for various non-islanding and islanding conditions

Sr. No.	Various conditions	Tested cases	Correctly classified cases	Wrongly classified cases	Accuracy(%)
Islanding conditions		61	60	1	98.36
Non-islanding conditions					
1	Faults on main feeder	40	40	0	100.00
2	Faults on adjacent feeder	15	14	1	93.33
3	Sudden load change on target DG bus	28	28	0	100.00
4	Sudden power factor of load change on target DG bus	20	20	0	100.00
5	Capacitive load switching at PCC of target DG	19	19	0	100.00
6	Induction motor load switching at PCC of target DG	19	19	0	100.00
7	Sudden addition of new DG with load	19	19	0	100.00
8	Sudden tripping of adjacent DG	19	18	1	94.74
9	Fault on adjacent feeder at various X/R ratio of target DG	31	31	0	100.00
10	Sudden load change with changed Network topology	19	18	1	94.74
All Non-islanding Events		229	226	3	98.69
Overall Result (Islanding + Non-islanding)		290	286	4	98.62

$$I_{Fuse} = I_R \text{ and } I_{DG} = 0 \tag{2.10}$$

Conversely, when the DG is connected, the current flowing from the lateral fuse is the sum of the current flowing from recloser and current supplied by the DG. This situation is described by (2.11):

$$I_{Fuse} \neq I_R \text{ but } I_{Fuse} = I_R + I_{DG} \tag{2.11}$$

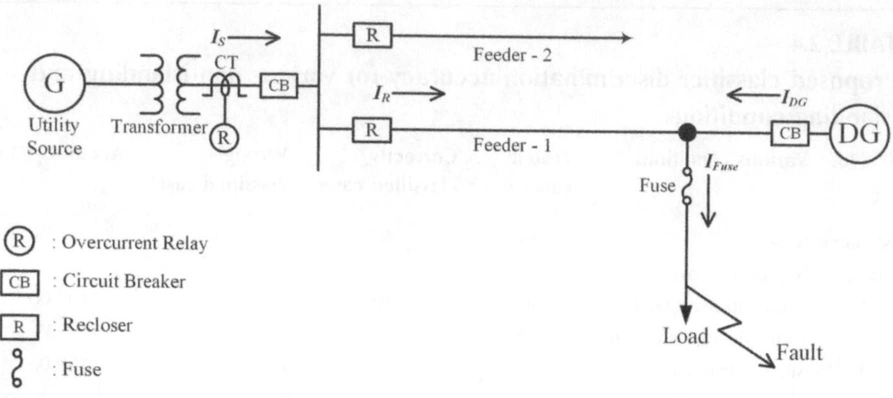

FIGURE 2.9 Typical radial distribution network with DG.

In the DN, the low rated fuse is coordinated with a higher capacity upstream fuse. The high-rated fuse is coordinated with the upstream recloser and the recloser is coordinated with the upstream OCR. Finally, the OCR is coordinated with another upstream OCR. In this way, the protection coordination covers the entire DN and provides primary and backup protection to every segment of the DN. The co-ordinated backup protection protects the network in case of failure of primary protection. Hence, the protection coordination problem in DN can be categorized as (i) fuse to fuse coordination, (ii) fuse to recloser coordination, (iii) recloser to OCR coordination, and (iv) OCR to OCR coordination. Among all coordination issues, the recloser–fuse coordination is very crucial [30], [31]. This is due to the fact that in case of a temporary fault on any of the lateral, the recloser has to operate faster than the fuse. Conversely, during a permanent fault, the fuse has to blow before the recloser. On contrary, in the presence of DG on the lateral, an additional fault current is supplied by the DG. This makes the fuse blow prior to the recloser for every temporary fault condition and hence, there is a permanent disconnection of the network. This is against the fuse-saving principle and makes an expensive fuse blow even for a temporary fault [32], [33]. During some cases, the loss of co-ordination remains undetected for a long time until a major disturbance occurs.

2.8.2 Application of AI Technique for Restoration of Protection Coordination

In order to restore the protection coordination among various protective devices, researchers have proposed various techniques based on acquired signals of currents, voltages, utlization of fault current limiter (FCL), use of superconducting fault current limiter (SFCL), change in the operating characteristic of OCR, and recloser. Recently, with the advancement in AI techniques, some of the mehtods have been proposed based on utilization of different AI techniques. These techniques utilize the acquired signals of either current or voltage in addition with the status of DG interconnection breaker. The variation in the acquired signals during DG interconection and disconnection and status of DG interconnecting breaker is observed by these AI-based techniques. The AI-based techniques would take a decision to update/change the recloser/backup relay operating characteristic to maintain proper coordination between relay to relay, relay to recoser, and recloser to fuse.

2.9 SUMMARY

In this chapter, a general overview on the distributed generation and its advantages over the conventional power system is given. Further, several technical and operting issues that arise due to incorporation of DG are also explained. In this chapter, the problems of DG interconnection and its probable solutions have been summarized. Furthermore, an application of the novel concept of artificail intelligence has been explained to sort out the protection issues of DG interconnection. In addition, with a case study and its results, it has been throughly explained how one of the proteciton issues (islanding detection) of a DG interconnection can be resolved. This chapter will be beneficial to the readers to carry out further research in the area of islanding detection and restoration of protection coordination in the presence of DG by utilizing various AI-based techniques.

REFERENCES

[1] C. Christopoulos and A. Wright, *Electrical power system protection*. Springer Science & Business Media, 2012.

[2] B. Bhalja, R. P. Maheshwari, and N. G. Chothani, *Protection and switchgear*. Oxford University Press, 2011.

[3] M. H. Bollen and F. Hassan, *Integration of distributed generation in the power system*, vol. 80. John Wiley & Sons, 2011.

[4] A.-M. Borbely and J. F. Kreider, *Distributed generation: the power paradigm for the new millennium*. CRC Press, 2001.

[5] IEEE Std 1547-2003, "IEEE Standard for interconnecting distributed resources with electric Power systems," *IEEE Standards*. pp. 1–16, Jul. 2003.

[6] IEEE Std 1547a-2014, "IEEE Standard for interconnecting distributed resources with electric power systems amendment," *IEEE Standards*. pp. 1–16, May 2014.

[7] T. Ackermann, G. Andersson, and L. Söder, "Distributed generation: a definition," *Electr. Power Syst. Res.*, vol. 57, no. 3, pp. 195–204, Apr. 2001.

[8] A. A. Memon and K. Kauhaniemi, "A critical review of AC Microgrid protection issues and available solutions," *Electr. Power Syst. Res.*, vol. 129, pp. 23–31, Dec. 2015.

[9] M. F. Akorede, H. Hizam, and E. Pouresmaeil, "Distributed energy resources and benefits to the environment," *Renew. Sustain. Energy Rev.*, vol. 14, no. 2, pp. 724–734, Feb. 2010.

[10] N. K. Roy and H. R. Pota, "Current status and issues of concern for the integration of distributed generation into electricity networks," *IEEE Syst. J.*, vol. 9, no. 3, pp. 933–944, Sep. 2015.

[11] B. J. Brearley and R. R. Prabu, "A review on issues and approaches for microgrid protection," *Renew. Sustain. Energy Rev.*, vol. 67, pp. 988–997, Jan. 2017.

[12] C. Li, C. Cao, Y. Cao, Y. Kuang, L. Zeng, and B. Fang, "A review of islanding detection methods for microgrid," *Renew. Sustain. Energy Rev.*, vol. 35, pp. 211–220, Jul. 2014.

[13] Y. Guo, K. Li, D. M. Laverty, and Y. Xue, "Synchrophasor-based islanding detection for distributed generation systems using systematic principal component analysis approaches," *IEEE Trans. Power Deliv.*, vol. 30, no. 6, pp. 2544–2552, Dec. 2015.

[14] B. Bahrani, H. Karimi, and R. Iravani, "Nondetection zone assessment of an active islanding detection method and its experimental evaluation," *IEEE Trans. Power Deliv.*, vol. 26, no. 2, pp. 517–525, Apr. 2011.

[15] A. Samui and S. R. Samantaray, "An active islanding detection scheme for inverter-based DG with frequency dependent ZIP–Exponential static load model," *Int. J. Electr. Power Energy Syst.*, vol. 78, pp. 41–50, Jun. 2016.

[16] D. Motter, J. C. M. Vieira, and D. V. Coury, "Development of frequency-based anti-islanding protection models for synchronous distributed generators suitable for real-time simulations," *IET Gener. Transm. Distrib.*, vol. 9, no. 8, pp. 708–718, May 2015.

[17] V. Menon and M. H. Nehrir, "A hybrid islanding detection technique using voltage unbalance and frequency set point," *IEEE Trans. Power Syst.*, vol. 22, no. 1, pp. 442–448, Feb. 2007.

[18] K. El-Arroudi, G. Joos, I. Kamwa, and D. T. McGillis, "Intelligent-based approach to islanding detection in distributed generation," *IEEE Trans. Power Deliv.*, vol. 22, no. 2, pp. 828–835, Apr. 2007.

[19] S. R. Samantaray, K. El-Arroudi, G. Joos, and I. Kamwa, "A fuzzy rule based approach for islanding detection in distributed generation," *IEEE Trans. Power Deliv.*, vol. 25, no. 3, pp. 1427–1433, Jul. 2010.

[20] A. Khamis, H. Shareef, A. Mohamed, and E. Bizkevelci, "Islanding detection in a distributed generation integrated power system using phase space technique and probabilistic neural network," *Neurocomputing*, vol. 148, pp. 587–599, Jan. 2015.

[21] V. L. Merlin, R. C. Santos, A. P. Grilo, J. C. M. Vieira, D. V. Coury, and M. Oleskovicz, "A new artificial neural network based method for islanding detection of distributed generators," *Int. J. Electr. Power Energy Syst.*, vol. 75, pp. 139–151, Feb. 2016.

[22] M. R. Alam, K. M. Muttaqi, and A. Bouzerdoum, "An approach for assessing the effectiveness of multiple-feature-based SVM method for islanding detection of distributed generation," *IEEE Trans. Ind. Appl.*, vol. 50, no. 4, pp. 2844–2852, Jul. 2014.

[23] N. W. A. Lidula and A. D. Rajapakse, "A pattern-recognition approach for detecting power islands using transient signals-Part II: Performance evaluation," *IEEE Trans. Power Deliv.*, vol. 27, no. 3, pp. 1071–1080, Jul. 2012.

[24] O. N. Faqhruldin, E. F. El-Saadany, and H. H. Zeineldin, "A universal islanding detection technique for distributed generation using pattern recognition," *IEEE Trans. Smart Grid*, vol. 5, no. 4, pp. 1985–1992, Jul. 2014.

[25] S. Kar and S. R. Samantaray, "Data-mining-based intelligent anti-islanding protection relay for distributed generations," *IET Gener. Transm. Distrib.*, vol. 8, no. 4, pp. 629–639, Apr. 2014.

[26] B. Matic-Cuka and M. Kezunovic, "Islanding detection for inverter-based distributed generation using support vector machine method," *IEEE Trans. Smart Grid*, vol. 5, no. 6, pp. 2676–2686, Nov. 2014.

[27] M. Tipping, "Sparse Bayesian learning and the relevance vector machine," *J. Mach. Learn. Res.*, vol. 1, pp. 211–244, 2001.

[28] M. Tipping, "Relevance Vector Machine," US Patent 6 633 857 B1, 2003.

[29] Y. Makwana and B. R. Bhalja, "Islanding detection technique based on relevance vector machine," *IET Renew. Power Gener.*, vol. 10, no. 10, pp. 1607–1615, Nov. 2016.

[30] P. H. Shah and B. R. Bhalja, "New adaptive digital relaying scheme to tackle recloser–fuse miscoordination during distributed generation interconnections," *IET Gener. Transm. Distrib.*, vol. 8, no. 4, pp. 682–688, Apr. 2014.

[31] S. M. Brahma and A. A. Girgis, "Development of adaptive protection scheme for distribution systems with high penetration of distributed generation," *IEEE Trans. Power Deliv.*, vol. 19, no. 1, pp. 56–63, Jan. 2004.

[32] A. F. Naiem, Y. Hegazy, A. Y. Abdelaziz, and M. A. Elsharkawy, "A classification technique for recloser-fuse coordination in distribution systems with distributed generation," *IEEE Trans. Power Deliv.*, vol. 27, no. 1, pp. 176–185, Jan. 2012.

[33] B. Hussain, S. M. Sharkh, S. Hussain, and M. A. Abusara, "An adaptive relaying scheme for fuse saving in distribution networks with distributed generation," *IEEE Trans. Power Deliv.*, vol. 28, no. 2, pp. 669–677, Apr. 2013.

3 An Intelligent Scheme for Classification of Shunt Faults Including Atypical Faults in Double-Circuit Transmission Line

Valabhoju Ashok[1], Anamika Yadav[1], Mohammad Pazoki[2], and Almoataz Y. Abdelaziz[3]

[1]Department of Electrical Engineering, National Institute of Technology, Raipur, C.G., India

[2]School of Engineering, Damghan University, Damghan, Iran

[3]Faculty of Engineering and Technology, Future University in Egypt, Cairo, Egypt

3.1 INTRODUCTION

The power transmission line is the furthermost vulnerable component owing to its large physical dimension in power system networks. The thermal, electrical, mechanical, and environmental eccentricities are the foremost reasons for faults on transmission lines, which can be demarcated as common shunt faults (CSF), cross-country faults (CCF), and evolving faults (EVF). A common shunt fault, as depicted in Figure 3.1, can be of LG, LLG, LL, LLL, or LLLG type that occurs at any one location. The EVFs as depicted in Figure 3.2 can be categorized as ground faults that occur on diverse phases of the same circuit at the same locations at different fault inception times. For example, EVF (A1G-B1G) occurs on phase "A1" at 15 km, and it is evolved to phase "B1" after ½-cycle delay at the same location of circuit-I, as shown in Figure 3.2. At present, in this unified power system network, the detection/location of EVFs is wearisome; none of the recent numerical distance relays are efficient enough. Generally, these transmission lines lying all the way through the forest area, during thunderstorms/cyclones, incidences of EVFs are most severe, and the EVFs' location predicted by digital fault locators is not precise, and it misinforms the line patrolling team. When the transmission line length is very

DOI: 10.1201/9780367552374-3

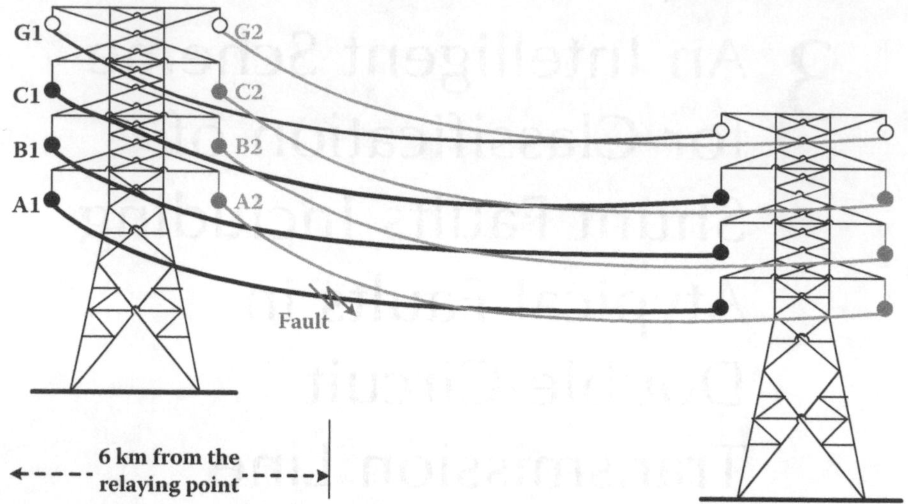

FIGURE 3.1 Common shunt fault in dual-circuit transmission line.

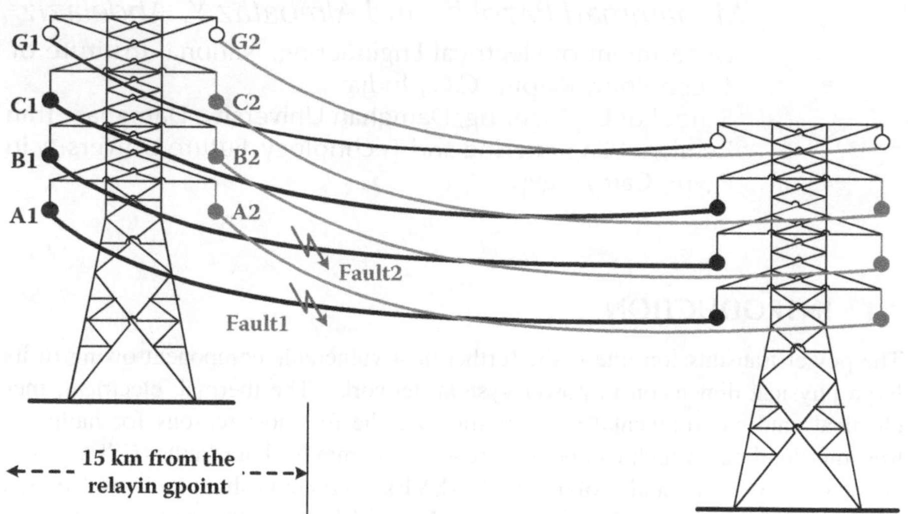

FIGURE 3.2 An evolving fault in a dual-circuit transmission line.

large, it becomes very tedious to patrol through the line. Similarly, there is another type of fault known as a cross-country fault that occurs in two different phases at different locations at the same time, as depicted in Figure 3.3.

Although the protection of double-circuit transmission line (DCTL) is more complex due to the effect of mutual coupling, the possibility of occurrence of CCFs, inter-circuit faults (ICF), and EVF is more. The ICFs, CCFs, and EVFs also result in ambiguity in the phase-selectivity for a single-pole tripping scheme because of zero-sequence current [1], and it is more severe where system stability is considered

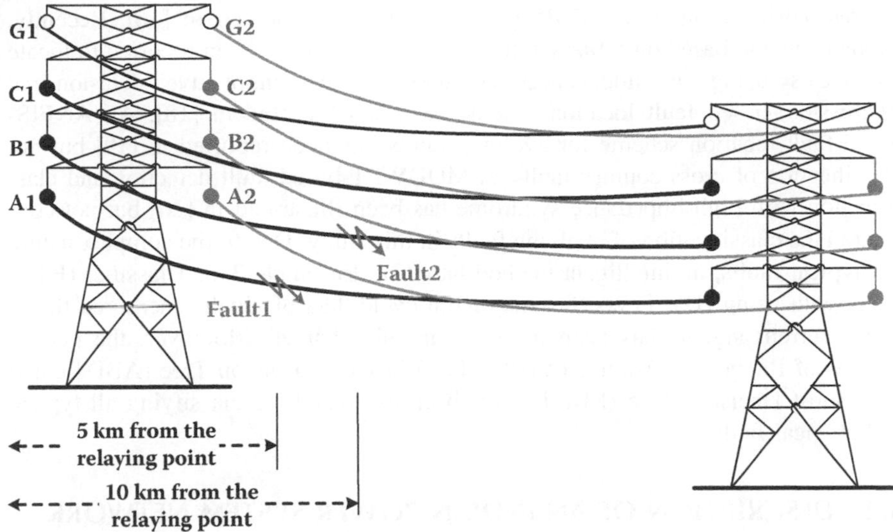

FIGURE 3.3 Cross-country fault in dual-circuit transmission line.

as most significant. Detection/classification of ICFs and CCFs using neural networks have been described in [2], and the investigation of the differential protection scheme for CCFs on transmission lines has been described in [3]. An effect of CCFs on the distance relaying scheme has been investigated on the 132-kV transmission line in [4]. The zone-I distance relaying algorithm for non-earthed CCFs in a parallel transmission line has been reported in [5], and CCFs with ground involvement has been proposed in [6]. The CCFs expose intricate characteristics and severely affect the efficacy of a distance relaying scheme since the CCF contains a combination of two faulty events that occur simultaneously at two diverse locations on different phases in the same circuit [7]. A discrete wavelet transform-artificial neural network (DWT-ANN)-based algorithm has been employed for the location of all CSFs, including CCFs and EVFs, without fault-type classification in a single-circuit transmission line in [8]. An innovative method for categorizing CCFs in a series compensated DCTL has been proposed in [9]. Further, a fault detection method for CCFs in medium distribution ring lines has been demonstrated in [10], where the mixed over headlines and cables have been studied from the Italian power distribution network. An equivalent circuit model has been featured for the assessment of CCF currents in a medium voltage distribution line in [11]. An ANN-based fault location scheme for multi-location fault in a series-compensated DCTL has been reported in [12]. The DWT ANN-based algorithm for detection/classification of multi-location faults and EVFs in a DCTL has been reported in [13,14] and Maximal Overlap DWT has been proposed in [15]. All the schemes [1–15] deal with only detection and classification of CCFs except [8], which also locates the CCF in a single-circuit transmission line. Further, the fault location assessment corresponding to common shunt faults in a DCTL has been carried out by using a discrete cosine transform and Bagged decision tree approach in [16]. Recently, another scheme based on a fuzzy inference system [17] has been reported to locate the CCFs using an impedance measurement. An innovative decision-tree (regression)-based fault location scheme has been reported in [18]. An ANFIS-based fault location scheme for evolving faults has been reported in [19] but not classification of cross-country faults. A MODWPT-based fault detection and classification with high-impedance syndrome has been illustrated in [20] but not considered the classification of evolving faults in this study. Due to the complex nature of atypical faults, an intelligent method based on Ensemble Tree Classifier (ETC) along with using (DWT) coefficients of full-cycle data of locally measured three-phase current signals has been proposed in this chapter. Moreover, the performances of Bagged Decision Tree (BGDT), AdaBoost Decision Tree (ABDT), and RUSBoost Decision Tree (RBDT) have been evaluated for classifying all typical and atypical faults.

3.2 DESCRIPTION OF AN INDIAN POWER SYSTEM NETWORK

A 400-kV and 50-Hz Indian power transmission network of Chhattisgarh state is shown in Figure 3.4 and corresponding equipment/network data is adopted from [15]. The transmission network comprises of two power generation units at bus-4 (KSTPS/NTPC), Unit-I with 4 × 500 MW and Unit-II with 3 × 210 MW. Here, the DCTL connected between bus-4 and bus-3 (i.e. between KSTPS/NTPC and Bhilai/Khedamara) of 198-km length is considered and its line configuration is adopted from [15]. This 400-kV Chhattisgarh state power transmission network is modeled

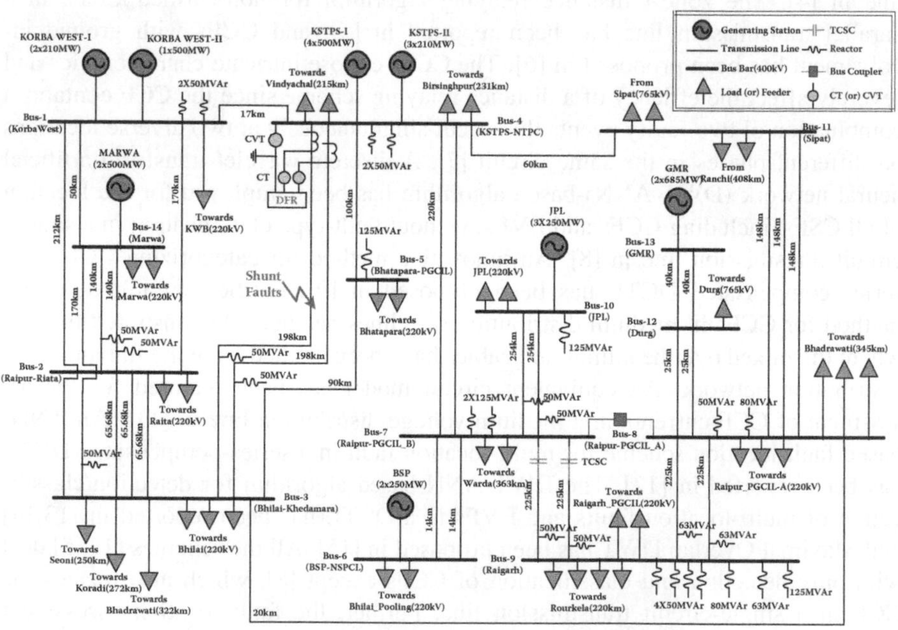

FIGURE 3.4 One-line diagram of 400-kV Chhattisgarh state power system network.

in the MATLAB®/Simulink environment and the proposed scheme (ETC-Model) is employed at bus-4 (KSTPS/NTPC) to classify all types of shunt faults including CSFs, CCFs, and EVFs at a wide range of variation in fault parameters.

3.3 ENSEMBLE TREE CLASSIFIER (ETC) MODEL FOR CLASSIFICATION OF CSFS, CCFS, AND EVFS

The proposed fault classification scheme based on the ETC model deals with shunt faults, which occur typically on different phases in the DCTL at different fault instants. Herein this ETC model incorporates four classifier modules (ETC-1, -2, -3, and -4) based on an ensemble of decision trees. These fault classifier modules are designed by considering three main types of shunt faults such as CSFs (LG, LLG, LL, and LLL), CCFs (LG-LG, LG-LLG, and LLG-LG), and EVFs (LG-LG, LG-LLG, and LLG-LG) which are most significant amongst all types of shunt faults. For instance, the CCFs/EVFs (A1G-B1G) is a combination of two faults, fault-1(A1G) and fault-2(B1G) which occurred subsequently on different phases at the same inception time/location on the same circuit-I. With the aid of the fault classifier modules (ETC-1, -2, -3, and -4), the classification of all types of shunt faults is possible using only one-end data of the DCTL. In the viewpoint of ETC-models, tree-based ensemble methods are normally used for prediction, classification, and regression objectives in numerous research domains for instant protection of transmission lines [16]. Figure 3.5 demonstrates the flow chart of the ETC-based scheme.

As shown in Figure 3.5, the flowchart comprises three stages: the first stage deals with data preprocessing followed by feature extraction, in the second stage, design of exclusive data sets have been done for training/testing of different ETC modules, and in the third stage, classification of all types of shunt faults has been carried out including CSFs, CCFs, and EVFs. The tree-based ensemble techniques such as Bagged Decision Tree (BGDT) and Boosted Decision Tree (BSDT) are outpacing to simple/conventional decision-tree-based techniques. The bagged decision tree and boosted decision tree are discussed in a further sub-section.

3.3.1 Designing of Exclusive Data Sets

An accurate and reliable fault classification scheme is a basic necessity of the contemporary interconnected transmission network. When a fault takes place, the fault current contains of several undesirable frequencies and the DC offset. To avoid an error in a subsequent signal processing due to signal aliasing (arising false frequencies in a signal) the voltage and current signals are passed through an anti-aliasing filter (second-order Butterworth filter with 400 Hz as cut off frequency). By using this second-order Butterworth filter, any undesired high frequencies (noise) are removed. Exclusive data sets are created by simulating several fault scenarios on a practical Chhattisgarh state power transmission network in MATLAB/Simulink software. Further, the voltage and current signals are recorded at a 1.0-kHz sampling frequency. Figure 3.6 illustrates the MATLAB/Simulink model with an anti-aliasing filter and DWT block for data acquisition and signal preprocessing and the

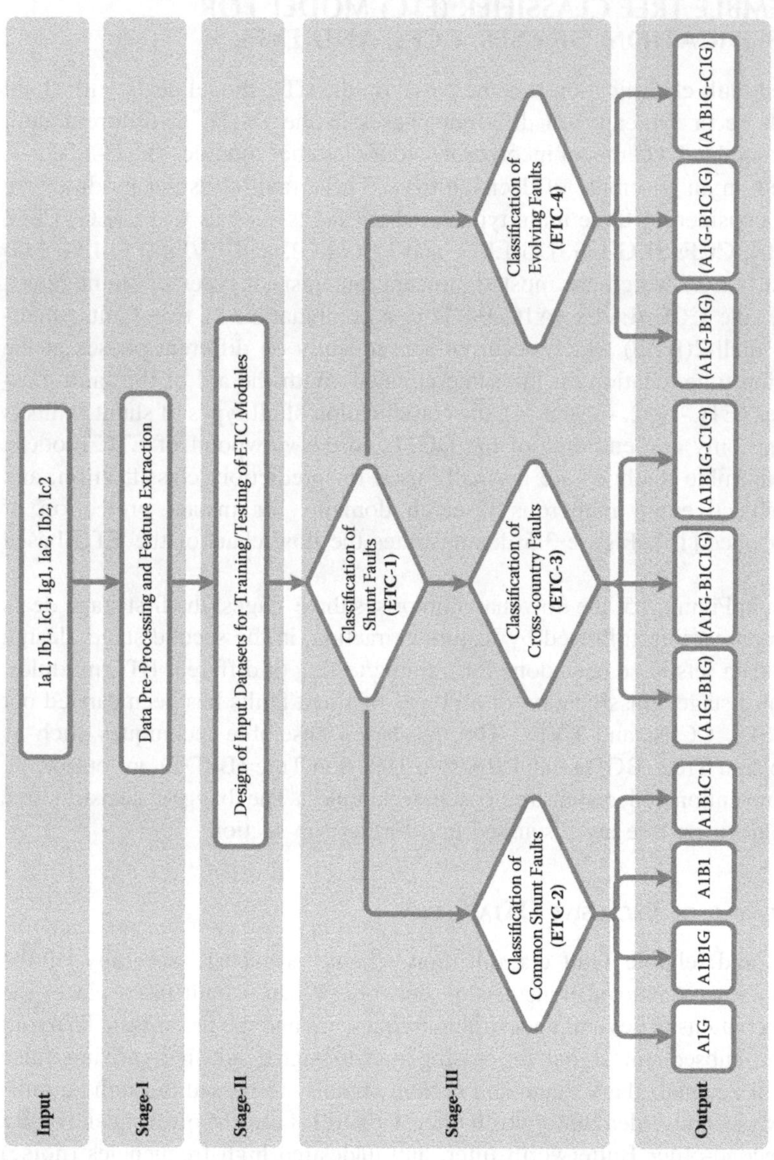

FIGURE 3.5 Proposed ETC-model for classification of shunt faults based on ensemble of decision trees.

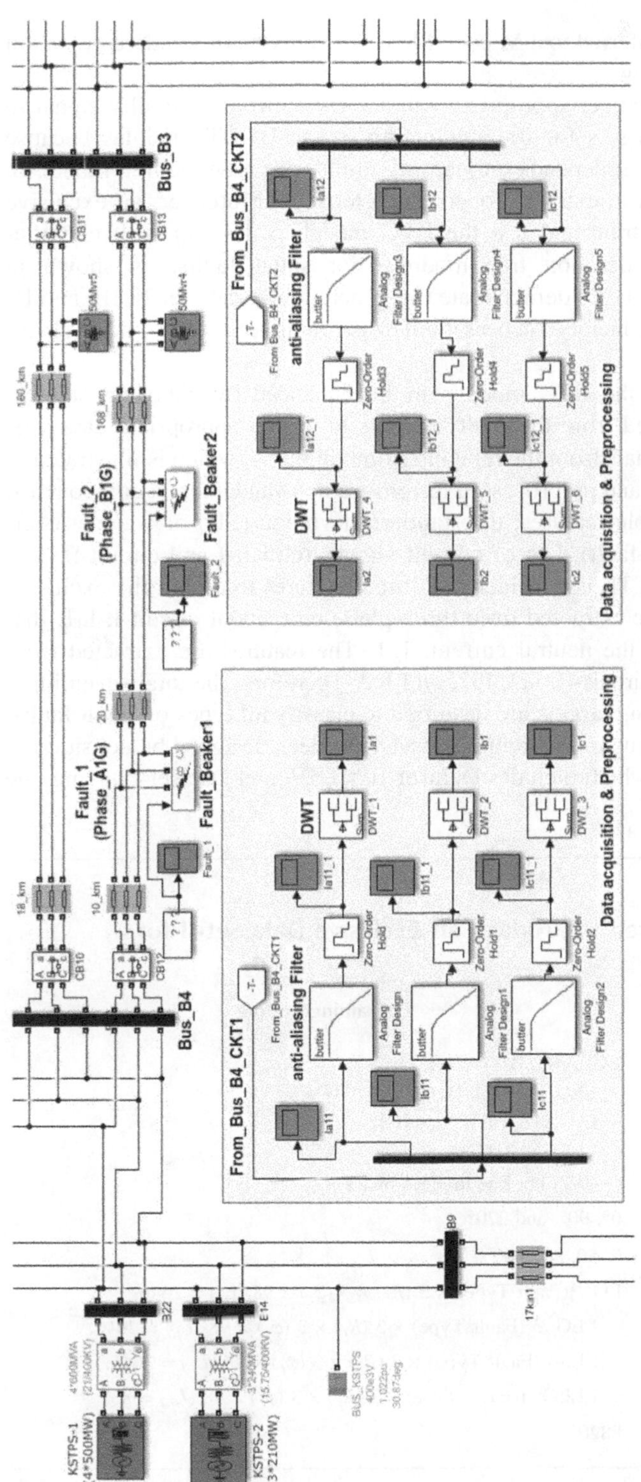

FIGURE 3.6 MATLAB/Simulink model for data acquisition and pre-processing with anti-aliasing filter and DWT block.

same output will be transferred to a M-code file for calculation of standard deviation value in a 1-cycle window.

It is very crucial to extract appropriate characteristics from the faulty signal to construct exclusive data sets for training/testing of the BGDT modules because efficacy of the ETC model depends on learning ability of ensemble tree model. In this work, DWT is used to extract proper characteristics. Moreover, an exclusive data set is provided to train and test the ETC model to develop fault classifier modules (ensemble of decision tree module) for shunt faults, as shown in Tables 3.1–3.3. Tables 3.1–3.3 demonstrate the generation of data set-II, III, and IV by altering several fault factors such as fault type, fault resistance, and fault inception angle.

The performance of the ETC model can be enhanced by choosing suitable features that are extracted from the faulty signals to design appropriate data sets. Therefore, the faulty signals from the relaying point (at bus-4, which is illustrated in Figure 3.4) are recorded and preprocessed to remove non-fundamental components. Further, to extract suitable features, the standard deviation (SD) of wavelet coefficients of 1-cycle (post-fault) data of current signals (circuit-I and circuit-II) and neutral current of the DCTL is considered as input features to design the exclusive data sets: the features are extracted from three-phase currents of circuit-I, Ia1, Ib1, Ic1, and extracted from the neutral current, Ig1. The features are extracted from three-phase currents of circuit-II, Ia2, Ib2, and Ic2. Therefore, the total seven input features and corresponding targets are designed to classify all types of shunt faults. From Figure 3.5, for example, the input Dataset-I has been designed by considering all types of shunt faults which includes Dataset-II, III, IV, and Dataset-I having one

TABLE 3.1
Various Constraints Used to Produce an Exclusive Data Set-II for Classification of CSFs

Parameter	Training/Testing
Fault Type	LG: A1G, B1G, C1G, LLG: A1B1G, B1C1G, A1C1G, LL: A1B1, B1C1, A1C1, LLL: A1B1C1
Fault Location (L_f)	(1–197) km line in steps of 1 km
Fault Inception Angle (ϕ_f)	0°, 90°, and 270°
Fault Resistance (R_f)	0, 50, and 100 Ω
Fault Type: No. of Fault Cases	LG: 3(Fault Type) × 3 (R_f) × 3 (ϕ_f) × 98 (L_f) = 2646, LLG: 3 (Fault Type) × 3 (R_f) × 3 (ϕ_f) × 98 (L_f) = 2646, LL: 3 (Fault Type) × 3 (R_f) × 3 (ϕ_f) × 98 (L_f) = 2646, LLLG: 1(Fault Type) × 3 (R_f) × 3 (ϕ_f) × 98 (L_f) = 882
Total No. of Fault Cases	8820

TABLE 3.2
Various Constraints Used to Generate an Exclusive Data Set-III for Classification of CCFs

Parameter	Training/Testing
Fault Location (L_f)	(1–197) km in steps of 5 km
Fault Inception Angle (ϕ_f)	0°, 90° and 270°
Fault Resistance (R_f)	0 Ω, 50 Ω and 100 Ω
Fault Type: No. of Fault Cases	(A1G-B1G): 3 (R_f) × 3 (ϕ_f) × 1921 (L_f) =17289
	(A1G-B1C1G): 3 (R_f) × 3(ϕ_f) × 1921 (L_f) = 17289
	(A1B1G-C1G): 3 (R_f) × 3(ϕ_f) × 1921 (L_f) = 17289
Total Number of Fault Cases	3 × 17289 = 51867

TABLE 3.3
Various Parameters Used to Produce an Exclusive Data Set-IV for Classification of EVFs

Parameter	Training/Testing
Fault Location (L_f)	(1–197) km in steps of 1 km
Fault Inception Angle (ϕ_f)	0°, 90°, and 270°
Fault Resistance (R_f)	0 Ω, 50 Ω, and 100 Ω
Fault Evolving Time (E_f)	10 ms, 20 ms, 40 ms, 60 ms, 80 ms, and 100 ms
Fault Type/No. of Fault Cases	(A1G-B1G) fault: 3(R_f) × 3(ϕ_f) × 6(E_f) × 197(L_f) = 10638
	(A1G-B1C1G) fault: 3(R_f) × 3(ϕ_f) × 6(E_f) × 197(L_f) = 10638
	(A1B1G-C1G) fault: 3(R_f) × 3(ϕ_f) × 6(E_f) × 197(L_f) = 10638
Total Number of Fault Cases	3 × 10638 = 31914

target vector with three classification labels, "1" for CSFs, "2" for CCFs, and "3" for EVFs. Further, Dataset-II having one target vector with ten classification labels, "1" for AG fault, "2" for BG fault, "3" for CG fault, "4" for ABG fault likewise all 10 types of single location common shunt faults are assigned from "1" to "10" as classification labels. The Dataset-III with one target vector with three classification labels, "1" for (A1G-B1G) fault, "2" for (A1G-B1C1G) fault, and "3" for (A1G-B1C1G) fault. Similarly, the Dataset-IV having one target vector with three classification labels, "1" for (A1G-B1G) fault, "2" for (A1G-B1C1G) fault, and "3" for (A1G-B1C1G) fault.

3.3.2 Discrete Wavelet Transform (DWT)

The DWT is a time-frequency-based multi-resolution analysis technique to decompose the signal into diverse levels. The original signal is decomposed into two components: approximation coefficient and detail coefficient [14]. A data window of 1-cycle is chosen from the fault instant and subsequently DWT is employed to extract detail coefficients at level-6 of the fault signal using the "db4" wavelet.

3.3.3 Bagged Decision Tree

The Bagged Decision Tree (BGDT) is built by randomization of abundant decision trees and thereafter accumulates their predictions. Each/every tree in the ensemble is grownup by separately derived bootstrap imitation of input data [21]. Observations not comprised in this imitation are "out of the bag" for this tree [22]. As shown in Figure. 3.7 the basic architecture of bagged decision tree, total four BGDT modules are designed/trained separately. The BGDT-1 is designed to train/test the Dataset-I which classifies the type of fault such as CSFs, CCFs, and EVFs. BGDT-2 is designed to classify CSFs such as LG, LLG, LL, and LLL. The BGDT-3 is designed to classify CCFs such as (A1-B1G) fault, (A1G-B1C1G) fault, and (A1B1G-C1G) fault. The BGDT-4 is designed to classify EVFs such as (A1G-B1G), (A1-B1C1G), and (A1B1G-C1G) which are most significant and non-repetitive among all possible fault cases. However, individual decision trees have a tendency to over-fit/under-fit. The BGDT combines the outcomes of numerous decision trees, thus upsetting the impacts of over-fitting and enhances generalization. The BGDT grows in the ensemble model using bootstrap examples of the input data. The BGDTs of ensemble models mainly explore unique functionality such as classification when the predicted outcome can be considered as an integer or binary value [23]. Bagging tree explores a good performance when the base decision tree process is not very stable. When minor variations in the training pattern

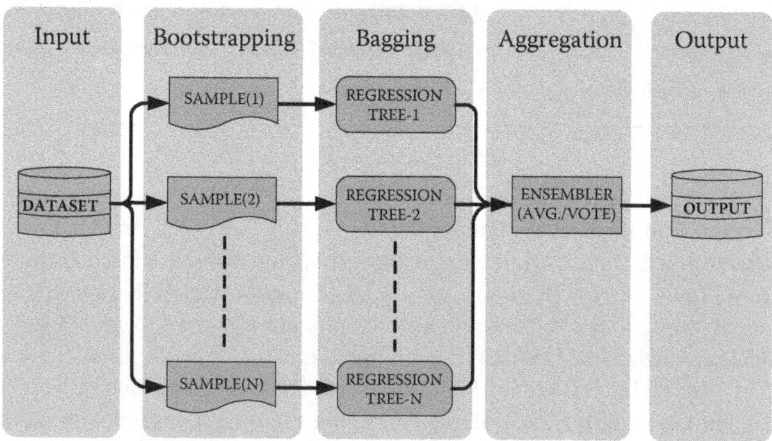

FIGURE 3.7 Basic Architecture of BGDT module.

repeatedly result in substantial discrepancies in the predictions attained, bagging can show a significant drop in average prediction error and it improves the accuracy of the ensemble model.

3.3.4 Boosted Decision Tree

The Boosted Decision Tree (BSDT) is a scheme that combines different decision trees, that are iteratively designed through weighted/biased versions of the training sample that were miscategorized in the preceding stage. The decisive predictions are attained by weighting the outcomes of the iteratively produced prediction. Dissimilar to bagging that works on simple averaging of outcomes to attain an overall decision, BSDT uses a weighted average of outcomes attained by applying a prediction method to several input patterns. Similarly, as shown in Figure 3.8, the basic architecture of boosted decision tree, a total of four BSDT modules are designed/trained separately for three main types of shunt faults such as CSFs, CCFs, and EVFS. In the boosting stage, the patterns used are the inaccurately predicted cases from a specified stage with improved weight/bias during the next stage [24]. Moreover, boosting is often rational to weak learners, not like bagging. Generally, this boosted decision-tree-based ensemble method with two different type of kernels such as AdaBoosts (ABSDT) and RUSBoost (RBSDT) is used. Boosting is preferred if the base classification technique is not stable. When minor variations in the training pattern repeatedly result in substantial discrepancies in the predictions attained, boosting can cause a considerable decrease in weighted prediction error and improves the accuracy of the ensemble model.

3.3.5 Training/Validation of Proposed ETC Model

Tables 3.1–3.3 show the generation of Dataset-II, III, and IV by changing various fault parameters such as fault type, fault resistance, and fault inception angle/time. For training/testing of BGDT-1 module for classification of three main types of

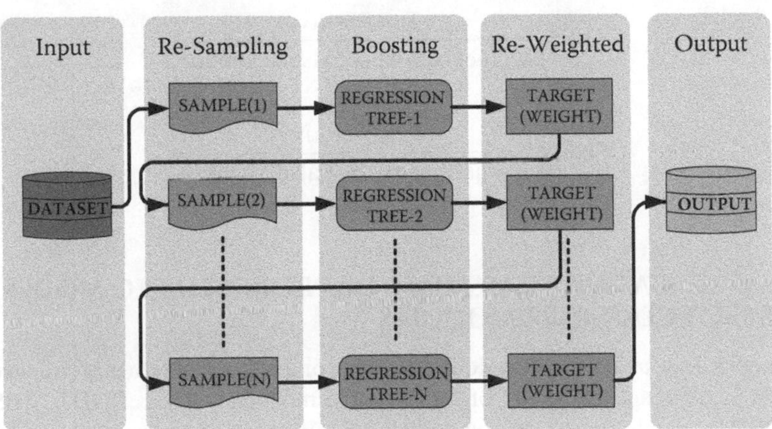

FIGURE 3.8 Basic Architecture of BSDT module.

faults, an exclusive Dataset-I is designed consisting of total no. of fault cases 92596 (3 fault types: CSFs, CCFs, and EVFs) and the final input data set size for BGRT-1 module is 8 × 92596 in which rows 1–7 represent input vectors and row 8 represents corresponding target vector/labels of actual type of shunt fault. Similarly, the input data sets for BGDT-2 (data set size: 8 × 8820), BGDT-3 (data set size: 8 × 51864), and BGDT-4 (data set size: 8 × 31914) modules are designed separately. Further, the proposed ETC model is trained and tested extensively to evaluate the generalized performance of fault classifier modules at the wide range of fault situations. In this context, a cross-validation method is used and fault classifier modules are trained/tested by performing various case studies using different combination of data sets for training and testing purpose correspondingly, such as case study-1 (90:10), case study-2 (80:20), case study-3 (70:30), case study-4 (60:40), and case study-5 (50:50). For example, a combination of (90:10) data set represents 90% of data is considered for training purposes and 10% of data set for testing purposes. Herein in the proposed scheme, four BGDT modules are developed to classify all types of shunt faults. Tables 3.4–3.7 provides a generalized performance assessment of different BGDT modules with their corresponding learning parameters and performance metrics such as Max. no. of splits (MS), No. of learners (NL), Training time, Prediction speed, Accuracy, True positive rate (TPR), and True negative rate (TNR), etc. To realize the different type of shunt faults occurred on a particular power system network, a comprehensive assessment has been demonstrated in terms of different performance metrics. These performance metrics give a kind of research insight to the protection/relaying engineers in a detailed manner and also it is very useful to the line patrolling crew so that they can identify the actual type of fault to diagnose and restore the power supply as early as possible.

$$\text{Accuracy}(\%) = \frac{\text{TruePositive} + \text{TrueNegative}}{\text{TruePositive} + \text{TrueNegative} + \text{FalsePositive} + \text{FalseNegative}} \times 100 \tag{3.1}$$

$$\text{TPR} = \frac{(\text{TruePositive})}{(\text{TruePositive} + \text{FalseNegative})} \tag{3.2}$$

$$\text{TNR} = \frac{(\text{TrueNegative})}{(\text{TrueNegative} + \text{FalsePositive})} \tag{3.3}$$

3.4 COMPARATIVE ASSESSMENT OF PROPOSED ETC MODEL BASED CLASSIFIER MODULES

The comparative assessment is done thereby comparing overall performance assessment of different ensembles of decision tree modules such as BGDT, ABSDT, and RBSDT, which elaborates comparison of outcomes of training/testing of different fault classifier modules. Herein this comparative assessment is done using

TABLE 3.4
Training/Testing Outcomes of BGDT-1 Module for Classification of all Types of Shunt Faults

Case Study	Learning Parameters			Performance metrics		
	MS/NL	Training time (sec)	Prediction speed (observation/sec)	Accuracy (%)	TPR	TNR
1	92565/30	59.696	44000	99.9	0.9989	0.9996
2	92565/30	45.582	47000	100	0.9998	0.9998
3	92565/25	31.067	91000	99.9	0.9995	0.9996
4	92590/31	035.958	78000	100	0.9997	0.9998
5	92585/42	51.007	50000	99.9	0.9995	0.9996

TABLE 3.5
Training/Testing Outcomes of BGDT-2 Module for Classification of CSFs

Case Study	Learning Parameters			Performance metrics		
	MS/NL	Training time (sec)	Prediction speed (observation/sec)	Accuracy (%)	TPR	TNR
1	8819/30	14.063	14000	100	1	1
2	8809/35	6.1226	21000	99.9	0.9994	0.9999
3	8810/25	3.9627	46000	99.8	0.9977	0.9998
4	8819/45	4.8964	30000	99.9	0.9988	0.9998
5	8811/36	3.8479	37000	99.8	0.9981	0.9998

TABLE 3.6
Training/Testing Outcomes of BGDT-3 Module for Classification of CCFs

Case Study	Learning Parameters			Performance metrics		
	MS/NL	Training time (sec)	Prediction speed (observation/sec)	Accuracy (%)	TPR	TNR
1	51861/30	22.097	45000	94.9	0.9485	0.9742
2	51861/36	23.822	51000	94.7	0.9469	0.8791
3	51861/50	40.476	39000	94.6	0.9460	0.973
4	51861/20	13.941	100000	94.4	0.9441	0.9716
5	51,861/33	19.05	69000	94.5	0.9450	0.9725

TABLE 3.7
Training/Testing Outcomes of BGDT-4 Module for Classification of EVFs

Case Study	Learning Parameters			Performance metrics		
	MS/NL	Training time (sec)	Prediction speed (observation/sec)	Accuracy (%)	TPR	TNR
1	31913/30	9.0768	53000	100	1	1
2	31913/40	12.615	43000	100	1	1
3	31913/34	9.8619	46000	99.9	0.9989	0.9994
4	31913/45	12.434	40000	100	0.9999	0.9999
5	31913/50	12.824	56000	100	1	1

different data set with 70:30 ratio for training phase and testing phase, respectively. For example, from Table 3.8, case study-1, BGDT-1 module is tested with 27778 fault cases. The test results from Table 3.8 confirms the BGDT is outperforming to the ABSDT and RBSDT with maximum accurateness for the fault classification in terms of performance metrics. Figure 3.9 shows the comparison of performance metrics of BGDT-1, ABSDT-1 and RBSDT-1 modules for three main types of shunt faults. Figure 3.10 exemplifies the comparison of performance metrics of BGDT-2, ABSDT-2, and RBSDT-2 modules for CSFs. Figure 3.11 shows the comparison of performance metrics of BGDT-3 ABSDT and RBSDT-3 modules for modules for CCFs. Figure 3.12 shows the comparison of performance metrics of BGDT-4, ABSDT-4, and RBSDT-4 modules for EVFs.

TABLE 3.8
Performance Comparison of Different Decision Tree Modules for Classification of Shunt Faults

Fault Type	Module	Performance metrics		
		Accuracy (%)	TPR	TNR
All Shunt Faults	BGDT-1	99.9	0.9995	0.9996
	ABSDT-1	99.9	0.9990	0.9993
	RBSDT-1	99.7	0.9977	0.9982
CSFs	BGDT-2	99.8	0.9977	0.9998
	ABSDT-2	93.3	0.9331	0.9925
	RBSDT-2	78.8	0.7876	0.9774
CCFs	BGDT-3	94.6	0.9460	0.973
	ABSDT-3	93.5	0.935	0.9674
	RBSDT-3	91.3	0.9126	0.9563
EVFs	BGDT-4	99.9	0.9989	0.9994
	ABSDT-4	99.9	0.9987	0.9993
	RBSDT-4	99.7	0.9872	0.9897

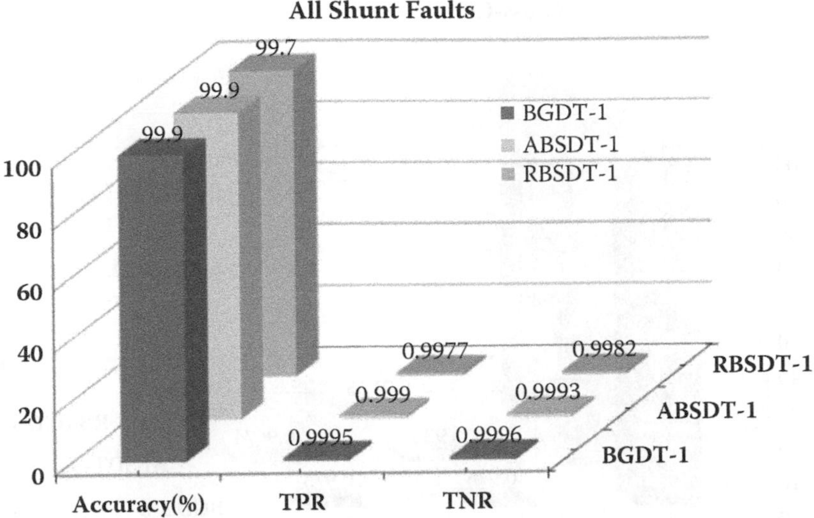

FIGURE 3.9 Comparison of performance metrics of fault classifier modules for all types of shunt faults.

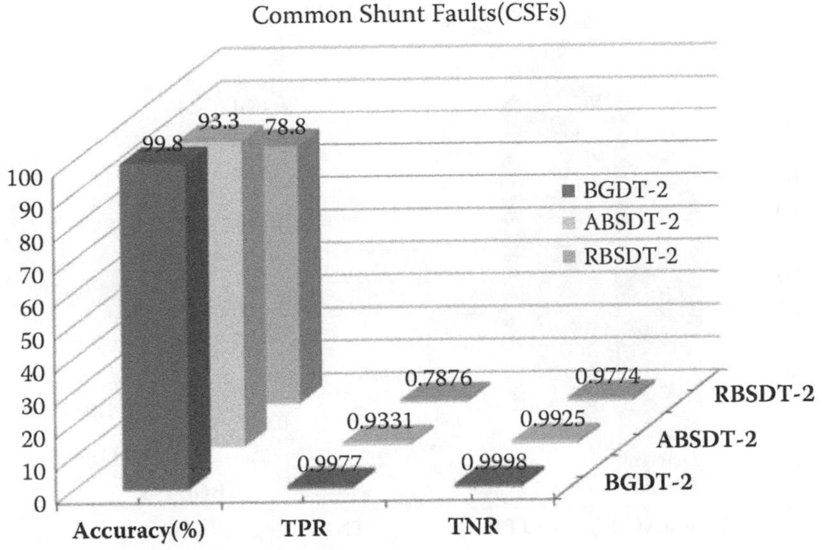

FIGURE 3.10 Comparison of performance metrics of fault classifier modules for CSFs.

3.5 RELATIVE ASSESSMENT OF PROPOSED SCHEME WITH OTHER AI TECHNIQUE-BASED FAULT CLASSIFICATION SCHEMES

The relative assessment has been done thereby comparing the performance metrics such as accuracy, training time and prediction speed of different AI

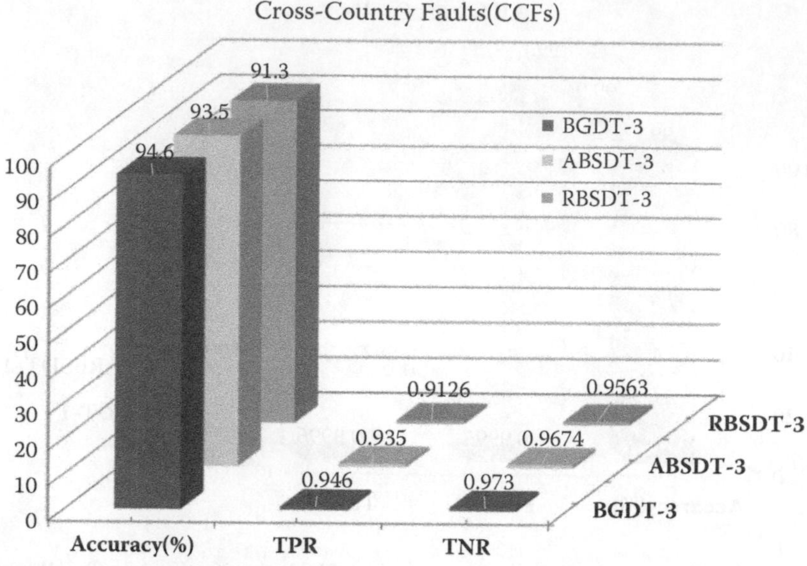

FIGURE 3.11 Comparison of performance metrics of fault classifier modules for CCFs.

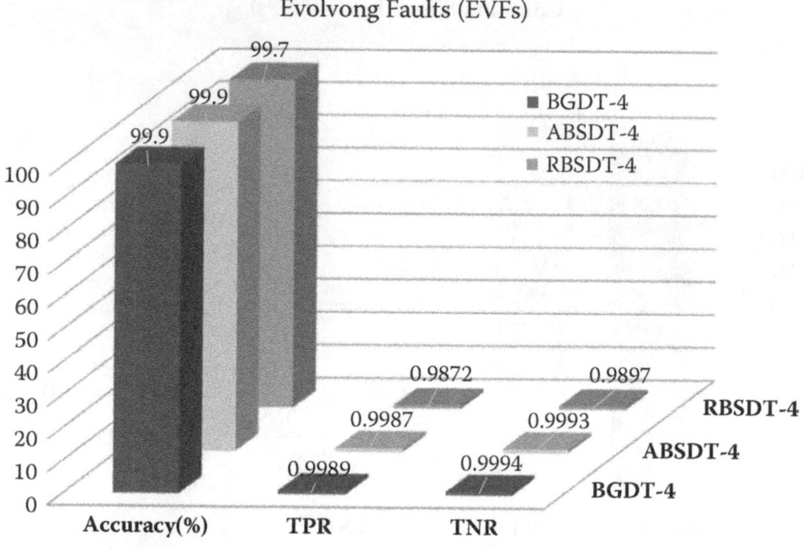

FIGURE 3.12 Comparison of performance metrics of fault classifier modules for EVFs.

technique base classification schemes such as BGDT, weighted KNN, Coarse KNN, and Coarse Gaussian SVM, etc. Herein this relative assessment has done using different data set with 70:30 ratio for training phase and testing phase, respectively. For example, from Table 3.9, case study-2, weighted KNN-2 module is tested with 27778 fault cases. The test results from Table 3.9 confirm

TABLE 3.9
Relative Assessment of Proposed Classification Scheme with Other AI Technique-Based Fault Classification Scheme

Fault Type	Module	Accuracy (%)	Training time (sec)	Prediction speed (observation/sec)
All Shunt Faults	BGDT-1	99.9	31.067	91000
	Weighted KNN-1	99.7	4.6786	38000
	Coarse KNN-1	93.8	133.55	5000
	Coarse Gaussian SVM-1	99.0	4299.6	450
CSFs	BGDT-2	99.8	3.9627	46000
	Weighted KNN-2	97.5	21.606	34000
	Coarse KNN-2	71.2	1.9555	11000
	Coarse Gaussian SVM-2	92.6	254.46	620
CCFs	BGDT-3	94.6	40.476	39000
	Weighted KNN-3	93.9	2.6609	72000
	Coarse KNN-3	91.8	53.127	13000
	Coarse Gaussian SVM-3	93.7	1351	760
EVFs	BGDT-4	99.9	9.8619	46000
	Weighted KNN-4	100	8.6918	62000
	Coarse KNN-4	96.7	2.8266	24000
	Coarse Gaussian SVM-4	99.6	290.08	2600

the BGDT is outperforming to other AI technique-based fault classification schemes with maximum accuracy with reliable training and prediction time for the fault classification.

3.6 EFFECT OF VARIATION IN SAMPLING RATE ON PERFORMANCE OF PROPOSED CLASSIFICATION SCHEME

The effect of variation in sampling rate on accuracy of proposed scheme has been reported in Table 3.10. Table 3.10 shows performance metrics of BGDT-3 for variation of sampling rate and its effect on accuracy, training, and prediction time for fault classification scheme. The total data set size is 8×28812, where 1–7 rows denotes feature vector and row 8 denotes target vector (class labels). Further, the 70% of data set used for training and 30% of data set is used for testing purposes.

TABLE 3.10
The Effect of Variation in Sampling Frequency on Performance of BGDT-3 Module

Fault Type/ Module	Sampling rate (kHz)	Accuracy (%)	Training time (sec)	Prediction speed (observation/sec)
CCFs/BGDT-3	1.0	94.6	40.476	39000
	1.2	92.5	101.28	5600
	5.0	91.1	10.013	77000
	10	91.7	8.4867	77000

3.7 CONCLUSION

In this chapter, a new fault classification scheme has been reported based on the ETC model using a bagged decision tree to classify three main types of shunt faults in the DCTL. The main contribution of this chapter is that here the bagged decision tree has been designed and developed for classification of shunt faults including CSFs, CCFs, and EVFs, which has been reported for the first time using an ensemble tree-based technique. Moreover, the standard deviation of wavelet coefficients of 1-cycle data (post-fault) of three-phase currents of the DCTL has been considered to design exclusive data sets to train/test the BGDT-1, 2, 3, 4 modules that erudite independently for the classification of different type of shunt faults. The proposed ETC-model-based fault classification scheme has been assessed for numerous fault situations considering variation in different fault parameters and in presence of mutual coupling between the two parallel circuits. However, accuracy of the proposed fault classification scheme has been evaluated in terms of performance metrics such as Accuracy (%), Sensitivity (TPR), and Specificity (TNR). The performance metrics obtained by the ensemble of bagged regression tree-based scheme are better than other schemes, which enhances the consistency of the power transmission system, thereby classifying the type of fault accurately and improves the fault diagnostic capabilities and decreases the outage time.

ACKNOWLEDGMENTS

The authors acknowledge the financial support of Central Power Research Institute, Bangalore for funding the project. no. RSOP/2016/TR/1/22032016, dated: 19.07.2016. The authors are grateful to the Head of the institution as well as Head of the Department of Electrical Engineering, National Institute of Technology, Raipur, for providing the research amenities to carry this work. The authors are indebted to the local power utility (Chhattisgarh State Power Transmission Company Limited) for their assistance in component/equipment data of real power system network.

REFERENCES

[1] D. J. Spoor and J. Zhu, "Inter circuit faults and distance relaying of dual-circuit lines," *IEEE Transactions on Power Delivery*, vol. 20, no. 3, pp. 1846–1852, July 2005, doi: 10.1109/TPWRD.2004.833899.

[2] A. Jain, A. S. Thoke, R. N. Patel and E. Koley, "Intercircuit and cross-country fault detection and classification using artificial neural network," 2010 Annual IEEE India Conference (INDICON), Kolkata, 2010, pp. 1–4, doi: 10.1109/INDCON. 2010.5712601.

[3] K. Solak and W. Rebizant, "Analysis of differential protection response for cross-country faults in transmission lines," 2010 Modern Electric Power Systems, Wroclaw, 2010, pp. 1–4.

[4] A. A. M. Zin, N. A. Omar, A. M. Yusof and S. P. A. Karim, "Effect of 132kV cross-country fault on distance protection system," 2012 Sixth Asia Modelling Symposium, Bali, 2012, pp. 167–172, doi: 10.1109/AMS.2012.17.

[5] Z. Y. Xu, W. Li, T. S. Bi, G. Xu and Q. X. Yang, "First-zone distance relaying algorithm of parallel transmission lines for cross-country nonearthed faults," *IEEE Transactions on Power Delivery*, vol. 26, no. 4, pp. 2486–2494, Oct. 2011, doi: 10.1109/TPWRD.2011.2158455.

[6] T. Bi, W. Li, Z. Xu and Q. Yang, "First-zone distance relaying algorithm of parallel transmission lines for cross-country grounded faults," *IEEE Transactions on Power Delivery*, vol. 27, no. 4, pp. 2185–2192, Oct. 2012, doi: 10.1109/TPWRD.2012. 2210740.

[7] A. Swetapadma and A. Yadav, "Improved fault location algorithm for multi-location faults, transforming faults and shunt faults in thyristor-controlled series capacitor compensated transmission line," *IET Generation, Transmission & Distribution*, vol. 9, no. 13, pp. 1597–1607, Oct. 10, 2015, doi: 10.1049/iet-gtd.2014.0981.

[8] A. Swetapadma and A. Yadav, "All shunt fault location including cross-country and evolving faults in transmission lines without fault type classification," *Electric Power Systems Research*, vol. 123, pp. 1–12, 2015, doi: 10.1016/j.epsr.2015.01.014.

[9] Sunil Singh and D.N. Vishwakarma, "A novel methodology for identifying cross-country faults in series-compensated double circuit transmission line", *Procedia Computer Science*, vol. 125, pp. 427–433, 2018, doi: 10.1016/j.procs.2017.12.056.

[10] A. Codino, F. M. Gatta, A. Geri, S. Lauria, M. Maccioni and R. Calone, "Detection of cross-country faults in medium voltage distribution ring lines," 2017 AEIT International Annual Conference, Cagliari, pp. 1–6, 2017, doi: 10.23919/AEIT.2017. 8240493.

[11] F. M. Gatta, A. Geri, S. Lauria and M. Maccioni, "An equivalent circuit for evaluation of cross-country fault currents in medium voltage (MV) distribution networks", *Energies*, 11(8), 1929, 2018, doi: 10.3390/en11081929.

[12] A. Swetapadma, and A. Yadav "An artificial neural network-based solution to locate the multilocation faults in double circuit series capacitor compensated transmission lines", *International Transactions on Electrical Energy Systems*, 28, e2517, 2018, doi: 10.1002/etep.2517.

[13] V. Ashok, A. Yadav, and V. K. Nayak, "Fault detection and classification of multi-location and evolving faults in double-circuit transmission line using ANN", In: J. Nayak, A. Abraham, B. Krishna, G. Chandra Sekhar, and A. Das (eds.), Soft Computing in Data Analytics. Advances in Intelligent Systems and Computing, vol. 758, pp. 307–317, 2019, Springer, doi: 10.1007/978-981-13-0514-6_31.

[14] V. Ashok, and A. Yadav "A protection scheme for cross-country faults and transforming faults in dual-circuit transmission line using real-time digital simulator: A case study of Chhattisgarh state transmission utility", *Iranian Journal of Science*

and Technology, Transactions of Electrical Engineering 43, pp. 941–967, 2019, doi: 10.1007/s40998-019-00202-w.

[15] V. Ashok, A. Yadav, and A. Y. Abdelaziz, "MODWT-based fault detection and classification scheme for cross-country and evolving faults", *Electric Power Systems Research*, vol. 175, 105897, October 2019, doi: 10.1016/j.epsr.2019.105897.

[16] V. Ashok and A. Yadav, "A novel decision tree algorithm for fault location assessment in dual-circuit transmission line based on DCT-BDT approach", In: A. Abraham, A. Cherukuri, P. Melin, and N. Gandhi (eds.), Intelligent Systems Design and Applications. ISDA 2018. Advances in Intelligent Systems and Computing, vol. 941, pp. 801–809, 2019, doi: 10.1007/978-3-030-16660-1_78.

[17] A. Naresh Kumar, C. Sanjay, M. Chakravarthy, "Fuzzy inference system-based solution to locate the cross-country faults in parallel transmission line", *The International Journal of Electrical Engineering & Education*, pp. 1–14, 2019, https://doi.org/10.1177/0020720919830905.

[18] A. Swetapadma and A. Yadav, "A novel decision tree regression-based fault distance estimation scheme for transmission lines", *IEEE Transactions on Power Delivery*, vol. 32, No. 1, pp. 234–245, February 2017.

[19] A. Naresh Kumar, P. Sridhar, T. Anil Kumar, T. Ravi Babu, and V. Chandra Jagan Mohan, "Adaptive neuro-fuzzy inference system based evolving fault locator for double circuit transmission lines", *IAES International Journal of Artificial Intelligence*, vol. 9, No. 3, pp. 448–455, September 2020, doi: 10.11591/ijai.v9.i3.pp448-455.

[20] V. Ashok and A. Yadav, "Fault diagnosis scheme for cross-country faults in dual-circuit line with emphasis on high-impedance fault syndrome," *IEEE Systems Journal*, 2020, doi: 10.1109/JSYST.2020.2991770.

[21] Breiman, L."Bagging predictors", *Machine Learning*, 24, 123–140, 1996, doi: 10.1023/A:1018054314350.

[22] R. Polikar, "Ensemble based systems in decision making," *IEEE Circuits and Systems Magazine*, vol. 6, no. 3, pp. 21–45, Third Quarter, 2006, doi: 10.1109/MCAS.2006.1688199.

[23] Wei-Yin Loh, "Classification and regression trees", Overview article in *WIREs Data Mining and Knowledge Discovery*, John Wiley & Sons Publication, vol. 01, pp. 14–23, 2011, doi: 10.1002/widm.8.

[24] Available online at https://in.mathworks.com/help/stats/treebagger-class.html#bvfstrb.

4 An Artificial Intelligence–Based Detection and Classification of Faults on Transmission Lines

Dalia Allam[1] *and Almoataz Y. Abdelaziz*[2]
[1]Department of Electrical Engineering, Faculty of Engineering, Fayoum University, Fayoum, Egypt
[2]Faculty of Engineering and Technology, Future University in Egypt, Cairo, Egypt

4.1 INTRODUCTION

Distance protection is the most complex part of the protection system as it tackles with all types of faults over a very long distance of the transmission line located in open air or in the ground. That's why the transmission lines are the most vulnerable part in the power system to be affected by the environmental conditions. Distance protection is a non-unit protection as an individual distance relay should be provided at each end of the transmission line. Therefore, there are two strategies for the detection and the classification of faults over the entire transmission lines. The first one depends on the existence of a communication channel which is able to communicate between the two terminal relays of the transmission line. This channel provides the (V, I) measurements from the remote end of the transmission line to provide accurate detection and classification of the faults on any part of the transmission line. The other strategy depends on the estimation of faults using (V, I) measurements available at just one terminal of the Transmission line to detect the faults and to classify their types. The first strategy needs an accurate communication channel as well as transmitters, receivers and signal processing devices which may need a large budget specially if the channel is private. Therefore, the second strategy based on estimation process is preferred. However, estimation of the fault existence and the fault type based on the measurements at one side of the transmission line needs very powerful techniques to provide accurate estimations with minimum errors.

DOI: 10.1201/9780367552374-4

Distance relays have passed through four generations of developments. The first one was the electromechanical relays that have been able to detect the faults by activating two coils using the voltage and current signals. The second one was the static relays that have utilized analog electronics for only fault detection based on detecting the phase shift between an operating voltage signal and a selected reference voltage signal named the polarizing signal. In the third generation, a microprocessor based digital system has been introduced where signal processing techniques such as Discrete Fourier Transform (DFT) and Discrete Wavelet Transform (DWT) in addition to coding by high-level programming algorithms that has been provided to calculate the circuit equations of the faults as shown in the block diagram in Figure 4.1. This generation has opened up the field of digital distance protection and has increased its capabilities to add more functions as classification of faults as well as an accurate determination of fault location to its main function of the fault detection. However, there are many factors that affects the voltage and current signals measured at the relay point and causes errors in the calculation of circuit equation within the program. Appropriate solutions of such errors should be provided individually to the digital relay system in both of the digital signal processing DSP stage and the stage of calculation of circuit equations within the algorithm. Therefore, there was a persistent need of introducing novel techniques to improve the performance of the distance relays and to overcome the problem of large dispersion of data as well as the effect of errors. This could be handled using Artificial Intelligence (AI) techniques accompanied by a preprocessing transformation tool for voltage and current signals. That's whythe fourth generation of AI-based smart distance relays has been introduced to overcome the most of the operating problems in addition to their features of high accuracy and low execution time.

The main purpose of using the preprocessing transformations for the phase voltage and the line current signals measured at the relay point is the decomposition of the voltage and the current signals and the extraction of the fundamental sinusoidal components to obtain the actual fault data needed as inputs to AI algorithms. The remaining harmonic components are just contaminations on the real signals and they should be eliminated to increase the accuracy of AI technique. There are three commonly used transformations in a fault diagnostic system. The first one is the Discrete Fourier Transform (DFT), the second is the Wavelet Transform (WT), and the third one is S-Transform (ST).

Fourier Transform (FT) is an effective tool for transformation of signal from time domain to frequency domain. It is the first generation of transformation used in the field of detection and classification of faults to decompose the voltage and

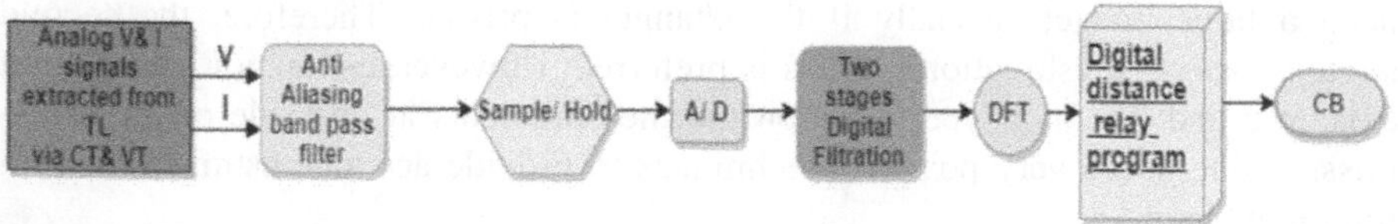

FIGURE 4.1 The block diagram of the digital distance relay.

current signals into their harmonic components and to extract the fundamental components. Discrete Fourier Transform (DFT) has been used to transform time domain discrete signals to frequency domain discrete signal within a specified window of time. In (Yu et al., 2001), DFT has been used to eliminate both of DC and harmonic components within two windows of one cycle and a half cycle.

DFT has been utilized within a half cycle for classification of faults in (Jamehbozorg et al., 2010). In (Hagh et al., 2007), Fast Fourier Transform (FFT) has been used within a full cycle.

Wavelet Transform (WT) is one of the most applicable transformations in the field of the fault diagnostic systems because it can provide frequency domain information as well as time domain information as in (Prasad et al., 2017). This can be considered a very good merit of WT over FT, where FT provides only the frequency domain information within a specified window of time and this isn't adequate in case of time variant signals of faults that need to be expressed in a time domain as well. Moreover, WT has the capability for extracting new features that may achieve better discrimination among various types of faults which is beneficial with non-linear problems that have large dispersion of data as fault analysis. A proper selections of the decomposition levels such as the Mother Wavelet (MW) and the frequency bands are important for the features extraction. WT has split the voltage and current signals for faults of overhead transmission lines into various frequency bands by using multi resolution analysis (MRA), as in (Youssef, 2001). MVs named Mexican hat and coif have been reported in (Ravindhranath Reddy et al., 2009) for fault classification within only a half cycle.

In (El Safty et al., 2009), the Wavelet entropy principle has been proposed for fault analysis. Two wavelet modules have been utilized in (Costa et al., 2009) to detect and classify the faults. Wavelets have played an important role in real-time detection of faults using the principle of the energy of the coefficients of the harmonics, as in (Costa et al., 2012). In (Lakshmana Nayak, 2014), DWT with daubechies eight MVs has been proposed for fault classification based on the availability of currents from both terminals of the transmission line. The energy of the current signals and its ratio of change as well as the maximum value of the detailed coefficient and the ratio of energy have been used for fault classification as proposed in (Jose et al., 2014). In (Prasad et al., 2016), DWT has been utilized to classify the faults using current signal at only one end of an overhead transmission line. The behavior of DFT coefficients has been studied for hybrid distribution system of overhead transmission lines and underground cables, as reported in (Pothisarn et al., 2019).

The third type of transformation is named S-Transform. It is a new avenue of time-frequency transformation that has been used in fault detection and classification as in (Dash et al., 2007). It has been derived from the continuous wavelet (CWT) and it has worked on sinusoidal function basis, which is physically closer in nature to the power signals. It has been able to decompose the voltage and current signals into their frequency components. Moreover, it could provide not only the amplitude of the frequency components as in case of DFT but also the time of their occurrence as well as in the case of DWT, as reported in (Pinnegar et al., 2003; 2004). Therefore, it has been a very efficient transformation when tackling time variant signals as voltages and currents during faults. Discrete S-Transform (DST) is more accurate than DWT. However, it has enormous numbers of calculations that may affect the practical real-time implementation on the protection system where the time of detecting the faults is a dominant factor and it is a crucial issue to be a minimum as in (Pei et al., 2011). Seeking a reduction of calculations of S-Transform, there were some attempts for increasing its speed at the frequencies associated with the power signals by merging an intelligent decision mechanism. This mechanism behaved numerically as a filter to decide the most significant frequencies in the signals and to reject the others in order to minimize the burden of the computations. Therefore, Fast Fourier Transform (FFT) has been added to DST to be developed to Fast Discrete S-Transform (FDST) where FFT has detected the significant frequencies that have a large effect on voltage and current signals and subsequently ST has been evaluated. Thereby, the mathematical computations would be decreased considerably and the time of computations would be reduced. That's why FDST is a very effective tool for real-time implementation in the relay operation and provides more accurate results than DWT, as in (Krishnanand et al., 2015).

Lately, another technique named Principal Component Analysis (PCA) has been merged with WT to reduce the dimensions of the data by mapping it and to extract new features that may achieve better accuracy in the fault-type selection problem, as in (Cheng et al., 2015). Furthermore, the Clarke Transformation (CT) may be used to transform a, b, and c to their equivalent modal components α, β, and 0. These modal components with the aid of the phase values have been used with WT and AI to improve the quality of classification of faults, as in (Zin et al., 2015).

Several AI techniques such as Artificial Neural Network (ANN), Fuzzy Logic (FL), Support Vector Machine (SVM), Support Vector Regression (SVR), Extreme Learning Machine (ELM), Classification and Regression Trees (CART), and K-Nearest Neighbors (K-NN) have been used for detection and classification of faults. However, the most commonly used AI techniques in detection and classification of faults are ANN and SVM. ANN needs large training data to provide accurate results compared to the amount of training data needed by SVM. Several types of ANN have been applied and tested on the problem of detecting and classifying faults, as in (Klomjit et al., 2020). However, Probabilistic Neural Networks (PNNs) have shown a superior behavior compared to various types of ANN in fault classification problems, especially when tackling a huge amount of data used for long transmission lines with consideration of many factors that may affect the relay operation. Integration between transformations as WT or ST with AI techniques has improved the accuracy of detection and classification of faults, as reported in (Raza et al., 2020). In (Mo et al., 1998), PNN has been implemented on the fault classification with an accuracy of 10% over that achieved by the other Feed Forward Neural Network. However, more accurate technique named adaptive PNN has been developed in (Rutkowski, 2004), where changes of PDF with time could be tracked and taken into account. In (Mishra et al., 2008), PNN has been used to classify 11 types of abnormalities in a power system using the voltage waveforms. The voltage signals have been preprocessed by implementing S-Transform and the covariance matrix named S-matrix has been calculated. The features' extraction is dependent on the standard deviation and the energy of maximum values of the columns as well as the standard deviation of the vector the maximum values of the rows of S-matrix. The selection of discriminative features is very essential to have better classification even with less available data. PNN integrated with DWT extracts the features for nine nodes, as in (Upendar et al., 2008). PNN has been used to discriminate between two events effectively as in case of internal fault and magnetizing inrush using an optimized factor, as in (Tripathy et al., 2010). In (Mirzaei et al., 2011), three types of ANN named FNN, PNN, and RBFN have been tested and PNN was recommended because of its higher accuracy and lower time of training.

As in (Ray et al., 2016), Support Vector Machine (SVM) is simply a classifier that uses an adaptive learning statistical based method where the input vectors are used for nonlinearly mapping into a feature space of high dimension. An optimal hyper plan, which is mainly a space gap, i.e. boundary between two nearby classes, created to be able to discriminate between the two classes. Features lying near the boundary are called support vectors. The training algorithm of an SVM fault classifier has utilized the training (Input/Target) pairs of the measured voltage and currents as inputs and the type of faults as targets. A radial basis function (RBF) with kernel parameter has been used for maximizing the gap between two classes to establish the optimal hyper plan. After training, the features of the real input data have been mapped into the same hyper plan and the prediction by the trained SVM was accomplished, as in (Parikh et al., 2010). SVM has shown a good convergence to the global solution and poor trapping in local minima, which has been considered as one of its best advantages in the fault classification process. Furthermore, proper identification of SVM parameters has led to higher accuracy and better performance, as reported in (Schittkowski, 2005). The optimal values of SVM parameters have been evaluated by the Particle Swarm Optimizer or by any other optimization algorithm, as in (Steinwart, 2003).

Moreover, one of the latest AI techniques that has been used in detection and classification of faults and has proved an effective behavior in this problem is the Convolution Neural Network (CNN). CNN has been used in conjunction with the WT signal processing technique, as proposed in (Fahim et al., 2020) and it has achieved promising results in the accuracy and in the operating time during fault detection and faulty phase selection.

In this chapter, the basic concepts of the distance protection as causes of faults, types of faults, relay characteristics, and sources of errors in voltage and current signals as well as in the fault point seen by the relay have been summarized. Moreover, signal processing techniques such as FT, WT, CT, and ST have been demonstrated and their roles in decomposition of signals as well as their feature extractions have been explained. Furthermore, several AI-based techniques for fault detection and classification have been clarified and three of them are PNN, SVM, and CNN, which have been selected to be explained in detail because of their importance and their efficient performance in the problem of fault diagnosis. Additionally, several combinations between signal processing transformations and AI methods have been illustrated to prove their robustness and efficiency in improving the accuracy of fault detection and classification and to make high-speed decisions in this problem.

The rest of the manuscript is organized as follows: Section 4.2 presents the basic concepts of distance protection. The AI-based fault diagnosis system is described in Section 4.3, while Section 4.4 presents the conclusion.

4.2 THE BASIC CONCEPTS OF DISTANCE PROTECTION

4.2.1 CAUSES OF CURRENT INCREASE UPON FAULT OCCURRENCE

If the transmission line is fed from only one terminal and the other terminal is loaded, this means that the current is limited by the source impedance, the transmission line impedance, and the load impedance. When the fault occurs at any point of the transmission line, the current is limited by the source impedance and only a part of the transmission line impedance, as in Figure 4.2. That's why the current drawn from the generator will increase as the generator output voltage is constant and the impedance seen by the relay is reduced according to the location of the fault point and the fault current is equal to the supply current. The most dangerous point of fault on the transmission line is at the relay point, where almost no impedance except the source impedance can limit the current drawn from the supply and this may cause serious damage in the supply. In case of TL of double feed, the fault current increases considerably as it is the vector sum of the two currents drawn from the two generators suppling the TL, as in Figure 4.3. The second reason that increases the current is the exponential component superimposed on the sinusoidal current due to the sudden switching of fault on the R-L circuit of the transmission line. The third factor that causes the increase of current is the harmonics added to the fundamental sinusoidal component of the current due to nonlinear equipment located in the power system as power electronic circuits and saturated transformers. The last factor is the value of the load current at the instant of fault.

4.2.2 CAUSES OF FAULTS

Short-circuit faults occur due to insulation failure and insulation may fail because of various reasons such as the following:

- Environmental conditions such as fall of snow, wind, change in temperature, existence of lightning strikes, or heavy rains
- Effects on the conductor structure as aging and stresses
- Occasional conditions such as birds landing, falling trees, and snakes wrapping around the transmission line as well as vehicles accidents with towers
- Atmospheric pollution, which leads to faster breakdown of insulation of transmission lines' conductors
- Switching surges

Some of the faults causes are shown in Figure 4.4.

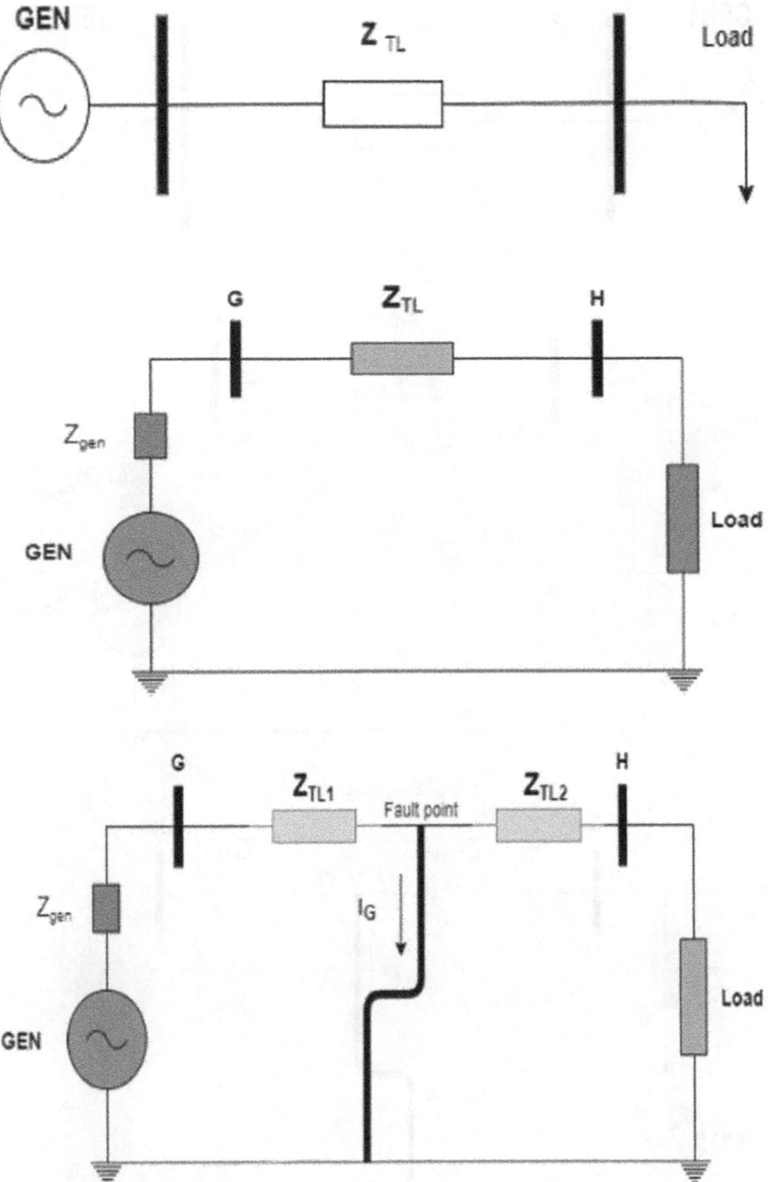

FIGURE 4.2 Circuit diagrams of one feed TL.

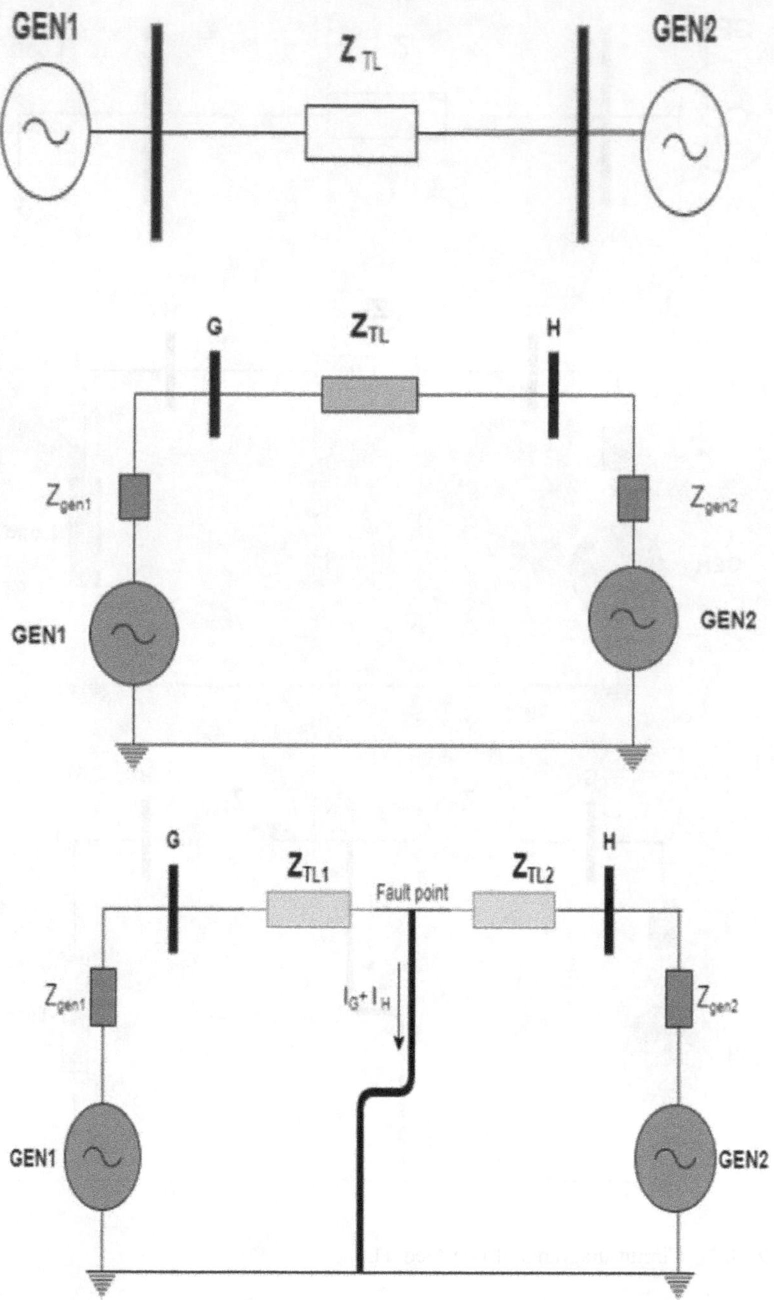

FIGURE 4.3 Circuit diagrams of double feed TL.

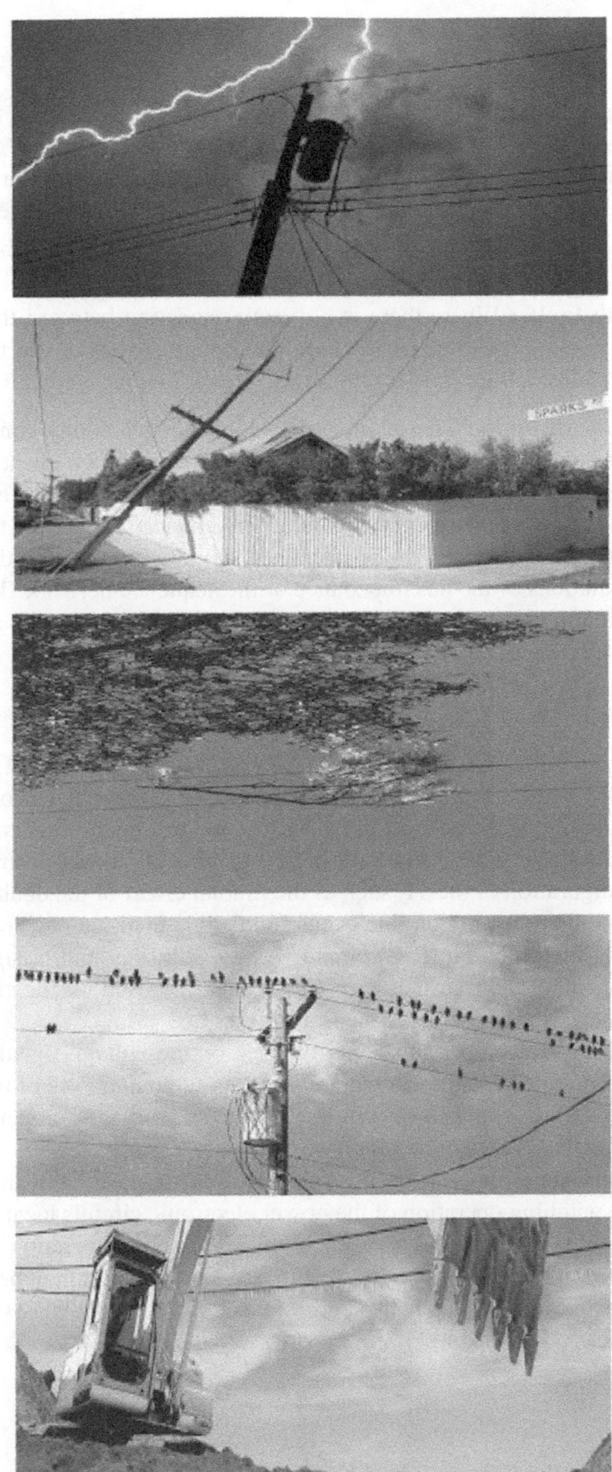

FIGURE 4.4 Causes of faults.

4.2.3 Types of Faults

- **Short-circuit faults:** These types of faults occur as a result of the insulation breakdown at any point on any phase of the transmission line and this point comes in contact with the ground or with any uninsulated point on any other phases. These faults can be divided into the following two categories:
  - **Phase faults:** At which there is a direct connection between the conductors of two or three phases of the transmission line
  - **Earth faults:** Where there is a direct connection between the conductors of one or two or three phases with the ground. Types of short circuit faults and their configuration on the transmission line are shown in Figure 4.5.
- **Open-circuit faults:** These types of faults occur when any conductor of the transmission line is broken due to storms or any unwanted event as well as its opening due to unbalanced operation of a circuit breaker. This causes an unbalance in the flow of currents in the three phases. However, these types of faults are not detected by the protection system but they can be detected using the calculations of the bus impedance of the sequence network. Types of open circuit faults and their configuration on the transmission line are shown in Figure 4.5. as well.

4.2.4 Sources of Errors in Detection and Classification of Faults

- The value of the fault resistance that may cause overreach or underreach depending on the phase shift between the two terminal currents injected into the TL results in seeing the fault before or after its actual position.
- The configuration of the TL such as the mutual effect of the double-circuit TL
- Faults on the transmission line occur suddenly which may be considered as a switching operation on (R-L) circuit. As a result, the current signal includes an exponential DC transient component superimposed on the sinusoidal AC steady state component as shown in Figure 4.6.
- High frequency components superimposed on the voltage signal measured at the relay point due to the interaction between the inductance of the TL and its capacitance for long TLs or with the capacitance of the compensators at the TL terminals, as shown in Figure 4.7.
- Harmonics superimposed on the current signals measured at the relay point due the switching operation of the power electronic circuits located at any part of the power system. As a result of this switching operation, the currents drawn from the supply by these circuits are distorted and they become almost a square wave rather than the pure sinusoidal wave shown as usual in the power network in Figure 4.8.

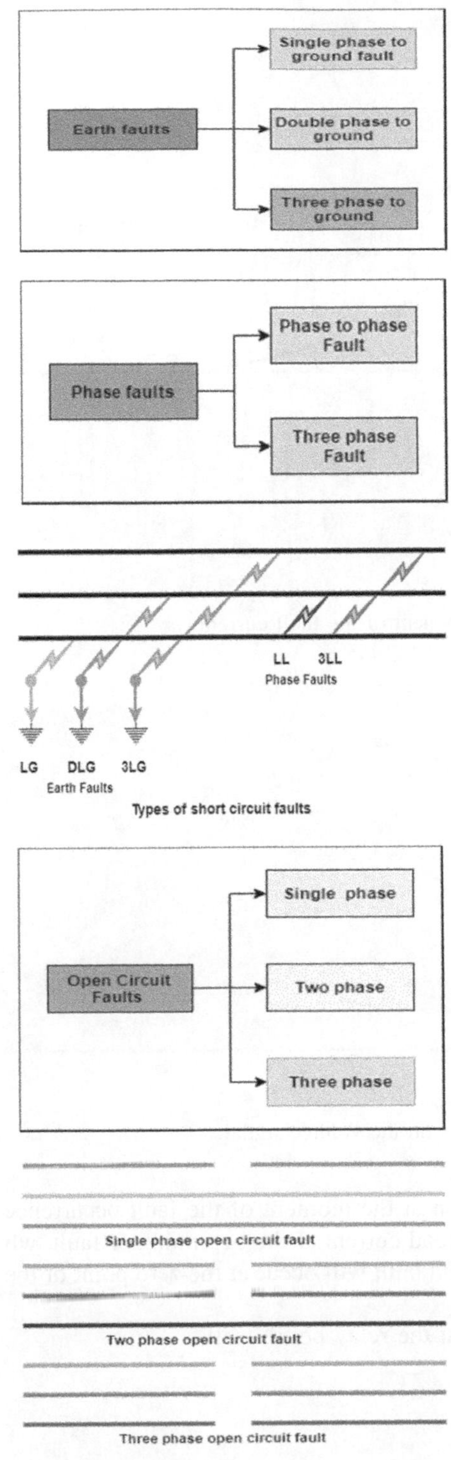

FIGURE 4.5 Short-circuit and open-circuit faults.

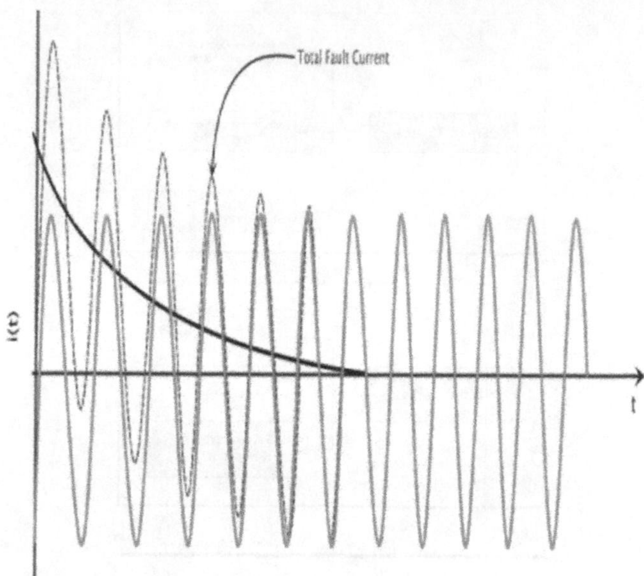

FIGURE 4.6 DC component of the fault current.

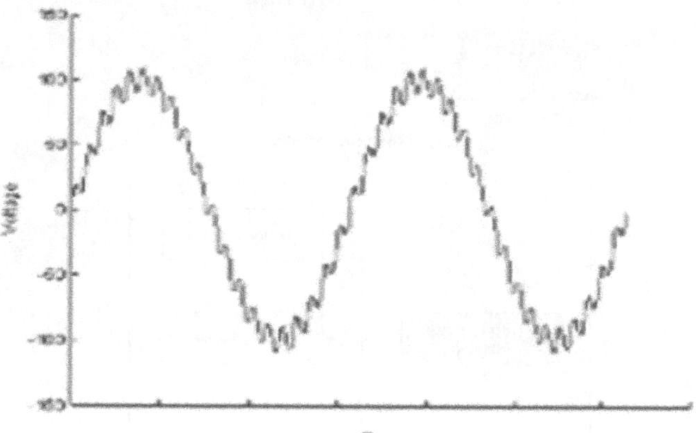

FIGURE 4.7 Harmonics on the voltage signal.

- Loading condition at the moment of the fault occurrence, which means that the value of the load current at the beginning of fault, where there is nothing guarantees that the fault will occur at the zero point of the sinusoidal wave of the load current.
- Close up faults at the relay point itself

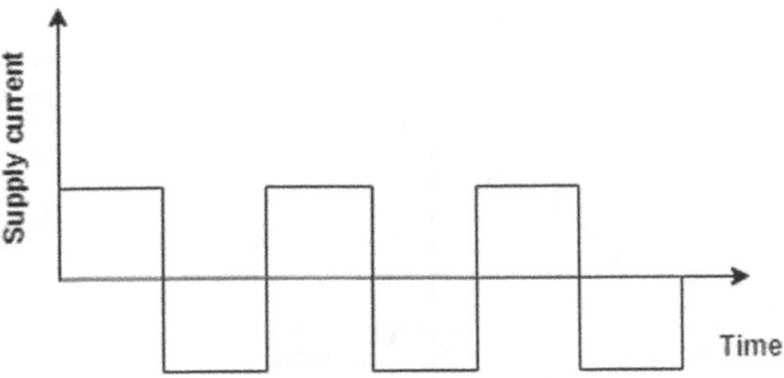

FIGURE 4.8 Harmonics on the current signal.

4.2.5 Distance Relay MHO Characteristic

There are two common characteristics used in distance protection:

- Impedance relay characteristic is a circle symmetrical around the origin of (R-X) plan and the relay setting is its radius. It has nondirectional characteristics and can respond to a fault from both sides of the relay location, as in Figure 4.9.
- MHO relay characteristics is a circle passing through the origin and it is a directional characteristic. The relay setting is its diameter, as in Figure 4.10. It can be considered a phase comparator that can detect the phase shift between an operating signal (V-IZ) and a reference signal, named a polarizing signal, such as the measured voltage in the self-polarized mho relay characteristic shown in Figure 4.11.

4.3 AI-BASED FAULT DIAGNOSIS SYSTEM

Creating a fault diagnosis system that is able to detect and classify different types of faults is a complicated problem because of its large dispersion and overlapping of the measured data for different types of faults as well as the influence of many sources of errors on this data. This problem can't be described by a definite function or criterion without many simplifications and assumptions that may affect the accuracy of the diagnostic system. Therefore, AI-based expert systems have been introduced to design robust fault diagnosis systems. The idea of AI is inspired by the human brain and it is accomplished by training a machine to be intelligent and to do several tasks as the human. This can be implemented using algorithms that enable the machine to learn based on an input data and to make decisions depending on this learning. Machine Learning (ML) is a subcategory of AL learning algorithms that can learn and understand by learning and parsing the data. It has three types of learning, named supervised, unsupervised, and reinforced learning. In supervised learning, there are training labels corresponding to the training samples. There are no training samples in the case of unsupervised learning and the algorithm

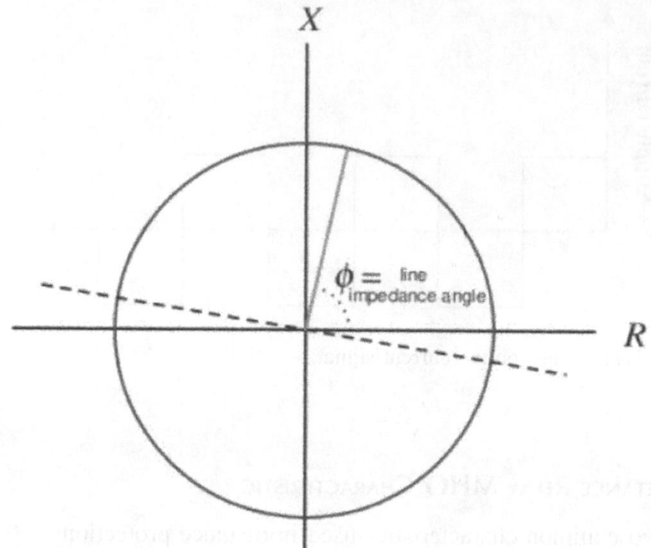

FIGURE 4.9 Impedance characteristic.

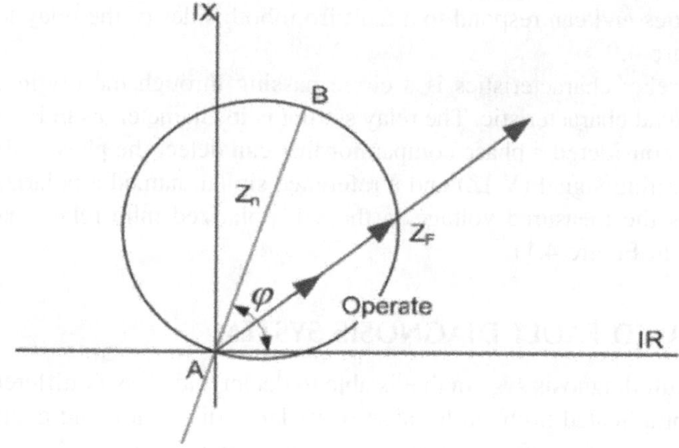

FIGURE 4.10 Mho characteristic.

itself makes a clustering process to reduce the dimensions of data. Reinforced learning has an agent that learns a behavior in an environment via taking actions. ML methods are support vector machines (SVMs), K-nearest neighbor (KNN), decision trees (DTs), K-means clustering, hidden Markov model, and Gaussian mixture model. Deep Learning (DL) is a subcategory of ML inspired by artificial neural networks (ANNs). It is named deep learning because it depends on stacking multiple hidden layers between the input and output layers and it is used for classification, regression, pattern recognition, feature extraction, data processing, and series prediction. The DL methods are CNN, generative adversarial network,

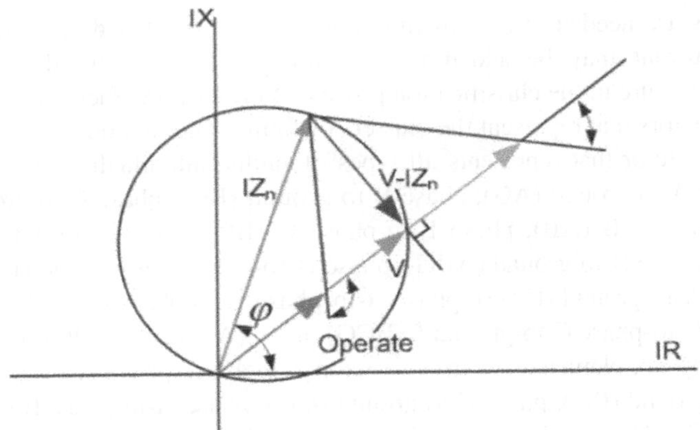

FIGURE 4.11 Self-polarized mho relay.

auto encoders, recurrent NN, and long- and short-term memory. Furthermore, feed forward neural networks (FFNNs) have been considered as one of the most powerful tools that are based on iterative and adaptive techniques to provide a fair description of the problems of distance protection. FFNN creates a transfer function named activation function and tries to adapt this function numerically by comparison between the actual output values and the desired target values obtained from the available measured or archived data until minimum errors are reached. Various types of FFNNs with different proposed activation functions have been implemented on this protective relay and their efficiency has been tested to prove their suitability for this application.

PNN, SVM, and CNN are selected to be demonstrated in details as examples of FFNN, ML, and DL, respectively, because they have been used widely in the field of detection and classification of faults on TLs. Moreover, integration between signal processing techniques such as FT or WT or ST and AI techniques speeds up the decision making in the problem of detecting and classifying faults.

4.3.1 Training Data for Artificial Neural Network: (Input/Target) Pairs

To design an ANN-based fault diagnostic system, archived data or measured data of the phase voltages and the line currents signals at the relay point during different types of faults as well as their corresponding types of faults should be available to obtain the training data required to train ANN. During the training process, the optimal values of the control parameters that provide the minimum errors between actual and target output are reached. The input vectors may be formed in frequency domains as magnitudes and angles or in time domains as time samples while the target vectors will be the corresponding clusters to these input vectors. In the case of using frequency domain data, each input vector will have 12 elements that represent the magnitudes and angles of the three phase voltages and the three line currents. Only six elements' column vectors represent the time samples of phase voltages and line currents are needed in case of time domain data input. Additionally, the zero sequence current may be added to the input vectors to provide another discriminative feature to the classification process. Furthermore, there are two types of the target vectors that represent the clusters of faults. The first one uses 12 elements per column vector that represents all types of faults individually, such as no fault (NF), phase A to ground (AG), phase B to ground (BG), phase C to ground (CG), phase A-to-phase B (AB), phase B-to-phase C (BC), phase C-to-phase A (CA), phase A-to-phase B-to-ground (ABG), phase B-to-phase C-to-ground (BCG), phase C-to-phase A-to-ground (CAG), phase A-to-phase B-to-phase C (ABC), and phase A-to-phase B-to-phase C-to-ground (ABCG), as in (Allam et al., 2007). The second one uses only six elements per target vector, which are phase A to ground (AG), phase B to ground (BG), phase C to ground (CG), phase A-to-phase B (AB), phase B-to-phase C (BC), and phase C-to-phase A (CA). Practically, the first form is more accurate and discriminative than the second one, where there may be a remarkable overlapping among different types of faults. Moreover, modal components α, β, and 0 can be used instead of the principle components A, B, and C to provide better results.

Example 4.1: This example demonstrates a sample of generating and preparing of data required to train the ANN via two stages of MATLAB® Simulink model. The first one is of a simple transmission line fed from its both sides. The transmission line length is 300 Km and it can be divided into sections of any required lengths to simulate all types of faults at different operating conditions on many locations of the transmission line. The three phase voltages and the three line currents at bus (B1) are measured. The lengths of the TL sections can be varied as well from 25 km to 15 km, as shown in Figure 4.12. The second stage of simulation is processing of Vs and Is signals using filters and DFT or DWT or any other transformations to be prepared as data inputs to ANN, as shown in Figure 4.13.

4.3.2 Feed Forward Artificial Neural Network

Artificial Neural Network (ANN) is a very efficient tool in building a mathematical model for a non- linear application that cannot be described by a specific equation. When there is a problem specified by measured or archived (X-Y) pairs and it has no mathematical criterion that can describe it, ANN will be an optimal solution in this case to provide a function that is able to fit these pairs with minimum errors. ANN is able to configure a mathematical model for this problem and uses these (X-Y) pairs as input and target vectors in its training process. The learning process of the ANN can be considered a closed-loop control system with a unity feedback where the controller is the adaptive weights, the plant is the transfer function, and the desired output is the target vector. There are several types of ANNs used in fault diagnosis expert systems, particularly in the detection and the classification of faults. FFNNs are the most commonly used type of the ANN as they are efficient and accurate in several non-linear applications. FFNN consists of an input layer, hidden layers, and an output layer. Its control parameters are named weights that

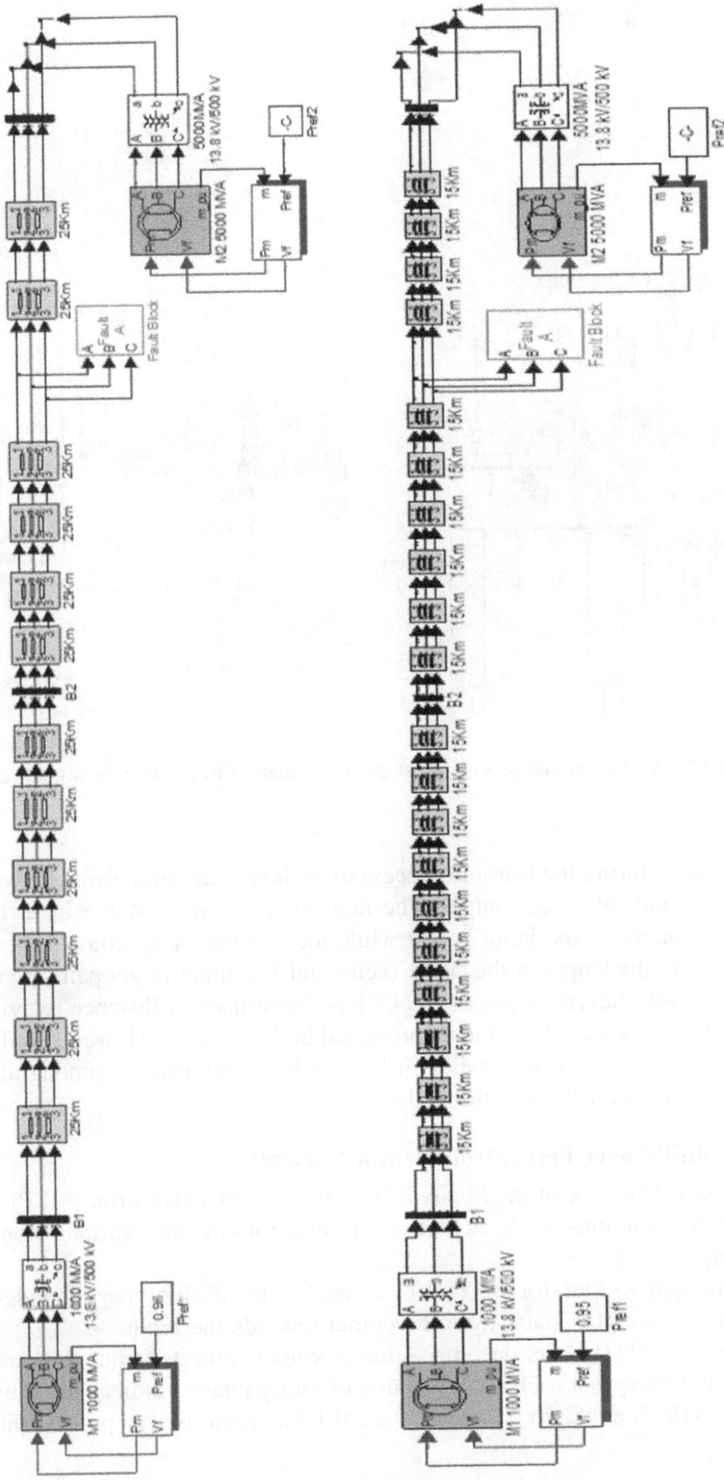

FIGURE 4.12 The first stage of simulation at 25 and 15 km lengths of the TL sections.

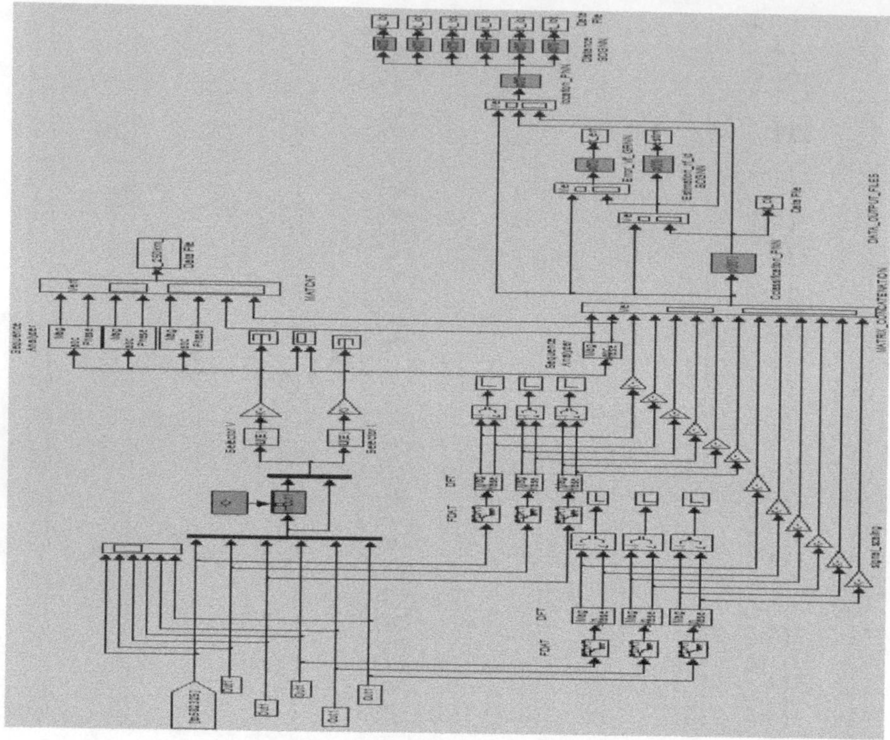

FIGURE 4.13 V and I signal processing stage of simulation to prepare the signals as inputs to the ANN stage.

can be adjusted during the learning process to achieve minimum errors between the actual output and the target output. The number of neurons of the input layer is equal to the length of the input vector while the number of neurons in the output layer is equal to the length of the target vector and The input/target pairs are used to train the network. Several types of FFNN have been used in distance relaying and some popular types are selected to be presented in this chapter. However, only PNN is chosen to be discussed in detail as it has the best performance among all other FFNNs in the classification of the faults.

4.3.2.1 Multi-Layer Perceptron Neural Network

The most prevalent type of the FFNN is the Multi-Layer Perceptron (MLP), which uses sigmoid functions such as activation function to suit various non-linear applications.

The training algorithm for this type of network is the back propagation algorithm where updating weights starts from the output towards the input.

In (Bo et al., 1997), After decomposition of voltage into its frequency bands and specifying the energy of each one, selection of faulty phase has been accomplished by MLP. In (Hagh et al., 2007), a modular MLP has been used to process the large

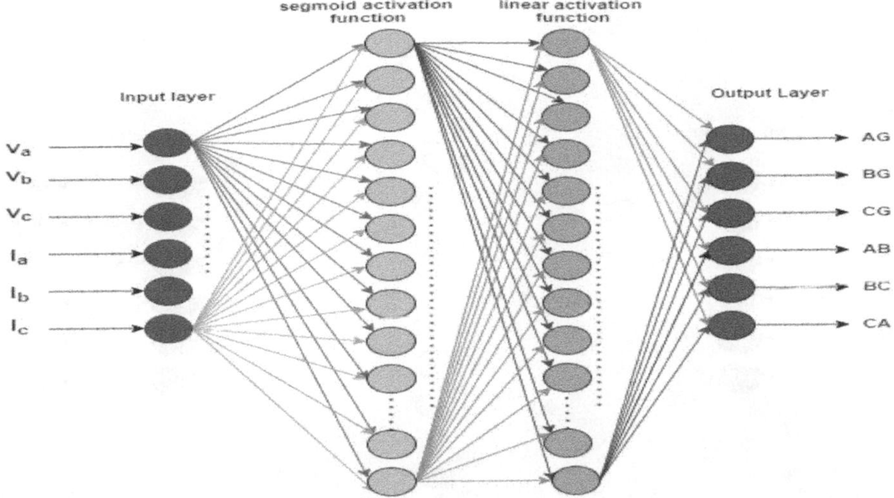

FIGURE 4.14 MLP architecture.

data of faults and the faulty phase selection has been carried out. Figure 4.14 shows MLP architecture.

4.3.2.2 Radial Basis Function Network

The Radial Basis Function Network (RBFN) is one type of FFNN. It utilizes the bell-shape Gaussian function as an activation function, which achieves a better performance when used as a classifier. Gaussian function is one of the most suitable functions when tackling random data, such as that available in the case of fault detection and classification. It has only one hidden layer with a larger number of neurons compared to the other standard FFNNs. However, its training time is a fractional of that of the standard networks. Moreover, it can withstand a large number of training vectors, which have improved the accuracy of the convergence, as in (Orr, 1996). In (Mahanty et al., 2004), RBFN has been used as a classifier of earth faults and phase faults.

The RBFN architecture is shown in Figure 4.15.

4.3.2.3 Chebyshev Neural Network

Functional expansion of a Chebyshev Neural Network (CHNN) is utilized to map the input vectors and only one layer is needed in the network, as shown in Figure 4.16, which has been illustrated in (Mall et al., 2017). This leads to adjusting only one parameter and saving remarkable processing time. Therefore, it has achieved good results in fault classification problems compared to other neural network techniques with lower training time, as in (Vyas et al., 2014).

4.3.2.4 Probabilistic Neural Network as a Detailed Example of FFNN

Probabilistic Neural Network (PNN) is a statistical memory-based technique that has been used in detection and classification of faults in many researches because of

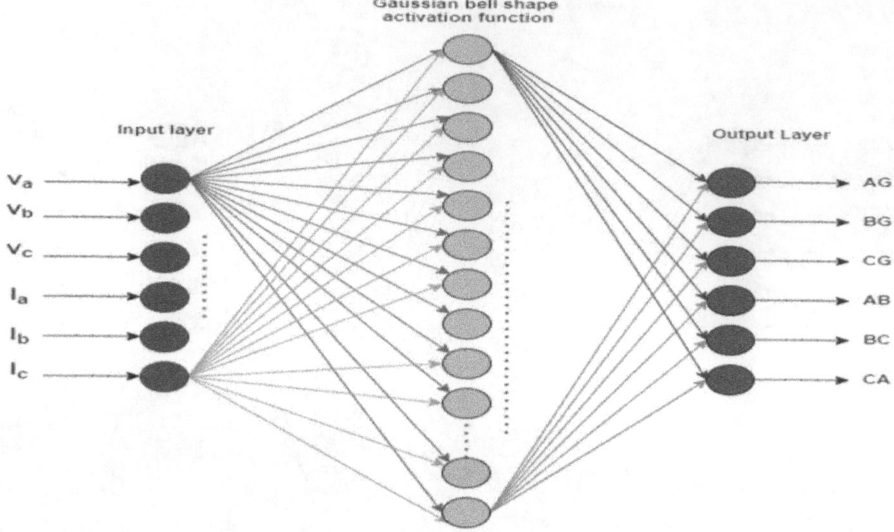

FIGURE 4.15 Radial basis function neural network.

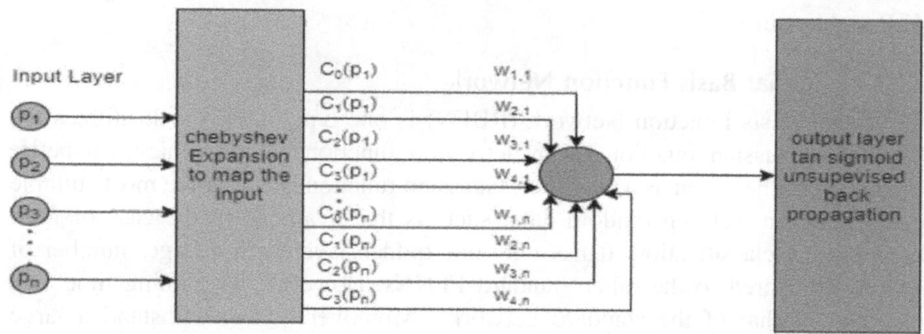

FIGURE 4.16 Chebyshev neural network architecture.

its efficiency as a classifier and its ability of tackling with large nonlinear data. In the standard structure of PNN there are four layers: input layer, radial basis layer (Gaussian uniform activation function), summation layer, and output layer. In PNN, approximation of the probability density function PDF has been performed via Parzen's criterion for each class. Bayes' Rule has been implemented to allocate class with the highest probability to the input in order to provide minimum error, as in (Mohebali et al., 2020). Moreover, kernels have been used to achieve better discrimination. Approximation of the probability density function PDF for each class is accomplished based on the training input/target pairs. Input/target pairs of the same class may be located far from each other, which may decrease in turn the probability of determining the right class. The flowchart shown in Figure 4.17 explains how PNN works. However, the performance of PNN has been improved

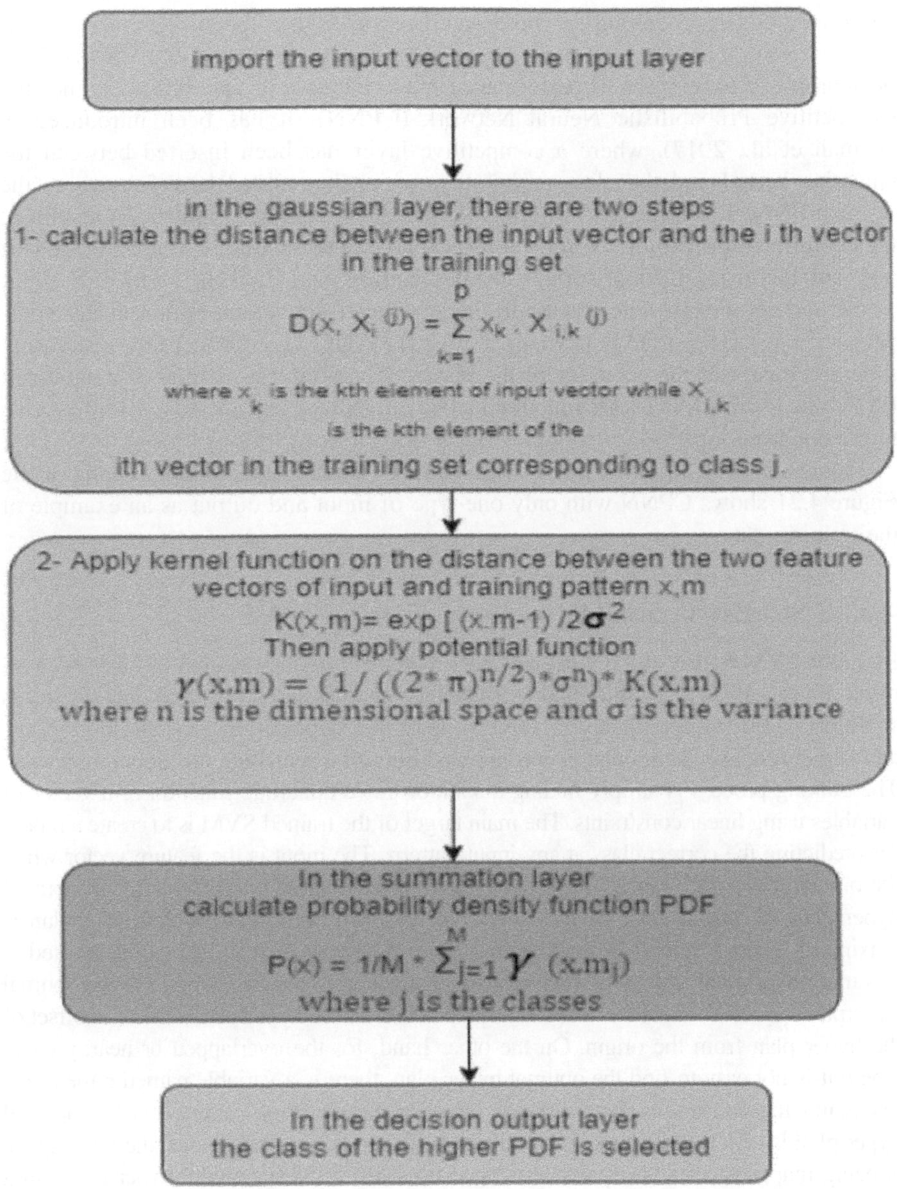

FIGURE 4.17 PNN flowchart.

via one out of three techniques. The first one is the processing of input data using WT or ST transformation before used as inputs to PNN. The second avenue is to modify PNN parameters such as the speed parameter using Local Decision Circles and Bilevel optimization. The third technique is formed by the integration between PNN and the other optimization techniques as firefly or with other types of ANN or with the fuzzy logic technique. This may lead to more accurate results of PNN in less operating time. Seeking for more accuracy and higher speed of convergence as well as lower storage requirements, there was a modified version of PNN that have been proposed to improve its performance and to increase its capabilities, named the Competitive Probabilistic Neural Network (CPNN). It has been introduced in (Zeinali et al., 2017), where a competitive layer has been inserted between the Gaussian kernel and the class probability, which has added new features to the original PNN. The selection of optimal kernels has been accomplished to estimate the conditional probability of each class. This novel technique has been able to sort and rank the magnitudes of kernels in a descending manner. That's why only a part of the kernels will be able to pass to the next layer after competition in this competitive layer, as reported in (Zeinali et al., 2017). This layer reduces the processing time and increases the accuracy of the results. The simplest method of obtaining a PNN code is via a MATLAB function named newpnn or via a MATLAB interactive tool named nntool.

Figures 4.18–4.20 show PNN with various inputs and outputs vectors while Figure 4.21 shows CPNN with only one type of input and output as an example of the training data.

4.3.3 Support Vector Machine as an Example of ML

Recently, SVM is one of the most prevalent methods for classification of faults because of its robustness as a classifier and its suitability for massive and sparse data due to its statistical-based learning theory. The SVM algorithm depends on generalization theory and kernel function. Its training is a convex optimization searching for global minimum. The training process is simply finding a solution for a quadratic function of a vector of variables using linear constraints. The main target of the trained SVM is to create a model for predicting the correct class at any input pattern. The input is the feature vector while the output is the class label. The goal of SVM is to separate two classes by an optimal hyper plan i.e. the distance of the closest vector from both sides of the hyper plan is maximum. There is a margin on the both sides of the hyper plan and the points located in this margin are named support vectors. The distance between the origin and the optimal hyperplane is $-b/ \| w \|$ where w is the normal vector to the hyper plan while b is offset of the hyper plan from the origin. On the other hand, for the overlapped or nearby cases where it is not easy to find the optimal hyper plan, there is a variable named ξ measures bypassing the constraints. The distance between classification failure and the optimal hyper plan is $-\xi/\| w \|$. In the case of non-linear separation of the classes, the input space is being mapped to feature space in order to establish a linear optimal hyper plan using kernel functions. Figure 4.22 illustrates a linear classifier, optimal hyper plan, as in (Welling, 2004). The optimal values of SVM parameters are obtained using Particle Swarm Optimizer (PSO) during the training process, the flowchart of optimization of SVM using PSO is shown in Figure 4.23, as illustrated in (Ray et al., 2016).

SVM has integrated with WT as a stage of preprocessing to improve its accuracy by extracting new features. The block diagram of a SVM classifier is shown in Figure 4.24, as in (Ray et al., 2016).

An example of a part of the MATLAB code of SVM established on the Math Works site is presented in this section.

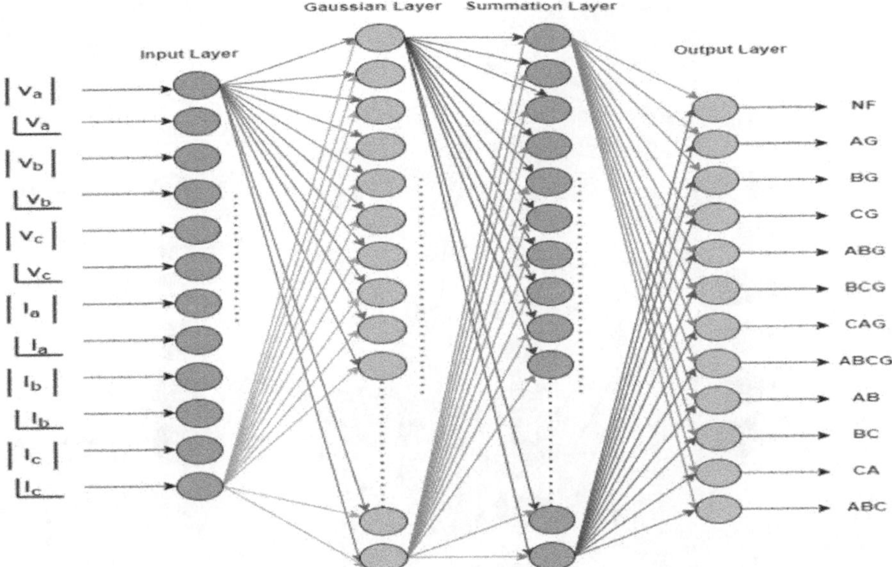

FIGURE 4.18 PNN with 12 input voltages and currents in phasor form and 12 output classes.

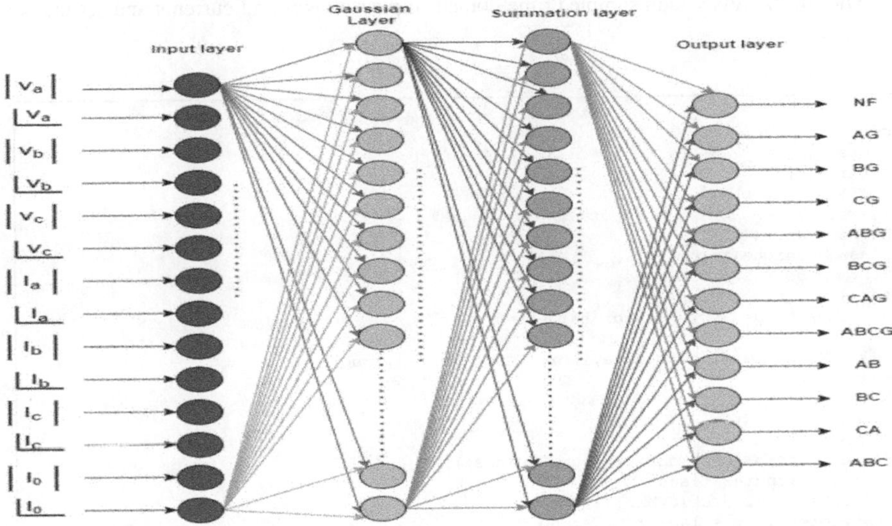

FIGURE 4.19 PNN with 14 inputs of voltages and currents magnitudes and angles including I_0 and 12 output classes.

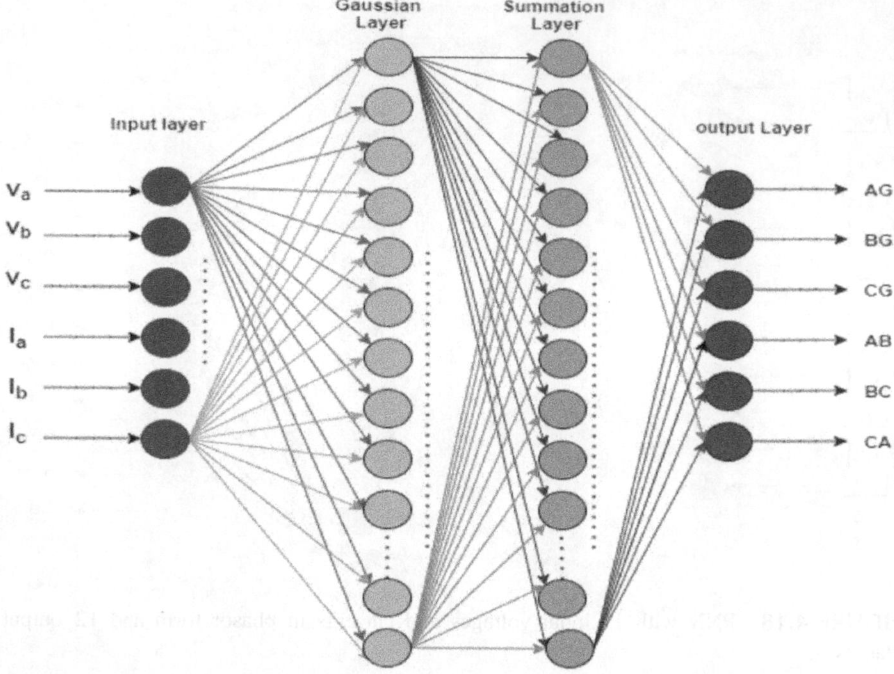

FIGURE 4.20 PNN with sampled time domain input voltages and currents and six classes of outputs.

```
%Example of using MATLAB SVM function from MATHWORKS site.
clc;
close all;
clear all;
 %% Training phase
% four classes and ten input vector length
SVMMS = cell(4,1);
Class = unique(target);
% SVM model is trained for each class
for j = 1:nume2(class)
    indx(target~=j)=-1;indx(target==j)=1; % Creating binary classes
% matlab function for SVM model is fitcsvm where the matlab code exists
    SVMMS{j} = fitcsvm(data,indx,'Standardize',true);
End
% saving of the trained SVMModels for testing phase.
save('SVMMS.mat','SVMMS');
% train performance
Scores = zeros(size(data,1),nume2(class));
 for j = 1:nume2(class)
    label = predict(SVMMS{j},data);
    Scores(:,j) = label; % Second column contains positive-class scores
end
Scores
[~,maxScore] = max(Scores,[],2)
 perf=sum(maxScore==target)/size(maxScore,1) % performance in the range of 0 to
1
 %% Testing phase
load('SVMMS.mat');
numclass = 4;
testdata = [10 0 0 0]; % check for a new unknown vector
for j = 1:nume2(classes)
    label = predict(SVMMs{j},testdata);
    Group(:,j) = label; % Second column contains positive-class scores
end
```

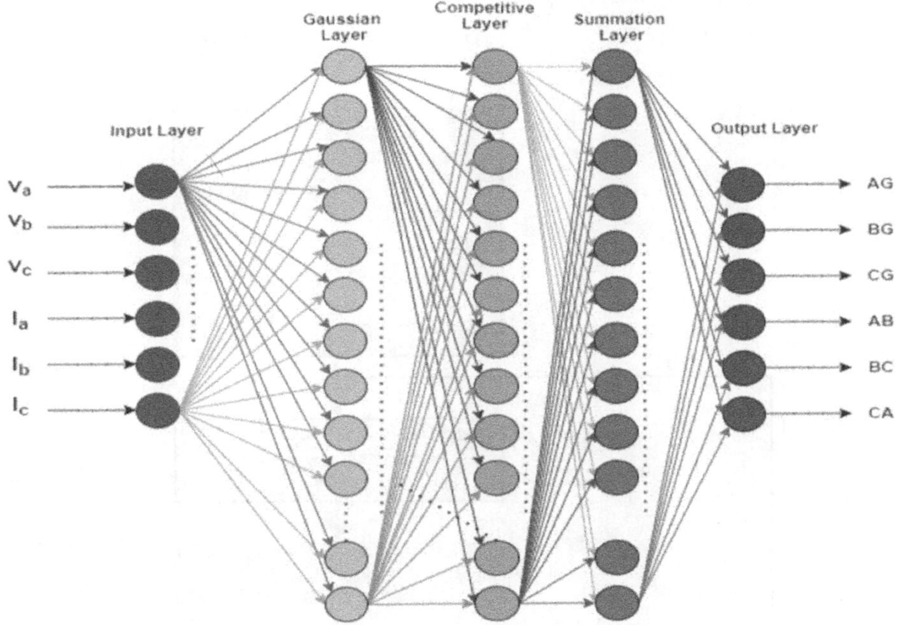

FIGURE 4.21 CPNN with sampled time domain input voltages and currents and six classes of outputs.

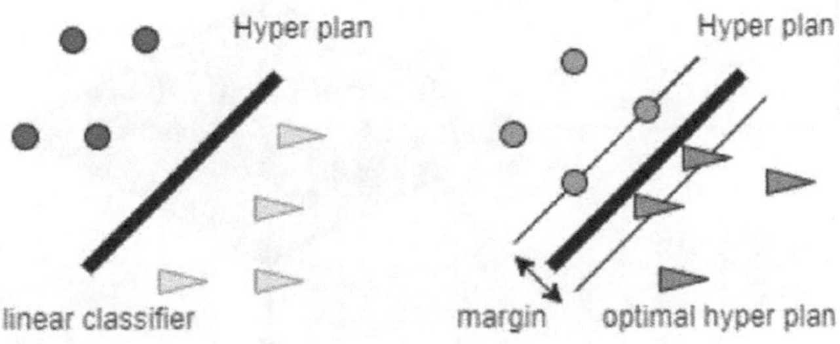

FIGURE 4.22 Linear classifier, optimal hyper plan.

4.3.4 Convolution Neural Network as an Example of DL

Lately, DL algorithms such as CNN have proved their efficiency in the classification field named image classification, where a vision algorithm has been used to convert signals into two-dimensional visual images. Therefore, representing the three-phase voltages and the three line currents measured at the relay point as images has opened up the field of visual fault classification, which has provided in turn high levels of fault features that cannot be reached by a one-dimension time series, as reported in (Yang et al., 2020). The time series voltages and currents have been encoded to images via Gramian Angular Field (GAF) technique, as in (Wang et al., 2015). Seeking for extraction of higher-level features, a robust DL technique

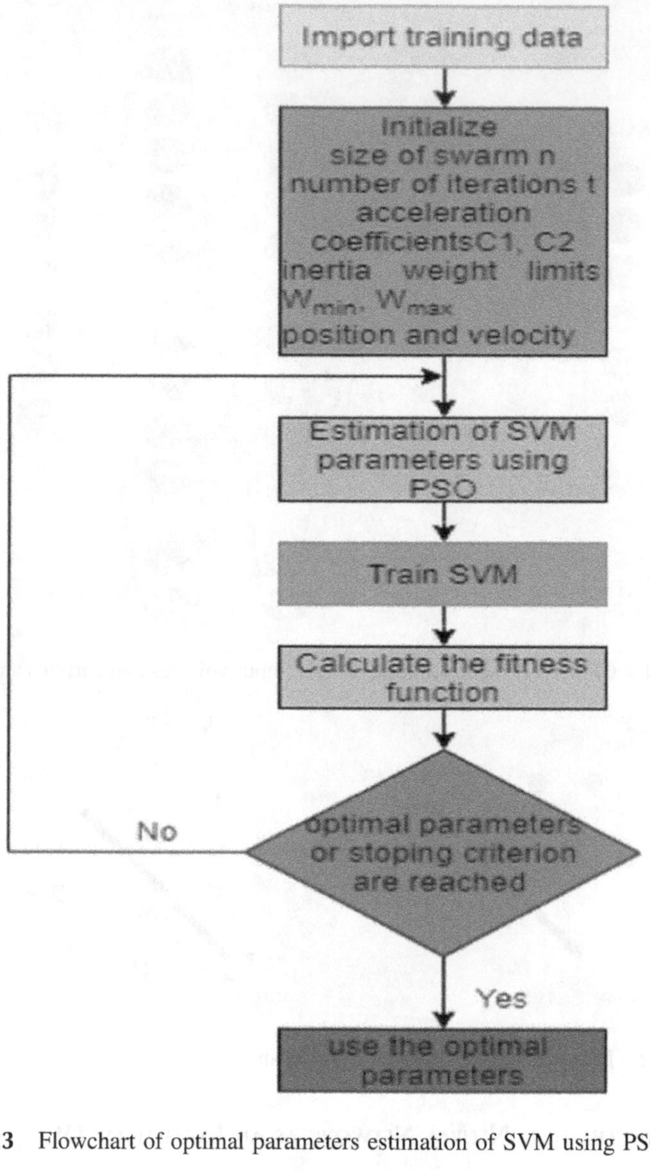

Import training data
Initialize
size of swarm n
number of iterations t
acceleration
coefficientsC1, C2
inertia weight limits
Wmin, Wmax
position and velocity
Estimation of SVM parameters using PSO
Train SVM
Calculate the fitness function
No
optimal parameters or stoping criterion are reached
Yes
use the optimal parameters

FIGURE 4.23 Flowchart of optimal parameters estimation of SVM using PSO.

Measured signals at the relay point
Decomposition of signals & Feature extraction
Best feature selection
Modular SVM classifier

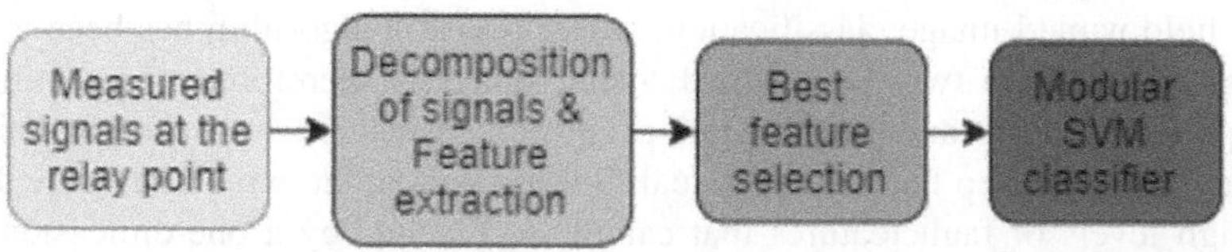

FIGURE 4.24 Block diagram of SVM fault classifier.

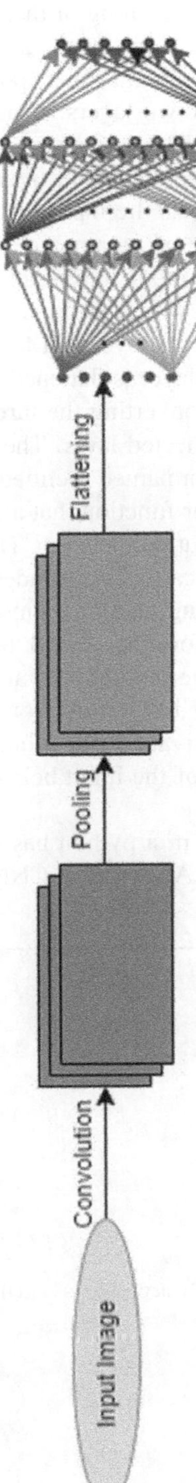

FIGURE 4.25 CNN structure.

named CNN has been selected to be used for learning of these novel features and a better classification has been achieved as in (Mei et al., 2017). CNN has several layers that have been categorized as convolutional layer, pooling layer, and fully connected layer. The main target of the first two layers is to reduce the number of parameters to be trained. The convolutional layer consists of an input matrix of three dimensions, kernels multiple filters, and an output three-dimensional matrix. The output matrix of the convolutional layer is the input of the pooling layer. The pooling layer is used to reduce the number of parameters of its input matrix by extracting representative features from the input tensor and thus the number of computations will be reduced considerably, which improves the efficiency of this technique. The third layer is the fully connected layer that forms the last few layers of the network. The output from the pooling layer is flattened before being the input of the fully connected layer, which means converting the three-dimensional output matrix into an input vector for the fully connected layer. The fully connected layer is simply a FFNN with an activation function named rectified linear unit activation function (ReLU). ReLU is a piece-wise linear function that allows only the positive input to have an output or otherwise the output will be zero. This activation function is used widely in DL neural networks because it provides easier training and achieves better performance. It is worth noting that the sigmoid and the hyperbolic tangent activation functions are not suitable for this network because they cannot be used in a network of many layers owing to the problem of vanishing gradient which can be overcome using ReLU. The last layer in the fully connected layer stage has an activation function named Softmax activation function. Softmax Activation function is used to provide the possibilities of the input being in a particular class. Figure 4.25 shows a CNN structure.

An example of a part of CNN algorithm in a python has been illustrated, as in (Lamadeu, 2018). Another complete MATLAB code of CNN has been created in (Mason, 2019).

```
# Example of CNN code in Python created by Antonio L. Amadeu, 2018.
# Building the CNN in Python
# Importing the Keras libraries and packages
from keras.models import Sequential
from keras.layers import Conv2D
from keras.layers import MaxPooling2D
from keras.layers import Flatten
from keras.layers import Dense
# Initialize  CNN
classf = Sequential()
# Convolution
classf.add(Conv2D(32, (3, 3), input_shape = (64, 64, 3), activation = 'relu'))
# Pooling
classf.add(MaxPooling2D(pool_size = (2, 2)))
# Flattening
classf.add(Flatten())
# Full connection
classf.add(Dense(units = 128, activation = 'relu'))
classf.add(Dense(units = 1, activation = 'sigmoid'))
# Compilation of CNN
classf.compile(optimizer = 'adam', loss = 'binary_crossentropy', metrics = ['accuracy'])
```

FIGURE 4.26 CNN block diagram for fault classification.

Recently, the performance of CNN has been enhanced by merging Self-Attention (SAT) mechanism illustrated in (Vaswani et al., 2017). This integration increases CNN classification capabilities and assists in tackling the new discriminative features. Thereby, A Self-Attentive Convolution Neural Network (SAT-CNN) technique is proposed to classify the fault types and the block diagram is shown in Figure 4.26, as in (Fahim et al., 2020).

4.4 CONCLUSION

Detection and classification of faults on the transmission lines are affected remarkably by many factors that contaminate the phase voltages and line current signals measured at the relay point as well as the errors that may be caused by the circuit elements. The protection system should respond to the faults efficiently despite these factors and should be able to discriminate between permanent faults and the other abnormalities of the power system. That's why there is a persistent need of using powerful techniques to ensure accurate operation of the distance relay. The first stage of these techniques was the transformations as FT, WT, CT, and ST. the first version of transformation has utilized FT to extract the fundamental components of voltages and currents in phasor form combined with the circuit equation. The next version of transformations has used WT to decompose the voltages and currents signals as well as to extract new features as coefficients of a detailed level to improve the discrimination among various types of faults. Thereafter, WT has been modified using many added techniques such as PCA, CT... etc. the last version of transformation was ST integrated with FFT. The second stage has used AI techniques such as PNN, CPNN, SVM ... etc. This stage improves the classification of faults considerably due to its ability in tackling parse and massive data as well as its strong capabilities as classifiers. The third category is integrating the transformations with AI techniques to create an accurate classifier based on principle components as well as novel feature extraction. This category has provided the best accuracy with a higher speed of execution. Lately, a new avenue of AI technique has been established to convert the voltage and current signals into images and to process the images for better classification of faults via an algorithm named CNN in conjunction with the other transformations. These techniques help to develop the distance protection system and provide accurate detection and classification of faults within a very short period of time, which improves the efficiency of the protection system in protecting the transmission lines.

REFERENCES

[1] A. Prasad, and J. B. Edward. "Application of wavelet technique for fault classification in transmission systems". *Procedia Computer Science*, 2016, vol. 92, pp. 78–83.

[2] A. Lamadeu. "Image_ classification_ CNN". Python code, github, 2018.

[3] A. Jamehbozorg, and S. M. Shahrtash. "A decision-tree-based method for fault classification in single-circuit transmission lines". *IEEE Transactions on Power Delivery*, 2010, 25, 2190–2196.

[4] A. Prasad, J. B. Edward, and K. Ravi. "A review on fault classification methodologies in power transmission systems: Part – I". *Journal of Electrical Systems and Information Technology*. 2017, 5, 48–60.

[5] A. Raza, A. Benrabah, T. Alquthami, and M. Akmal. "A review of fault diagnosing methods in power transmission systems". *Applied Science*, 2020, 10, 1312.

[6] A. Vaswani, N. Shazeer, N. Parmar, N. Uszkoreit, L. Jones, A.N. Gomez, L. Kaiser, and I. Polosukhin. "Attention is all you need". *Advances in Neural Information Processing Systems*, 2017, pp. 5998–6008.

[7] A. A. M. Zin, M. Saini, M. W. Mustafa, and A. R. Sultan. "New algorithm for detection and fault classification on parallel transmission line using DWT and BPNN based on Clarke's transformation". *Neuro Computing*, 2015, 168, 983–993.

[8] B. Lakshmana Nayak. "Classification of transmission line faults using wavelet transformer". *International Journal of Engineering Sciences & Research Technology*, February 2014, 3(2), pp. 568–574.

[9] B. Mohebali, A. Tahmassebi, A. Meyer-Baese, and A. H. Gandomi. "Probabilistic neural networks: a brief overview of theory, implementation, and application". *Handbook of Probabilistic Models*, 2020, CH 14, Elsevier, doi: 10.1016/B978-0-12-816514-0.00014-X.

[10] B. Ravindhranath Reddy, M. Vijay Kumar, M. Surya Kalavathi, and Y. Venkata Raju. "Detection & localization of faults in transmission lines using wavelet transforms (Coif Let & Mexican Hat)". *Journal of Theoretical and Applied Information Technology*, 2009, 8, 2, 99–104.

[11] B. Y. Vyas, B. Das, and R. P. Maheshwari. "Improved fault classification in series compensated transmission line: Comparative evaluation of Chebyshev neural network training algorithms". *IEEE Transactions on Neural Networks and Learning Systems*, 2014, 27, 1631–1642.

[12] C. Pothisarn, and C. Jettanasen. "The study on wavelet coefficient behavior of simultaneous fault on the hybrid between overhead and underground distribution system". *International Journal of Smart Grid and Clean Energy*, 2019, 8, 367–371.

[13] C. R. Pinnegar, and L. Mansinha. "Time local Fourier analysis with a scalable, phase modulated analyzing function: the S-transform with a complex window". *Signal Processing*, 2004, 84, no. 7, 1167–1176.

[14] C. R. Pinnegar, and L. Mansinha, "The S-transform with windows of arbitrary and varying window". *Geophysics*, 2003, 68, no. 1, 381–385.

[15] C.-L. Yang, Z.-X. Chen, and C.-Y. Yang. "Sensor classification using convolutional neural network by encoding multivariate time series as two-dimensional colored images". *Sensors*, 2020, 20, 168.

[16] D. Allam, M. Gilany, A. Elnagar, and M. Hassan. "A new successive ANN for fault classification and estimation of combined fault resistance and loading conditions". Power Tech Conference, 2007, 663.

[17] F. B. Costa, B. A. Souza, and N. S. D. Brito. "A wavelet-based method for detection and classification of single and cross country faults in transmission lines". International Conference on Power Systems Transients, June 2009, pp. 1–8.

[18] F. B. Costa, B. A. Souza, and N. S. D. Brito. "Real-time classification of transmission line faults based on maximal overlap discrete wavelet transform". Transmission and Distribution Conference and Exposition, IEEE/PES, 2012, vol. 1, pp. 1–8.

[19] F. Mo, and W. Kinsner. "Probabilistic neural networks for power line fault classification". In Proceedings of the IEEE Canadian Conference on Electrical and Computer Engineering, Waterloo, ON, Canada, 25–28 May 1998, pp. 585–588.

[20] H. Mason. "Matlab _ Convolutional _ neural_net". Matlab code, github, 2019.

[21] I. Steinwart. "On the optimal parameter choice for m-support vector machines". *IEEE Transactions on Pattern Analysis and Machine Intelligence*, 2003, 25, 1274–1284.

[22] J. Klomjit, and A. Ngaopitakkul. "Comparison of artificial intelligence methods for fault classification of the 115-kV hybrid transmission system". *Applied Science*, 2020, 10, 3967.

[23] J. Upendar, C.P. Gupta, and G.K. Singh. "Discrete wavelet transform and probabilistic neural network based algorithm for classification of fault on transmission systems". In Proceedings of the 2008 Annual IEEE India Conference, Kanpur, India, 11–13 December 2008, pp. 206–211.

[24] L. Cheng, L. Wang, and F. Gao. "Power system fault classification method based on sparse representation and random dimensionality reduction projection". In Proceedings of the 2015 IEEE Power & Energy Society General Meeting, Denver, CO, USA, 26–30 July 2015, pp. 1–5.

[25] K. R. Krishnanand, P. K. Dash, and M. H. Naeem. "Detection, classification, and location of faults in power transmission lines". *Electrical Power and Energy*, 2015, 67, 76–86.

[26] K. Schittkowski. "Optimal parameter selection in support vector machines". *Journal of Industrial and Management Optimization*, 2005, 1, 465–476.

[27] L. Rutkowski."Adaptive probabilistic neural networks for pattern classification in time-varying environment. "*IEEE Trans on Neural Network*, 2004, 15, 811–827.

[28] M. T. Hagh, K. Razi, and H. Taghizadeh. "Fault classification and location of power transmission lines using artificial neural network". In Proceedings of the 2007 International Power Engineering Conference (IPEC), Singapore, 3–6 December 2007, pp. 1109–1114.

[29] M. Mirzaei, M. Z. Kadir, H. Hizam, and E. Moazami. "Comparative analysis of probabilistic neural network, radial basis function and feed-forward neural network for fault classification in power distribution systems". *Electric Power Components and Systems*, 2011, 39, 1858–1871.

[30] M. J. Orr. "Introduction to radial basis function networks". Technical Report, *Center for Cognitive Science*, University of Edinburgh, Edinburgh, UK, 1996.

[31] M. Tripathy, R. P. Maheshwari, and H. Verma. "Power transformer differential protection based on optimal probabilistic neural network". *IEEE Transactions on Power Delivery*, 2010, 25, 102–112.

[32] M. Welling. *"Support Vector Machines"*. Toronto: University of Toronto, 2004.

[33] O. A. S. Youssef. "Fault classification based on wavelet transforms". Transmission and Distribution Conference and Exposition, IEEE/PES, 2001, vol. 1, pp. 531–536.

[34] P. K. Dash, S. R. Samantaray, G. Panda, and B. K. Panigrahi. "Time-frequency transform approach for protection of parallel transmission lines". *IET Generation, Transmission and Distribution*, 2007, 1, no. 1, 30–38.

[35] P. Jose, and V. R. Bindu. "Wavelet-based transmission line fault analysis". *International Journal of Engineering and Innovative Technology*, February 2014, vol. 3, no. 8, pp. 55–60.

[36] P. Ray, and D. P. Mishra. "Support vector machine based fault classification and location of a long transmission line". *Engineering Science and Technology, an International Journal*, 2016, 19, 1368–1380.

[37] R. N. Mahanty, and P. B. Gupta. "Application of RBF neural network to fault classification and location in transmission lines". In *IEE Proceedings-Generation, Transmission and Distribution*, IET, London, UK, 2004, pp. 201–207.

[38] S. El Safty, and A. El-Zonkoly. "Applying wavelet entropy principle in fault classification," *Electrical Power and Energy Systems*, Elsevier, 2009, pp. 604–607.

[39] S. R. Fahim, Y. Sarker, S. K. Sarker, R. Islam Sheikh, M. D. Sajal, K. Das. "Self attention convolutional neural network with time series imaging based feature extraction for transmission line fault detection and classification". *Electric Power Systems Research*, 2020, 187, 106437.

[40] S. Mall, and S. Chakraverty. "Single layer Chebyshev neural network model for solving elliptic partial differential equations". *Neural Processing Letters*, 2017, 45, pp. 825–840.

[41] S. Mei, J. Ji, J. Hou, X. Li, and Q. Du. "Learning sensor-specific spatial-spectral features of hyperspectral images via convolutional neural networks". *IEEE Transactions on Geoscience and Remote Sensing*, 2017, 55, 4520–4533.

[42] S. Mishra, C. Bhende, and B. Panigrahi. "Detection and classification of power quality disturbances using s-transform and probabilistic neural network". *IEEE Transactions on Power Delivery*, 2008, 23, 280–287.

[43] S. Pei, P. Wang, J. Ding, C. Wen. "Elimination of the discretization side-effect in the S-Transform using folded Windows". *Signal Processing*, 2011, 91, no. 6, 1466–1475.

[44] S. Yu, and J. C. Gu. "Removal of decaying DC in current and voltage signals using a modified Fourier filter algorithm". *IEEE Transactions on Power Delivery*, 2001, 16, 372–379.

[45] U. B. Parikh, B. Das, and R. Maheshwari. "Fault classification technique for series compensated transmission line using support vector machine". *International Journal of Electrical Power and Energy Systems*, 2010, 32, 629–636.

[46] Y. Zeinali, and B. A. Story. "Competitive probabilistic neural network". *Integrated Computer-Aided Engineering*, 2017, 1, 1–14.

[47] Z. Q. Bo, R. K. Aggarwal, A. T. Johns, H. Y. Li, and Y. H. Song. "A new approach to phase selection using fault generated high frequency noise and neural networks". *IEEE Transactions on Power Delivery*, 1997, 12, 106–115.

[48] Z. Wang, and T. Oates. "Imaging time-series to improve classification and imputation". Twenty-Fourth International Joint Conference on Artificial Intelligence, 2015.

5 Intelligent Fault Location Schemes for Modern Power Systems

Tamer A. Kawady, Mahmoud A. Elsadd, and Nagy I. Elkalashy
Electrical Engineering Department, Menoufia University, Shebin El-Kom, Egypt

5.1 INTRODUCTION

Owing to the increasing complexities of recent power systems, improving the existent protection functions and developing more advanced ones have got much attention. This is mainly to enhance the performance of the overall power system. Transmission and distribution systems are experiencing commonly different faults due to different causes including storms, lightning, snow, rain, insulation breakdown, … etc. Severe short circuits may also occur by birds or other external objects. In most cases, these electrical faults manifest in mechanical damage resulting in disconnecting the power supply to some customers. For these faults, the power system can be restored, when the fault position is estimated with acceptable accuracy. Although the transient faults may cause minor damage to be easily visible on inspection, their effects may be exaggerated later. Hence, fault location schemes can help effectively for assisting early repair plans to prevent recurrence and consequent major damages. This corroborates the effectiveness and importance of fault locators in power systems [1–3].

The schemes of fault location and distance protection are closely related. However, some major important differences arise between both of them. Fault locators are used for pinpointing the fault position accurately rather than indicating the faulted area (defined by the predetermined protective zones). Both the measurement and the decision-making procedures of protective relays are commonly performed in an on-line regime [4,5]. The high speed of operation of protective relays appears as a crucial requirement imposed on them. In contrast, the calculations of fault location schemes are typically performed in an off-line mode. Fault location information, including the position of the fault and its type, are then fed into maintenance teams. This implies that the required speed for fault location procedures can be performed in some seconds or even in minutes as well [6–8]. Including the fault location function as an additional function of microprocessor-based relays is commonly used in practice. In this case, high computational

DOI: 10.1201/9780367552374-5

capability and communication with remote sites of modern relays are utilized at little or almost no additional cost. Also, digital fault recorders enable easy and not costly incorporation of the fault-location function. In turn, stand-alone fault locators are applied in the case of using sophisticated fault location algorithms and under the condition that higher cost of the implementation is accepted [9–11].

Fault location methods can be categorized into three types including the phasor measurement schemes using either single or double end data, traveling wave-based techniques, and non-conventional techniques using AI and advanced signal analysis tools. On the other hand, modern smart grid concepts in distribution networks encourage using more sophisticated fault location schemes based on recent hardware elements and broad measurements. In spite of the reliability and simplicity of the first category, the increasing complexities of modern network topologies utilizing distributed generators and multi-terminal feeders seriously affects their performance. Also, utilizing the second traveling wave-based scheme have their own difficulties regarding the implementability and dependability as well. On the other hand, the increasing development of communication and hardware tools facilitates realizing more practical, broad, and versatile intelligent schemes for fault location purposes. Such schemes may represent great solutions to provide electric utilities with more powerful fault locators [12,13]. As seen in the literature, different AI tools were successfully utilized for these purposes such as Expert Systems, Artificial Neural Network, Support Vector Machine, Fuzzy Logic, and Genetic Algorithms.

The main goal of this chapter is to emphasize a visualization of the most recent developments for fault location schemes for transmission and distribution systems using AI tools. AI currently represents an ideal alternative for performing those complex and unsolved problems that can not be processed using the conventional tools. First, the conventional fault location methodologies are shortly reviewed. Then, contributions with AI for fault location purposes are covered. Finally, smart grid requirements and their contributions for fault location tasks are visualized.

5.2 CONVENTIONAL FAULT LOCATION REVIEW

As described in Figure 5.1, conventional fault location methods can be classified into two basic groups: traveling wave fault locators and impedance measurement-based ones [2]. Traveling wave schemes can be used either with injecting a certain traveling wave from the locator position or with analyzing the generated transients due to the fault occurrence. Impedance measurement schemes depend on either the data from one or both line ends. Each category can be then classified according to the considered line model during the derivation method using either simpler (lumped) models or detailed (distributed parameters) ones [14–17].

5.2.1 Traveling Wave-Based Fault Locators

Due to the fault occurrences, voltage and current transients travel toward the line terminals. These transients continue to bounce back and forth between the fault point and both line terminals for the faulted line until the post-fault steady state

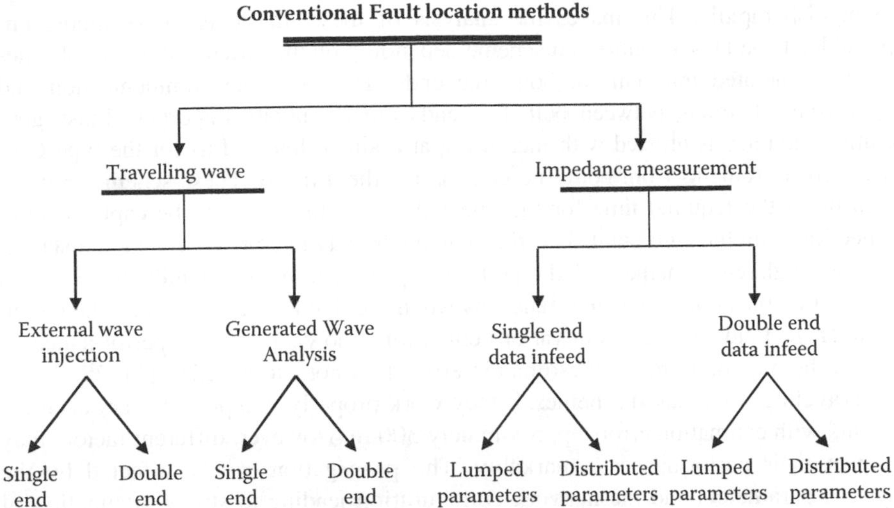

FIGURE 5.1 Classification of conventional fault location methods.

condition is reached [18]. For a fault at unknown distance from the sending end, an abrupt injection appears at the fault point. Then, the injected wave travels as a surge along the line segment in both directions between the fault point and both line terminals until the post-fault steady state condition is reached. These traveling waves can be then captured to be utilized for locating these faults. These methods rely on calculating the required time of the captured waves to reach the end of the line. This is achieved by comparing the wave arrival time difference at each end of the line to determine the distance to the fault point. Since the wave moves at the speed of light, this requires extremely accurate timing for calculating the fault distance precisely using either voltage or current signals. The voltage portion of the traveling waveform tends to be reduced as a result of buses with lower impedance. On the other hand, the current waveform tends to double as the result of a constant current source. Thus, the required fault location can be correctly calculated using the precise wave-arrival times at each end of the faulted line [19]. Technically, different equipment is required including accurate time-stamping devices such as GPS devices on both ends of the line, appropriate sensors to detect the voltage or current, and communication circuits.

Employing traveling wave phenomena for fault location purposes for both underground cables and overhead lines was reported since 1931. In 1951, Lewis classified traveling wave-based fault location schemes into different four types A, B, C, and D according to their modes of operation using the traveling voltage waves [20]. Types A and D depend on analyzing the resulting transients from the fault itself needing no further pulse-generating circuitry. Type A is a single end one capturing the transients only at one end depending on the generated transients from the arcing flashover during the fault. However, this assumption regarding getting the generated transients at the line ends cannot be always satisfied. Moreover, the arc itself may extinguish rapidly. This makes the analysis of these transients to be almost impossible. Type D is a double-end scheme depending on the difference in arrival times of the generated transients at both line ends. However, the communication and synchronized timing between both line ends are essentially important. Pulse generating circuitry is utilized with measuring at a single line end as for the type C or using measurements from both line ends as for the Type B. These schemes rely on measuring the required time for the injected pulses to go and to be captured after reflection from the fault point. This time can be directly interpreted as a fault distance. A new single-end scheme, called type E, was proposed in 1993. Unlike the previous types, it employs the generated transients when the line is re-energized by the circuit breaker. Its field test records in various conditions show a promising performance, in which the maximum resulted estimation error does not exceed 2.7% [21–25].

Traveling wave-based schemes, if they work properly, can provide very accurate results with estimation errors approximately 300 m. However, different factors may impact their performances remarkably. The propagation can be affected by the system parameters and the network configuration leading to strong attenuation of the waves. Another difficulty arises for those faults occurring near to the buses or for faults at or near zero-voltage inception angles. Moreover, the reflected waves can be seriously affected by the line discontinuities such as branches, tapped loads, and cable sections. Also, the complexity of developing their simulation, especially with considering the frequency dependency of system parameters, is another real difficulty [26]. Also, the economical factor is an essential disadvantage of such schemes due to the extra required hardware including wave sending equipment and capturing instruments as well as the communication and timing synchronization tools for double-end ones. Actually, different requirements are required for realizing an accurate performance of such schemes. Time stamping must be very precise. Thus, GPS-based traveling-wave fault location schemes, with the availability of time-stamp information, show accurate performance. Also, utilizing data from both line terminals allows timing from the initiation of the short circuit, which effectively assists to realize a precise estimation of the fault distance. More accurate current and voltage transformers are required to provide reasonable reproduction of transients [27–29].

5.2.2 Impedance Measurement-Based Fault Locators

These schemes provide another applicable alternative for fault location purposes. Figure 5.2 shows the one-line diagram of a three-phase and double-fed faulted transmission line. A line-to-ground fault occurred on phase A at point F through a resistance R_F at a distance x from the locator position. Then, the fault current I_F is comprised of both components I_{Fs} and I_{Fr} flowing from sending and receiving ends, respectively. The essential task of the fault location algorithm is to estimate the fault distance x as a function of the total line impedance Z_L using the sending end measurements (for single-end algorithms) or both end measurements (for double-end algorithms) with the most possible accuracy. Both schemes are briefly described as follows.

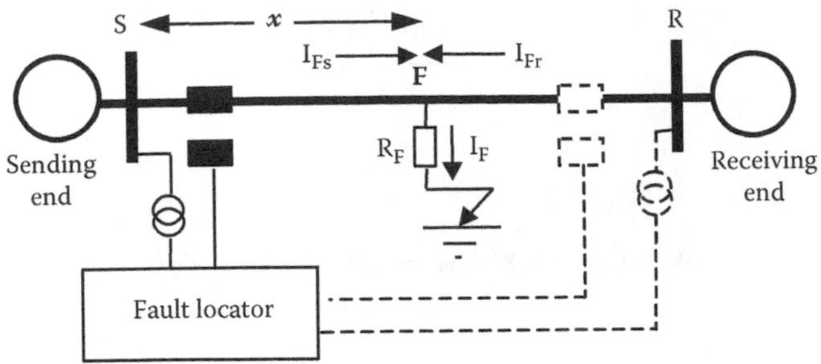

FIGURE 5.2 One-line diagram of a faulted transmission line [1].

Certainly, the direct and most accurate way to calculate the fault distance is to depend on the measuring voltage and current quantities at both line ends using V_S, I_S and V_R, I_R at both sending and receiving ends for the faulted phase, respectively. Related to Figure 5.2, the voltage of the fault point V_F can be written as a function of both sending and receiving end voltages as:

$$V_F = V_S - I_{Fs}\,xZ_L \tag{5.1}$$

$$V_F = V_R - I_{Fr}(1 - x)Z_L \tag{5.2}$$

Equating both equations and rearranging yield:

$$x = \frac{\frac{V_S - V_R}{Z_L} + I_{Fr}}{I_{Fs} + I_{Fr}} \tag{5.3}$$

Then, the formulas for other fault types (double phase, double phase to ground, and three-phase faults) can be derived similarly [30–35]. In spite of the simple and direct derivation of the algorithm equations, its performance is remarkably questionable due to the simple line model (basic lumped resistance model) neglecting the capacitive currents and mutual coupling [36–41].

The same algorithm was reformed using the distributed parameter line model aiming to realize a more accurate performance [42]. For this purpose, equations (5.1) and (5.2) are modified using the line characteristic impedance Z_0 and line propagation constant γ as:

$$V_F = \cosh(x\gamma)V_s - Z_0 \sin(x\gamma)I_{Fs} \tag{5.4}$$

$$V_F = cosh((L - x)\gamma)V_R - Z_0 sin((L - x)\gamma)I_{Fr} \tag{5.5}$$

The unknown fault distance x can be then written as:

$$x = \frac{\tanh^{-1}\left(\frac{-B}{A}\right)}{\gamma} \tag{5.6}$$

where

$$A = Z_0 \cosh(\gamma L) I_{Fr} - \sin(\gamma L) V_R + Z_0 I_{Fs} \tag{5.7}$$

and

$$B = \cosh(\gamma L) V_R - Z_0 \sin(\gamma L) I_{Fr} - V_S \tag{5.8}$$

On the other hand, the extra requirements for communication and accurate timing purposes at both line ends are real disadvantages of all double-end algorithms. This problem was partially solved by utilizing unsynchronized and parameter-less fault location computation, as seen in [43–46].

Another double-end algorithm was introduced utilizing the Global Positioning System (GPS) in conjunction with the Phase Measurement Unit (PMU) for fault location purposes [47–49]. Employing the GPS enables to ensure the accurate timing between both of the line ends, whereas the PMU is employed for phasor estimation purposes as described in Figure 5.3. The algorithm core is basically similar to the main double-end algorithm equations by equating the fault point voltage from both line ends using a distributed parameter line modeling. All evaluation tests of the performance revealed higher accuracy as reported in their relevant references. However, the economical factor arises here as a basic disadvantage due to the required cost for the GPS synchronization systems as well as for the communication requirements.

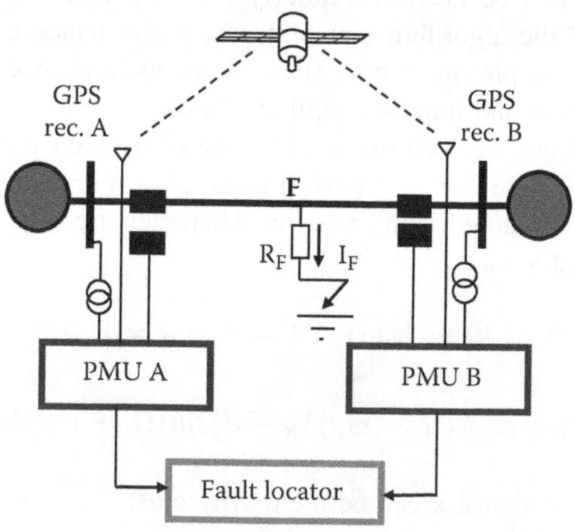

FIGURE 5.3 PMU-based fault location schematic.

Due to the extra costs for double-end algorithms, single end ones attract an increasing attention and consequently they have the superiority from the commercial point of view [50–56]. Referring to Figure 5.2, equation (5.1) can be rewritten as:

$$V_S = xZ_L I_{Fs} + R_F I_F \tag{5.9}$$

Then, the unknown fault distance x can be directly computed by equating the imaginary parts of both equation sides as:

$$x = \frac{\mathrm{Im}\{V_S - R_F I_F\}}{\mathrm{Im}\{Z_L I_{Fs}\}} = \frac{\mathrm{Im}\left\{\frac{V_S}{I_{Fs}}\right\} - \mathrm{Im}\left\{R_F\left(\frac{I_F}{I_{Fs}}\right)\right\}}{\mathrm{Im}\{Z_L\}} \tag{5.10}$$

To solve the aforementioned equation, the unknown R_F should be excluded, considering here some proper simplifying assumptions. Assuming both the sending end and fault currents (I_{Fs} and I_F) are in phase, the term containing R_F vanishes as its imaginary part equal to zero. This yields the final form for fault distance x as [57–60].

$$x = \frac{\mathrm{Im}\left\{\frac{V_S}{I_{Fs}}\right\}}{\mathrm{Im}\{Z_L\}} \tag{5.11}$$

This derivation summarizes the basic feature of single end algorithms, in which the lower amount of available data (as compared with the other two end ones) leads to simplifying the fault location equations with some assumptions. It consequently affects the overall accuracy of the calculation process remarkably. The research efforts aim therefore to improve these algorithms in order to get the possible highest accuracy [56–59].

5.2.3 Requirements for Fault Location Process

Figure 5.4 illustrates a general explanation of the basic requirements for fault location schemes. As described before, the fault locator works in off-line mode after performing the relaying action. Once the fault is detected and the faulty phase(s) are successfully classified, the fault locator is enabled to find out the estimated fault distance. The recorded data by the available Digital Fault Recorder (DFR) is then passed through the locator input manipulator to the fault locator block. A few seconds or minutes (according to the locator speed) later, the fault distance is estimated and then the maintenance crews can be sent to the fault position. For those locators that use double terminal information, an extra data communication link is fitted between both line ends. Also, traveling wave-based locators require extra hardware for generating and capturing the resulting waves. Practical fault locators

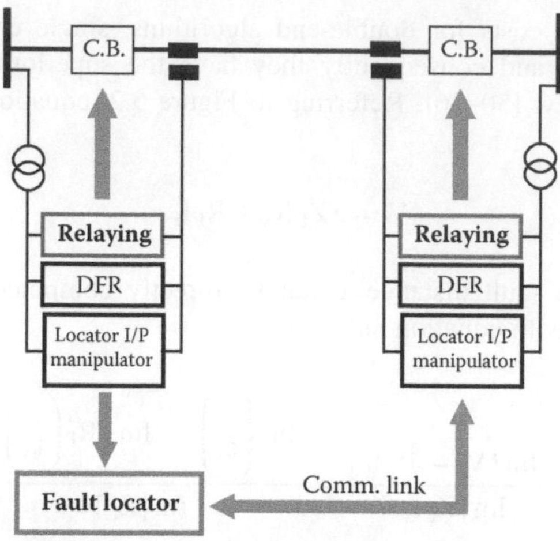

FIGURE 5.4 General requirements for fault location schemes.

may be standalone devices in the substations or included as parts of the modern multi-function protection equipment and distance relays for overhead lines, which is the most economical and common protection tools for transmission networks recently [60–62].

5.3 AI-BASED FAULT LOCATION SCHEMES

AI is a branch of computer science concerning with investigating how to mimic human beings' actions by machines. As seen in the literatures, AI has shown great potential for a broad variety of applications in electrical engineering. AI tools possesses great characteristics such as fast learning, fault tolerance, and ability to produce correct output when fed with partial input [1]. Therefore, it can be adapted to recognize learned patterns of the measured quantities in electric systems to be utilized in various applications in power networks. For protection engineering, in particular, different difficulties arise with transmission and distribution networks due to different factors such as the heterogeneity of transmission lines as a result of geographical nature [60], the insertion of series capacitors through the transmission lines [63,64], the insertion of AC/DC converter [65], and new deregulation and free marketing polices that lead to an increase in utilizing distributed energy resources including wind energies and PVs. Different drawbacks of integrating these green power sources into the power network arise including voltage fluctuations, frequency stabilization, and power quality issues. Another drawback is the oscillating nature of the renewable sources that are highly dependable on the environment aspects and weather conditions as well. Moreover, the two-way power flow, presence of load taps and laterals, the existence of short and heterogeneous lines, and uncertainties in line and system parameters impact the accuracy of the corresponding fault detection and location tools using conventional methods. [66,67]. Instead of the normal mathematical derivation, non-conventional fault location algorithms were introduced, depending on other processing platforms such as Wavelet Transform (WT), Artificial Neural Network (ANN), Fuzzy Logic (FL), or Genetic Algorithm (GA) tools. These methods have their own problems that result from the line modeling accuracy, data availability, and the method essences. The contributions of such tools are reviewed as follows.

5.3.1 ANN-Based Fault Location Computation

Artificial Neural Network (ANN) provides a promising tool for classification and non-linear mapping problems. For power system purposes, many successful applications have been proposed for different purposes such as load forecasting, security assessment, control,… etc. This is well covered in the published literature [68–71]. For protection purposes, different applications were developed covering a wide range of protection purposes such as faulted area estimation, fault direction discrimination, generator protection, transformer protection,… etc. However, almost all of these applications mainly use the ANN as a simple discriminator having only the outputs of 1 or 0 using voltage and current samples directly. This simple topological use of ANN may reach the aimed accuracy with the proper training. However, employing ANN for fault location purposes requires the ANN to perform more advanced calculations in order to predict the fault impedance seen from the locator location. The training sets should be prepared properly, covering all situations that can happen in real situations. Thus, the ANN efficiency essentially depends on the properly selected network design as well as the sufficient amounts of training data. Some successful applications employing the ANN for fault location purposes were published, as seen in [72–76].

The ANN is a collection of neurons (or simply nodes) that are connected in a certain configuration to allow communication among these nodes [1]. Each neuron receives the weighted sum of the received signals coming from the former layer (or from the network general inputs) and propagates the weighted outputs to the next nodes as shown in Figure 5.5. For each node, the weighted inputs to the node are aggregated with a dedicated activation function resulting in the output of the node. Many activation functions are available.

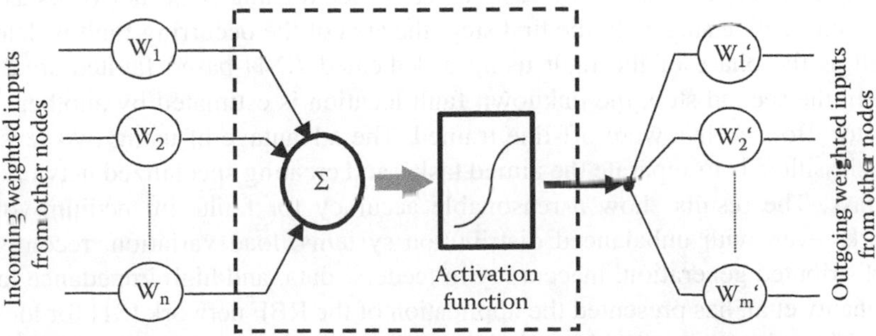

FIGURE 5.5 A single node basic construction.

The node output is transmitted to other nodes through the pre-defined selected weights. The ANN nodes can be connected in different ways according to the selected ANN architecture. Each of them is described with certain characteristics. It is consequently applicable for certain applications. One of the most common ANN structures is realized by adjusting the nodes into sequential layers. Each layer includes a certain number of nodes. Practically one or two hidden layers are sufficient for almost all applications. The wiring of connectivity and its direction among the different layers detect the type of the network and refer consequently to the suitable training paradigm. Among the available network architectures, Multi-Layer Feedforward (MLF), Radial Basis Functions (RBF), and Elman networks arise as examples of the most known and common ANN architectures. RBF network can be considered as a special feedforward network utilizing basis functions in its hidden layer rather than using the sigmoid activation as in normal MLF networks. Each of these hidden neurons has a symmetrical response around its selected center vector. Each of the output nodes accumulates the weighted sum of all hidden neurons. The output layer combines linearly the content in this new space, where the weights connecting both layers are adjusted through the training process using the Least Square optimization. This simple topological structure and fast training paradigm provide a powerful tool for various applications including function approximation and classification purposes. Recurrent networks, on the other hand, employ one or more feedback from a certain layer to the foregoing one(s). Unlike the normal MLF networks, recurrent ones can distinguish the temporal patterns due to the memory action of the context inputs. A popular example of this category is the Elman network [77,78].

As seen in [79], feeder protection was successfully developed using the feedforward ANN based on the voltage signals, circuit breaker status, real power of the feeder during the normal condition, and real power of the feeder during the short circuit, operator experience to train the neural network. In the training process, additional data were used during and after the fault, such as voltage angle, X/R ratio during the fault, short circuit current angle, … etc. In the training process, additional data were used during and after the fault, such as voltage angle, X/R ratio during the fault, short-circuit current angle, etc. As described in the relevant reference, this combination of the selected inputs guarantees an acceptable efficiency for fault location computation.

In [80], the fault location process was constructed using MLF networks as well in two consecutive stages. In the first step, the area of the occurring fault is detected as well as the phase of the fault using a dedicated ANN-based faulted-area classifier. In the second step, the unknown fault location is estimated by another ANN estimator. Both ANNs were off-line trained. The advantage of using two levels of ANN classifiers is to separate the aimed tasks, and creating specialized networks for each task. The results show a reasonable accuracy for faults in medium-voltage networks even with unbalanced distribution systems, load variation, reconfiguration, distributed generation, inaccuracy in feeders' data, and high impedance faults.

Mahanty et al. has presented the application of the RBF network [81] for locating faults. The advantage of this technique is that it has two ANNs (ANN-I and ANN-II) for every type of fault (two ANNs for L-G fault, L-L fault, L-L-G fault, and L-L-L fault). Joorabian et al. proved that an ANN-based fault location scheme can be useful for real-time systems as well [74]. They have used a separate RBF network for each type of fault to identify the fault location in extra high voltage (EHV) transmission lines. Another method for fault location using Elman Recurrent Network (ERN) has been implemented, as seen in [82]. This network is an alternative method to overcome the difficulties in feedforward back propagation network MLF and RBF to identify the location of faults. A 380-kV, 360-km transmission line was used for simulation purposes. The authors in [82] got better results with ERN compared with MLF and RBF networks for the location of faults in transmission lines. A Complex Adaptive Linear Neural Network (CADALINE) algorithm [83] has been proposed to identify the aimed fault location in overhead lines. Field Programmable Gate Array (FPGA) implementation using ANN [84] has been proposed for the location of faults as well. The authors have presented a methodology and a prototype for fault location using FPGA implementation process for fast protective relaying. A hybrid frame work for location of faults with the help of Adaptive Structural Neural Networks (ASNNs) has been proposed in [85]. The results shown in these papers are interesting and simulation results verified with developed hardware results as well. A comparison of different neural network approaches for fault location has been presented in [86–88], where the authors compared three different ANN structures. These results corroborated the efficacy of the ANNs with a variety network structures have a superior performance for fault location purposes.

Instead of generating the unknown fault distance directly by the constructed ANN, the conventionally computed fault distance was successfully corrected by a dedicated ANN fault location corrector, as described in Figure 5.6 [1]. The

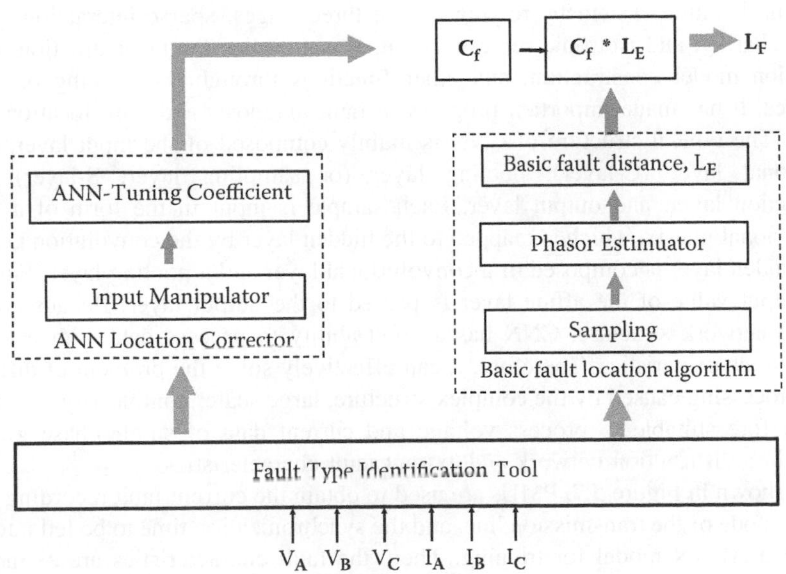

FIGURE 5.6 Schematic diagram of the ANN-based tuning mechanism for fault location [1].

generated correction factor, C_f, is utilized to compensate the accumulated errors in the computed basic fault distance, L_E. Then, the final fault distance, L_F, can be found as:

$$LF = Cf * LE \tag{5.12}$$

As seen in [1], different network structures and different training paradigms were considered for designing the most optimized ANN-based tuner, including MLF, RBF, and Elman networks. Then, the performance of each constructed ANN architecture was investigated thoroughly, raising a perfect tuning performance for fault location purposes by compensating the overall computational errors accompanied with the conventional computed algorithm.

On the other hand, recent ANN constructions and training mechanisms were successfully utilized for fault location purposes such as deep learning modules and Support Vector Machines (SVMs) [89]. As seen in the literature, SVM was successfully used in classification and regression problems by gaining popularity among the available various intelligent techniques due to its performance [90]. The number of support vectors for the SVM mechanism is determined by SVM algorithm itself. In neural network, the number of hidden layers is determined by trial-and-error method. This facility makes SVM a better classification algorithm than ANN. Also, SVM does not require any training effort like the neural network for good performance. The advantage of SVM is that it is faster even for large-sized problem and requires fewer heuristics. The main features of SVM are the upper bound on the generalization error does not depend on the dimension of the space. As seen in [91], Convolution Neural Network (CNN) improves the traditional machine learning system by relying on the three stages: sparse interaction, parameter sharing, and isotropic representation. It can realize feature extraction, classification model construction, and other functions through the training of input samples. It has made important progress in fault diagnosis and fault location purposes. The typical structure of CNN is mainly composed of the input layer, convolutional layer (C-layer), pooling layer (or sampling layer, S-layer), full connection layer, and output layer. Each sample is input in the form of a two-dimensional matrix, which is mapped to the hidden layer by the convolution kernel. The hidden layer is composed of a convolutional layer and a pooling layer. Finally, the output value of the affine layer is passed to the output layer. Because of its unique network structure, CNN has a good ability to process data with network structure characteristics. Therefore, it can effectively solve the problem of difficult data processing caused by the complex structure, large scale, nonlinearity,.... etc. It is therefore suitable to process voltage and current data of single-phase ground faults in a distribution network and extract fault characteristics.

As shown in Figure 5.7, PMUs are used to obtain the current fault recording data of each node of the transmission line, and the synchronization time to be fed into the designed ACNN model for training. Then, the fault characteristics are extracted. During the training of the network, the model can automatically extract the features of the input data through the C-layer and the adaptive S-layer. The number of

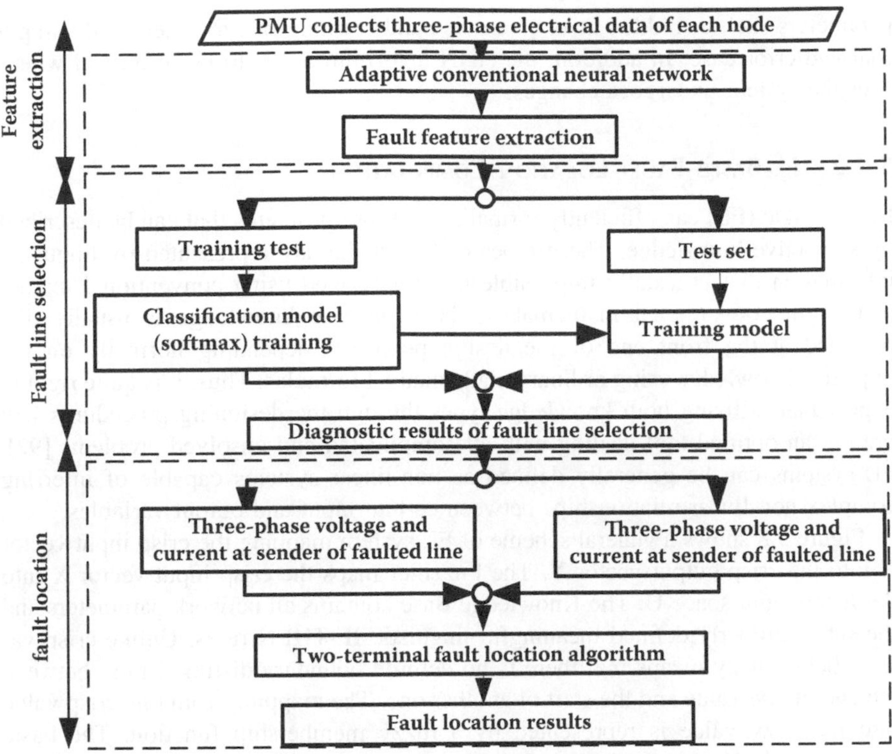

FIGURE 5.7 Schematic of the CNN-based fault locator [91].

hidden layers of ACNN is set according to the actual needs, through the analysis of the above input matrix dimensions and considering the dimensionality reduction effect of convolution and pooling processes. Therefore, in the process of model design, the number of hidden layer network layers is four, comprising two C-layers and two S-layers. In the second part, the current phasor sampling data of each node of the line after feature extraction is divided into the training sample set and the test set. Then, CNN network parameters with high accuracy after training are saved. In the third part, the fault current data of each node uploaded by PMU in real time are input into the trained ACNN model for fault line selection. When the fault line is determined, the fault record data of both ends of the fault line with the synchronization mark is called from the background, and the fault location algorithm is applied to locate the fault accurately.

The main advantage of the ANN is its simplicity in implementation resulting from its simple and straightforward mathematical basis. However, its disadvantage is that it is highly dependent on the amount and quality of the trained data in the preparation stage. A limited amount of information will therefore affect the performance of the method. This problem happens for distribution systems with limited information resulting from an insufficient number of monitoring devices. The other disadvantage of ANN is that the training process has slow convergence. Also, the parameters such as hidden layers, neurons, and learning rate are identified using a trial-and-error case. In addition, the ANN algorithm needs to be re-trained whenever the system undergoes changes.

5.3.2 FL-Based Fault Location Computation

Fuzzy Logic (FL) can efficiently formalize all those problems that can be described by subjective knowledge. These types of knowledge are represented by linguistic information that is usually impossible to be described using conventional mathematics. In conventional mathematics, the subjective knowledge is usually disregarded at the front end of the design processes depending normally on the objective knowledge using ordinary mathematical formulas. Thus, it is quite right to expect that utilizing both knowledge types through the designing procedures will present an optimal tool dealing with all complicated and unsolved problems [92]. FL systems can be generally defined as non-linear systems capable of inferring complex non-linear relationships between certain input and output variables.

Figure 5.8 shows a general scheme of FL system mapping the crisp input vector X into the crisp output vector Y. The Fuzzifier maps the crisp input vector X into the fuzzy input space U. The Knowledge Base contains all network parameters and the set of properly defined meaningful linguistic IF-THEN rules. Unlike crisp values, being fuzzy means that there is no definite boundary distinguishing between the end of one value and the start of another one. The mapping from this crisp value into its fuzzy value is represented by a fuzzy membership function. The basic function of the defuzzifier is to map the fuzzy outputs from the resulted output space vector V from the inference stage into the crisp output vector. On the other hand, designing a FL system can be described as a process for approximating a non-linear function or fitting a complex surface in a properly high-dimensional space [92]. First, a preliminary design of the FL network can adopt considering the arbitrary initial rules and membership functions describing the corresponding behavior of the considered problem. Building meaningful rules is an essential step in order to develop a successful fuzzy system. These rules are developed based on the available knowledge about the required task and its relation to the selected inputs. Thus, the importance of the successful selection of these input features among all of the

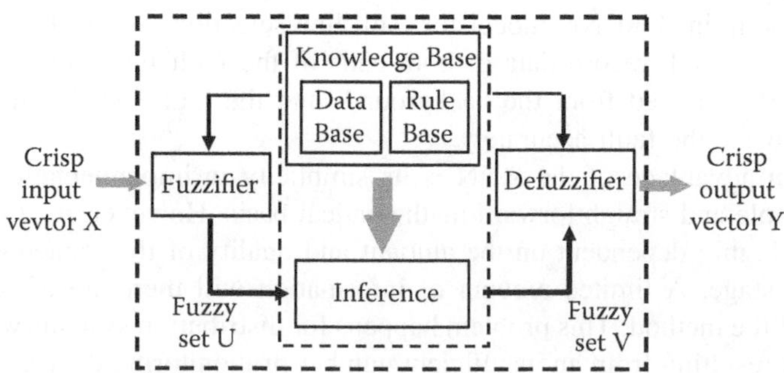

FIGURE 5.8 FL system schematic diagram.

available ones is obvious. Then, the related parameter of these networks should be optimized through a pre-prepared set of input-output pairs via a proper simulation. Different training paradigms can be employed using different methods such as Least Squares, GA-based, or BP training. The training efficiency mainly depends on the characteristics of the training method, the amount of training data, and the complexity of the required task.

FL system in addition to synchronized phasor measurements was used for fault location as seen in [93] to identify the fault type and fault location in double circuit transmission lines. The method classifies successfully series, shunt, and simultaneous series-shunt faults. Test results show that the fault location error percentage is within 1% for series faults and is up to 5% for shunt faults. Another approach for fault detection, classification, and location using FL and Programmable Automation and Control technology was presented in [94]. Although fuzzy logic–based scheme is quite satisfactory, the main drawback of fuzzy logic is in determining the global minimum using fuzzy membership functions. Also, feature definition and extraction have to be enhanced for the classification algorithms. Hence, the need for optimizing FL systems is clear.

Among the different available methods for fuzzy logic system training, the Adaptive Network Fuzzy Inference System (ANFIS) routine arises as a direct and powerful tool [95–98]. It is, therefore, included in the fuzzy logic toolbox in MATLAB®. However, it is characterized by a few limitations. The available version of ANFIS supports only first- or zero-order Sugeno-type fuzzy systems with multi-inputs and one output network. In Sugeno-based fuzzy systems, all inputs are described by a set of selected overlapping membership functions, while the outputs are described as a linear function of the inputs. This was performed via a dedicated five-layer network describing the FL system, as shown in Figure 5.9. Details for ANFIS construction and training were found in [99]. It has, therefore, a wide variety of successful applications for power system protection purposes as seen in [100–103].

As seen in [1], the ANFIS-based tuning mechanism was successfully developed and tested similarly to that utilized by the ANN in the preceding section. A promising performance was reordered even for locating transmission line faults even with high impedance faults.

5.3.3 GA-Based Fault Location Computation

Genetic Algorithm (GA) is a heuristic search technique that tries to mimic the process of natural evolution to decide which solutions should be preserved and allowed to be reproduced. Then, the best solutions can be precisely detected. This is conducted via a dedicated fitness function (related to the considered problem) based on some emulated steps including crossover and mutation procedures. Recent researches proved that GA has many advantages. It doesn't need a continuous search space as it jumps from point to point in search space, which allows escaping from the local optima, in which other algorithms may fall. It can also deal with complex and nonlinear functions without linearization efficiently [104].

For fault location purposes, in particular, the fault can be estimated by solving the derived optimization problem defined based on objective function and

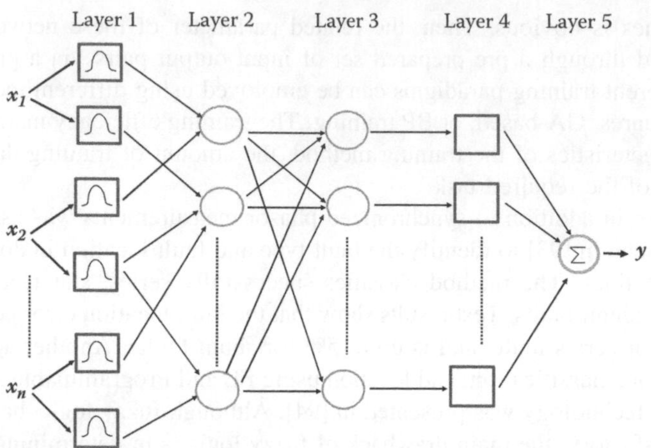

FIGURE 5.9 ANFIS network structure.

depending on the mathematical representation of the faulted power network [105]. As illustrated in Figure 5.10, local voltage and current inputs are first sampled. Then, the associated phasors are computed with a dedicated signal analysis tool such as DFT algorithm, while the magnitudes of the remote voltages and currents are only retrieved. A GA mechanism is utilized for optimizing the generated objective function for each phase. Three different variables are defined here as the unknown optimization parameters; the fault distance x (as a function of the line length L) and the angles δ_{ri} and δ_{rv} for the remote end current and voltage, respectively. Upon the occurring fault type, the corresponding GA mechanism(s) to the participated phase(s) are initialized. The total error sum is then minimized in order to get the corresponding unknowns. Three sequential steps have been carried out in order to develop the proposed GA fault locator. First, the fault location problem itself should be computationally formulized as a typical minimization problem to cope with the nature of GA procedure. Then, the primary GA mechanism is established by constructing the procedure details as well as the associated objective function. Finally, the parameters of each utilized GA mechanism should be optimized to realize the highest accuracy as described below. For a line to ground fault occurring on phase (A) at point F is assumed at a distance x from the sending end, the calculated voltage V_{SF} for the point F, seen from the sending end side, can be described considering distributed line parameters as:

$$V_{SF} = \cosh(\gamma x) V_{SA} - Z0 \sinh(\gamma x) I_{SA} \tag{5.13}$$

where V_{SA} and I_{SA} are the sending end voltage and current for phase (A), respectively. γ and Z_0 are the propagation constant and the surge impedance for the line. Both quantities are calculated as a function of line impedance and admittance as follows:

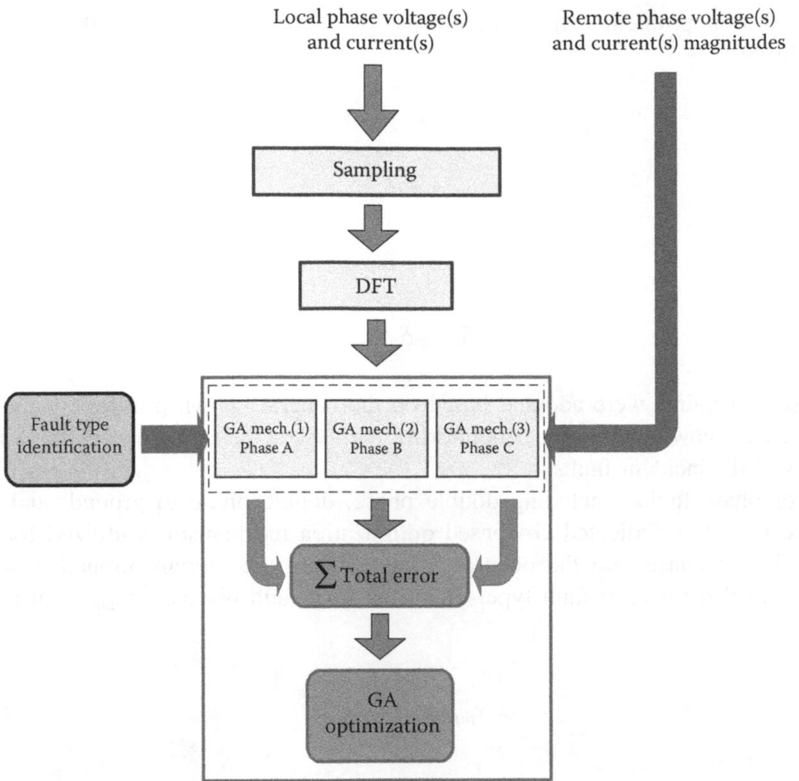

FIGURE 5.10 GA-based un-synchronized double-end fault location scheme [105].

$$\gamma = \sqrt{Z\ Y} \tag{5.14}$$

$$Z_0 = \sqrt{\frac{Z}{Y}} \tag{5.15}$$

Similarly, the fault voltage point, V_{RF}, seen from the receiving end can be written as a function of the receiving end voltage and current as:

$$V_{RF} = \cosh(\gamma(L - x))V_R - Z_0 \sinh(\gamma(L - x))I_R \tag{5.16}$$

Searching for the fault distance x is carried out by minimizing the computed fault location error, e_{FA}, defined as follows:

$$e_{FA} = |V_{SF}| - |V_{RF}| \tag{5.17}$$

The target of the GA mechanism is to search the complete space for the relevant unknowns including x, δ_{ri} and δ_{rv} to realize the global minimum as a typical optimization problem. These unknowns are constrained through the following conditions:

$$0 < x < 1 \tag{5.18}$$

$$-\pi < \delta_{rv} < \pi \tag{5.19}$$

$$-\pi < \delta_{ri} < \pi \tag{5.20}$$

$$|\delta_{si} - \delta_{ri}| < \pi/2 \tag{5.21}$$

These constraints were adopted based on the general known profiles for the considered unknowns as well as monitoring the system behavior during the various types of the incident faults.

For phase faults (including double phase, double phase to ground, and three phase faults), a dedicated GA-based optimization mechanism is utilized for each participated phase into the occurring fault, where the aforementioned equations were profiled for each fault type. Then, the final fault distance, x_{final}, can be calculated as:

$$x_{final} = \frac{1}{n_p} \sum_{k=1}^{n_p} x_{F_k} \tag{5.22}$$

where n_p is number of the participated phase(s) in the fault according to the fault type as well as the associated phases and x_{Fk} is its related estimated fault distance. Further details and evaluation tests are documented in its aforementioned reference.

Similar to the genetic optimization, other natural inspired optimization tools like the Ant Colony and the Artificial Bee Colony (ABC) were utilized as well for fault location purposes. As seen in [106], ABC for fault location in the transmission system utilized only the post-fault measurements. This method was, however, only valid for asymmetric faults and used a short line model to locate the fault location. An Artificial Bee colony is an optimization technique that is based on the natural phenomena to reach an optimal solution as a metaheuristic technique based on swarm intelligence. It has no centralized controller and self-organizing techniques. In BCO, bees use Path Integration and use direct path to come back to hive instead of back tracking their original route. Bee colony optimization algorithm is very efficient as it takes a lower number of steps when finding and collecting the food. BCO algorithm requires less computation time to complete task so it is more scalable [107,108].

5.3.4 WT-Based Fault Location Computation

Wavelet Transform (WT) provides an advanced tool for signal analysis purposes. Unlike the conventional signal processing tools such as Fourier analysis, the wavelets are not only localized in the frequency domain but also in the time domain as well. This localization enables to detect the occurrence times of even fast disturbances such as fault transients. These transients are generated by the fault and travel continuously between the fault point and the line terminals until the post fault steady state is reached. Thus, processing these signals with WT reveals their travel times between the fault point and the locator position. These times can directly refer to the unknown fault distance [109,110].

As a pre-analysis tool, WT can be utilized for preparing the captured signals before performing the decision-making procedure. A combination of wavelet transform and modular multilayer feed forward neural network was presented in [111], which identified the fault type and the location. Test results show that the method identified the faulty phase and fault type within one cycle time from the inception of the fault. Fault distance was identified with a maximum error percentage of 0.688%.

A hybrid approach using WT and SVM was suggested in [112,113] for precise fault location in transmission lines based on the voltage and current signals at one end of the transmission system in three consecutive stages. First, the WT was employed to extract the high frequency components of the extracted voltage and current signals. Then, the aimed fault type was identified using SVM classification. Finally, the fault distance was identified using SVM regression analysis. The accuracy of fault type identification is 1% and 0.7% for fault distance. The limitation of the method is that it does not consider the faulty phase in the system.

A smart fault location technique was proposed by [114] combining RTUs readings, WT, and ANN/ANFIS regimes, as illustrated in Figure 5.11. The fault currents have been acquired from the ends of the transmission line using RTUs located on buses. The GPS technology facilitates the process of synchronizing the acquired currents and communicates with the Transmission Control Center (TCC) using high-speed broadband technology. In the TCC, the WT is applied to synchronized currents to extract the features. These features with computational intelligent techniques are used for fault location to complete necessary maintenance action. The inputs for the WT are the synchronized currents measured from remote telemetry units (RTUs) using GPS technology on different buses. This smart fault-location technique is based on the WT analysis to extract the features of the transient current signals based on the generated harmonics at the instant of the fault occurrence due to the abrupt changes of currents in the three-phase transmission line. These extracted features, with such computational intelligence techniques as an adaptive neuro-fuzzy inference system (ANFIS) and artificial neural network (ANN), lead the grid toward smarter strategies for locating the fault distance.

5.4 RECENT TRENDS IN DISTRIBUTION NETWORK AND SMART GRID REQUIREMENTS

The exponential increase in the demand for electricity production and distribution in parallel with technological progress is a major challenge. These challenges are mainly concerning energy sustainability and enhancing both efficiency, reliability, and sustainability of distribution networks under these burden conditions. To

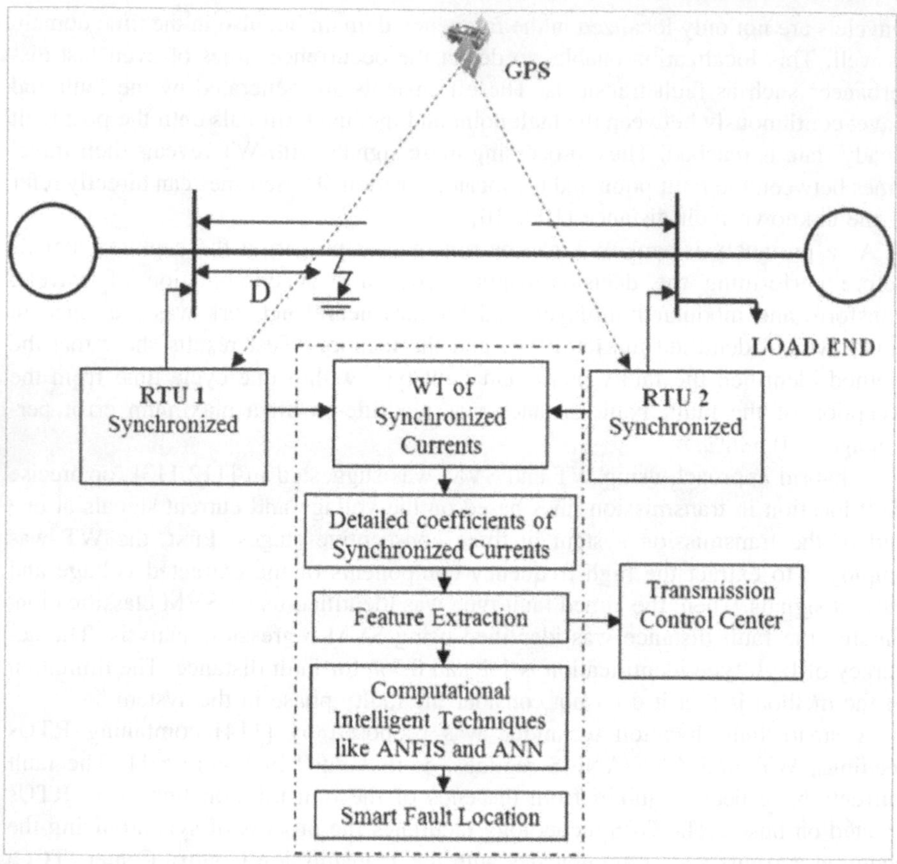

FIGURE 5.11 Schematic of the smart fault location technique using WT, RTUs, and ANNs [114].

implement these requirements, the power grid is heading towards the smart grid. In this section, a general discussion of the main parts and advantages of the smart grid is introduced, with a focus on the potential for enhancing the distribution system. The generated difficulties in locating all fault types caused by the evolution of the grid have been also described.

First, the basic parts of the smart grid are summarized from the technical viewpoint into three major systems that are smart infrastructure system, smart management system, and smart protection system as shown in Figure 5.12. The smart infrastructure system consists of smart energy, smart information, and smart communication subsystems. The smart energy subsystem is responsible for the generation, delivery, and consumption of advanced electricity generation where it can generate electricity using renewable energy, solar panels, or wind turbines. It supports the two-way flow of electricity where electricity can be delivered to users and can also be fed the energy back to the grid by distributed generations at medium voltage level or at customers. This backward flow is very helpful in a micro grid, which has been islanded due to a certain fault. The smart information subsystem provides advanced information metering and monitoring, where the intelligent electronic devices (IEDs) are now widely used [115–117]. These computer-based devices can record and store a huge amount of data with a periodicity according to the intended purpose of the device. The IEDs can provide higher sampling rates than those provided by remote terminal units (RTUs) and with higher accuracy [115].

These IEDs are installed all over the system, from the substation until the customer point. Examples of the IEDs are circuit breaker monitor (CBM), digital fault recorder (DFR), automatic meter reading (AMR) installed at customer sites, power quality meters (PQM) installed at strategic locations, low-cost current sensor, and low-cost voltage sensor. An example of the variety of IEDs applications in the distribution network is illustrated in Figure 5.13, where the solid wires are the overhead distribution lines and the dotted lines are the available communication media. The smart communication subsystem provides media to communicate information between all systems and devices. The second system is the smart management system. It provides advanced management and control services using the advantage of the smart infrastructure. The third one (last system) is the smart protection system that can be divided into two main targets. The first target is system reliability and failure protection, whereas the second one is security and privacy protection services containing information metering, measurement, and information transmission [118–123]. The failure protection mechanism contains failure prevention by predicting some possible faults as well as fault detection, diagnosis, and restoration. Also, it contains micro grid protection. The failure diagnosis and restoration in distribution grid is a crucial task in the smart grid taking into consideration problems resulting from the availability of energy flow in two ways and exploiting the advantages of the smart grid. Thus, fault location determination is a vital requirement.

The smart grid supremacy to the conventional one can be summarized as in Table 5.1. These merits are summarized from the viewpoint of both availabilities of power flow and tools. From the power flow viewpoint, smart grid can transfer power in more efficient ways by providing distributed advanced energy resources. Also, additional tools are available in the smart grid such as more sensors (RTU, IED, … etc.), as illustrated in Figure 5.13, concentrating on the distribution systems, two-way communication instead of one-way communication, distributed control instead of central control. However, there are different challenges for engineers in implementing the requirements of the smart grid that are the remote operation, adaptive protection and islanding, and self-healing [125–127]. Hence, the smart grid can be regarded as an electric system that uses information, two-way communication technologies, and computational intelligence in an integrated system across electricity generation, transmission, substations, distribution, and consumption. This is to achieve a system that is clean, safe, secure, reliable, flexible, efficient, and sustainable. Consequently, the smart grid may respond to events that happen in any region through the grid, such as power generation, transmission, distribution, and consumption. For example, when a fault occurs in a feeder in the distribution grid, it is normally de-energized by opening the associated circuit

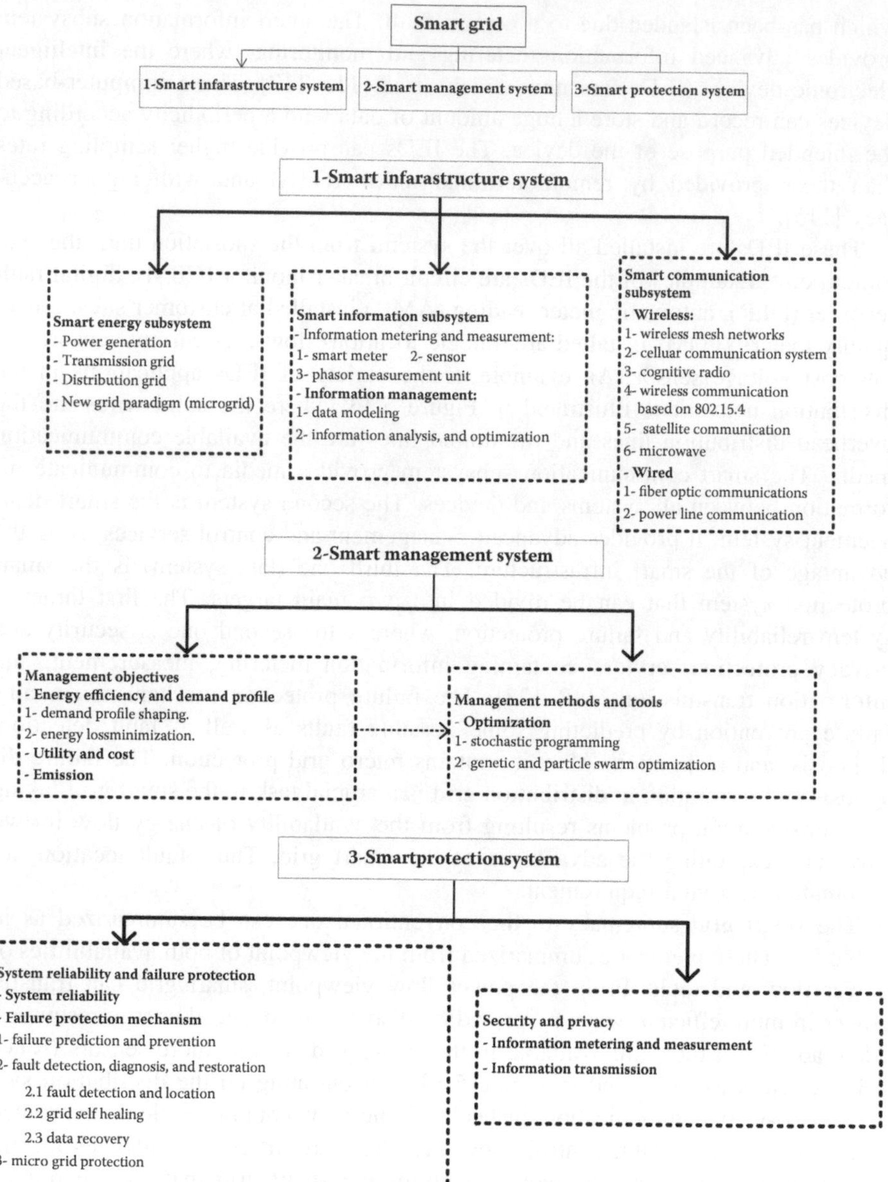

FIGURE 5.12 The construction of smart grid from the technical viewpoint.

breaker in a fast and selective way using adaptive protection. This event will result in service interruption for the customers. The duration of the interruption should be as short as possible to improve the reliability of the system. This is verified by both fast identifying the faulted section, isolating the faulted section using switching devices (sectionalizers that are normally closed switches) with the aid of a suitable

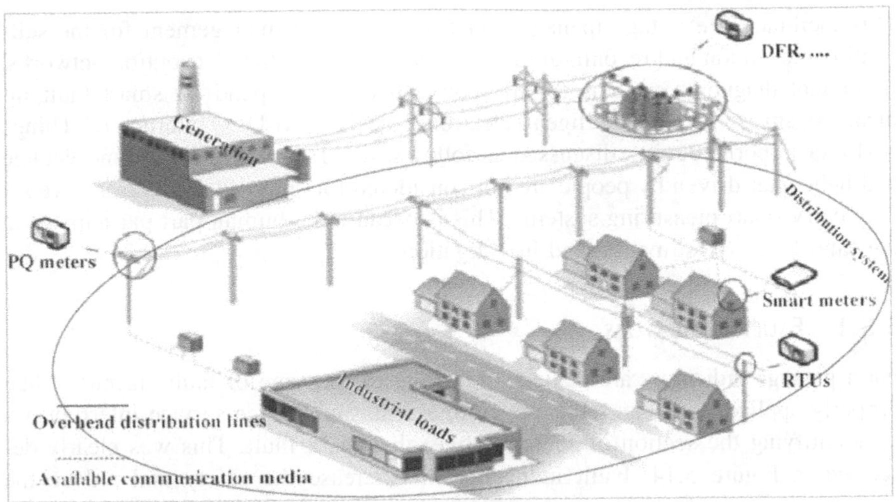

FIGURE 5.13 Locations of some IEDs in a distribution network [123, 124].

distributed control scheme, restoring the main power supply of the feeder (reclosing the associated circuit breaker), and reconfiguring the system using tie switching (normally open) to feed the regions losing service. If the fault location, isolation, and network reconfiguration are fully automated, this is sometimes denoted as a "self-healing" grid.

5.5 SMART FAULT LOCATION TECHNIQUES

Smart fault diagnosis estimation is attained in the field toward the detection and location of the proactive faults as well as the condition monitoring assessment. Accordingly, the smart fault diagnosis includes the smart fault locations as well.

TABLE 5.1
A Brief Comparison Between the Conventional and the Smart Grids [118]

Comparison Items	Conventional Grid	Smart Grid
Power flow	Centralized generation system	Distributed generation system
	Radial distribution network	Meshed distribution network
Available tools	Few sensors used	More sensors will be used
	One-way communication	Two-way communication
	Central control	Distributed control
Requirements	Manual operation	Remotely operation
	Failures and blackouts	Adaptive protection and islanding
	Manual restoration	Self-healing

This facilitates the outage management and restoration management for the self-healing realization and reconfiguration implementation of the distribution networks. Intelligent diagnosis systems for the power networks depend on smart fault indicators, smart meters, intelligent electronic devices (IEDs), Internet of Things (IoT), or airborne device, discussed as follows. In this chapter, the ground vehicle and helicopter driven by people are not considered for smart fault locations even if they carry smart measuring systems. This is because the human part participates in the fault location estimation and investigation.

5.5.1 Fault Indicators

Fault passage indicators are devices indicating the passage of fault current. When properly applied, they can reduce operating costs and reduce service interruptions by identifying the section of cable (or line) that has a fault. This was clearly described in Figure 5.14. Fault indicators can increase the safety and reduce the equipment damage by reducing the need for hazardous fault chasing procedures. To provide the greatest benefit, the fault indicator must indicate reliably when fault current passes through the cable to which the fault indicator is mounted. Misapplication or improper selection of the fault indicator can reduce reliability. Fault indicators are currently commonly utilized in most distribution networks in the distribution feeders.

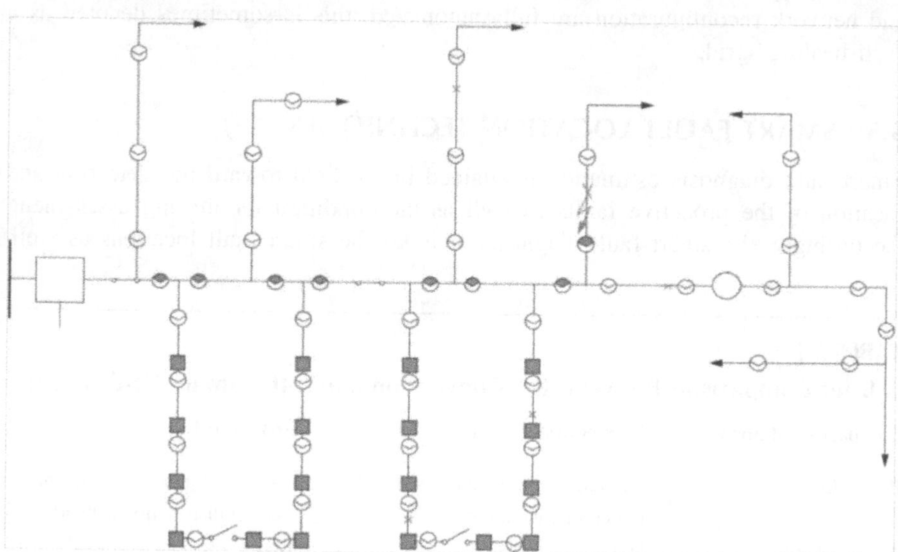

FIGURE 5.14 Fault passage indicator in distribution networks.

5.5.2 Distributed Smart Meters

The smart meters are distributed in the distribution networks with the characteristics of data processing and communication features. This provides the advantages of evaluating the measurements and communicating with the measured data in order to pinpoint the faults in the distribution networks as these networks are complicated in their structures, operations, and disturbances.

The distributed smart meters are applied and considered for implementing the fault location estimation based on voltage sag measurements, as addressed in [128]. The voltage sage estimation capability enhanced the fault location process concerning the bus impedance matrix to estimate the fault current that is assumed to be at different points. First, the voltage deviation due to the fault event is calculated by:

$$\Delta V_i^{(abc)} = V_i^{(abc)p} - V_i^{(abc)f} \tag{5.23}$$

where $V_i^{(abc)p}$ and $V_i^{(abc)f}$ are respectively the bus i voltage of pre-fault and during fault for phases a, b, and c. Assuming the fault is at bus k, then the fault current $\hat{I}_{\text{fault}_{ik}}^{(abc)}$ for the assumed fault event based on the distributed measured voltage deviation is [128]:

$$\hat{I}_{\text{fault}_{ik}}^{(abc)} = (Z_{ik}^{(abc)})^{-1} \cdot \Delta \hat{V}_i^{(abc)} \quad (\text{for } i \neq k) \tag{5.24}$$

where $Z_{ik}^{(abc)}$ is a submatrix of the bus impedance matrix $\mathbf{Z}_{bus}^{(abc)}$. For accurate network representation, the distributed loads are represented in the impedance matrix. This assumption of the fault point is done concerning all available buses in the networks. By comparing between the measured and calculated currents to estimate the error, the minimum error indicates to the faulted bus using the fault-location index by:

$$\delta_k = \sum_{ph}^{a,b,c} \sum_{i=1}^{N_{fm}} \left(\left| \hat{I}_{\text{fault}_{ik}}^{ph} - \overline{\hat{I}_{\text{fault}_k}^{ph}} \right| \right) = \sum_{ph}^{a,b,c} \sum_{i=1}^{N_{fm}} d_{ik}^{ph} \tag{5.25}$$

where $\hat{I}_{\text{fault}_{ik}}^{ph}$ is the calculated fault current using measurements for each phase at bus i, and $\overline{\hat{I}_{\text{fault}_k}^{ph}}$ is the average current calculated using the voltage measurements at bus k, the minimum δ_k. The advantage of this algorithm is that it depends on the measured distributed voltage sags as a magnitude, in which the fault event is used to differentiate between pre- and post-fault measurements. The disadvantage of this fault location technique is that it indicates to the faulted zone not to the fault location point. Accordingly, it can be used for the faulted zone isolation in the smart distribution networks.

5.5.3 IoT for Data Collections

IoT enhances the real-time data collections concerning all boundary conditions related to the power systems. The collected data are acquired for the supervisory and control system with the aid of widely distributed smart sensors. Intelligent techniques are applied to manipulate the big data in order to predict the fault and assess the line condition monitoring. Due to the widely distributed smart sensors, big data communication and collections, and intelligent processing methods, this smart fault diagnosis system is expensive.

Wide measurements are attained using the distributed phasor measurement unit (PMU) in the distribution networks for faulted section estimation and fault distance determination [129]. The IoT architecture is shown in Figure 5.15. It includes three layers that are sensing, networking, and application layers. In the sensing layers, the distributed smart sensors and signal collector system attain the unified data collection system for the distribution networks. In the networking layer, the collected measurement data are transmitted and stored with the aid of different communication technologies. In the application layer, the data management, data analyses,

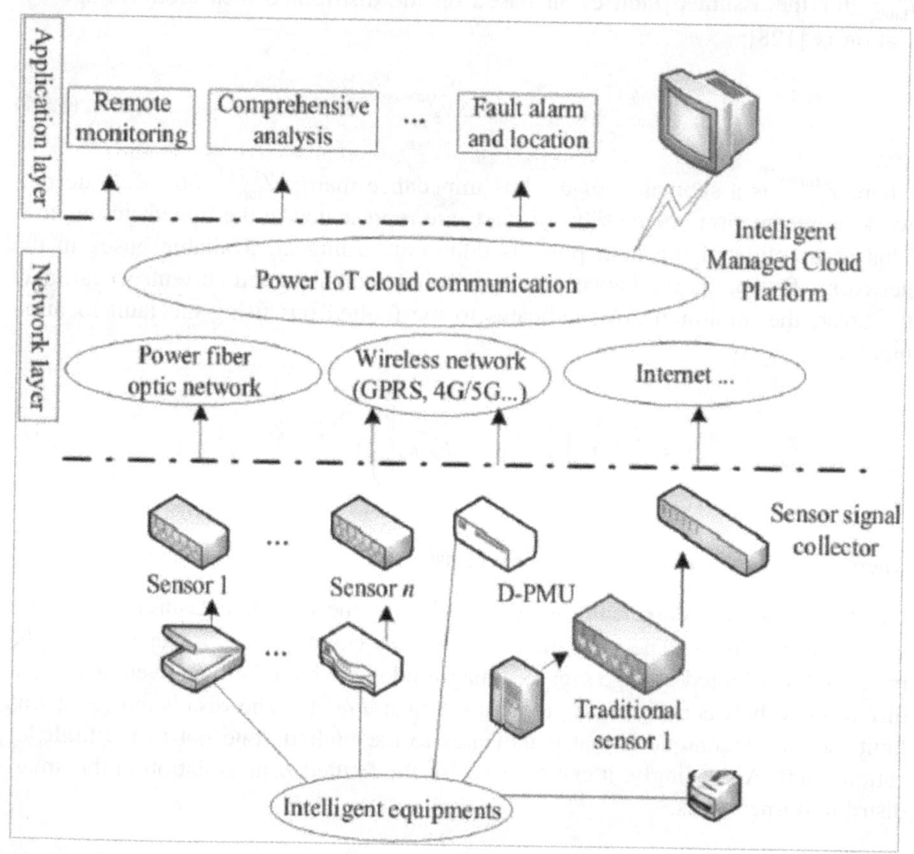

FIGURE 5.15 Architecture of IoT as reported in [129].

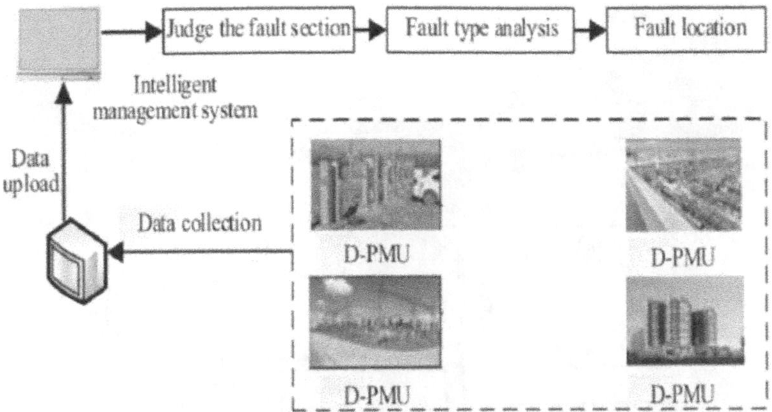

FIGURE 5.16 Fault analyses and location modules using distributed PMU as presented in [129].

and decision making are processed for attaining faulted section estimation and inspection plan consideration.

As shown in Figure 5.16, the distributed PMUs are utilized to facilitate and widely estimate the faulted section, fault type, and fault location. Using this platform, the fault diagnosis can be done using an accurate fault diagnosis system such as analyzing the phase difference between the positive and negative sequences [129]. In this system, the smart fault indicators are used to indicate to the faulted section using voltage and current measurements and then initiate alarms to declare their fault indication. Then, the management system finds the faulted section and isolate it before the system restoration process. During the fault distance estimation process, different fault location algorithms can be applied either single-end or double-end measurement methods.

5.5.4 Unmanned Aerial Vehicles (Drones)

The unmanned aerial vehicles (drones) are used for implementing smart fault diagnosis and location for the power transmission lines as reported in [130–133]. The mobility of the unmanned aerial vehicles provides advantages of using a single device for the online monitoring and condition assessment for the transmission system and, therefore, save cost and downtime of the transmission system. The unmanned aerial vehicle system includes ultraviolet and thermal cameras to capture images analyzed to investigate the transmission line insulation status. This investigation is applied during that the transmission system is in the service. For example, the camera captures images that are digitally processed for detecting the partial discharge in order to enhance the condition monitoring and proactive fault detection and location in the power transmission system. Therefore, the line defect points are detected and planned toward the maintenance task before the damage stage.

Figure 5.17 shows the unmanned aerial vehicle device including flight controller, ultraviolet cameras, thermal cameras, gimbal controller, microcomputer system,

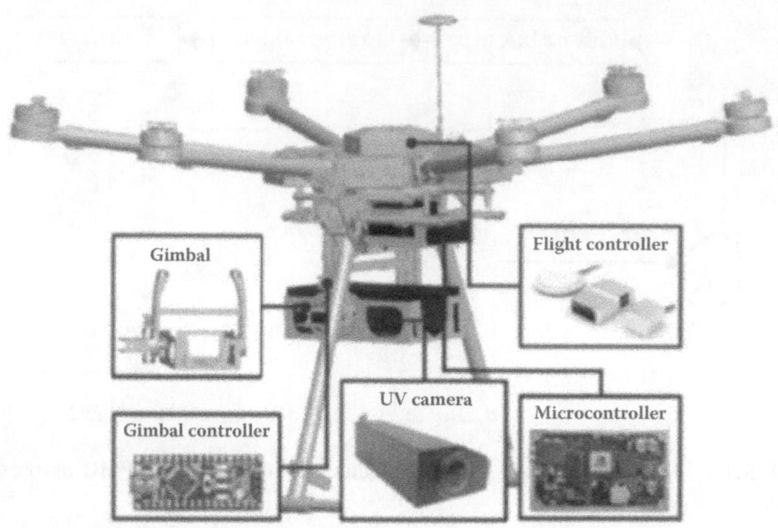

FIGURE 5.17 Unmanned aerial vehicles system reported in [130].

communication system, tracking system, energy storage system, servo motor system, electric drive system, etc. The unmanned aerial vehicle tracks the power line based on the image processing method. The unmanned aerial vehicles are designed in a small size, high controllability, and high observability, in which these features attain the vast and fast geographical accessibility throughout the transmission line conductors and towers.

The unmanned aerial vehicle includes two main approaches that are automatic line tracking and consistency estimation. The automatic line tracking approach objectively attains the line targeting continuously with the aid of GPS, UV camera, and digital signal processing. The consistency estimation approach is designed to effectively monitor and identify the line fault conditions as well as to highlight the alarms with the information of PD and location.

Fortunately, such unmanned aerial vehicles-based condition monitoring systems can be applied during disaster times of places close to the power networks either transmission, sub-transmission, and distribution systems. The catastrophic causes are such as fires, storms, … etc. During these circumstances, the unmanned aerial vehicles are useful to reach, classify, and estimate the damage. On the other hand, and during the normal atmosphere conditions, these vehicles enhance the online condition monitoring such as against vegetations in the forest countries. The unmanned aerial vehicles-based fault location diagnosis can be applied not only to detect the partial discharges through the transmission systems but also to investigate the fault point physically based on image processing methods. The smart inspection of the fault point using the unmanned aerial vehicles provide accurate information about the fault point location and damage percentage with the aid of suitable photos. Accordingly, it increases the safety of the maintenance man, reduces the maintenance time attained by the maintenance crew, and enhances the transmission system reliability.

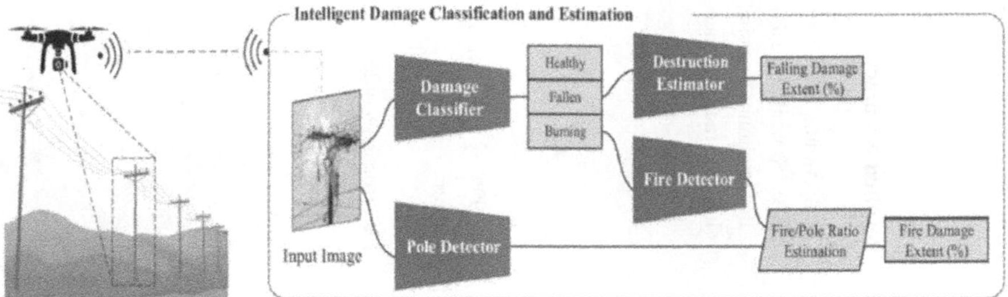

FIGURE 5.18 Architecture as reported in [131].

In [131], the pole fault damages are quantified and located using the unmanned aerial vehicles for the distribution networks. The structure of the intelligent damage classification and estimation system is shown in Figure 5.18. As shown in the figure, the unmanned aerial vehicle sends the information of each pole consequently to the control center using the cellular communication link. The vehicle can attain portable satellite communication as a backup communication system to achieve reliable communication with the control center. By investigating the image of the power line poles, the faulted pole is precisely located and diagnosed using the intelligent image processing system. Then, the quantifying process intelligently indicates the type of damage (falling or burning) and the percentage of damage. Therefore, the faulted feeder is visualized by estimating the faulted poles during the fault period, in which the fault can be more than one damaged pole. The fault damage estimation model is designed based on deep conventional neural networks, in which it is trained, validated, and tested using 1615 images.

5.6 CONCLUDING REMARKS

In spite of the accepted performance of these non-conventional applications with simulation testing, their reliability and dependability are still questionable. Therefore, almost none of these proposed applications are commercially developed until the present. However, the wide capabilities of these modern tools may motivate to improve the employment of these tools so that they will fit the practical requirements for real usage. Recent surveying was introduced in [134,135] comparing the performance of most recent AI-based fault location tools in the literature, as summarized in Table 5.2. This comprehensive study analyzed the performance of various fault location techniques considering the algorithm employed, input, test system, complexity level, features, and results. Complexity is defined by considering the number of inputs and rules involved in algorithm development as simple, medium, and complex. Further, the selection of the complexity level is achieved based on feature training and testing time, accuracy, convergence, and variance along with data required. More details regarding the performance of the selected algorithms and their related references are available in the aforementioned reference. Finally, the unmanned aerial vehicles are found a promising intelligent

TABLE 5.2
Comparative Study of the Performance of AI-based Fault Location Schemes [134]

Algorithm	Inputs	Selected Test System	Features	Complexity	Resulted Accuracy
ANN	Pre-fault current and voltage samples	380 kV, 189.3 km long TL, Spanish power system (50 Hz)	• FALNEUR software is used to train network data. • Training time varies from 5 s to 2.5 min to accomplish the mentioned error level. • BP based on Levenberg–Marquardt optimization technique is selected. • The "ANSIG" is selected as a transfer function for the hidden layer, and the linear function for the output layer.	Medium	The maximum error noted is 0.7% while 0.12% is the minimum error in locating fault distance
Least Square Error	Current and voltage magnitudes	Length of TL is 100 km, 400 kV, 50 Hz	• The sampling frequency is 6400 Hz. • 20 ms is the duration of the data window.	Simple	0.0099% is the relative error
Impedance-Based Algorithm (IBA)	Voltage profile	500 kV, 200 miles TL, 50 Hz	• Shunt capacitance is neglected of the TL which is desirable for online applications. • Data synchronization is not required.	Simple	1% error is recorded for IBA
Neuro-Fuzzy Systems and WT	Current and voltage profiles	Hybrid transmission system: 6.06-km cable and 14 km TL with 154 kV operating voltage	• DC o_set is removed via FIR. • Db4 mother wavelet is used and decomposed into three levels. • Back-propagation is used for learning and 228 various faults created for analysis. • Post-fault time is a half-cycle.	Medium	-

WT	Current samples	50 Hz, 60 km, 400 kV	• Db5 mother wavelet is used and decomposed into three Levels. • The sampling frequency is 3840 Hz with 64 samples/cycle. • The fault is located within 1 cycle via A3 components.	Simple	-
ANN and wavelet packet transform (WPT)	Current and voltage samples	360 km, 380 kV, and 50 Hz	• Db4 mother wavelet is employed and dissolved up to three levels by WPT. • The 10 kHz is the sampling frequency. • The computation burden is reduced as it is a reduction technique. • Pre- and post-fault is a half-cycle.	Complex	Minimum and maximum errors in finding fault location are 0.06% and 1.67%, respectively
RBF-based SVM and scaled conjugate gradient (SCALCG)-based NN approach	Positive sequence voltage and line currents	150 km double circuit TL, 400 kV is operating voltage	• The 5 kHz is selected as the sampling rate. • The 2e-004s is time to locate the fault. • RBF kernel is used to extract principal eigenvectors of the feature space and to remove noise from the signal.	Complex	Maximum fault error observed is 1.852 km while 7.874e-003 km is minimum
Nelder–Mead simplex	Post-fault voltage phasors	320 km, 500 kV, 50 Hz	• 960 Hz is the sampling rate. • Current transformer (CT) errors are avoided by not using post-fault current.	Complex	2.7% error is expected with _5% error in post-fault voltages
ANN and WPT	Current samples	360 km, 380 kV, 50 Hz	• Wavelet entropy and energy features are extracted from the decomposed signal. • Db4 is the mother wavelet. • 10 kHz is the sampling rate.	Complex	FL finding error is less than 2.05%

(Continued)

TABLE 5.2 (Continued)
Comparative Study of the Performance of AI-based Fault Location Schemes [134]

Algorithm	Inputs	Selected Test System	Features	Complexity	Resulted Accuracy
ANFIS	Zero and fundamental components of three-phase currents	Hybrid transmission system: 10-km cable and 90 km TL. Operating voltage is 220 kV	• ANFIS is trained for 2132 patterns, where 1520 patterns are for TL and rest for cable. • During training, the maximum percentage error of 0.031% and 0.0109% is observed for TL and cable, respectively. • During the testing process, the maximum % error of 0.0277% and 0.039% are observed for TL and cable, respectively.	Medium	The maximum error in finding FL is expected below than 0.07%.
ANNs with FPGA	Pre-fault current and voltage samples from one end	L 380 kV, 189.3 km long TL, Spanish power system (50 Hz)	• SARENEUR tool is used to run ANN. • Hardware is also implemented. • FPGA is designed for 60 MHz and consumes less power.	Complex	Error in finding fault location is 0.03%

machine in order to scan the faulted transmission line, diagnose the faulted point, and transmit the data to the control center.

REFERENCES

[1] T. Kawady, "Fault Location Estimation in Power Systems with Universal Intelligent Tuning", Ph.D. Thesis, TU-Darmstadt, Germany, 2005.

[2] M. Saha, J. Izykowski, and E. Rosolowski, *"Fault Location on Power Networks"*, Springer, London, 2010.

[3] D. M. Welton (Editor), *"Transmission Lines: Theory, Types and Applications"*, Nova Publishers, USA, 2010, ISBN: 978-1-61761-423-1, USA.

[4] T. A. Kawady, G. Sowilam, and R. Shalwala, "Improved Distance Relaying for Double-Circuit Lines Using Adaptive Neuro-Fuzzy Inference System", *Arabian Journal for Science and Engineering*, 45(3): 1969–1984, 2020. https://doi.org/10.1007/s13369-020-04369-x.

[5] V. Cook, *"Analysis of Distance Protection"*, Research Studies Press Ltd., John Wiley & Sons, Inc., New York, 1985.

[6] M.S. Sachdev, "Advancement in Microprocessor Based Protection and Communication", IEEE Tutorial Course, IEEE PES, IEEE Catalog Number: 97TP120-0, 1997.

[7] J. Izykowski, E. Rosolowski, and M.M. Saha, "Post-Fault Analysis of Operation of Distance Protective Relays of Power Transmission Lines", *IEEE Transactions on Power Delivery*, 22(1), 74–81, 2007.

[8] T. W. Stringfield, D. J. Marihart, and R. F. Stevens, "Fault Location for Overhead Lines", *AIEE Transactions,* Vol. 76, No. 3, pp. 518–526, Aug.1957.

[9] J. Blackburn and T. Domin, *"Protective Relaying: Principles and Applications"*, CRC Press – Taylor & Francis Group, New Tork, USA, 4th edition, 2014.

[10] Electricity Training Association, "Power System Protection", Institution of Electrical Engineers, 1997.

[11] W. Elmore, "Protective Relaying: Theory and Applications", CRC Marcel Dekker, 2nd edition, 2003.

[12] S. Horowitz and A.G. Phadke, "Power System Relaying", Wiley Press, New York, USA, 4th edition, 2014.

[13] C. Russel Masson, "The Art of Science of Protective Relaying", *GE*, https://www.gegridsolutions.com

[14] T. Kawady and J. Stenzel, "Investigation of Practical Problems for Digital Fault Location Algorithms based on EMTP simulation", IEEE/PES Transmission and Distribution Conference and Exhibition: Asia-Pacific, Yokohama, Japan, 6–10 Oct. 2002, pp. 118–123.

[15] F. Gaugaz, F. Krummenacher, and M. Kayal, "Implementation Guidelines for a Real-Time Fault Location System in Electrical Power Networks", 15th Biennial Baltic Electronics Conference (BEC), 2016, pp. 47–50,.

[16] F. Gaugaz, F. Krummenacher, and M. Kayal, "Ultra High-Speed Hardwaree Emulator for Real-Time Fault Location in Multi-Conductor Power Systems", 16th IEEE International New Circuits and Systems Conference (NEWCAS), 2018, pp. 313–316, IEEE.

[17] X. Wang et al., "Location of Single Phase to Ground Faults in Distribution Networks Based on Synchronous Transients Energy Analysis", *IEEE Transactions on Smart Grid*, Vol. 11, No. 1, pp. 774–785 11 (1), 2020.

[18] R. Jalilzadeh and H. Livani, "A Recursive Method for Traveling-Wave Arrival-Time Detection in Power Systems", *IEEE Transactions on Power Delivery*, 34 (2), pp. 710–719, 2019.

[19] H. Shu et al. "Speeded-Up Robust Features Based Single-ended Travelling Wave Fault Location: A Practical Case Study in Yunnan Power Grid of China", *IET Generation, Transmission & Distribution*, 12, 2018, pp. 886–894.

[20] L. Lewis, "Travelling Wave Relations Applicable to Power System Fault Locators", *AIEE Transactions*, 1951, Vol. 70, issue 2, pp. 1671–1680.

[21] M. Sneddom and P. Gale, "Fault Location on Transmission Lines", IEE Colloquium on Operational Monitoring of Distribution and Transmission Systems (Digest No. 1997/050), 28 Jan. 1997, pp. 2/1–2/3.

[22] F. Gale et al., "Fault Location Based on Travelling Waves", 5th International Conference on Developments in Power System Protection, 1993, pp. 54– 59.

[23] F. Gale, J. Stokoe, and P. Crossley, "Practical Experience with Travelling Wave Fault Locators on Scottish Power's 275 & 400 kV Transmission System", 6th International Conference on Developments in Power System Protection, 25–27 Mar. 1997, pp. 192–196.

[24] Z. Bo, G. Weller, F. Jiang, and Q. X. Yang "Application of GPS based Fault Location Scheme For Distribution System", International Conference on Power System Technology, Proceedings. POWERCON '98, 18–21 Aug. 1998, 1, pp. 53–57.

[25] Z. Q. Bo, A. T. Johns, and R. K. Aggarwal, "A Novel Fault Locator Based on the Detection of Fault Generated High Frequency Transients", Sixth International Conference on Developments in Power System Protection(Conf. Publ. No. 434), 1997, pp. 197–200.

[26] A. Di Tomasso G. Invernizzi, and G. Vielmini, "Accurate Single-End and Double-End Fault Location by Traveling Waves: A Review with Some Real Applications", 2019 AEIT International Annual Conference (AEIT), pp. 1–6., 2019.

[27] G.B. Ancell and N.C. Pahalawaththa, "Effects of Frequency Dependence and Line Parameters on Single Ended Travelling Wave Based Fault Location Schemes", *IEE Proceedings C*, 139, pp. 332–342, 1992.

[28] P. Gale, P. Taylor, P. Naidoo, C. Hitchin, and D. Clowes, "Travelling Wave Fault Locator Experience on Eskom Transmission Network", Seventh International Conference on Developments in Power System Protection, Apr. 9–12, 2001, pp. 327–331.

[29] F. Gale, J. Stokoe, and P. Crossley, "Practical Experience with Travelling Wave Fault Locators on Scottish Power's 275 & 400 kV Transmission System", 6th International Conference on Developments in Power System Protection, 25–27 Mar. 1997, pp. 192–196.

[30] G. Rockefeller, "Fault Protection with a Digital Computer", *IEEE Transactions on Power Apparatus and Systems*, PAS-88, pp. 438–461, 1969.

[31] B. Mann and I. Marrison, "Digital Calculation of Impedance for Transmission Line Protection", *IEEE Transactions on Power Apparatus and Systems*, PAS-90, pp. 270–279, 1971.

[32] G. Gilchrist, G. Rochefeller, and E. Udren, "High Speed Distance Relaying using a Digital Computer, Part I - System Description", *IEEE Transactions on Power Apparatus and Systems*, PAS-91, pp. 1235–1243, 1972.

[33] G. Gilchrist, G. Rochefeller, and E. Udren, "High Speed Distance Relaying using a Digital Computer, Part II", *IEEE Transactions on Power Apparatus and Systems*, PAS-91, pp. 1244–1258, 1972.

[34] P. McLaren and M. Redfern, "Fourier Series Techniques Applied to Distance Protection", *Proceedings of IEE*, 122, pp. 1295–1300, 1975.

[35] A. Johns and M. Martin, "Fundamental Digital Approach to the distance Protection of EHV Transmission Lines", *Proceedings of IEE*, 125, pp. 377–384, 1978.

[36] A. Phadke, J. Thorp, and M. Adamiak, "A New Measurement Technique for Tracking Voltage Phasors, Local System Frequency and Rate of Change of Frequency", *IEEE Transactions on Power Apparatus and Systems*, PAS-102, pp. 1025–1038, May 1983.

[37] A. Phadke and J. S. Thorp, *"Computer Relaying for Power Systems"*, John Wiley & Sons, 1988.

[38] A. Johns and S. Salman, *"Digital Protection for Power Systems"*, IEE Power Series 15, The Institution of Electrical Engineering, London, UK, 1995.

[39] T. Funabashi, H. Otoguro, Y. Mizuma, L. Dube, M. Kizilcay, and A. Ametani, "Influence of Fault Arc Characteristics on the Accuracy of digital Fault Locators", *IEEE Transactions on Power Delivery*, 16 (2), pp. 195–199, Apr. 2001.

[40] IEEE-Power System relaying Committee (PSRC), "IEEE Guide for Protective Relay Applications to Transmission Lines", IEEE Std C37.113-1999.

[41] B. Jeyasura and M. A. Rahman, "Accurate Fault Location of Transmission Lines Using Microprocessors", Fourth International Conference on Developments in Power System Protection, pp. 13–17, 1988.

[42] A. Johns and S. Jamali, "Accurate Fault Location Technique for Power Transmission Lines", *IEE Proceedings*, 137 Pt. C (6), pp. 395–402, 1990.

[43] N. I. Elklashy and T. Kawady, "Evaluation of Sequence Components-Based Parameterless Fault Location for Overhead Transmission Lines", International Middle East Power System Conference (MEPCON2012), Alexandria, Egypt, 23–25 Dec., pp. 1–6, 2012.

[44] N. I. Elkalashy, T. Kawady, W. M. Khater, and A.-M. I. Taalab, "Unsynchronized Fault-Location Technique for Double-Circuit Transmission Systems Independent of Line Parameters", *IEEE Transactions on Power Delivery*, 31 (4), pp. 1591–1600, Aug. 2016.

[45] D. Novosel, D. Hart, E. Udren, and J. Garitty, "Unsynchronized Two-terminal Fault Location Estimation", *IEEE Transactions on Power Delivery*, 11 (1), pp. 130–138, Jan. 1996.

[46] M. A. Elsadd, N. I. Elkalashy, T. A. Kawady, and A.-M. I. Taalab, "Earth Fault Location Determination Independent of Fault Impedance for Distribution Systems", *International Transactions on Electrical Energy Systems (ETEP)*, 27, (5), 1–16, May 2017, Wiley, DOI: 10.1002/etep.2307.

[47] A. A., Girgis, D. G. Hart, and W. L. Peterson, "A New Fault Location Technique for Two and Three Terminal Lines", *IEEE Transactions on Power Delivery*, 7 (1), pp. 98–107, Jan. 1992.

[48] J. Jiang, J. Yang, Y. Lin, C. Liu, and J. Ma, "An Adaptive PMU based Fault Detection/Location Technique for Transmission Lines. Part I: Theory and Algorithms", *IEEE Transactions on Power Delivery*, 15 (2), pp. 486–493, Jan. 2000.

[49] J. Jiang, J. Yang, Y. Lin, C. Liu, and J. Ma, "An Adaptive PMU Based Fault Detection/Location Technique For Transmission Lines. I. Theory and Algorithms", *IEEE Transactions on Power Delivery*, 15 (4), pp. 1136–1146, Oct. 2000.

[50] T. Kawady and J. Stenzel, "A Practical Fault Location Approach for Double Circuit Transmission Lines Using Single End Data", *IEEE Transactions on Power Delivery*, 18 (4), pp. 1166–1173, Oct. 2003.

[51] M. S. Sachdev and R. Agarwal, "A Technique for Estimating Line Fault Locations from Digital Impedance Relay Measurements", *IEEE Transactions on Power Delivery*, 3 (1), pp. 121–129, Jan. 1988.

[52] T. Adu, "A New Transmission Line Fault Locating System", *IEEE Transactions on Power Delivery*, 16 (4), pp. 498–503, Oct. 2001.

[53] T. Takagi, Y. Yamakoshi, J. Baba, K. Uemura, and T. Sakaguchi, "A New Algorithm for EHV/UHV Transmission Lines: Part I-Fourier Transform Method", *IEEE Transactions on Power Apparatus and Systems*, PAS-100, pp. 1316–1323, 1981.

[54] T. Takagi, Y. Yamakoshi, J. Baba, K. Uemura, and T. Sakaguchi, "A New Algorithm for EHV/UHV Transmission Lines: Part-II Laplace Transform Method", IEEE PES Summer Meeting 81, SM 411-8.

[55] T. Takagi, Y. Yamakoshi, M. Yamaura, R. Kondow, and T. Matsushima, "Development of a New Type Fault Locator Using the One-Terminal Voltage and Current Data", *IEEE Transactions on Power Apparatus and Systems*, PAS-101, pp. 2892–2898, 1982.

[56] A. Wiszniewski, "Acuurate Fault Impedance Locating Algorithm", *IEE Proceedings*, 130 (Pt. C), pp. 331–314, 1993.

[57] L. Eriksson, M. Saha, and G. D. Rockefeller,"An Accurate Fault Locator with Compensation for Apparent Reactance in the Fault Resistance Resulting from Remote End Infeed", *IEEE Transactions on Power Apparatus and Systems*, PAS-104 (2), pp. 424–436, Feb. 1985.

[58] A. Girgis and E. Makram, "Application of Adaptive Kalman Filtering in Fault Classification, Distance Protection and Fault Location Using Microprocessors", *IEEE Transactions on Power Systems*, 3 (1), pp. 301–309, Feb. 1988.

[59] T. Kawady and J. Stenzel, "A New Single End Approach for Transmission Line Fault Location Using Modal Transformation", 13th International Conference on Power System Protection, Bled, Slovenia, Sep. 25–27, 2002, pp. 98–103.

[60] A. D. Zahran, N. I. Elkalashy, M. A. Elsadd, T. A. Kawady and A. I. Taalab, "Improved Ground Distance Protection for Cascaded Overhead-Submarine Cable Transmission System", 2017 Nineteenth International Middle East Power Systems Conference (MEPCON), Cairo, 2017, pp. 778–758.

[61] F. V. Lopes et al., "Phasor-Based Fault Location Challenges and Solutions for Transmission Lines Equipped with High-speed Time-domain Protective Relays", *Electric Power Systems Research*, 189, pp. 1–8, Dec. 2020.

[62] A. Mouco and A. Abur, "Improving the Wide-Area PMU-Based Fault Location Method Using Ordinary Least Squares Estimation", *Electric Power Systems Research*, 189, Dec. 2020, pp. 1–8.

[63] H. Abd el-Ghany, M. A. Elsadd, and E. S. Ahmed, "A Faulted Side Identification Scheme-based Integrated Distance Protection for Series-compensated Transmission Lines", *International Journal of Electrical Power & Energy Systems*, 113, pp. 664–673, Dec. 2019.

[64] A. Adly, Z. Ali, M. A. Elsadd, H. M. Abdel-Mageed, and S. H. E. Abdel Aleem, "An integrated Scheme for a Directional Relay in the Presence of a Series-compensated Line", *International Journal of Electrical Power & Energy Systems*, 120, 2020, Art. no. 106024.

[65] M. Elgeziry, M. Elsadd, N. Elkalashy, T. Kawady, and A. Taalab, "AC Spectrum Analysis for Detecting DC Faults on HVDC Systems", Nineteenth International Middle East Power Systems Conference (MEPCON), Cairo, 2017, pp. 708–715.

[66] J. J. Q. Yu, D. J. Hill, A. Y. S. Lam, J. Gu, and V. O. K. Li, "Intelligent Time-Adaptive Transient Stability Assessment System", *IEEE Transactions on Power Systems*, 33 (1), pp. 1049–1058, 2018.

[67] N. B. Hartmann, R. C. dos Santos, A. P. Grilo, and J. C. M. Vieira, " Hardware Implementation and Real-Time Evaluation of an ANN-Based Algorithm for Anti-Islanding Protection of Distributed Generators", *IEEE Transactions on Industrial Electronics*, 65 (6), pp. 5051–5059, 2018.
[68] O. A. Alimi, K. Ouahada, and A. M. Abu-Mahfouz, "A Review of Machine Learning Approaches to Power System Security and Stability", *IEEE Access*, 8, pp. 113512–113531, 2020.
[69] B. K. Bose, "Artificial Intelligence Techniques in Smart Grid and Renewable Energy Systems—Some Example Applications", *Proceedings of the IEEE*, 105 (11), pp. 2262–2273, 2017.
[70] H. A. Darwish, A. I. Taalab, and T. Kawady, "Development and Implementation of an ANN- Based Fault diagnosis Scheme for Generator Winding Protection", *IEEE Transactions on Power Delivery*, 16 (2), pp. 208–214, Apr. 2001.
[71] A. I. Taalab, H. A. Darwish, and T. A. Kawady, "ANN-Based Novel Fault Detector for Generator Windings Protection", *IEEE Transactions on Power Delivery*, 14 (3), pp. 824–830, Jul. 1999.
[72] N. Saravanan and A. Rathinam, "A Comparative Study on ANN Based Fault Location and Classification Technique for Double Circuit Transmission Line", 2012 Fourth International Conference on Computational Intelligence and Communication Networks, IEEE, pp. 219–224.
[73] M. Tawfik and M. Morcos, "ANN-Based Techniques for Estimating Fault Location on Transmission Lines Using Prony Method", *IEEE Transactions on Power Delivery*, 16 (2), pp. 219–224, Apr. 2001.
[74] M. Joorabian, "Artificial Neural Network Based Fault Locator for EHV Transmission System", 10th Mediterranean Electromechanical Conference (MELECON 2000), 29–31 May 2000, vol. 3, pp. 1003–1006.
[75] A. J.Mazon, "Fault Location System on Double Circuit Two-terminal Transmission Lines Based on ANNs", PowerTech 2001, Porto, Portugal, 10–13Sep. 2001, pp. 1–5.
[76] H. Fathabadi, "Novel Filter Based ANN Approach for Short-circuit Faults Detection, Classification and Location in Power Transmission Lines", *International Journal of Electrical Power & Energy Systems*, 74, pp. 374–383, Jan. 2016.
[77] J. M. Keller, D. Liu, and D. B. Fogel, *"Fundamentals of Computational Intelligence: Neural Networks, Fuzzy Systems, and Evolutionary Computation (IEEE Press Series on Computational Intelligence)"*, 1st Edition. Wiley-IEEE Press, Jul. 13, 2016, IEEE Press, USA.
[78] P. Kim, *"MATLAB Deep Learning: With Machine Learning, Neural Networks and Artificial Intelligence"*, 1st Edition, Apress, Jun. 15, 2017. Apress, USA
[79] S. A. M. Javadian, A. M. Nasrabadi, M.-R. Haghifam, and J. Rezvantalab, "Determining Fault's Type and Accurate, Location in Distribution Systems with DG Using MLP Neural Networks", International Conference on Clean Electrical Power (ICCEP), IEEE, 2009, pp. 284–289.
[80] J. J. G. Ledesma, K. Nascimento, L. Araujo, and D. R. Ribeiro Penido, "A Two-Level ANN-Based Method Using Synchronized Measurements to Locate High-impedance Fault in Distribution Systems", *Electric Power Systems Research*, 189, Dec. 2020, Art. no. 106576.
[81] R. N. Mahanty and P. B Dutta Gupta, "Application of RBF Neural Network to Fault Classification and Location in Transmission Lines", *IEE Proceedings - Generation, Transmission and Distribution*, 151 (2), Mar. 2, 2004, pp. 201–212.
[82] S. Ekici, . ppS. Yildirim, and M. Poyraz, "A Transmission Line Fault Locator Based on Elman Recurrent Networks", *Applied Soft Computing*, 9 (1), pp. 341–347, Jan. 2009.

[83] I. Sadinezhad et al., "An Adaptive Precise One-end Power Transmission Line Fault Locating Algorithm Based on Multilayer Complex Adaptive Artificial Neural Networks", International Conference on Industrial Technology, IEEE, 2009.

[84] J. Ezquerra et al, "Field Programmable Gate Array Implementation of A Fault Location System in Transmission Lines Based on Artificial Neural Networks", *IET Generation, Transmission & Distribution*, 5(2), pp. 191–198, 2011.

[85] J.-A. Jiang et al., "A Hybrid Framework for Fault Detection, Classification, and Location—Part I: Concept, Structure, and Methodology", *IEEE Transactions on Power Delivery*, 26 (3), 1989–1998, 2011.

[86] A. Sanad Ahmed et al. "Modern Optimization Algorithms for Fault Location Estimation in Power Systems", *Engineering Science and Technology, an International Journal*, 20 (5), pp. 1475–1485, Oct. 2017.

[87] S. S. Gururajapathy, H. Mokhlis, and H. A. Illias, "Fault Location and Detection Techniques in Power Distribution Systems with Distributed Generation: A Review", *Renewable and Sustainable Energy Reviews*, 74, pp. 949–958, Jul. 2017.

[88] D. P. Mishra and P. Ray, "Fault Detection, Location and Classification of a Transmission Line", *Neural Computing and Applications*, 30, pp. 1377–1424, 2018.

[89] P. Ray and D. P. Mishra, "Support Vector Machine Based Fault Classification and Location of a Long Transmission Line", *Engineering Science and Technology, an International Journal*, 19 (3), pp. 1368–1380, Sep. 2016.

[90] X. Deng et al., "Fault Location in Loop Distribution Network Using SVM Technology", *International Journal of Electrical Power & Energy Systems*, 65, pp. 254–261, Feb. 2015.

[91] J. Liang et al. "Two-Terminal Fault Location Method of Distribution Network Based on Adaptive Convolution Neural Network", *IEEE Access*, 8, pp. 54035–54043, 2020.

[92] I. Altas, *"Fuzzy Logic Control in Energy Systems with Design Applications in MATLAB/Simulink"*, The Institution of Engineering and Technology, IET, London, UK, Oct. 6, 2017.

[93] A. Swetapadma and A. Yadav, "Fuzzy Inference System Approach for Locating Series, Shunt, and Simultaneous Series–Shunt Faults in Double Circuit Transmission Lines", *Computational Intelligence and Neuroscience*, Hindawi, pp. 1–13, 2015.

[94] S. Adhikari, N. Sinha, and T. Dorendrajit, "Fuzzy Logic Based On-line Fault Detection and Classification in Transmission Line", *Springerplus*, pp. 1–14, 2016.

[95] T. Kari et al., "An Integrated Method of ANFIS and Dempster-Shafer Theory for Fault Diagnosis of Power Transformer", *IEEE Transactions on Dielectrics and Electrical Insulation*, 25, pp. 360–370, Mar. 2018.

[96] M. Reddy and D. Mohanta, "Adaptive-Neuro-fuzzy Inference System Approach for Transmission Line Fault Classification and Location Incorporating Effects of Power Swings", *IET Generation, Transmission & Distribution*, 10 (2), pp. 235–244, 2018.

[97] J. Sadeh and H. Afradi, "A New and Accurate Fault Location Algorithm for Combined Transmission Lines Using Adaptive Network-Based Fuzzy Inference System", *Electric Power Systems Research*, 129 (11), pp. 1538–1545, 2009.

[98] H. Khorashadi-Zadeh and Z. Li, "Transmission Line Distance Protection Using ANFIS and Positive Sequence Components", iREP Symposium - Bulk Power System Dynamics and Control - VII. Revitalizing Operational Reliability,pp. 1–9, 2007.

[99] F. Belhachat and C. Larbes, "Global Maximum Power Point Tracking Based on ANFIS Approach for PV Array Configurations under Partial Shading Conditions", *Renewable and Sustainable Energy Reviews*, 77, pp. 875–889, Sep. 2017.

[100] T. S. Kamel, and M. A. Moustafa Hassan, "Using a Combined Artificial Intelligent Approach in Distance Relay for Transmission Line Protection", 5th International Conference on Soft Computing, Computing with Words and Perceptions in System Analysis, Decision and Control, USA, pp. 1–6, 2009.

[101] M. S. Aziz, M. A. Hassan, and Abo-Elzahab, "Applications of ANFIS in High Impedance Faults Detection and Classification in Distribution Networks", *International Journal of System Dynamics Applications*, 1 (2), 2012.

[102] H. Khorashadi-Zadeh and Z. Li, "Transmission Line Distance Protection Using ANFIS and Positive Sequence Components", iREP Symposium - Bulk Power System Dynamics and Control - VII. Revitalizing Operational Reliability, USA, 2007, pp. 1–9.

[103] M. Sanaye-Pasand and P. Jafarian, "An Adaptive Decision Logic to Enhance Distance Protection of Transmission Lines", *IEEE Transactions on Power Delivery*, 26 (4), pp. 2134–2144, 2011.

[104] M. Gen and R. Cheng, *"Genetic Algorithms and Engineering Design (Engineering Design and Automation"*, 1st Edition, Wiley-Interscience, New York, USA, 1997.

[105] T. Kawady, "A Genetic Algorithm-Based Fault Locator Using Unsynchronized Double End Data", 15th International Conference on Power System Protection, Bled, Slovenia, Sep. 6–8, 2006, pp. 275–280.

[106] Y. Liao and S. Elangovan, "Unsynchronized Two-Terminal Transmission-line Fault-Location Without Using Line Parameters", *IET Generation, Transmission and Distribution*, 153 (6), pp. 639–643, 2006.

[107] A. Kaur and S. Goyal, "A Bee Colony Optimization Algorithm for Fault Coverage Based Regression Test Suite Prioritization", *International Journal of Advanced Science and Technology*, 29, pp. 2786–2793, 2011.

[108] M. K. Bedi and S. Singh, "Comparative Study of Two Natural Phenomena Based Optimization Techniques", *International Journal of Scientific and Engineering*, 4 (3), pp. 1–4, 2013.

[109] A. R. Adly, S. H. E. Abdel Aleem, M. A. Elsadd, and Z. M. Ali, "Wavelet Packet Transform Applied to a Series-Compensated Line: A Novel Scheme for Fault Identification", *Measurement*, 151, 2020, Art. no. 107156. https://doi.org/10.1016/j.measurement.2019.107156

[110] P. J. Van Fleet, *"Discrete Wavelet Transformations: An Elementary Approach with Applications"*, 2nd Edition, Wiley, New York, USA, 2019.

[111] E. Koley, K. Verma, and S. Ghosh, "An Improved Fault Detection Classification and Location Scheme Based on Wavelet Transform and Artificial Neural Network for Six Phase Transmission Line Using Single End Data Only", *SpringerPlus* 4, 2015, Art. no. 551. https://doi.org/10.1186/s40064-015-1342-7

[112] S. Ekici, "Support Vector Machines for Classification and Locating Faults on Transmission Lines", *Applied Soft Computing*, 12 (6), pp. 1650–1658, Jun. 2012.

[113] X. Deng et al., "Fault Location in Loop Distribution Network Using SVM Technology", *International Journal of Electrical Power & Energy Systems*, 65, pp. 254–261, Feb. 2015.

[114] M. Jaya Bharata Reddy et al., "Smart Fault Location For Smart Grid Operation Using RTUs and Computational Intelligence Techniques", *IEEE Systems Journal*, 8 (4), Dec. 2014.

[115] PSRC WG C2 Report, "Role of Protective Relaying in the Smart Grid", 2010.

[116] R. Moghaddass and J. Wang, "A Hierarchical Framework for Smart Grid Anomaly Detection Using Large-Scale Smart Meter Data", *IEEE Transactions on Smart Grid*, 9 (6), pp. 5820–5830, Nov. 2018.

[117] M. S. Mahmoud and Y. Xia, "Smart Grid Infrastructures", *Networked Control Systems*, pp. 315–349, 2019.

[118] O. G. Abood, M. A. Elsadd, and S. K. Guirguis, "Investigation of Cryptography Algorithms Used for Security and Privacy Protection in Smart Grid", 2017 Nineteenth International Middle East Power Systems Conference (MEPCON), Cairo, 2017, pp. 644–649.

[119] O. Abood, M. Elsadd, and S. Guirguis, "Incorporating Deoxyribonucleic Acid in AES Scheme for Enhancing Security and Privacy Protection", *Journal of Theoretical and Applied Information Technology*, 97 (2), pp. 349–360, Jan. 2019.

[120] J. M. Junior et al., "Data Security and Trading Framework for Smart Grids in Neighborhood Area Networks", *Sensors*, 20, Feb. 2020, Art. no. 1337.

[121] Y. Xiang, L. Wang, and Y. Zhang, "Adequacy Evaluation of Electric Power Grids Considering Substation Cyber Vulnerabilities", *International Journal of Electrical Power & Energy Systems*, 96, pp. 368–379, Mar. 2018.

[122] A.O. Otuoze, M.W. Mustafa, and R.M. Larik, "Smart Grids Security Challenges: Classification by Sources of Threats", *Journal of Electrical Systems and Information Technology*, 5, pp. 468–483, Dec. 2018.

[123] M. Kezunovic, "Smart Fault Location for Smart Grids", *IEEE Transactions on Smart Grid*, 2, pp. 11–22, 2011.

[125] A. Esteban, L. Rivas, and T. Abrao, "Faults in Smart Grid Systems: Monitoring, Detection and Classification", *Electric Power Systems Research*, 189, pp. 1–26, Dec. 2020.

[126] N. G. Tarhuni, N. I. Elkalashy, T. A. Kawady, and M. Lehtonen, "Autonomous Control Strategy for Fault Management in Distribution Networks", *Electric Power Systems Research*, 121, pp. 252–259, Apr. 2015.

[124] S. Howell, Y. Rezgui, J.-L. Hippolyte, B. Jayan, and H. Li, "Towards the next Generation of Smart Grids: Semantic and Holonic Multi-Agent Management of Distributed Energy Resources", *Renewable & Sustainable Energy Reviews*, 77, pp. 193–214, Mar. 2017.

[127] M. A. Elsadd, N. I. Elkalashy, T. A. Kawady, A.-M. I. Taalab, and M. Lehtonen, "Incorporating Earth Fault Location in Management Control Scheme for Distribution Networks", *IET Generation, Transmission and Distribution*, 10 (10), pp. 2389–2398, 2016.

[128] X. Kong, Y. Xu, Z. Jiao, D. Dong, X. Yuan, and S.H. Li, "Fault Location Technology for Power System Based on Information About the Power Internet of Things", *IEEE Transactions on Industrial Informatics*, 16 (10), pp. 6682–6692, Oct. 2020.

[129] F. Trindade, W. Freitas, and J. Vieira "Fault Location in Distribution Systems Based on Smart Feeder Meters", *IEEE Transactions on Power Delivery*, 29 (1), pp. 251–260, Feb. 2014.

[130] S. Kim, D. Kim, S. Jeong, J.-W. Ham, J.-K. Lee and K.-Y. Oh "Fault Diagnosis of Power Transmission Lines Using a UAV-Mounted Smart Inspection System", *IEEE Access*, 8(2), pp. 149999–150010.

[131] M. Hosseini, A. Umunnakwe, M. Parvania, and T. Tasdizen, "Intelligent Damage Classification and Estimation in Power Distribution Poles Using Unmanned Aerial Vehicles and Convolutional Neural Networks", *IEEE Transactions on Smart Grid*, 11 (4), pp. 3325–3334, Jul. 2020.

[132] M. Ostendorp, "Innovative Airborne Inventory and Inspection Technology for Electric Power Line Condition Assessments and Defect Reporting", IEEE 9th International Conference on Transmission and Distribution Construction, Operation and Live-Line Maintenance Proceedings, ESMO 2000 Proceedings, Global ESMO 2000, Montreal, Quebec, Canada, pp. 123–128, 8–12 Oct. 2000.

[133] C. H. Deng, S. H. Wang, Z. Huang, Z. Tan, and J. Liu, "Unmanned Aerial Vehicles for Power Line Inspection: A Cooperative Way in Platforms and Communications", *Journal of Communications*, 9 (9), pp. 687–692, Sep. 2014.

[134] A. Raza et al., "A Review of Fault Diagnosing Methods in Power Transmission Systems", *Applied Sciences*, 10, 2020, Art. no. 1312, DOI:10.3390/app10041312.

[135] D. P. Mishra and P. Ray, "Fault Detection, Location and Classification of A Transmission Line", *Neural Computing and Applications*, 30, pp. 1377–1424, 2018.

6 An Integrated Approach for Fault Detection, Classification and Location in Medium Voltage Underground Cables

M. Karthikeyan[1] *and R. Rengaraj*[2]
[1]Department of Electrical and Electronics Engineering, Velammal Engineering College
[2]Department of Electrical and Electronics Engineering, Sri Sivasubramaniya Nadar College of Engineering

6.1 INTRODUCTION

Electrical power transmission and distribution companies utilize underground cables to provide reliable power supply to its customers. Underground cable systems require a high installation cost. But the operational cost and outages are less in underground cables compared to overhead systems. Also, protection is a challenging task in underground cables as the cables are laid underground. To protect the underground cables from fault, protective devices are used. Protective devices detect the fault and send information to the power system operator. When a fault occurs, the power system operator interprets this information and takes necessary steps to restore power supply. For doing this, the power system operator needs the following information: fault type and its exact location in underground cable. The power supply restoration time depends on the accuracy of information received by the power system operator. Power system engineers look for innovative methods to detect, classify, and locate the faults. To accomplish this task, researchers in power system protection carryout developments in three methods: impedance measurement, traveling wave and hybrid methods.

The impedance-based protection measures the impedance between the source end and fault point. Here, the operating parameter of the relay is the impedance calculated by taking the ratio of voltage and current measured at the source end. This method requires data from a single terminal. When the calculated impedance is

DOI: 10.1201/9780367552374-6

less than the set value, the relay trips. The impedance-based protection tells us whether the fault has occurred or not but it does not provide data about exact location of the fault. In the second method, protective device acts based on reflection of traveling waves. The traveling wave gets reflected when there is a change in impedance. The fault point causes the incident wave to reflect as it creates discontinuities. The traveling wave-based protection records the time taken by the reflected traveling wave to calculate the location of the fault; but this method does not provide fault classification information. The hybrid method uses signal processing and machine learning algorithms to design the protection scheme. Power system operator retrieve accurate information from the hybrid methods like fault type and exact location of the fault. The hybrid methods use transforms like Fourier transform, discrete Wavelet transform, and S-transform to extract the features from the current and voltage signal. The machine learning algorithms require these extracted features as input. The algorithms designed in classification mode categorizes the type of fault. The algorithms designed in regression mode give the actual location of fault in kilometer. The performance of the machine learning algorithms depends on the features chosen. As the hybrid methods provide all valuable information to the power system operator, researchers are focusing on designing hybrid protection schemes. Table 6.1 shows a few of the significant methods proposed by researchers in the past, addressing the faults in underground cables.

From the literatures, we understood the demand for an integrated approach to complete the three tasks: fault detection, classification, and location in underground cables. Completion of these three tasks quickly is mandatory to operate power system back to normal condition. This chapter presents such an integrated approach utilizing model-based signal processing tool autoregressive signal modeling and extreme learning machine. The model-based approach provides good frequency resolution as compared to transforms, while the extreme learning machine has better generalization capability and faster learning rate. Both of these will help the power system operator during the fault by providing reliable information for necessary action to restore the power supply.

6.2 AUTOREGRESSIVE MODELING

Researchers are using signal processing methods to extract features from the signal. Two broad categories of signal processing methods are transforms and model-based approaches. Fast Fourier transform estimates spectral density inaccurately, whereas discrete Wavelet transform fails for noisy signals. Researchers used the model-based approach Autoregressive Modeling (AR) in power system engineering like fault diagnosis in induction motor [16], oscillation estimation [17], and detecting islanding condition [18]. Performance of model-based approach depends on the model chosen for representing the signal [19,20]. Autoregressive (AR) model features the signal spectral and energy properties by representing them as time series. AR coefficients provides information of signal dynamics. When a fault or disturbance occurs in a signal, its dynamics change. Therefore, the AR coefficients also changes. Thus, AR coefficients can be used as a feature for fault detection in

TABLE 6.1
Significant Methods Available for Fault Detection, Classification, and Location Scheme

Year	Objectives	Operating quantity	Methods	Remarks	Reference
2006	Fault location	Three-phase voltage and current from both ends	Discrete Fourier transform (DFT)	The maximum percentage error is 1.643%. DFT provides good frequency resolution.	[1]
2008	Fault classification	Three phase current and voltage for full cycle duration	Fast Fourier transform (FFT) – Neural network-based module to classify fault	Fault classification accuracy is 99.92%. Sampling frequency = 800 Hz.	[2]
2009	Fault detection, classification and location	Three-phase current	Wavelet transform	Sampling frequency = 1600 Hz. Selection of mother wavelet requires time. Wavelets are sensitive to signals with harmonics.	[3]
2011	Fault location	Three-phase current and voltage	Direct circuit analysis	Maximum error = 2.88%	[4]
2012	Fault location	Three-phase current	Discrete Wavelet transform and support vector machine (SVM)	Quarter cycle samples alone used. Negligible average error in fault location. Performance of SVM is affected to data set with similar target classes.	[5]
2014	Fault classification and location	Three-phase and zero-sequence current	Wavelet transform and adaptive network fuzzy inference system	Sampling frequency used is 200 kHz. Percentage error of less than 3%. To improve the accuracy, the number of rules increases exponentially.	[6]
2014	Fault location	Three phase current	Visual inspection and Wavelet transform	Sampling frequency of 10 MHz is used.	[7]

(Continued)

TABLE 6.1 *(Continued)*
Significant Methods Available for Fault Detection, Classification, and Location Scheme

Year	Objectives	Operating quantity	Methods	Remarks	Reference
2014	Fault location	Three phase current and voltage	Arc voltage algorithm	Sampling frequency of 1600Hz is used.	[8]
2014	Fault detection and location	Three-phase current and voltage	Fourier transform, artificial neutral network, time domain reflectometry (TDR)	TDR gives precise measurement but requires high frequency signals.	[9]
2015	Fault location	Three-phase current and voltage	S-Transform and extreme learning machine	Sampling frequency of 10 kHz is used. Error is less than 0.3%. S-Transform requires high computational time.	[10]
2016	Fault classification	Three-phase current, voltage and zero-sequence current	Artificial neural network	Neural network requires large data set with larger training time.	[11]
2016	Fault detection, classification, and location	Three-phase current and voltage	Discrete wavelet transform	Sampling frequency of 100 kHz is used. Maximum percentage error is 5.1%.	[12]
2018	Fault classification	Three-phase voltage	Deep belief network	Fault detection accuracy is 97.8%.	[13]
2020	Fault location	Cable resistance	Murray loop test, Varley loop test, and Box's optimization	Maximum percentage error is 0.41%.	[14]
2020	Fault impedance and location	Three-phase current and voltage	Discrete wavelet and artificial neural network	Sampling frequency required is 4 kHz. Maximum error is 0.296 km.	[15]

underground cables. The spectral characteristics of current under normal and fault conditions vary. Hence AR coefficients can be used to classify the fault. The spectral characteristics of the fault current vary at each location in underground cable. Therefore, AR coefficients can be used to locate fault in underground cables.

This makes AR modeling as a proper feature extraction tool for protecting underground cables. Equation (6.1) represents the AR modeling of the current signal $i(t)$ with white noise $e(t)$:

$$i(t) = -\sum_{j=1}^{p} a_j i(t-j) + e(t) \tag{6.1}$$

where a_j is the AR coefficients, $i(t-j)$ is the delayed current signal (j samples), and p is the number of poles. Equation (6.2) gives the z-transform of (6.1):

$$I(z) = -I(z) \sum_{j=1}^{p} a_j z^{-j} + E(z) \tag{6.2}$$

where $I(z)$ and $E(z)$ are the z-transforms of $i(t)$ and $e(t)$, respectively. Equation (6.3) gives the system function $H(z)$:

$$H(z) = \frac{I(z)}{E(z)} = \frac{1}{1 + \sum_{j=1}^{p} a_j z^{-j}} \tag{6.3}$$

We calculated AR coefficients a_j using the Burg algorithm. This algorithm used three parameters: forward prediction error, backward prediction error, and reflection coefficient. Equations (6.4)–(6.6) defines these three parameters:

$$\hat{e}_{f,p}(t) = i(t) + \sum_{m=1}^{p} \hat{a}_{p,m} i(t-m), \quad t = p+1, \ldots, N \tag{6.4}$$

$$\hat{e}_{b,p}(t) = i(t-p) + \sum_{m=1}^{p} \hat{a}^{*}_{p,m} i(t-p+m), \quad t = p+1, \ldots, N \tag{6.5}$$

$$\hat{k}_p = \frac{-2\sum_{t=p+1}^{N} \hat{e}_{f,p-1}(t)\hat{e}^{*}_{b,p-1}(t-1)}{\sum_{t=p+1}^{N} \left[|\hat{e}_{f,p-1}(t)|^2 + |\hat{e}_{b,p-1}(t-1)|^2\right]} \tag{6.6}$$

Equation (6.7) defines the AR coefficients in terms of reflection coefficient and earlier estimates:

$$\hat{a}_{p,m} = \begin{cases} \hat{a}_{p-1,m} + \hat{k}_p \hat{a}^{*}_{p-1,p-m}, & m = 1, \ldots, p-1 \\ \hat{k}_p, & m = p \end{cases} \tag{6.7}$$

The Burg algorithm

Step1 Initialize $\hat{e}_{f,0}(t) = \hat{e}_{b,0}(t) = i(t)$.
Step2 For $p = 1, .. n$,

a. Calculate $\hat{e}_{f,p-1}(t)$ and $\hat{e}_{b,p-1}(t)$ for $t = p + 1, \ldots, N$ from (6.4) and (6.5).
b. Calculate reflection coefficient $\hat{k}_p$ from (6.6).
c. Calculate $\hat{a}_{p,m}$ for $m = 1, \ldots, p$ from (6.7).

Then $\hat{\theta} = [\hat{a}_{p,1}, \ldots, \hat{a}_{p,p}]^T$ is the vector of calculated AR coefficient.

6.3 EXTREME LEARNING MACHINE

Guang-Bin Huang proposed the extreme learning machine (ELM), which is a single hidden layer feed forward neural network [21,22]. ELM learns faster and has better generalization characteristics than a gradient-based algorithm like backpropagation. ELM does not have problems like accurate learning rate selection, local minima, and overfitting [23]. ELM works for all kinds of activation and kernel functions in electrical engineering applications [24–26]. Figure 6.1 shows the structure of a ELM binary classifier that performed fault classification. The input for the ELM binary classifier is the AR coefficients, as shown in Figure 6.1. The ELM binary classifier has two output nodes: one node representing fault current (class 1) and other for normal current (class 0). Out of two outputs, one output will have a higher value than the other, according to system conditions. Accordingly, the ELM classifier output was class 1 or class 0. ELM selects the weight between input nodes and hidden nodes arbitrarily and then adjusts the weight between the hidden and output nodes analytically.

Equation (6.8) gives the mathematical model of an ELM structure for N samples $\{(x_k, t_k)\}_{k=1}^{N}$, where input vector $x_k = [x_{k1}, x_{k2}, \ldots, x_{kn}]^T$, target vector $t_k = [t_{k1}, t_{k2}, \ldots, t_{km}]^T$, $\tilde{N}$ hidden neurons, and activation function $g(x)$. Here, $w_i \cdot x_k$ denotes the inner products of w_i and x_k:

$$\sum_{i=1}^{\tilde{N}} \beta_i g(w_i \cdot x_k + b_i) = O_k, \quad k = 1, 2, \ldots, N \tag{6.8}$$

Weight vectors: (1) $w_i = [w_{i1}, \ w_{i2}, \ldots, w_{in}]^T$ connects i^{th} hidden node and the input nodes.

(2) $\beta_i = \left[\beta_{i1}, \beta_{i2}, \ldots, \beta_{im}\right]^T$ connects i^{th} hidden node and the output nodes.

Output vector: $O_k = [O_{k1}, O_{k2}, \ldots, O_{kn}]^T$.

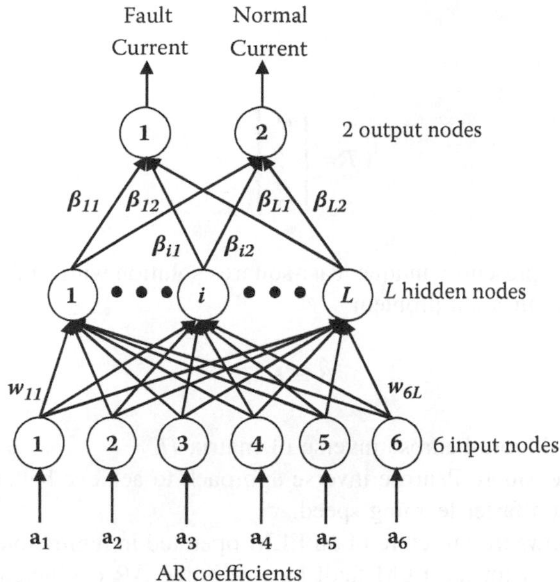

FIGURE 6.1 ELM binary classifier.

Bias: b_i – bias of the ith hidden neuron.

The other version of Equation (6.8) is as follows:

$$H\beta = O \tag{6.9}$$

where

$$H = \begin{bmatrix} g(w_1 \cdot x_1 + b_1) & \dots & g(w_{\tilde{N}} \cdot x_1 + b_{\tilde{N}}) \\ \vdots & \dots & \vdots \\ g(w_1 \cdot x_N + b_1) & \dots & g(w_{\tilde{N}} \cdot x_N + b_{\tilde{N}}) \end{bmatrix}_{N \times \tilde{N}} \tag{6.10}$$

$$\beta = \begin{bmatrix} \beta_1^T \\ \vdots \\ \beta_{\tilde{N}}^T \end{bmatrix}_{\tilde{N} \times m} \quad and \quad O = \begin{bmatrix} O_1^T \\ \vdots \\ O_N^T \end{bmatrix}_{N \times m} \tag{6.11}$$

To minimize the training error, Guang-Bin Huang formulated the minimization problem as shown below by substituting target in equation (6.9) instead of output:

$$\min_{\beta} \|H(w_1, \dots, w_{\tilde{N}}, b_1, \dots, b_{\tilde{N}})\beta - T\|^2 \tag{6.12}$$

where target:

$$T = \begin{bmatrix} t_1^T \\ \vdots \\ t_N^T \end{bmatrix}_{N \times m} \tag{6.13}$$

Equation (6.14) represents a unique least-squares solution with a minimum norm for the previous minimization problem:

$$\hat{\beta} = H^{\dagger} T \tag{6.14}$$

where $H^{\dagger}$ is the Moore-Penrose inverse of matrix H.

ELM uses the Moore-Penrose inverse approach to achieve better generalization characteristics at a faster learning speed.

Figure 6.2 shows the structure of an ELM operated in regression mode to locate the fault. The input for the ELM fault locator is the AR coefficients, as shown in Figure 6.2. There exists a difference between classification and regression mode in the number of output nodes. For binary classification, there will be two output nodes, whereas for regression there is only one output node. In binary classification, the output is class number like class 0 or class 1, whereas in regression mode, the output is a continuous value. Design procedures are same in both the modes. Since the fault distance is a continuous value, we designed the ELM in regression mode to locate the fault.

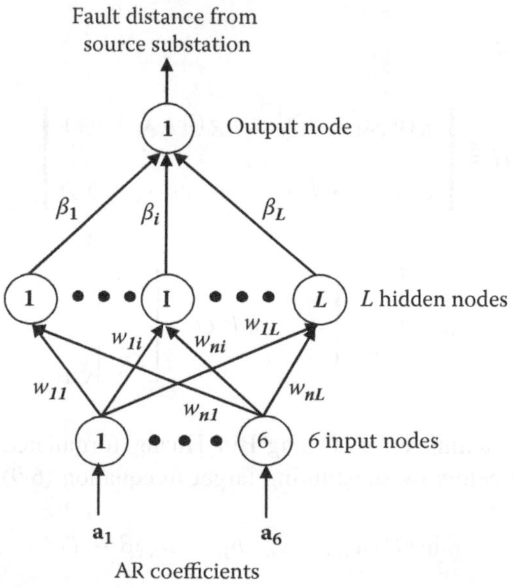

FIGURE 6.2 ELM fault locator.

6.3.1 Training Extreme Learning Machine

Figure 6.3 shows the training process followed in designing ELM for classification and regression mode. The input data for a training process were AR coefficients and target output. The target outputs were fault type in classification mode and fault distance (km) in regression mode. The two main stages in the training process was cross-validation and validating it. The training process fragmented the input training samples in to five-folds in cross-validation. To reduce the effect of random data fragmentation in cross-validation, the training process repeated cross-validation for 10 times. Training accuracy (TA) or root mean square error (RMSE) decided the completion of training process. Training the ELM finalized the number of nodes in a hidden layer for desired performance in a particular mode.

6.4 INTEGRATED APPROACH OF THE PROTECTION SCHEME

The three stages of the integrated approach are fault detection, classification, and location. The fault detection scheme used a three-phase current measured at one end of the line with a sampling frequency of 1 kHz. Figure 6.4 shows the flowchart of the fault detection scheme. The fault detection algorithm used samples with a sliding window length equal to half cycle fundamental frequency. When a new sample entered the window, the detection scheme discarded the old sample. Half-cycle data contains 10 samples for each phase for a system frequency of 50 Hz. We modeled 30 samples of a three-phase current using AR signal modeling and obtained its poles using equation (6.7). We selected the model order using the common rule that AR order should be around one-third of the data window size [19]. The critical pole is the pole which moves closely to the unit circle. We identified the third pole as the critical pole and calculated the power of the critical pole. If power of the critical pole exceeds the threshold value, the detection scheme declared the current signal as the disturbance signal. Otherwise the detection scheme discarded the oldest sample and a new sample entered the window. To discriminate between the fault and inrush current, the detection scheme used a counter i. The maximum duration of the transient or inrush current is two cycles, whereas the fault current exists for more than two cycles' duration. When the counter value is greater than the duration of inrush current (N), the detection scheme made a decision that the fault occurred in the system and generated the trip signal.

After fault detection, the next stage is fault classification. Fault classification provides the information about the status of each phase in the system that the power system operator requires. The second stage of the integrated approach is the fault classification scheme, as shown in Figure 6.5, that identified the status of each phase. We modeled half-cycle current samples of each phase using AR signal modeling. The input for each ELM classifier is the AR coefficient. We designed the ELM classifier as a binary classifier, where class 1 denotes fault condition and class 2 normal condition. The ground detector helps to find the ground fault [27]. We calculated the ground index using equation (6.15) and used a threshold of 0.05 to detect the ground fault. The fault classification scheme classifies the fault by using ELM binary classifiers and ground detectors:

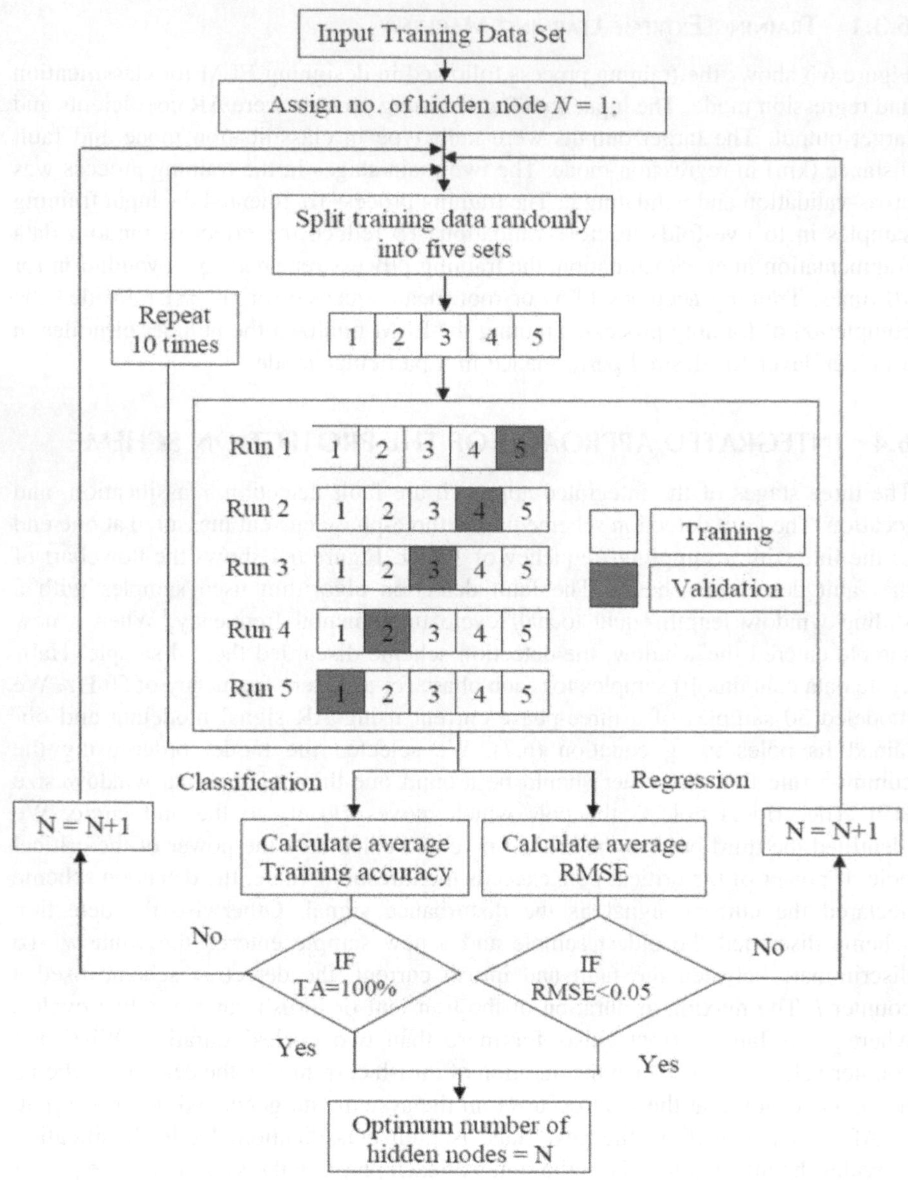

FIGURE 6.3 Flowchart for training process of two modes of ELM.

$$Index = \frac{|\bar{I}_a + \bar{I}_b + \bar{I}_c|}{median\,(|\bar{I}_a|,\ \ |\bar{I}_b|,\ |\bar{I}_c|)} \tag{6.15}$$

The last stage in the integrated approach is locating the fault. According to the type of fault classified, the fault location structure picks one fault locator as shown in Figure 6.6. This scheme finds the exact km at which the fault occurred. We designed the ELM in regression mode to find the exact km of fault. The input to the

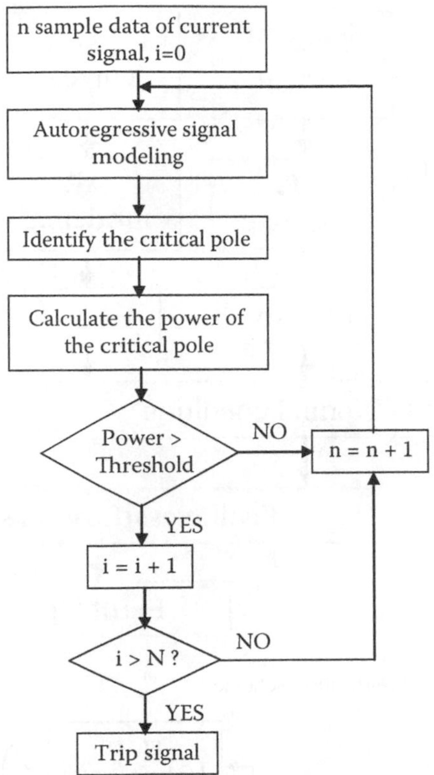

FIGURE 6.4 Flowchart of the fault detection scheme.

ELM in regression mode is the AR coefficient. The accuracy of this part of the integrated approach helps the power system operator to a greater extent. The higher the accuracy, the power system operator investigates a small portion of the underground cable. The entire integrated approach not only detects the fault but also provides status of each phase and exact km data to the power system operator, thus making the protection system smarter and more reliable.

6.5 TEST SYSTEM

Figure 6.7 shows the test system simulated using PSCAD/EMTDC, where a virtual operator will be available in a source substation. A voltage source of 11 kV, 50 Hz represented the substation in simulation. Figure 6.7 shows two circuit breakers that are available at both ends of the underground cable to protect it. Table 6.2 shows the specifications of the underground cable. Figure 6.8 shows the cable configuration of underground cable. The integrated approach processed the three-phase current and implemented the schemes using MATLAB® programs available in the source substation. The fault detection scheme sends a trip signal to the circuit breakers when the fault occurs to protect the underground cable. With the data provided by

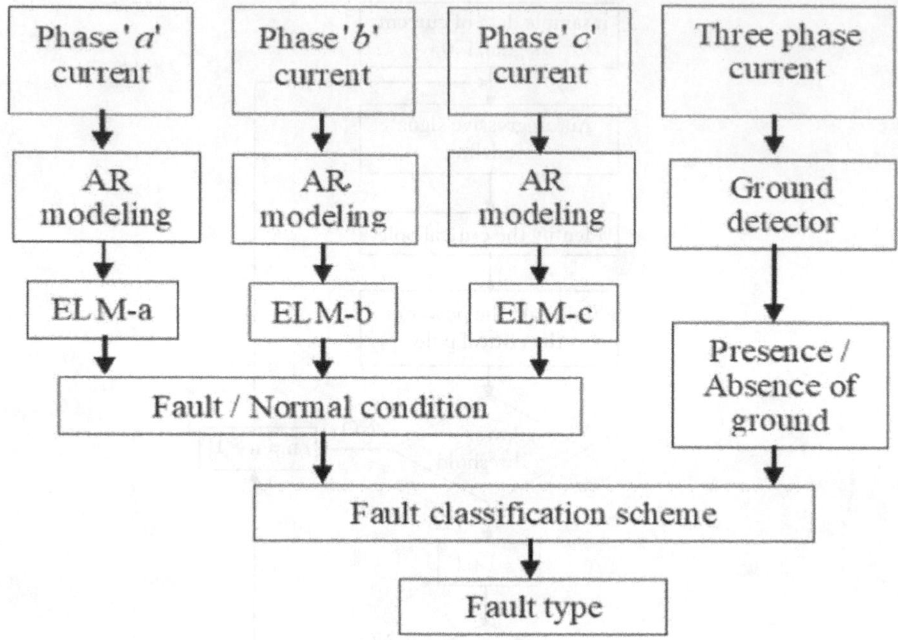

FIGURE 6.5 Fault classification scheme.

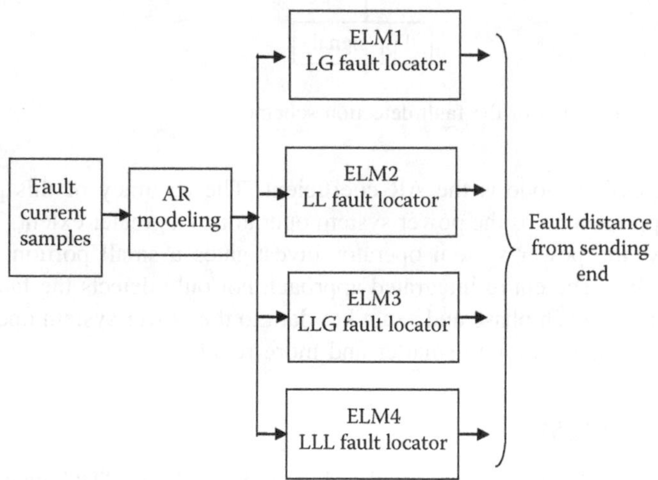

FIGURE 6.6 Fault location scheme.

the integrated approach, the virtual operator rectifies the fault. After clearing the fault, power supply will be restored to the load by closing the circuit breaker.

Figure 6.9 shows a three-phase current of the test system considered with a double line to ground fault simulated at 0.4 s. Figure 6.10 shows the AR coefficients of the respective fault phases that clearly distinguished the fault condition from

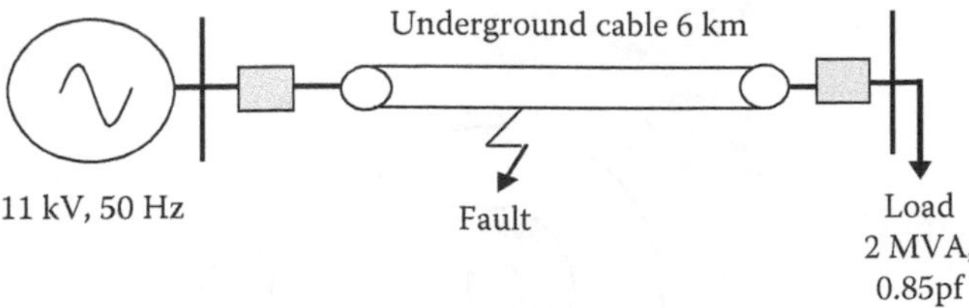

FIGURE 6.7 Test system.

normal condition. The integrated approach used this discriminating feature of AR coefficients to design the protection system.

6.5.1 Simulation Parameters for Training and Testing

To design and test the protection system using an integrated approach, we simulated 1260 fault scenarios for training and 10,800 fault scenarios for testing the protection system. Table 6.3 shows the details of simulation parameters. We recorded the three-phase current waveforms of all the fault scenarios in PSCAD/EMTDC and processed the waveforms in MATLAB.

6.6 FAULT DETECTION

Figure 6.11 shows the three-phase current waveform for *a-b-g* fault at a 1-km distance during 0.4 to 0.5 seconds. The fault detection scheme calculated the power of third pole by capturing samples from the three-phase current using a sliding window of half cycle duration. Figure 6.12 depicts the pole power calculated at every instance of time; it shows pole power is above the threshold value of 0.5 per unit during fault, whereas it is less than the threshold for normal operating condition of the system. When the pole power was above the threshold, the detection scheme started incrementing the counter *i*. Once *i* was greater than *N*, the scheme altered the status of the circuit breaker as open.

TABLE 6.2
Specifications of underground cable

Specification of Medium Voltage Underground Cable (XLPE Stranded Copper Conductor – 6 km – Bergeron Model)	
Radius (mm)	$r_1 = 6.75$, $r_2 = 10.15$, $r_3 = 12.05$, $r_4 = 12.2$, $r_5 = 13.8$
Resistivity and permeability of core conductor	$\rho = 1.7e^{-8}\,\Omega m$, $\mu = 1.0$
Relative permeability and permittivity of insulation	$\mu = 1.0$, $\varepsilon = 2.7$
Resistivity and permeability of sheath	$\rho = 2.5e^{-8}\,\Omega m$, $\mu = 1.0$

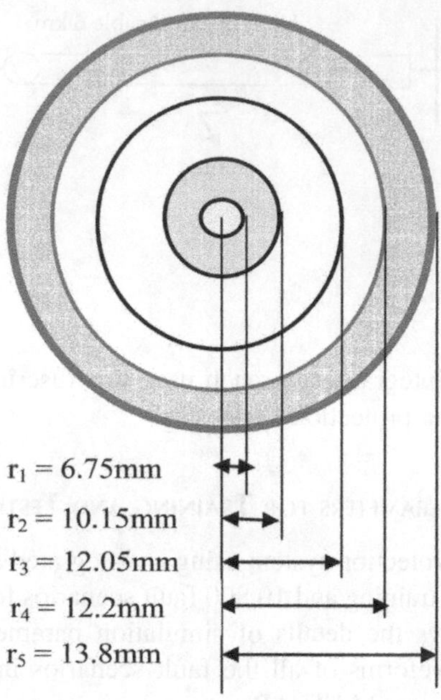

FIGURE 6.8 Cable configuration.

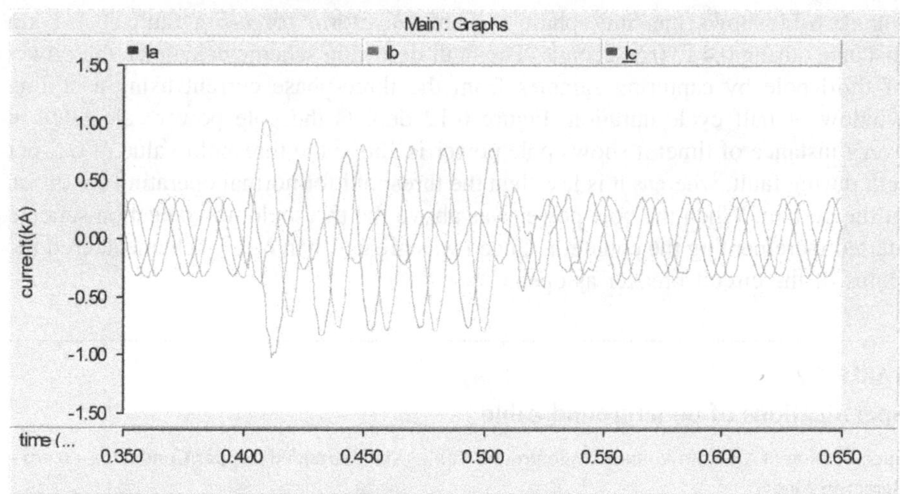

FIGURE 6.9 Three-phase current during *a-b-g* fault in underground cable.

6.7 FAULT CLASSIFICATION

The next step is fault classification. We trained the ELM classifiers with 1260 fault scenarios using the sigmoid activation function. Figure 6.13 shows training accuracy for each increment of nodes in a hidden layer during the training process. The

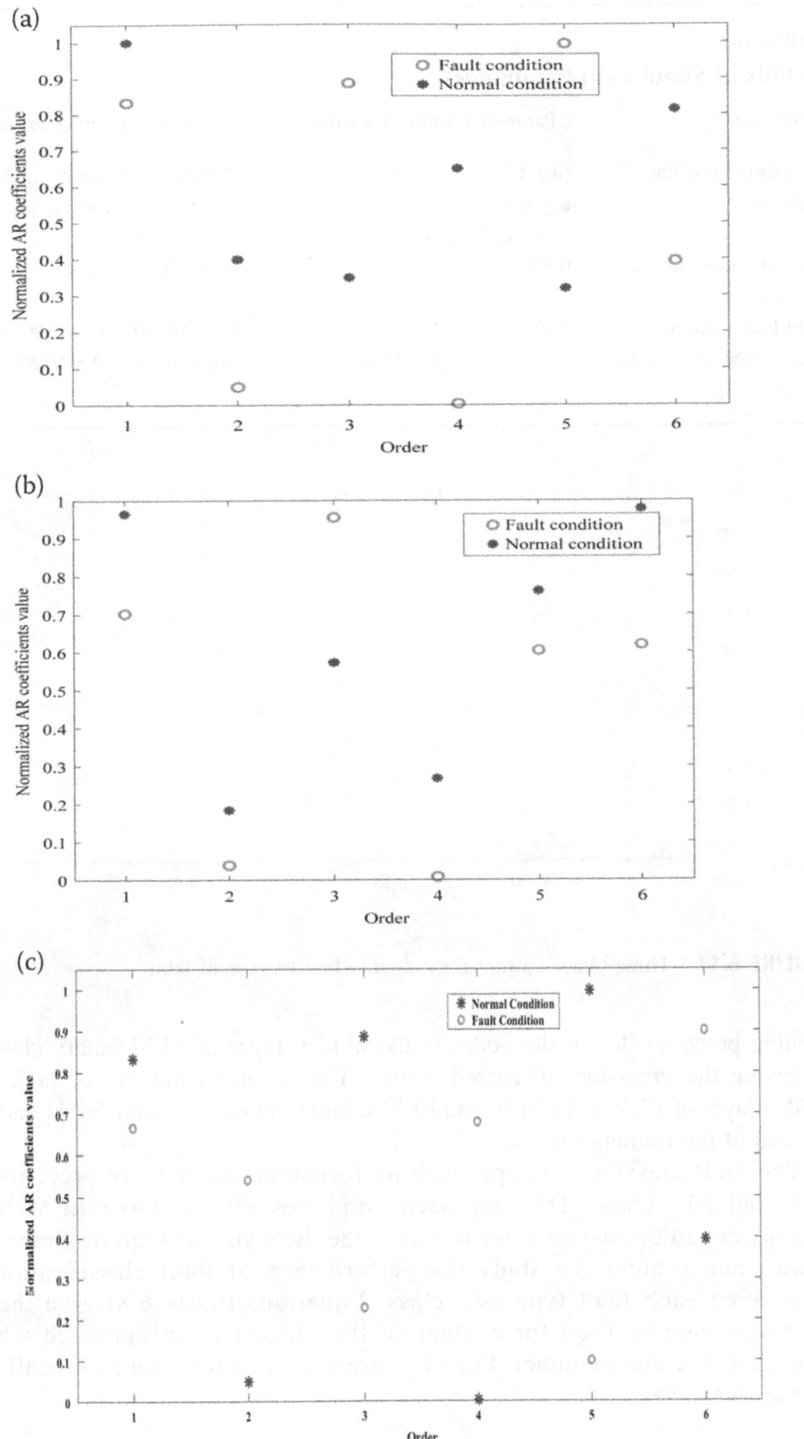

FIGURE 6.10 AR coefficients (a) "*a*" phase current; (b) "*b*" phase current; (c) "*c*" phase current.

TABLE 6.3
Details of Simulation Parameters

Parameters	Parameter values for training	Parameter values for testing
Fault distance in km	[0.5, 0.75, 1, ..., 5.5]	[0.625, 0.875, 1.125, ..., 5.375]
Fault type	a-g, b-g, c-g, a-b, a-c, b-c, a-b-g, a-c-g, b-c-g, a-b-c	a-g, b-g, c-g, a-b, a-c, b-c, a-b-g, a-c-g, b-c-g, a-b-c
Fault inception angle in degrees	0, 90	0, 18, 36, 54, 72, 90
Fault Resistance in Ω	0, 80, 100	0, 5, 10, 30, 50, 80, 100, 150, 200
Total number of simulated cases	21 × 10 × 2 × 3 = 1260	20 × 10 × 6 × 9 = 10,800

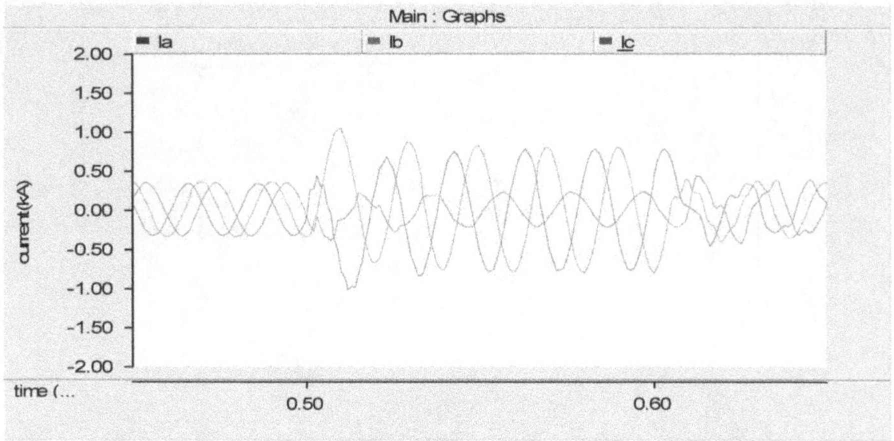

FIGURE 6.11 Three-phase current for a double line to ground fault.

training process selected the nodes in the hidden layer of ELM binary classifiers following the procedure discussed earlier. The optimum number of nodes in a hidden layer of ELM-a, ELM-b, and ELM-c binary classifiers were 39,41 and 84 at the end of the training process.

The fault classification approach performance metrics are precision, recall, and F1- score. This approach combines all the three ELM binary classifiers and ground detector to recognize the status of fault of all the three phases and ground. To study the performance of fault classification, we considered each fault type as a class. Equations (6.6)–(6.8) give the performance metrics used for evaluating the classification approach where i represents the class number. For F1-score average precision and recall of all classes was used:

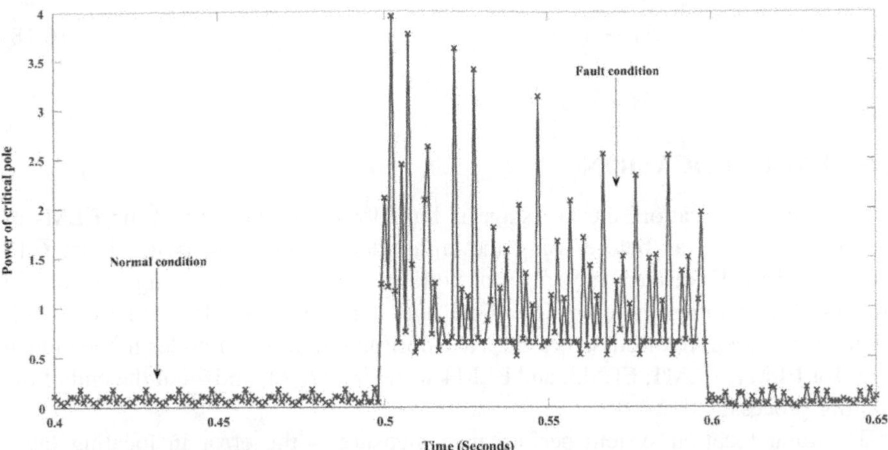

FIGURE 6.12 Power of critical pole.

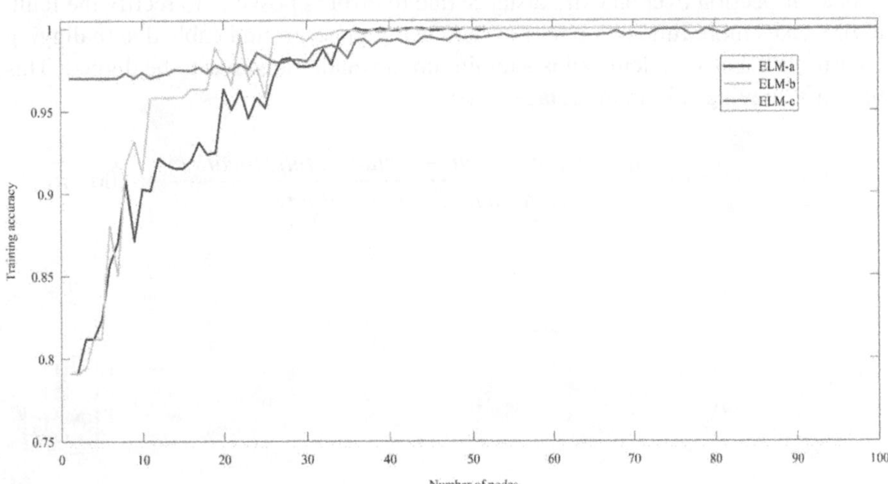

FIGURE 6.13 Training accuracy of ELM binary classifiers.

$$Precision_i = \frac{True\ \ positive_i}{True\ \ positive_i + False\ \ positive_i} \tag{6.16}$$

$$Re\,call_i = \frac{True\ \ positive_i}{True\ \ positive_i + False\ \ negative_i} \tag{6.17}$$

$$F1 - score = \frac{2 * \text{Precision} * \text{Re}\,call}{\text{Precision} + \text{Re}\,call} \tag{6.18}$$

6.8 FAULT LOCATION

After fault classification, the last stage is fault location. We trained the ELMs of four fault locators as before using the sigmoid activation function. Figure 6.14 shows the RMSE for each increment of node in a hidden layer during the training process. The training process selected the nodes in the hidden layer of ELMs following the procedure discussed earlier. The optimum number of nodes in the hidden layer for ELM1, ELM1, ELM3, and ELM4 were 78, 77, 74, and 74 at the end of the training process.

The fault location system performance measure is the error in locating fault. Equation (6.19) gives the formula used for calculating the error. This error should be a minimum that helps the power system operator to restore the system back to normal condition with minimum outage time. In the overhead transmission line, physical inspection over an extra distance due to error is possible to rectify the fault. But this takes more time in a power system using underground cable due to digging the earth. If the error calculated is a minimum, a smaller area has to be digged. This helps in achieving minimum outage time:

$$\%\ error = \frac{|Fault\ \ locator\ \ output - Actual\ \ fault\ \ location|}{Total\ \ length\ \ of\ \ the\ \ line} \times 100 \tag{6.19}$$

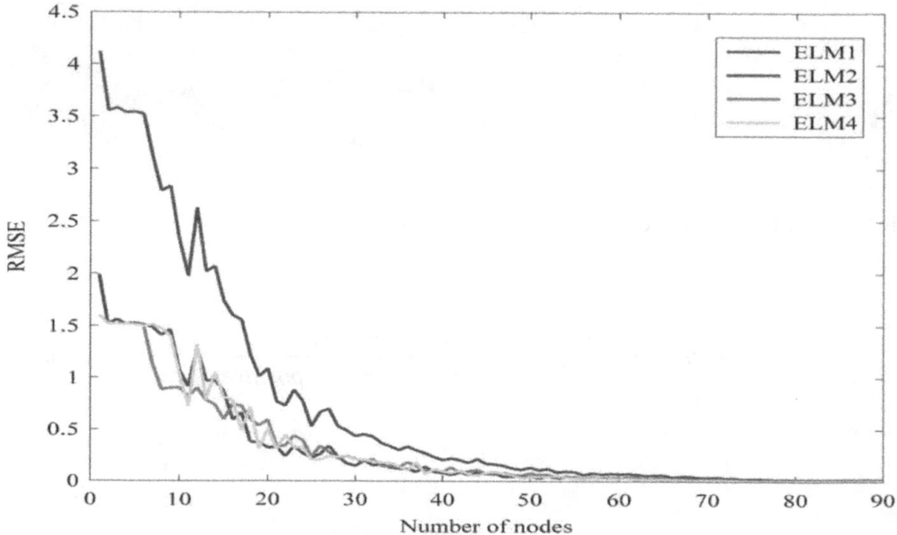

FIGURE 6.14 RMSE of ELM fault locators.

6.9 RESULTS AND DISCUSSION

The training process ended with finalizing the structure of ELMs. Then we tested the ELM binary classifiers and ELM in regression mode with 10,800 fault scenarios. Figure 6.15 shows the confusion matrix of the fault classification scheme. The diagonal elements of the confusion matrix correspond to true classification, whereas the off-diagonal elements correspond to false negatives and false positives, as shown in Figure 6.15. The column at the right end of Figure 6.15 shows the precision of the fault classification approach for all the fault types. The row at the bottom of Figure 6.15 shows the recall of the fault classification approach for all 10 faults. The F1-score of the fault classification scheme was 0.99 (99% classification accuracy) for the tested 10,800 fault scenarios. Table 6.4 shows the output of ELM binary classifiers for the misclassified fault scenarios. The target fault type was $a - c - g$, whereas the output fault type of the fault classification approach was $a - g$. This was because the ELM-c classifier provided incorrect values for output nodes, leading to false negative of the particular classifier. The fault resistance and actual fault location of the false negative test samples were 200 Ω and 90% of cable length. Therefore, high fault resistance and faraway fault from the measuring end affected the performance of the fault classification scheme to a slight extent. But overall performance of the fault classification for this test system was better to integrate with the fault location. Also, the noise interference effect was evaluated by adding a noise level of 20 dB in the original signals. The classification accuracy drops by a percentage of 1%, which doesn't affect the overall performance of the proposed scheme.

The protection scheme classified the fault accurately that helped in choosing the correct ELM fault locator for each test sample. The ELM in regression mode estimated the location of fault for all the 10,800 test samples. The percentage error in locating the fault was calculated from equation (6.19) for each test sample. Figure 6.16 shows the maximum error, mean error, and standard deviation calculated for all the fault types. Standard deviation values are close to the mean error, as shown in Figure 6.16. This indicates that percentage error of most of the test fault scenarios were around the mean error, while the percentage error of a few samples was far away from the mean % error.

Figure 6.17 shows the box plot of percentage error with the number of outliers for each fault type. Outliers represent the number of test samples for which the percentage error differed significantly from the remaining test samples. We found 724 outliers out of 10,800 tested fault scenarios. Therefore, the percentage error of 94.3% of tested fault scenarios was in a similar range, excluding outliers (6.7% samples) that proved the robustness of the fault location approach.

Figure 6.18 shows the percentage error range and it indicates 101 samples have a percentage error of greater than 0.05%. So, the fault location approach located the fault for 99% of tested samples with less than 0.05% error. Even a fault location error of 0.05% corresponds to a length of 3 meters for a 6 km length cable. Hence, the power system operator has to locate the fault in less than 3 meters in distance estimated by the fault location approach. This is a good acceptable range for locating the fault in the underground cable considered. From the previous discussions,

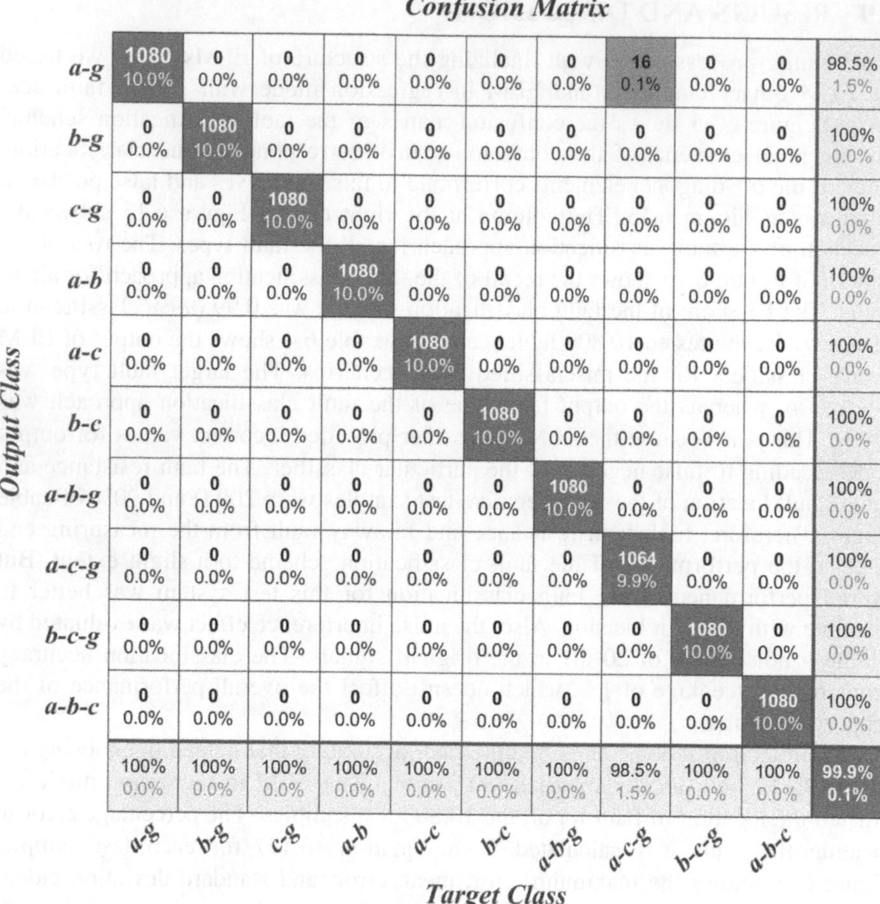

Confusion Matrix
Output Class
a-g
b-g
c-g
a-b
a-c
b-c
a-b-g
a-c-g
b-c-g
a-b-c
1080 10.0% | 0 0.0% | 0 0.0% | 0 0.0% | 0 0.0% | 0 0.0% | 0 0.0% | 16 0.1% | 0 0.0% | 0 0.0% | 98.5% 1.5%
0 0.0% | 1080 10.0% | 0 0.0% | 0 0.0% | 0 0.0% | 0 0.0% | 0 0.0% | 0 0.0% | 0 0.0% | 0 0.0% | 100% 0.0%
0 0.0% | 0 0.0% | 1080 10.0% | 0 0.0% | 0 0.0% | 0 0.0% | 0 0.0% | 0 0.0% | 0 0.0% | 0 0.0% | 100% 0.0%
0 0.0% | 0 0.0% | 0 0.0% | 1080 10.0% | 0 0.0% | 0 0.0% | 0 0.0% | 0 0.0% | 0 0.0% | 0 0.0% | 100% 0.0%
0 0.0% | 0 0.0% | 0 0.0% | 0 0.0% | 1080 10.0% | 0 0.0% | 0 0.0% | 0 0.0% | 0 0.0% | 0 0.0% | 100% 0.0%
0 0.0% | 0 0.0% | 0 0.0% | 0 0.0% | 0 0.0% | 1080 10.0% | 0 0.0% | 0 0.0% | 0 0.0% | 0 0.0% | 100% 0.0%
0 0.0% | 0 0.0% | 0 0.0% | 0 0.0% | 0 0.0% | 0 0.0% | 1080 10.0% | 0 0.0% | 0 0.0% | 0 0.0% | 100% 0.0%
0 0.0% | 0 0.0% | 0 0.0% | 0 0.0% | 0 0.0% | 0 0.0% | 0 0.0% | 1064 9.9% | 0 0.0% | 0 0.0% | 100% 0.0%
0 0.0% | 0 0.0% | 0 0.0% | 0 0.0% | 0 0.0% | 0 0.0% | 0 0.0% | 0 0.0% | 1080 10.0% | 0 0.0% | 100% 0.0%
0 0.0% | 0 0.0% | 0 0.0% | 0 0.0% | 0 0.0% | 0 0.0% | 0 0.0% | 0 0.0% | 0 0.0% | 1080 10.0% | 100% 0.0%
100% 0.0% | 100% 0.0% | 100% 0.0% | 100% 0.0% | 100% 0.0% | 100% 0.0% | 100% 0.0% | 98.5% 1.5% | 100% 0.0% | 100% 0.0% | 99.9% 0.1%
a-g
b-g
c-g
a-b
a-c
b-c
a-b-g
a-c-g
b-c-g
a-b-c
Target Class

FIGURE 6.15 Confusion matrix of fault classification.

we found that this protection scheme operated with good accuracy in locating the fault for most of the fault scenarios tested.

6.9.1 Comparative Evaluation

Table 6.5 shows the comparative evaluation of the proposed integrated approach with three other schemes reported in the literature. These methods used full cycle current and voltage signal samples for fault classification and location, whereas the proposed scheme used half-cycle single end current samples only. We designed the approach using lesser training samples with the same range of fault inception angles and fault resistance values as in the other schemes. On the other end, we tested larger testing samples. This "less training and larger testing" proved the robustness of the integrated approach with better maximum and mean error compared to other schemes considered.

TABLE 6.4
ELM Binary Classifiers Outputs for Misclassified Fault Scenarios

Sl.No	ELM – 1 (Phase 'a')		ELM – 2 (Phase 'b')		ELM – 3 (Phase 'c')		Ground index	Target fault type	Output fault type
	Node 1	Node 2	Node 1	Node 2	Node 1	Node 2			
1.	0.94	–0.94	-0.96	0.96	–0.47	0.47	0.09	$a-c-g$	$a-g$
2.	0.67	–0.67	-0.34	0.34	–0.30	0.30	0.07	$a-c-g$	$a-g$
3.	1.00	–1.00	–1.55	1.55	–1.07	1.07	0.1	$a-c-g$	$a-g$
4.	1.13	–1.13	–1.15	1.15	–1.51	1.51	0.07	$a-c-g$	$a-g$
5.	1.31	–1.31	–0.99	0.99	–1.67	1.67	0.06	$a-c-g$	$a-g$
6.	1.40	–1.40	–0.67	0.67	–1.22	1.22	0.09	$a-c-g$	$a-g$
7.	2.19	–2.19	–1.58	1.58	–1.94	1.94	0.1	$a-c-g$	$a-g$
8.	0.44	–0.44	–0.78	0.78	–0.86	0.86	0.09	$a-c-g$	$a-g$
9.	0.03	–0.03	–0.04	0.04	–0.56	0.56	0.08	$a-c-g$	$a-g$
10.	0.72	–0.72	–0.50	0.50	–0.20	0.20	0.07	$a-c-g$	$a-g$
11.	0.71	–0.71	–0.20	0.20	–0.59	0.59	0.07	$a-c-g$	$a-g$
12.	0.65	–0.65	–1.22	1.22	–0.26	0.26	0.07	$a-c-g$	$a-g$
13.	1.14	–1.14	–0.85	0.85	–0.30	0.30	0.08	$a-c-g$	$a-g$
14.	1.47	–1.47	–1.28	1.28	–1.26	1.26	0.09	$a-c-g$	$a-g$
15.	1.74	–1.74	–0.31	0.31	–0.52	0.52	0.07	$a-c-g$	$a-g$
16.	0.32	–0.32	–0.32	0.32	–0.56	0.56	0.07	$a-c-g$	$a-g$

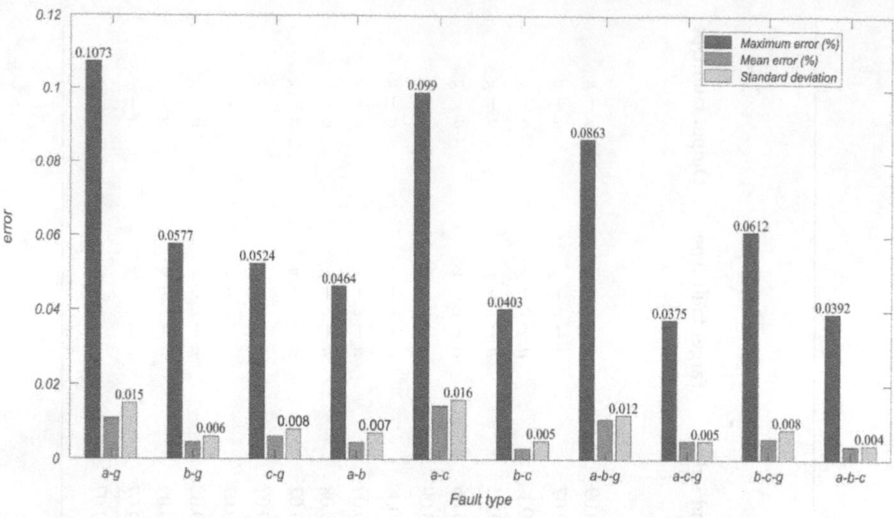

FIGURE 6.16 Percentage error for 10 types of faults.

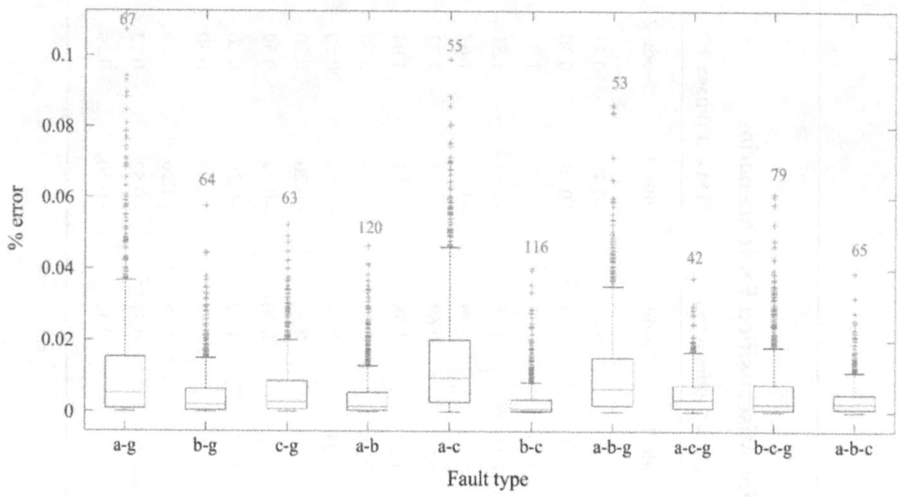

FIGURE 6.17 Boxplot for percentage error for 10 types of faults.

6.10 SUMMARY

This chapter presented an integrated approach for fault detection, classification, and location in medium voltage underground cables. The integrated approach used three phase current measured at single terminal of the system which was modeled using AR signal modeling. All three stages of the protection scheme used AR coefficients extracted from a three-phase current. The fault detection approach detected the fault using the critical pole calculated from AR coefficients. The fault classification approach used ELM binary classifiers and ground index to know the fault type. Fault location approach used ELM in

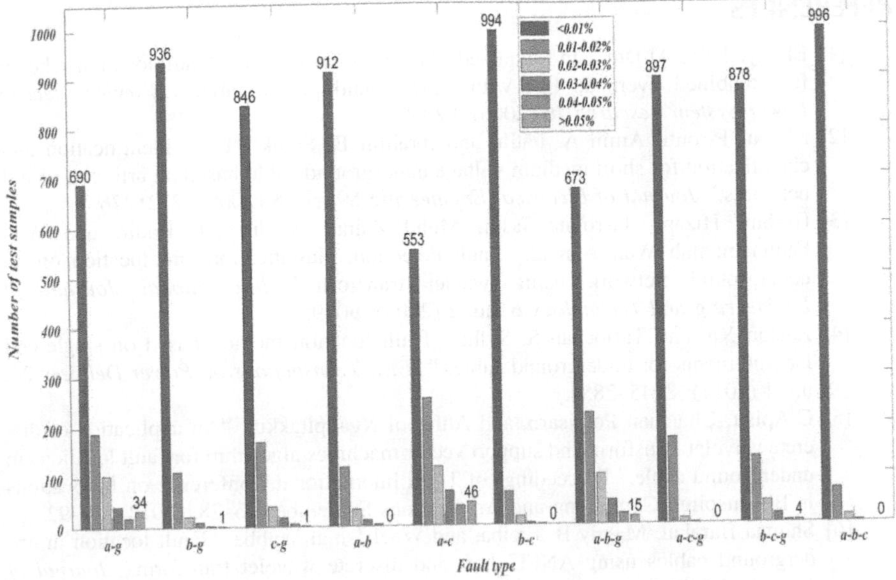

FIGURE 6.18 Percentage error distribution for 10 types of fault.

TABLE 6.5
Comparative Evaluation of Proposed Integrated Approach

References	[6]	[12]	[10]	Proposed integrated approach
Schemes → Parameters ↓	DWT - ANFIS	DWT	S-Transform - ELM	AR-ELM
Number of fault scenario trained	2520	-	2100	1260
Number of fault scenario tested	540	-	4000	10,800
Maximum error (%)	3	5.1	0.15	0.10
Mean error (%)	-	-	0.08	0.01

Note: "-" indicates not considered/mentioned

regression mode to find the fault distance. The training process of fault classification and location of ELMs helped in fixing the structure of ELMs for the integrated approach. Trained ELMs are tested to evaluate the performance metrics of the integrated approach. Test results confirm the robustness of the proposed integrated approach to protect the medium voltage underground cable. Comparative evaluation shows the proposed integrated approach outperforms the other schemes considered. Also, the proposed integrated approach provides reliable data to the power system operator during a fault to restore the supply with less outage time.

REFERENCES

[1] El Sayed Tag Al Din, and Mohamed Mamdouh Abdul Aziz. "Fault location scheme for combined overhead line with underground power cable." *Elsevier Electric Power System Research* 76 (2006): 928–935.

[2] Marouf Pirouti, Amin A. Fatih, and Ibrahim B. Sadik. "Fault identification and classification for short medium voltage underground cable based on artificial neural networks." *Journal of Electrical Engineering* 59, no. 5 (2008): 272–276.

[3] Hashim Hizam, Jasronita Jasni, Mohd Zainal Abidin Ab Kadir, and Wan Fatinhamamah Wan Ahmad. "Fault detection, classification and location on an underground network using wavlet transform." *International Journal of Engineering and Technology* 6, no. 2 (2009): 90–95.

[4] Zhihan Xu, and Tarlochan S. Sidhu. "Fault location method based on single-end measurements for underground cables." *IEEE Transactions on Power Delivery* 26, no. 4 (2011): 2845–2854.

[5] C Apisit, Chaichan Pothisarn, and Atthapol Ngaopitakkul. "An application of discrete wavelet transform and support vector machines algorithm for fault locations in underground cable." Proceedings of Third International Conference on Innovations in Bio-Inspired Computing and Applications September 26–28, (2012): 89–92.

[6] Shimaa Barakat, Magdy B. Eteiba, and Wael Ismal Wahba. "Fault location in underground cables using ANFIS nets and discrete wavelet transform." *Journal of Electrical Systems and Information Technology* 1 (2014): 198–211.

[7] Christian Flytkjær Jensen, and O. M.K.Kasun Nanayakkara. "Online fault location on AC cables in underground transmission systems using sheath currents." *Elsevier Electric Power System Research* 115 (2014): 74–79.

[8] Saurabh Kulkarni, Surya Santoso, Thomas A. "Incipient fault location algorithm for underground cables." *IEEE Transactions on Smart Grid* 5, no.3 (2014): 1165–1174.

[9] Kunal Hasija, Shelly Vadhera, Abhishek Kumar and Anurag Kishore. "Detection and location of faults in underground cable using Matlab/Simulink/ANN and OrCad." 6th IEEE Power India International Conference (2014): 1–5.

[10] Papia Ray, and Debani Mishra. "Application of extreme learning machine for underground cable fault location." *International Transactions on Electrical Energy Systems* 25, no. 12 (2015): 3227–3247.

[11] Ankita Nag and Anamika Yadav. "Fault classification using Artificial Neural Network in combined underground cable and overhead line," 2016 IEEE 1st International Conference on Power Electronics, Intelligent Control and Energy Systems (2016): 1–4.

[12] D Prabhavathi, D.R.M. Surya Kalavathi, and K Prakasam. "Detection, classification and location of faults in underground cable by wavelet technique." *International Journal of Advanced Research in Electrical, Electronics and Instrumentation Engineering* 5, no. 2 (2016): 1160–1171.

[13] Xuebin Qin, Yizhe Zhang, Wang Mei, Gang Dong, Jun Gao, Pai Wang, Jun Deng, and Hongguang Pan, "A cable fault recognition method based on a deep belief network." *Computer and Electrical Engineering* 71 (2018): 452–464.

[14] Ankita Nag, Anamika Yadav, A. Y. Abdelaziz, and Mohammad Pazoki. "Fault Location in Underground Cable System Using Optimization Technique," 2020 First International Conference on Power, Control and Computing Technologies (2020): 261–266.

[15] Kanendra Naidu, Mohd Syukri Ali Ab Halim Abu Bakar, Chia Kwang Tan, Hamzah Arof, and Hazlie Mokhlis. "Optimized artificial neural networkto improve the accuracy of estimated fault impedances and distances for underground distribution system." *PLoS ONE* 15 (2020): 1–22.

[16] Suguna Thanagasundram, Sarah Spurgeon, and Fernando Soares Schlindwein. "A fault detection tool using analysis from an autoregressive model pole trajectory." *Journal of Sound and Vibration* 317, no. 3 (2008): 975–993.

[17] Richard Wires, John W. Pierre, and Daniel J. Trudnowski. "Use of ARMA block processing for estimating stationary low-frequency electromechanical modes of power systems." *IEEE Transactions on Power System* 18, no. 1 (2003): 167–173.

[18] Biljana Matic-Cuka, and Mladen Kezunovic. "Islanding detection for inverter-based distributed generation using support vector machine method." *IEEE Transactions on Smart Grid* 5, no. 6 (2014): 2676–2686.

[19] Petre Stoica, and Randolph Moses. *"Spectral Analysis of the Signals."* Upper Saddle River, NJ, USA: Prentice Hall, 2005.

[20] M. Kemal Kiymikb, Mehmet Akinc, and Ahmet Alkanb "AR spectral analysis of EEG signals by using maximum likelihood estimation." *Computers in Biology and Medicine* 31, no. 6 (2001): 441–450.

[21] Guang-Bin Huang, Qin-Yu Zhu, and Chee-Kheong Siew. "Extreme learning Machine: Theory and applications." *Neurocomputing* 70 (2006): 489–501.

[22] Guang-Bin Huang, Qin-Yu Zhu, and Chee-Kheong Siew. "Universal approximation using incremental constructive feedforward networks with random hidden nodes." *IEEE Transactions on Neural Networks* 17, no. 4 (2006): 879–892.

[23] Guang-Bin Huang, H. Zhou, X. Ding, and R. Zhang. "Extreme learning machine for regression and multiclass classification." *IEEE Transactions on Systems Man, and Cybernetics – PartB: Cybernetics* 42, no. 2 (2012): 513–529.

[24] R. Dubey, S. R. Samataray, and B.K. Panigrahi. "An extreme learning machine based fast and accurate adaptive distance relaying scheme." *International Journal of Electrical Power and Energy Systems* 73(2015): 1002–1014.

[25] Rahul Dubey, Subhransu Rajan Samataray, Bijay Ketan Panigrahi, and Vijendran G. Venkoparao. "Extreme learning machine based adaptive distance relaying scheme for static synchronous series compensator based transmission lines." *International Journal of Electric Power Components and Systems* 44, no. 2 (2016): 219–232.

[26] Yi Chen, Enyi Yao, and Aridam Basu. "A 128 channel extreme learning machine based neural decoder for brain machine interfaces." *IEEE Transactions on Biomedical Circuits and Systems* 10, no. 3 (2016): 679–692.

[27] Magnus Akke, and James T. Thorp "Some improvements in the three phase differential equation algorithm for fast transmission line protection" *IEEE Transactions of Power Delivery* 13, no. 1 (1998): 66–72.

7 A New High Impedance Fault Detection Technique Using Deep Learning Neural Network

M. M. Eissa[1,2], *M. H. Awadalla*[1,3] *and A. M. Sharaf*[4]

[1]Department of Electrical and Computer Engineering Sultan Qaboos University, Oman

[2]Department of Electrical and Power Engineering, Faculty of Engineering at Helwan, Helwan University

[3]Department of Computer Engineering, Faculty of Engineering at Helwan, Helwan University

[4]Sharaf Energy Systems Inc, NB, Canada

7.1 INTRODUCTION

High Impedance Arc-Type Faults (HIFs) are characterized by an irregular nature of the arc-type and low-level fault currents on meshed electrical distribution utilization networks. Typical Electrical faults in distribution networks are found by traditional relays over current or ground relays running on a solid-state or electromechanical basis. It leaves the traditional relays inadequate for high impedance fault detection. Artificial intelligence techniques can provide solutions for this critical and still uncovered area of electrical protection systems, viz., High Impedance (HIF) Faults Detection. A high impedance fault is a significant threat to health. When left undetected, it can cause arcing and a possible electric fire [1]. High-impedance-low-current faults, luckily, show some harmonic properties that could be used to detect high impedance [2]. A method using fast Fourier transform and examining the auto and cross spectrum is examined [3]. Wavelet transformation–based algorithm for the detection of stochastic high impedance faults is also discussed in [4]. The topic of interest is to develop a robust and accurate relay, which can recognize these unique harmonic patterns using the capabilities of neural networks for pattern classification. The sequence of operations will be as follows after the relay is

DOI: 10.1201/9780367552374-7

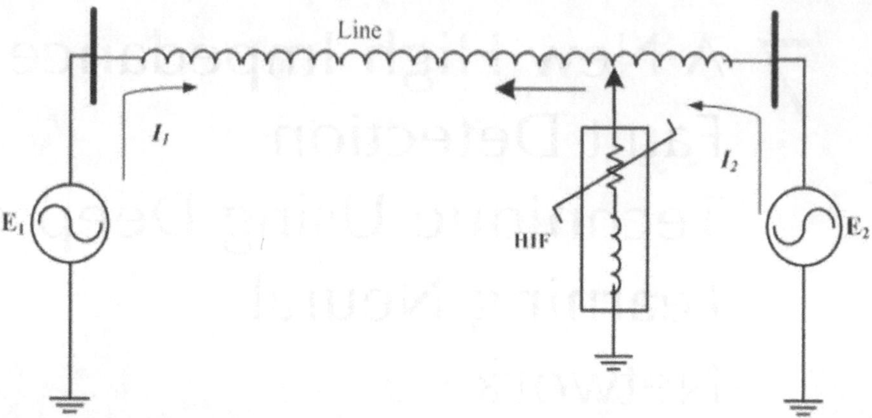

FIGURE 7.1 Single line diagram of a radial transmission line.

mounted in the protection-relaying scheme: Using one cycle FFT, the instantaneous voltage and current values at the feeder substation bus can be converted into a frequency domain. To obtain that function, the FFT-harmonic vector is processed. The neural network is then fed a vector with voltage and current signals, which activates the correct output neurons. Figure 7.1 demonstrates the radial sample method used to produce all the data for preparation, validation, and testing.

7.2 FAULT MODEL

Training data is obtained by applying nonlinear and linear faults at different locations (x) to the circuit model given in Figure 7.2. The model was simulated, as shown in Figure 7.1. The device consists of a 138 kV. X is to be taken in length per line. Data for the verification of the proposed technique were created using the MATLAB®/Simulink model to model the selected method. In this chapter's Appendix, the nonlinear arc model and the parameters of the transmission line are given. The relays are located at the respective sending and receiving ends (S and R). The standard power flows from the end of sending to the end of receiving. Only the effects of S relay responses are shown. The same results for R relay are obtained.

7.3 THE PROPOSED DEEP LEARNING APPROACH

Machine learning research has a new field called deep learning that has the aim of making machine learning to approach its original goals: Artificial Intelligence. In addition, it is a specific machine learning subset and to predict things, this kind of algorithms works well and a significant performance can be achieved. To classify images and texts, machine learning has been used for decades, but the accuracy is still behind a certain level that is crossed using deep learning.

The accuracy of classifying and labeling images using deep learning outperforms any other traditional algorithms and also get remarkable results that cannot be

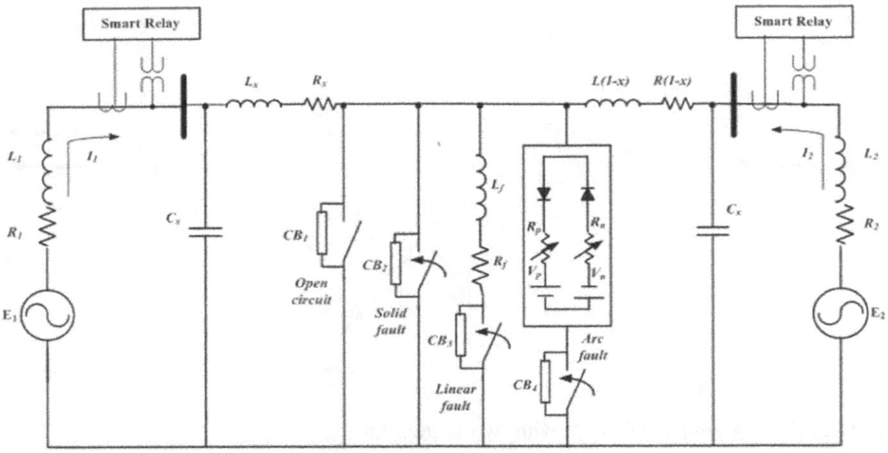

FIGURE 7.2 The fault circuit used in the digital simulation.

obtained using humans. Deep learning is currently used by Google to manage and control the energy at data centers of the company. They reduced 40% of their need of the energy for cooling. That leads to saving hundreds of millions of dollars for the company and achieving more than 15% in power.

Robust open-source software packages have been widely spred due to the advances in practical applications of deep learning. In addition, it urges the people to push the development forward.

Artificial Neural Networks (*ANN*) are one of the most powerful machine-learning algorithms that is used to mimic the processing of the information by the human brain [5]. To manipulate the regression and classification problems in a very flexible way, artificial neural networks are used without the determination of the input and output variables relationships. Generally, neural networks are arranged in three types of layers, as shown in Figure 7.3. The first layer, input layer, where the inputs are submitted to this layer, is based on the number of available input features; this layer neurons' number is developed. The last layer, the output layer, where the number of different classes the network should provide, determines the number of neurons in this layer. The intermediate layers are called hidden layers. The hidden layers and the output layer receive their inputs from the output of the preceding layers. The output of each neuron in the preceding layer is multiplied with learned weights of the interconnected links and then added to the learned biases, if necessary. The result is fed into the neurons' activation function to determine the input for the next layer. The number of hidden layers, the number of neurons per each layer, and the type of the activation function are determined based on the learning phase and the tackled application. Network processing consists of two main phases: forward and backward phase. In the forward phase, each neuron sums its weighted inputs with bias and then applies an activation function on that sum and propagates its value to the following layer [5] (see Figure 7.3).

In the backward phase, the error issues from the difference between the network predicted output and the actual output is passed backward to modify the weights of

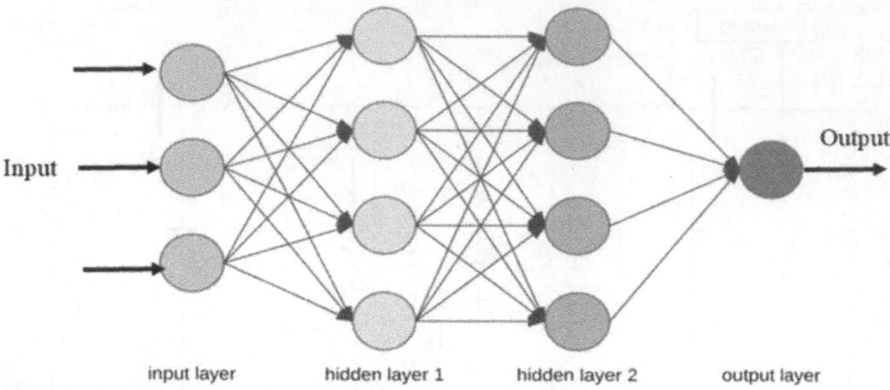

FIGURE 7.3 A simple MFANN with two hidden layers.

the links and the biases. The Activation function is usually used to perform non-linear transformation of inputs. Most commonly functions that are used as activation functions are sigmoid function and hyperbolic tangent function. Backward propagation stage starts by calculating the value of the loss function and updating the neurons weights in such a way to reduce the value of the loss function [6]. Convolutional Neural Network (CNN) or ConvNet) is a class of artificial neural networks. These networks are able to perform relatively complex tasks with images, sounds, texts, and videos. CNN networks share their parameters that do not need to feed them with hand-crafted features [6–11]. CNN is made of different types of neuron layers such as Convolutional Layers, Pooling Layers, and Fully-Connected Layers (see Figure 7.4). Figure 7.4 shows the general architecture of a Convolutional Neural Network. The deep architecture of a Convolutional Neural Network (CNN) reflects its power for extracting a set of discriminating feature representations at multiple levels of abstraction. In the application of face recognitions, CNN recently has been widely applied due to its outstanding achieved performance. The ability of CNN to learn the set of features of rich images is the

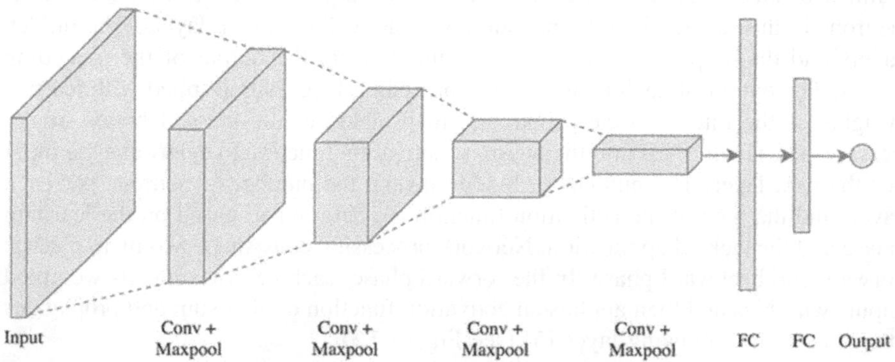

FIGURE 7.4 General architecture of a convolutional neural network [12].

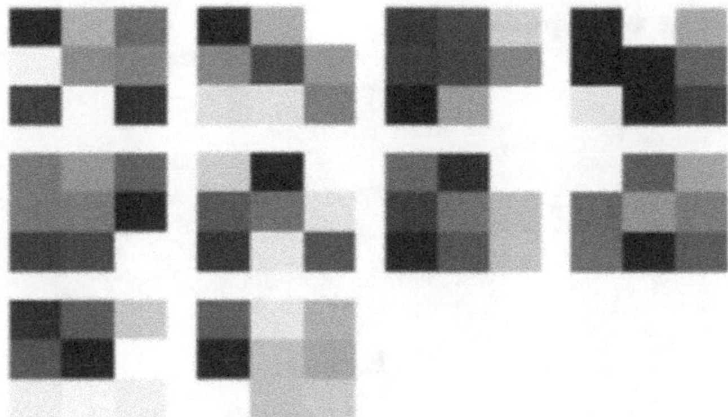

FIGURE 7.5 Initial convolutional filters generated by uniform distribution [12].

main contribution to its success. However, the huge number the parameters of network to be learned, millions, of labeled data sets is required.

The main building block of CNN is the convolutional layer. It contains a set of filters whose parameters need to be learned. Figure 7.5 shows an example of convolution layer filters, which contains 10 different initial filters.

A filter passes over the image, scanning a few pixels at a time and creating a feature map. Computing an activation map made of neurons, the input volume is convolved with ach filter. In other words and as shown in Figure 7.6, the dot products between the input and filter are computed at every spatial position of the filter by sliding the filter across the width and height of the input. The convolution operation extracts different features. Usually, the first convolution layers are used to extract low-level features such as edges, lines, end-points, and corners. Then, those features are combined by the successive layers to learn higher-level features [13].

Several parameters must be set before using the CNN, such as the number and size of the filters for each convolutional layer, the stride that defines the amount by which the filter shifts each time, and padding that indicates the number of rows and columns added around the border of the input 3D-matrix. The spatial size of the output matrix can be calculated as a function of the size of the input matrix (W), the size of the Convolution Layer filters (F), the size of the padding (P), and the stride (S), as follows [12] in equation 7.1:

$$O = \frac{W - F + 2P}{S + 1} \tag{7.1}$$

Padding is the process of adding layers of zeros around the image to avoid shrinking the size of the output from the current convolutional layer, as shown in in Figure 7.7. The reduction in the output dimension can become a problem with many successive convolutional layers, since some area is lost at every convolution. For many layers convolution network, zero padding can shrink this effect [12].

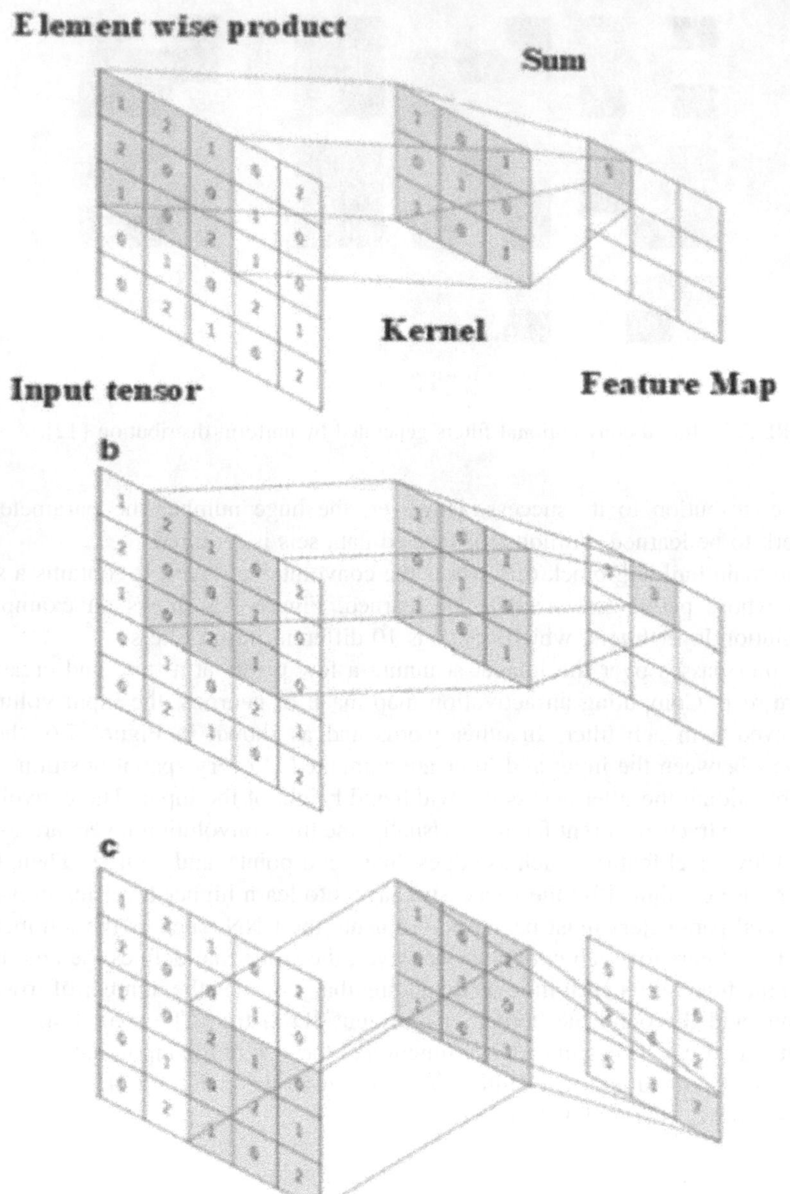

FIGURE 7.6 3 × 3 kernel size used in the convolution operation (S = 1, and no padding) [12].

The number of pixels to be shifted over the input matrix is determined by the value of the stride S. Convolution with zero padding and S > 1 is shown in Figure 7.8. The activation function that is commonly used for CNN is the Rectified Linear Units (ReLU) defined as follows [13] in equation 7.2:

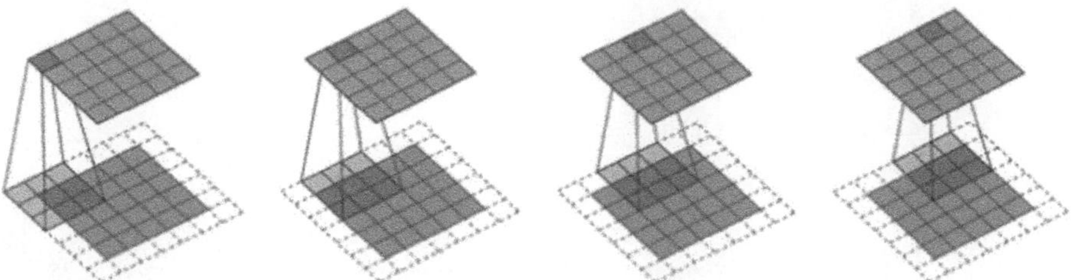

FIGURE 7.7 The effect of zero padding on the output size reduction [12].

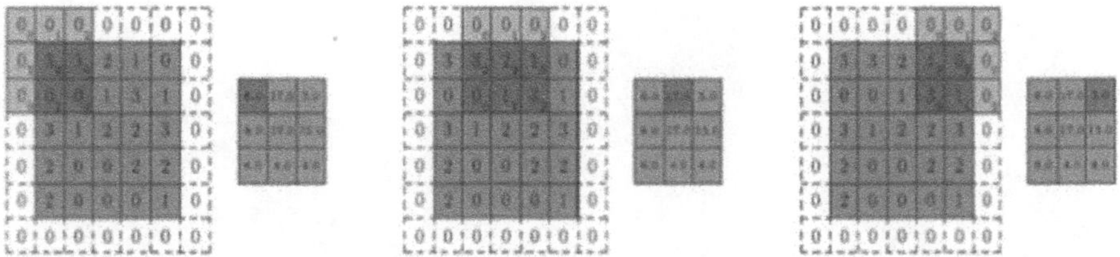

FIGURE 7.8 Convolution with zero padding and stride > 1 [15].

$$ReLu(x) = Max(0, x) \tag{7.2}$$

It is fast learning and resistance to over-fitting [14] that characterizes it. The ReLU function is used to change the values of all negative pixel in the feature map to zero. There are different activation functions such as the function *tanh* [15] and the sigmoid function [16], as shown in Figure 7.9. The activation function that is given by (7.3) affects the output value $s_k^{(l)}$ of a neuron k of the layer l:

$$s_k^{(l)} = f\left(\sigma_k^{(l)}\right) \tag{7.3}$$

with f as is the activation function and $\sigma_k^{(l)}$*is* the neuron k value. The nature of the problem affects the choice of the activation function.

To reduce the number of network parameters and network computation, the Pooling Layer, called subsampling or down sampling layer, is used [16]. Reducing each rectified feature map dimensions and maintains the most important information can be achieved through the implementation of the pooling layer. The pooling layer is inserted between layers and operates on each feature map independently. There are several nonlinear functions to implement pooling, such as max pooling, min pooling, and average pooling. The most common approach used in pooling is max pooling (see Figure 7.10). After conducting the pooling operation, the achieved feature map is defined as in equation 7.4:

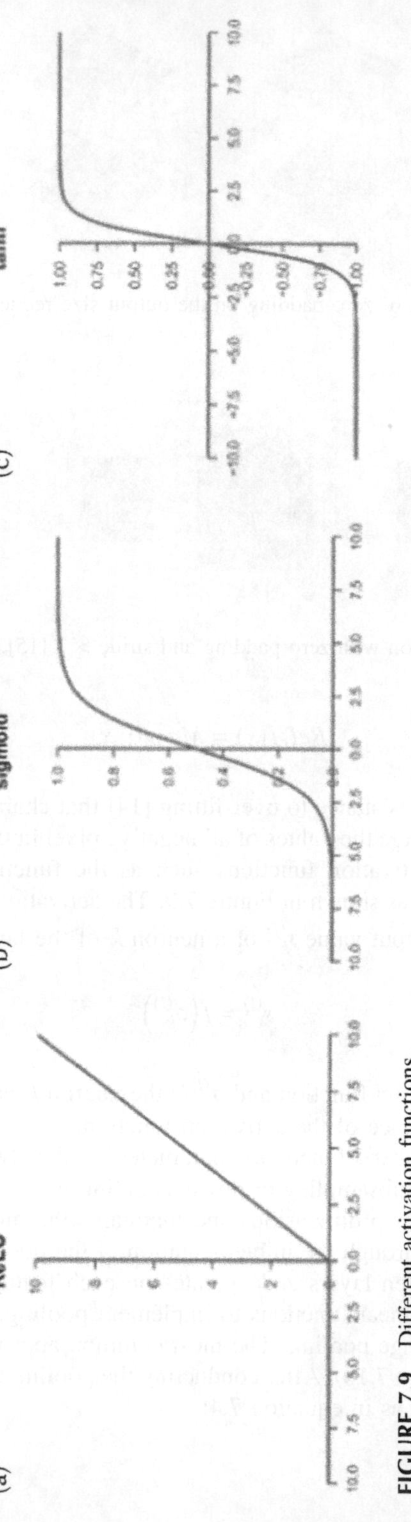

FIGURE 7.9 Different activation functions.

Single depth Slice

1	1	3	5
5	6	7	8
3	2	1	0
1	2	2	5

Max pool 2X2 filters & Stride 2

6	8
3	4

FIGURE 7.10 Pooling layer.

$$I_k^{(l)} = pool\,(s_k^{(l)}) \tag{7.4}$$

where $I_k^{(l)}$ *is* the feature map of the layer l, the *pool* () is pooling operation, and $s_k^{(l)}$ *is* the output value of the neuron k of the layer l.

The achieved image features vector from both of the convolutional and pooling layers of a CNN are fed to the fully connected layer to categorize them into different classes based on a labeled training data set. The fully connected layer is defined as a regular layer. It receives its input from the preceding layer, and then it calculates the scores of the classes. Finally, it gives 1-D array output and the size of the array equals the number of classes [16].

The Softmax function [17] is the commonly used function, which produces a probability distribution between the different classes.

The softMax function is usually applied to the output of the last fully connected layer to predict the probability distribution over m different classes. It is defined as [17] in equation 7.5:

$$\sigma\,(z)_j = \frac{e^{z_j}}{\sum_{k=1}^{k} e^{z_j}}\, for\; j = 1,\, \ldots,\, k \tag{7.5}$$

Training a network is to determine the weights of the filters in convolution layers and weights in fully connected layers, which minimizes the difference between the predicted outputs and the given ground truth labels on a training dataset. During the forward propagation and under particular kernels and weights, the predicted output is calculated. Using an optimization algorithm such as back propagation, and according to the loss value, the learnable parameters, kernels, and weights are updated. The commonly used back propagation algorithm is the gradient decent, which is a partial derivative of the loss function related to each learnable parameter, and equation 7.6 gives a single update of that parameter:

$$w = w - \alpha \frac{\partial L}{\partial w} \tag{7.6}$$

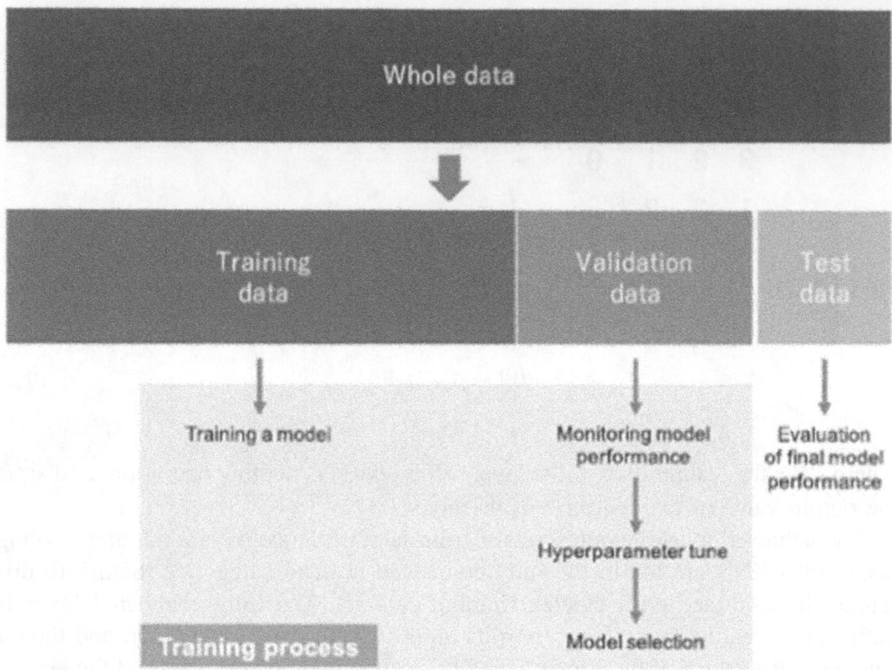

FIGURE 7.11 The data are divided into a training, validation, and testing set [12].

where w is the learnable parameter, α is the learning rate, and L is the loss function. Practically, one of the most important hyper-parameters to set before the training starts is the learning rate.

The key factor behind the success of deep learning is the collection of data set and ground truth labels that are used for training the network. These data are portioned for training, validating, and testing the developed model, as shown in Figure 7.11.

7.4 THE SIMULATED EXPERIMENTS AND DISCUSSIONS

The technique developed is based on a novel low frequency (the diagnostic vector for the third and fifth harmonic feature). Using one Fast Fourier Transform FFT loop, the instantaneous current and voltage values in feeder substation buses, shown in Figure 7.1, are captured and transformed into a frequency domain. The FFT-harmonic vectors extraction [$i3$], [$v3$], [$i5$], and [$v5$] are processed to obtain feature vectors. Current patterns are classified with the CNN. A general CNN figure is shown in Figure 7.12. The parameters for the network structure built are set according to Tables 7.1 and 7.2.

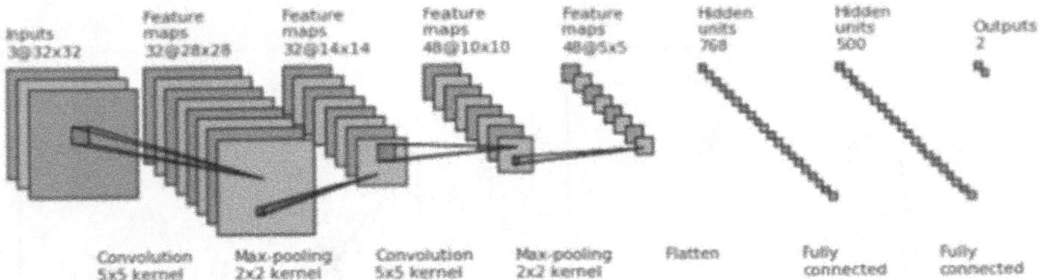

FIGURE 7.12 The developed convolutional neural network [15].

TABLE 7.1
The Parameters of the Proposed CNN

Layer	Parameter	Hyper-parameters
Convolution layer	Kernels	The size of the filter, the number of filters, the size of the stride, the padding, activation function
Pooling layer	None	The method of pooling
Fully connected layer	Weights	The layer weights and their activation function
Others		The value of the learning rate, what is the loss function, the initializations of the weights

TABLE 7.2
Layer-Wise Input and Output Dimensions

Layer Number	Layer Prototype	Layer Name	Layer Specification
1	"inputs"	Input current and voltage	227×227×3 images with "zero center" normalization
2	"conv1"	Convolution	96 11×11 convolutions with stride [4 4] and padding [0 0 0 0]
3	"relu1"	ReLU	Rectified linear unit
4	"norm1"	Batch Normalization	Batch Normalization
5	"pool1"	Max Pooling	3×3 max pooling with stride [2 2] and padding [0 0 0 0]
6	"conv2"	Convolution	256 5×5 convolutions with stride [1 1] and padding [2 2 2 2]
7	"relu2"	ReLU	Rectified linear unit
8	"norm2"	Batch Normalization	Batch Normalization
9	"pool2"	Max Pooling	3×3 max pooling with stride [2 2] and padding [0 0 0 0]
10	"fc23"	Fully Connected	2 fully connected layer, (2 classes)
11	"prob"	Softmax	Softmax
12	"classifier"	Classification Output	cross entropy

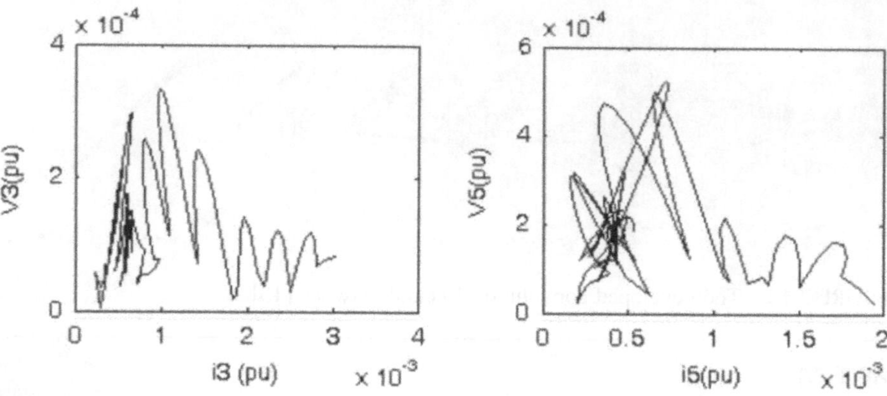

FIGURE 7.13 An internal fault at F1 (phase-a-to-ground), linear fault at x = 10%.

7.5 CASE STUDY

The device shown in Figure 7.1 used MATLAB/Simulink software to produce data to test the output of the proposed technique was subjected to different types of faults. The suggested solution has been validated through training described above by the CNN. CNN threshold limits were set to define the linear and non-linear impedance flaws. The performance of the proposed technique was tested for different kinds of internal faults. A broad range was investigated in fault sites, source impedances, near fault, and fault resistances. The relay output shown in Figure 7.2 for a phase-a-to-ground fault on the transmission line, Figure 7.1. Source capacities are updated at S and R. The fault is situated from the receiving end at 10% of the transmission line length, as shown in Figure 7.1. The developed CNN has been trained for different computed input vectors [***i3::v3***] and [***i5::v5***] and their corresponding outputs/classes as shown in Figures 7.13 to 7.18. For example, the corresponding output for the exposed input in Figure 7.13 is zero value against samples (class 0), as shown in Figure 7.15. This indicates that the fault is a linear fault. Figure 7.14 shows different fault conditions for a linear fault at $x = 50\%$. Again, the output of the relay for a phase-a-to-ground fault on the transmission line is zero, as given in Figure 7.15. Figures 7.16 and 7.17 show the non-linear faults that applied to the CNN and their output is one (class 1), as shown in Figure 7.18.

The performance of the neural network for differentiation between linear and non-linear faults is determined by one or zero. The qualified patterns are 200, which had been obtained using MATLAB's toolboxes by simulating the proposed method. Figure 7.19 and Table 7.3 display the progress of the learning process through the different iterations.

The deep learning neural network performance will effectively differentiate between multiple cases of linear and non-linear faults. The relay works efficiently to classify nonlinear faults as seen in the figures. The identification error rate in all test simulation was nearly 0%. The validation test cases covered all the distance of one form of single line to ground fault on the transmission line. Successfully, the proposed CNN addresses all of the cases reviewed.

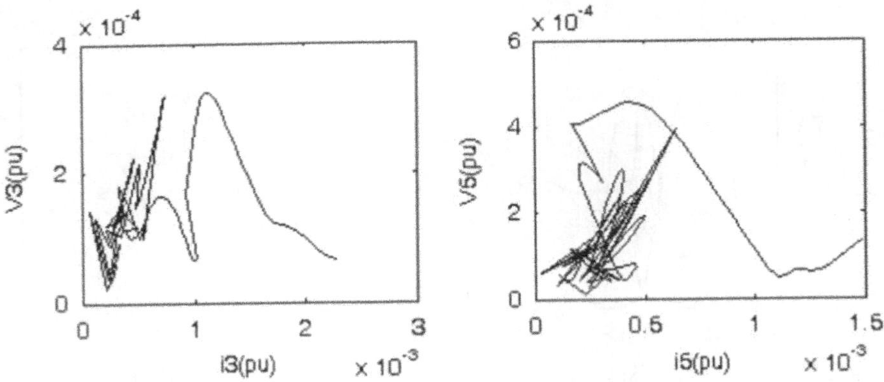

FIGURE 7.14 An internal fault at F1 (phase-a-to-ground), linear fault at x = 50%.

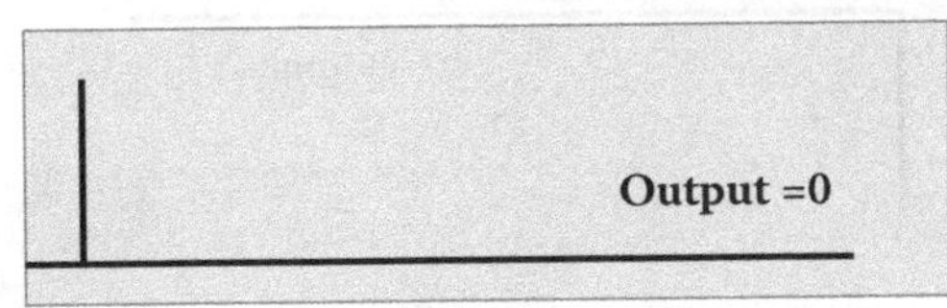

FIGURE 7.15 Response to an internal fault at F1 (phase-a-to-ground), linear fault at x = 10% and 50%.

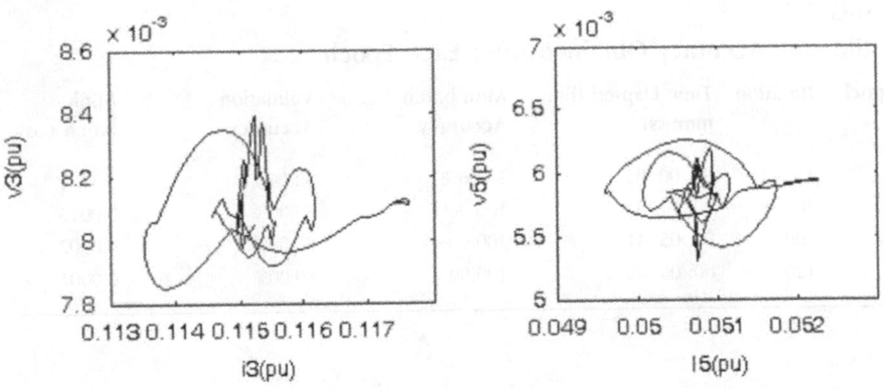

FIGURE 7.16 An internal fault at F1 (phase-a-to-ground), non-linear fault at x = 10%.

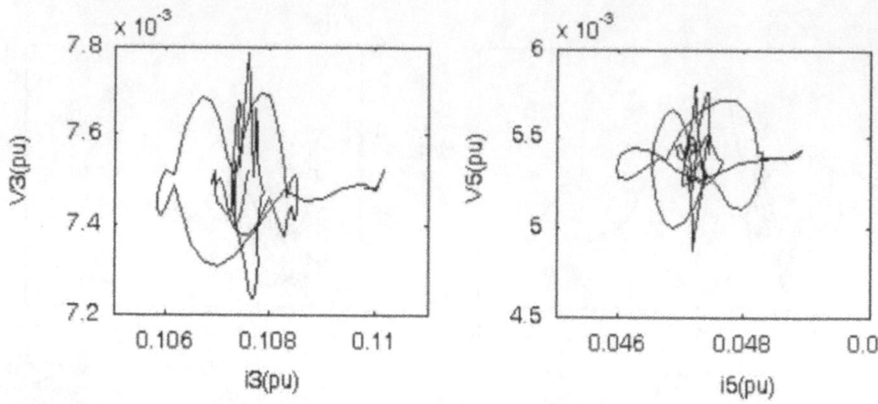

FIGURE 7.17 An internal fault at F1 (phase-a-to-ground), non-linear fault at x = 50%.

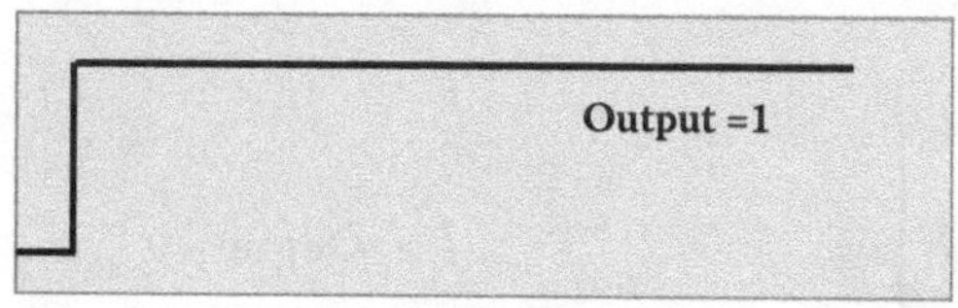

FIGURE 7.18 Response to an internal fault at F1 (phase-a-to-ground), non-linear fault at x = 10% and 50%.

TABLE 7.3
Validation Accuracy Obtained After Each Epoch

Epoch	Iteration	Time Elapsed (hh: mm: ss)	Mini-batch Accuracy	Validation Accuracy	Mini-batch Loss
1	1	00: 00: 02	14.06%	34.00%	5.3171
9	50	00: 00: 44	100.00%	85.00%	0.0013
17	100	00: 05: 41	100.00%	88.00%	0.0002
20	120	00: 06: 48	100.00%	89.00%	0.0001

7.6 CONCLUSIONS

This chapter introduced a novel low-order harmonic current pattern for *High Impedance Fault Arc* detection and discrimination. The technique is based on the present pattern form analysis. The power signals are analyzed using the *FFT* to obtain the vector [i3 v3 v5 i5] in harmonic form. The pattern is classified using the deep learning neural network technique. A lot of data is found in this study which is

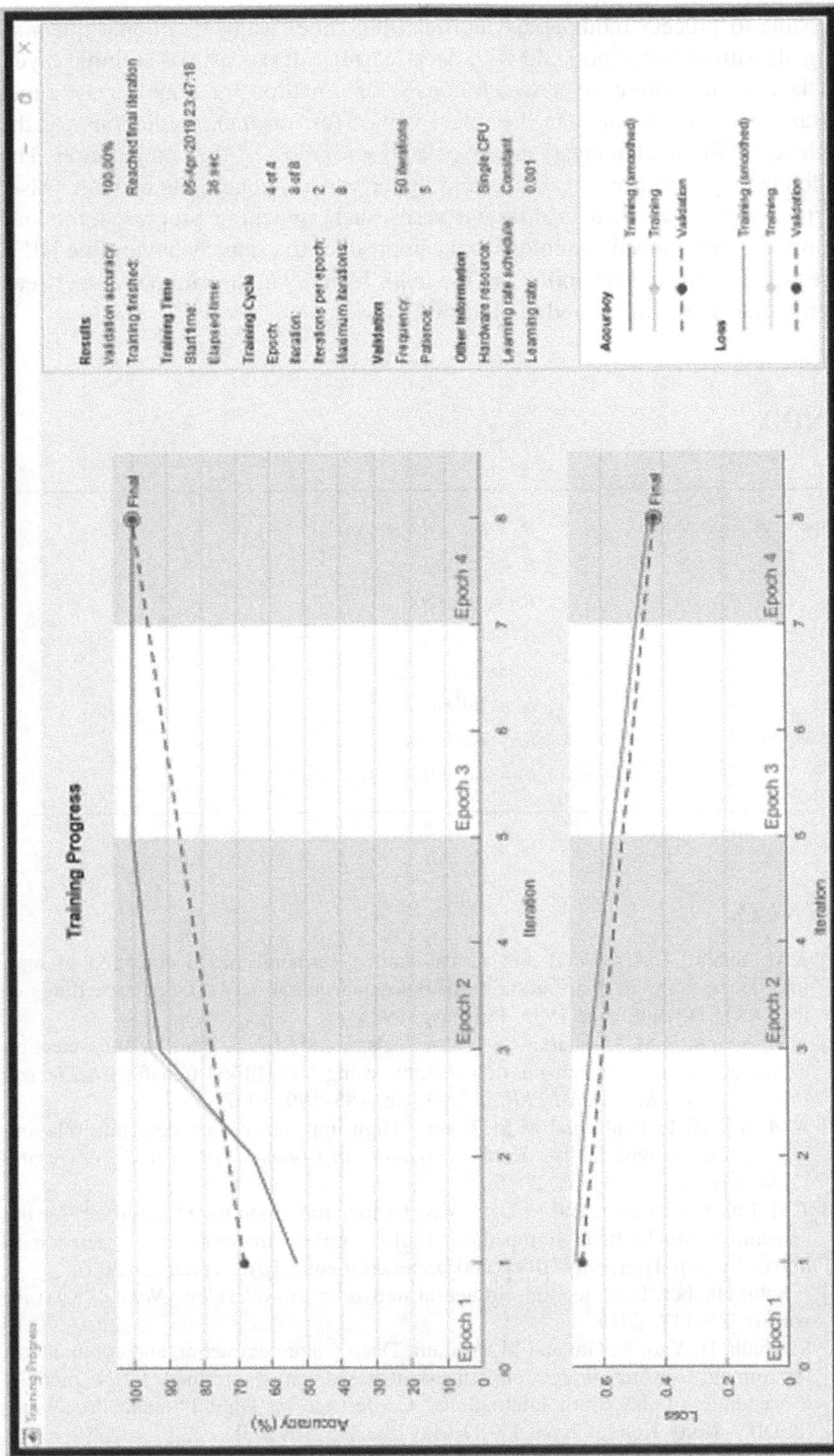

FIGURE 7.19 Validation accuracy versus number of iterations.

not possible to process (complexity and learning time) using traditional machine learning algorithms but is possible with deep learning. It consists of an input layer, hidden layers, and output layer which consist of a neuron (or node) array. Input layers take an array of numbers (i.e. pixel values for images), predictions for the output layer, while hidden layers are associated with most of the computation. The methodology suggested has been checked under various conditions of fault. Also, the performance is satisfied in different cases such switching process, light load switching, and unbalanced conditions that can produce the same behavior like HIFs. The great selectivity and reliability are the main features in discrimination between linear, non-linear arc faults, and similar other cases.

APPENDIX

High Impedance Fault Model	$R_f = R_{f0} + R_{f1} \propto (i_f/i_{f0})^{\beta}$
	$R_{f0} = 100\Omega,$
	$R_{f1} = 50to150\Omega$
	i_{f0}=70 A
	$\propto$ =0. 3*to*0. 7, $\beta = 2.$
	$L_f = 1to5mH$
Linear Fault model	$R_f = 100to150\Omega$
	$L_f = 1to5mH$

REFERENCES

[1] A.M. Sharaf, L.A. Snider, and K. Debnath, "Harmonic based detection of high impedance faults in distribution networks using neural networks", Proceedings of the IASTED conference 1993, Pittsburg, PA.

[2] A.M. Sharaf, R.M. El-Sharkawy, H.E.A. Talaat, and M.A.F. Badr. "Fault detection on radial and meshed transmission systems using fast Hilbert transform", *Electric Power Systems Research-EPSR*, vol. 41, pp. 185–190, 1997.

[3] A.M. Sharaf, L. Paull, and M.M. Eissa, "High impedance fault detection relaying scheme for arc type faults", *Recent Advances in Communication and Networking Technology*, vol. 4, 90–94, 2015.

[4] T.M. Lai, L.A. Snider, and E. Lo, "Wavelet transformation based algorithm for the detection of stochastic high impedance faults," in Proc. International Conference on Power System Transients (IPST 2003), New Orleans, LA, September 2003.

[5] J. Schmidhuber, Deep learning in neural networks: An overview. *Neural Network*, vol. 61, 85–117, 2015.

[6] S. Chaib, H. Yao, Y. Gu, and M. Amrani, Deep feature extraction and combination for remote sensing image classification based on pre-trained CNN models, Proceedings of the Ninth International Conference on Digital Image Processing (ICDIP), Hong Kong, China, 19–22 May 2017, p. 104203D.

[7] Y. LeCun, Y. Bengio, and G. Hinton, Deep learning. *Nature*, vol. 521, 2015, pp. 436–444.

[8] M. He and D. He, "Deep learning based approach for bearing fault diagnosis", *IEEE Transactions on Industry Applications*, vol. 53, no. 3, pp. 3057–3065, May–June 2017.

[9] Y. Tamura, S. Ashida, and S. Yamada, "Fault identification tool based on deep learning for fault big data", International Conference on Information Science and Security, pp. 1–4, 2016.

[10] D. Lee, V. Siu, R. Cruz, and C. Yetman, "Convolutional neural net and bearing fault analysis", Proceedings of the International Conference on Data Mining, pp. 194–200, 2016.

[11] J. Lu, G. Wang, and J. Zhou, Simultaneous feature and dictionary learning for image set based face recognition. *IEEE Transactions on Image Processing*, vol. 26, 2017, pp. 4042–4054.

[12] R. Yamashita, M. Nishio, R. Gian Do, and K. Togashi, Convolutional neural networks: overview and application in radiology. *Insights Imaging*, vol. 9, 2018, pp. 611–629.

[13] G. Hu, X. Peng, Y. Yang, T.M. Hospedales, and J. Verbeek, Frankenstein: Learning deep face representations using small data. *IEEE Transactions on Image Processing*, vol. 27, 2018, pp. 293–303.

[14] R. Yamashita, M. Nishio, R.K.G. Do, and K. Togashi. Convolutional neural networks: An overview and application in radiology. *Insights into Imaging*, vol. 9, 2018, pp. 611–629.

[15] J. Murphy. *An Overview of Convolutional Neural Network Architectures for Deep Learning*. Microway, Inc., Fall 2016.

[16] B. Xu, N. Wang, T. Chen, and M. Li, *Empirical Evaluation of Rectified Activations in Convolutional Network*, 2015, arXiv 1505.00853v2.

[17] A.W. Dumoulin, and F. Visin, *A Guide to Convolution Arithmetic for Deep Learning*, 2016, arXiv:1603.07285v1.

8 AI-Based Scheme for the Protection of Multi-Terminal Transmission Lines

Bhavesh Kumar R. Bhalja
Department of Electrical Engineering, Indian Institute of Technology Roorkee, Uttarakhand, India

ABBREVIATIONS:

ANN	Artificial neural network
CT	Current Transformer
DUTT	Direct underreach transfer tripping
DWT	Discrete Wavelet Transform
FDRM	Fault distance ratio matrix
FIA	Fault inception angle
FIS	Fuzzy interference system
GPS	Global positioning system
RBF	Radial basis function
TT	Traveling time
SLD	Single line diagram
SONET	Synchronous optical network
SVM	Support vector machine
WTC	Wavelet-transformation coefficient

8.1 INTRODUCTION TO MULTI-TERMINAL TRANSMISSION LINE

A multi-terminal line is formed, as shown in Figure 8.1, when a two-terminal line is tapped with one or more power sources. As shown in Figure 8.1(a), the most common configuration of the multi-terminal transmission line is a three-terminal line. In this three-terminal line, the line section connecting the tap point (M) and the third terminal (T) is called a tapped line. Similarly, the line connecting two existing terminals (S and R) is known as a main line. Further, Figure 8.1(b) shows an example of a multi-terminal transmission line with several line sections connected at

DOI: 10.1201/9780367552374-8

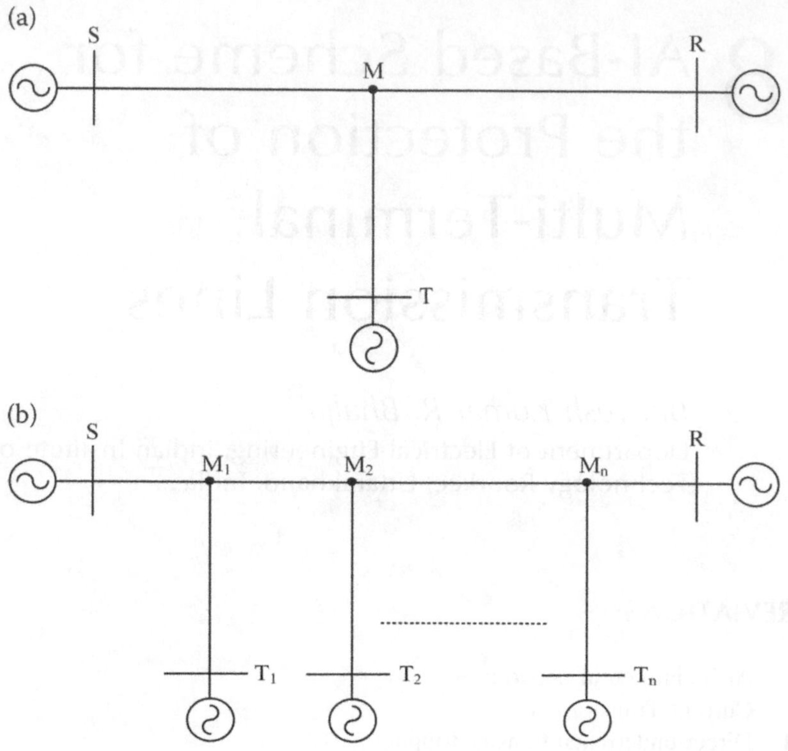

FIGURE 8.1 SLD of (a) three-terminal line, and (b) multi-terminal line.

different tap points (M_1, M_2 … M_n where "n" is the number of tap points) on the main line.

8.2 NEED OF A MULTI-TERMINAL TRANSMISSION LINE

Electricity demand is set to soar all across the globe. It is because the developed countries continue to consume vast amounts of energy, while the demand is increasing for the developing countries. However, the key in meeting this increased power demand load will be to match the generation capacity with adequate power transmission. Further, it also demands to upgrade the existing transmission network/lines. Constructing the new transmission line is not necessarily the best option due to escalating property prices and right-of-way problems. A right-of-way involves permission for additional ground space, which can take several months or years to negotiate in addition to the time needed to upgrade to higher transmission voltage. Besides, the approvals for adding new lines are difficult to meet due to environmental concerns because of the depreciation of forest and agricultural cover. Hence, an upgrade of a two-terminal transmission line into a multi-terminal line is a quick and economical solution.

8.2.1 Benefits of a Multi-Terminal Transmission Line

The multi-terminal transmission line has the following benefits:

1. There is an economic advantage with a multi-terminal line as it reduces the expenditure of all or part of a substation.
2. It reduces right-of-way requirements of new lines and stations, as it is not often straightforward to construct new facilities due to environmental considerations or limited area.
3. The configuration of a multi-terminal line can minimize the risk of overloading due to single contingency events.

8.2.2 Limitations of a Multi-Terminal Transmission Line

A multi-terminal line faces several limitations in its implementation because of the protection and technical challenges. The protection challenge is due to the presence of an infeed current. Similarly, the requirement of a communication channel for its protection is a big technical challenge. These issues are discussed in detail in the following section.

8.2.3 Protection and Other Technical Issues with Multi-Terminal Transmission Line

The existing transmission line protection schemes are well suited for a two-terminal line, as most of the transmission lines are of two-terminal configuration. However, the protection of multi-terminal lines faces several challenges in comparison to the two-terminal transmission line:

1. The under reach effect due to infeed current and fault resistance (R_F) reduces the reach of the relay during internal faults.
2. The settings, typically needed to protect the multi-terminal line, would be much greater than the settings necessary without the infeed currents. These settings may reach several multiples of the protected line's actual impedance, resulting in a decrease in line loading capability unless some form of load blinder or encroachment logic is used.
3. With a greater number of terminals, the associated communication system will increase system complexity and cost.

Hence, the protection of a multi-terminal transmission line requires careful design and application to ensure the overall system reliability.

8.3 CONVENTIONAL PROTECTION SCHEMES

A well-designed protection scheme of a multi-terminal line must take into account the particular topology of the multi-terminal line, corresponding settings that meet the necessary clearing times, reliability, and security of the system. The conventional protection schemes for multi-terminal transmission lines can be classified mainly into single-end schemes and multi-ended schemes. Single-ended schemes utilize only the local end measured voltage and/or current data (for example distance relay). Hence, this does not require any communication facility. Conversely, multi-end schemes use the local as well as remote end voltage and current data (for example differential relay). Thus, it requires a reliable communication facility between local and remote end relays. The distance and differential schemes are discussed in the following sections for a three-terminal line, as it is the simplest form of a multi-terminal line.

8.3.1 Distance Protection Scheme

The conventional distance relay works based on the apparent impedance seen by the relay. The apparent impedance seen by a relay installed at terminal S (as shown in Figure 8.2) can be expressed as (8.1). The zone-1 of each relay in the three-terminal line is set up to 80%–90% of the line section's impedance connected between the relay terminal to the nearest terminal:

$$Z_{seen} = \frac{I_S Z_{SM} + (I_S + I_T) Z_{MF} + I_F R_F}{I_S} = Z_{SM} + Z_{MF} + \frac{I_T}{I_S} Z_{MF} + \frac{I_F}{I_s} R_F \quad (8.1)$$

where I_S, I_T, and I_R are the currents flowing toward the tap-point from terminals S, T, and R, respectively. The terms Z_{SM} and Z_{MF} are the line-section impedance connecting the tap point to terminal S and fault point F, respectively. I_F and R_F are the total fault current and fault resistance, respectively.

The last two components of (8.1) represent the effect of infeed current and R_F. The sum of these two components causes a positive difference between actual and apparent impedance. This causes the misinterpretation of a fault in reach of the relay as an out-of-zone fault. This is known as the relay's underreach effect, where the relay does not operate for an in-zone fault.

Practically, the tap point location and the length of the three line sections of a three-terminal line can vary for each design. In any case, each relay should not respond for faults external to the protected section (zone-1). Generally, a DUTT

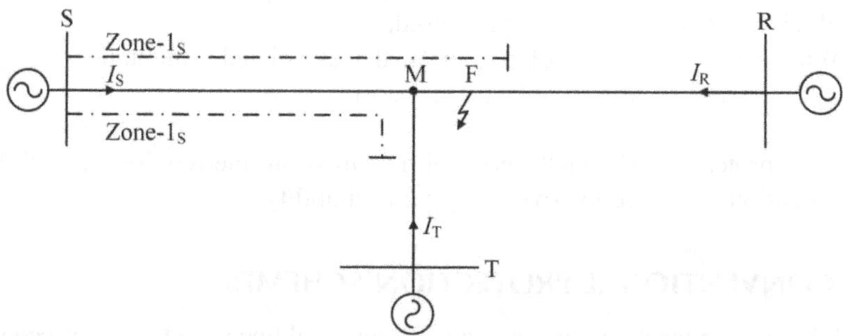

FIGURE 8.2 Zone-1 reach of the distance relay located at terminal S for a three-terminal line.

scheme is utilized with distance relays for high-speed fault clearance [1]. With the equal length of line section, each relay's reach setting would be beyond the tap point, as marked in Figure 8.3. In such a system, for a fault anywhere on the line section connected between terminal S and the tap point, the relays connected at terminal R and T would underreach due to infeed current. Simultaneously, the relay at terminal S does not have any infeed current effect for the mentioned fault case. Hence, the fault is detected at least at terminal S, which is enough to clear the fault at three terminals protected with the DUTT scheme. However, this is not the case for a three-terminal system with a different length of line sections.

With unequal section lengths, problems may be evident during relay setting of the longest branch's relay. In some cases, the overlapping of relay zone-1 reach may be minimal, which vanishes due to the underreach effect caused by infeed current. For the configuration shown in Figure 8.4, the shaded portion in-line section RM is not covered in zone-1 of any relay. This causes underreach problem for faults in the shaded portion and this underreach problem is resolved by extending the relay reach or by using the zone-2 element of the permissive overreaching scheme. The permissive zone-2 setting is usually set at 120–130% of the line impedance. However, this extended setting of the relay limits the load-carrying capacity and increases the probability of undesired operation during stable power swings.

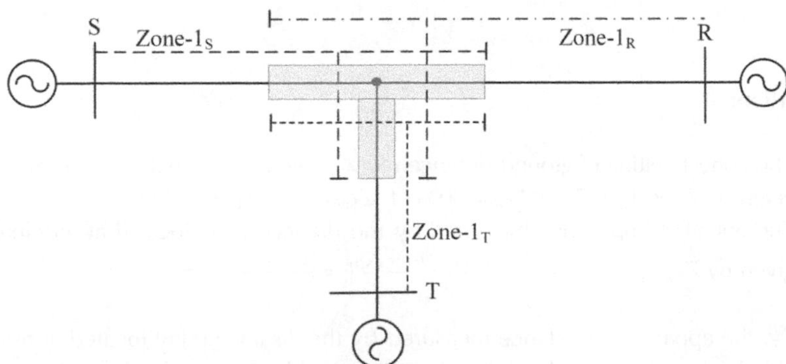

FIGURE 8.3 Overlapping region of relay reach near tap point for a line having equal line lengths.

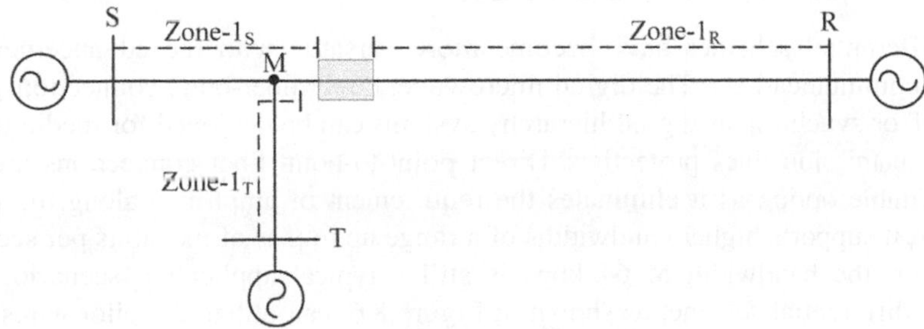

FIGURE 8.4 SLD of a three-terminal line with an unequal length of line section.

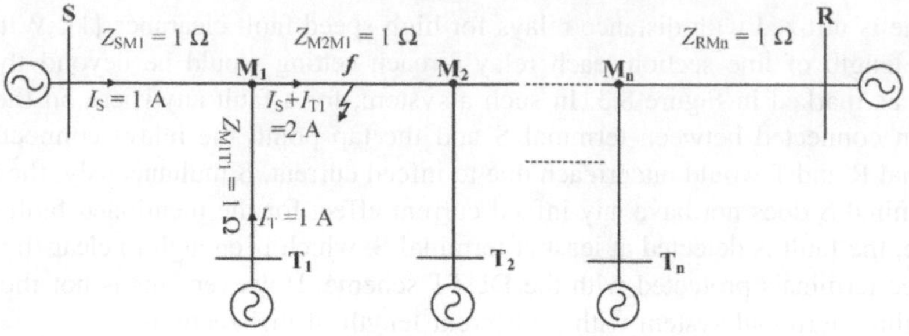

FIGURE 8.5 SLD of a multi-terminal line for example 8.1.

Example 8.1: Figure 8.5 shows SLD of a multi-terminal line. The impedance of each line section is 1 Ω and the current contribution from each end is mentioned in Figure 8.5. The first zone setting of all the distance relays is 80% of the line length.

a. Calculate zone-1 setting of ground distance relay installed at terminal S and terminal T1 for a line to ground fault at point "*f*."
b. Also, calculate the apparent impedance measured by the distance relays located at terminals S and T_1 for a three-phase fault in the middle of line section M_1M_2 (at point "*f*").

Solution:

a. The zone-1 setting of ground distance relay located at terminal S and terminal T is given by Zone-1_S = Zone-1_{T1} = 80% of $(Z_{SM1} + Z_{M1T1}) = 1.6\ \Omega$.
b. The apparent impedance measured by the distance relay located at terminal S is given by $Z_{apparent} = \frac{V_A}{I_A} = \frac{I_S \times Z_{SM1} + (I_S + I_{T1}) \times Z_{M1f}}{I_S} = \frac{1 \times 1 + 2 \times 0.5}{1} = 2\ \Omega$.

Similarly, the apparent impedance measured by the distance relay located at terminal T_1 would also be 2 Ω. It is to be noted that the actual impedance between the relaying point (S or T) and the fault point is 1.5 Ω.

8.3.2 Current Differential Scheme

The differential schemes have become more versatile with the advancement in digital communications. The digital microwave, direct fiber-optic connections, and SONET or synchronous digital hierarchy systems can be deployed for medium and long transmission lines protection. Direct point-to-point fiber connections are the best suitable option as it eliminates the requirement of amplifiers along the path. Further, it supports higher bandwidths of a range up to tens of megabits per second. However, the bandwidth of 64 kbps is still a typical application scenario. The current differential scheme, as shown in Figure 8.6, can utilize the pilot wires as a communication medium for short lines.

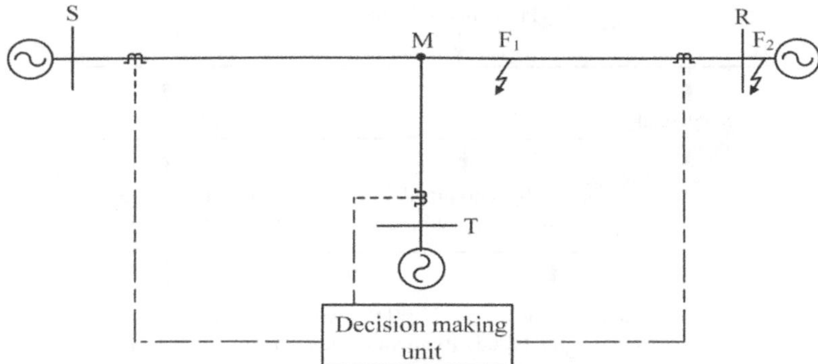

FIGURE 8.6 Schematic diagram indicating current differential principle for a three-terminal line.

The current differential principle does not need to contend with the power swings. However, the performance of line current differential protection scheme may be affected for internal line faults in case of an outfeed condition and high charging current. Besides, it may mal-operate in case of an external fault during CT saturation condition. To overcome the drawbacks of distance and current differential schemes, many advanced communication-based (multi-end) protection schemes are proposed in literature and these schemes are explained in the following sub-sections.

8.4 ADVANCED MULTI-END PROTECTION SCHEMES

The multi-end protection schemes can be categorized based on (i) the type of synchronized and unsynchronized measurements and (ii) the type of frequency component (fundamental or transient frequency) and AI or knowledge. Detailed explanation of these schemes is given in the following sub-sections (Figure 8.7).

8.4.1 Synchronized and Unsynchronized Measurement-Based Schemes

A circuit diagram considering the synchronized measurement is shown in Figure 8.8. Signals from both local and remote end terminals will be processed in this algorithm, making use of a greater volume of information. This helps to increase the effectiveness of communication-based multi-end algorithms compared to the single-end approaches. The SLD of a multi-end (three-terminal) transmission line is given in Figure 8.8, which demonstrates the synchronized measurement utilizing the satellite GPS. A synchronized measurement system gathers voltage and/or current data from various substations with respect to common reference. The signal data from different locations are stamped with time. Hence, the time axis of all measurements must be aligned with a common time reference. In most cases, the time reference is derived from GPS. The GPS capability to provide a time reference has been broadly accepted and has tremendous potential for power system applications. Today, the GPS satellite

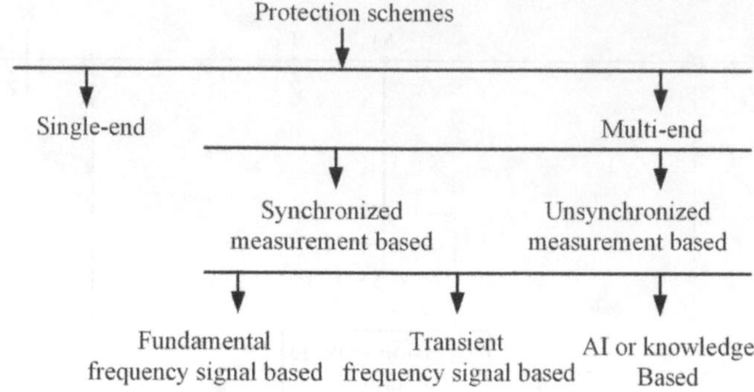

FIGURE 8.7 Classification of algorithms.

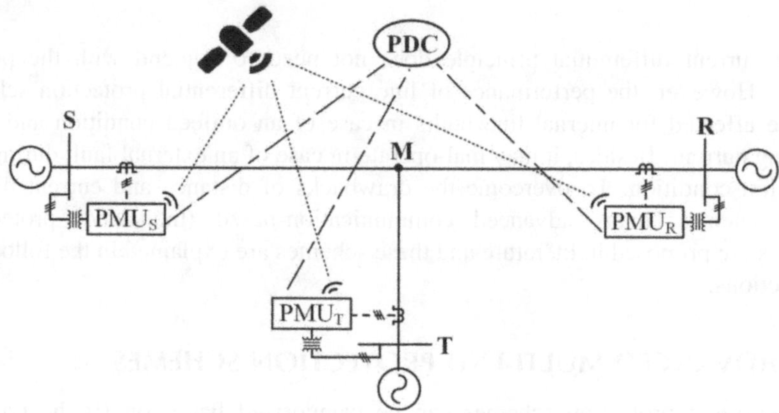

FIGURE 8.8 Three-terminal transmission line equipped with facility of synchronized measurement.

is the most common synchronization mean. GPS satellites are located in such a manner that four or more of them can be seen at any point on the earth. The satellites are supposed to maintain the coordinated universal time with a precision of ±0.5 μs. The accuracy of GPS is usually less than ±0.5 μs, which is equivalent to (±1/40,000) × 360° = ±0.009°) for a fundamental frequency of 50 Hz.

In case of unsynchronized measurement or loss of the GPS signal, the measured data is only tagged with the local time, which is completely asynchronous in nature. Generally, the local time axis, $t^S = 0$ (at the terminal S), $t^T = 0$ (at the terminal T), and $t^R = 0$ (at the terminal R) does not coincide, as shown in Figure 8.9(a). In such a case, the measurement at one end is to be assumed as reference or base, while measurement of the other two ends has to be synchronized analytically. As per the pictorial description given in Figure 8.9, the time axis at terminals R and T has a phase difference with reference to terminal S that is equivalent to time $t_{\delta R}$ (or angle δ_R) and $t_{\delta T}$ (or angle δ_T), respectively. Thus, the two additional unknowns, δ_R and δ_T, appear in unsynchronized measurement compared to synchronized measurement. Here, the

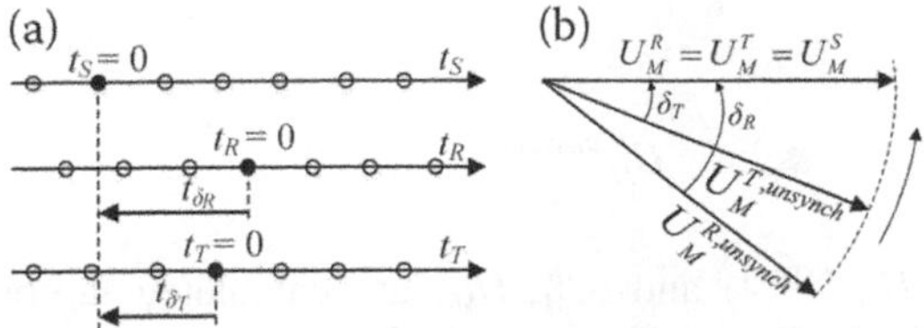

FIGURE 8.9 Diagram showing (a) local time axis at terminal S, R, and T; (b) tap point voltage in respect to synchronized and unsynchronized voltage measurement.

electrical signals of terminals S, R, and T can be expressed as per (8.2)-(8.4), respectively.

$$U_S = |U_S|e^{j\theta_{US}},\ I_S = |I_s|e^{j\theta_{IS}} \tag{8.2}$$

where ($|U_S|$, θ_{US}) and ($|I_S|$, θ_{IS}) are the phasor values (magnitude and phase angle) of voltage and current at terminal S, respectively:

$$\left.\begin{aligned} U_R^{unsynch} &= |U_R|e^{j(\theta_{UR}-\delta_R)} = U_R e^{-j\delta_R} \\ I_R^{unsynch} &= |I_R|e^{j(\theta_{IR}-\delta_R)} = I_R e^{-j\delta_R} \end{aligned}\right\} \tag{8.3}$$

$$\left.\begin{aligned} U_T^{unsynch} &= |U_T|e^{j(\theta_{UT}-\delta_T)} = U_T e^{-j\delta_T} \\ I_T^{unsynch} &= |I_T|e^{j(\theta_{IT}-\delta_T)} = I_T e^{-j\delta_T} \end{aligned}\right\} \tag{8.4}$$

where ($U_R^{unsynch}$, $I_R^{unsynch}$) and (U_R, I_R) are the un-synchronized and synchronized values of voltage and current at terminal R. Similarly, ($U_T^{unsynch}$, $I_T^{unsynch}$) and (U_T, I_T) are the un-synchronized and synchronized values of voltage and current at terminal T. Further, (θ_{UR}, θ_{IR}) and (θ_{UT}, θ_{IT}) are the synchronized values of voltage and current phase angle at terminals R and T, respectively. δ_R and δ_T are the synchronization phase angle operators of un-synchronized measurements at terminal S, corresponding to terminals R and T, respectively.

Synchronization angles of tap-point voltages $U_M^S U_M^{R,unsynch}$ and $U_M^{T,unsynch}$ with reference to terminals S, R, and T, respectively, can be calculated using (8.5), (8.6), and (8.7), correspondingly:

$$U_M^S = \cosh(\gamma_{SM} l_{SM}) U_S - Z_{C,SM} \sinh(\gamma_{SM} l_{SM}) I_S \tag{8.5}$$

where U_M^S is the calculated tap-point voltage corresponding to terminal S:

$$\begin{aligned} U_M^{R,unsynch} &= \cosh(\gamma_{RM} l_{RM}) U_R^{unsynch} - Z_{C,RM} \sinh(\gamma_{RM} l_{RM}) I_R^{unsynch} \\ &= \{\cosh(\gamma_{RM} l_{RM}) U_R - Z_{C,RM} \sinh(\gamma_{RM} l_{RM}) I_R\} e^{-j\delta_R} = U_M^R e^{-j\delta_R} \end{aligned} \tag{8.6}$$

Similarly:

$$U_M^{T,unsynch} = U_M^T e^{-j\delta_T} \tag{8.7}$$

where ($U_M^{R,unsynch}$, $U_M^{T,unsynch}$) and (U_M^R, U_M^T) are calculated tap-point voltage with reference to terminals R and T, respectively, by utilizing unsynchronized and synchronized measurements, in that order.

The U_M^S, U_M^R, and U_M^T would be equal during pre-fault conditions as observed in Figure 8.9(b). Hence, the expression of δ_R and δ_T is derived from (8.5)–(8.7) that is given as (8.8) and (8.9), respectively:

$$\delta_R = -\arg(U_M^{R,unsynch} / U_M^S) \tag{8.8}$$

$$\delta_T = -\arg(U_M^{T,unsynch} / U_M^S) \tag{8.9}$$

8.4.2 Fundamental and Transient Frequency-Based Schemes

The protection algorithms of multi-end transmission lines are classified on the basis of fundamental frequency and transient frequency. Some algorithms utilize only the fundamental frequency component of data, whereas the others utilize the high-frequency component (which is present only during transients) of data.

8.4.2.1 Fundamental Frequency-Based Schemes

The fundamental frequency-based algorithm mostly utilizes the phasor quantity or sequence component of voltage/current data with synchronized or unsynchronized measurement. Figure 8.10 shows SLD of a three-terminal transmission line with hypothetical fault locations. Many fundamental frequency-based algorithms have been proposed in literature for the protection of three-terminal lines. Mainly, the protection scheme of a three-terminal line has three primary tasks: fault detection, faulty section identification, and fault distance calculation. Protection schemes described in [2–10] have utilized several indices to accomplish the above-mentioned tasks. These algorithms assume three hypothetical fault locations, each in separate line sections, i.e. F_S, F_T, or F_R, as shown in Figure 8.10. The distance of each hypothetical fault point from the particular terminals S, T, and R is denoted by d_S, d_T, and d_R, respectively. These three hypotheses are utilized to identify the faulty section. Then, the indices corresponding to faulty section are used to determine the actual fault distance.

It is crucial to determine the faulty section of the line before determining the fault distance. The technique started with a hypothetical fault in line section SM at distance d_S (pu) from terminal S. With reference to Figure 8.10, the three-terminal line is virtually converted into two terminals with tap point as a remote terminal.

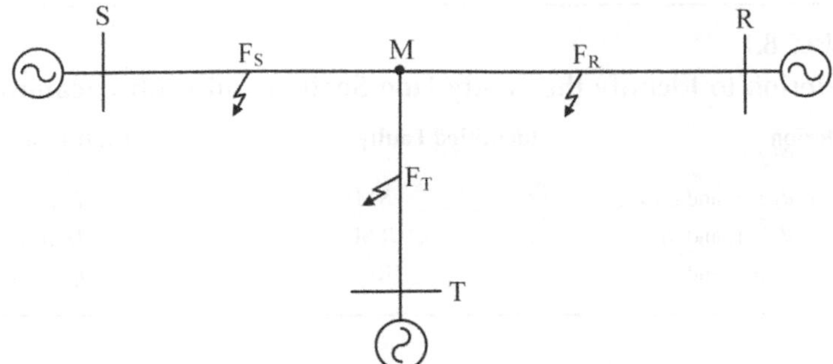

FIGURE 8.10 Hypothetical fault location on three-terminal line.

The voltage and current of the tap point can be calculated with reference to terminals T and R as per (8.10), (8.11):

$$U_M = \cosh(\gamma_{RM} l_{RM}) U_R - Z_{C,RM} \sinh(\gamma_{RM} l_{RM}) I_R \tag{8.10}$$

$$\begin{aligned} I_M = &-\frac{\sinh(\gamma_{RM} l_{RM}) U_R}{Z_{C,RM}} + \cosh(\gamma_{RM} l_{RM}) I_R \\ &- \frac{\sinh(\gamma_{TM} l_{TM}) U_T}{Z_{C,TM}} + \cosh(\gamma_{TM} l_{TM}) I_T \end{aligned} \tag{8.11}$$

Now, the distance (d_S) to hypothetical fault F_S is calculated using signals from terminal S (U_S, I_S) and the tap point (U_M, I_M) obtained after analytical transfer using (8.10) and (8.11).

Then, the voltage equation at fault point F_S and the fault distance (d_s) can be expressed as (8.12) and (8.13), respectively:

$$\begin{aligned} U_{FS} &= \cosh(\gamma_{SM} d_S l_{SM}) U_S - Z_{C,SM} \sinh(\gamma_{SM} d_S l_{SM}) I_S \\ &= \cosh(\gamma_{SM} (1 - d_S) l_{SM}) U_M - Z_{C,SM} \sinh(\gamma_{SM} (1 - d_S) l_{SM}) I_M \end{aligned} \tag{8.12}$$

$$d_S = \frac{\tanh^{-1}\left(\dfrac{U_M \cosh(\gamma_{SM} l_{SM}) - Z_{C,SM} I_M \sinh(\gamma_{SM} l_{SM}) - U_S}{U_M \sinh(\gamma_{SM} l_{SM}) - Z_{C,SM} I_M \cosh(\gamma_{SM} l_{SM}) - Z_{C,SM} I_S} \right)}{\gamma_{SM} l_{SM}} \tag{8.13}$$

Analogous to (8.13), the other two fault indices, d_R and d_T, are for an imaginary fault in the transmission line section RM and TM, respectively, are calculated. The faulty line segment and the fault location are identified based on the criterion shown in Table 8.1.

The previous algorithm was discussed for the three-terminal line. However, in an actual power system, one can find lines with more than one tap. The previously discussed technique requires specific modification before applying to a multi-terminal line. Figure 8.11 shows the five-terminal line that includes three tapped

TABLE 8.1
Criterion to Identify the Faulty Line Section and Fault Location

Criterion	Identified Faulty Line Section	Fault Distance
$d_S < 1, d_T > 1$, and $d_R > 1$	SM	$d_S \times l_{SM}$
$d_S > 1, d_T < 1$, and $d_R > 1$	TM	$d_T \times l_{TM}$
$d_S > 1, d_T > 1$, and $d_R < 1$	RM	$d_R \times l_{RM}$

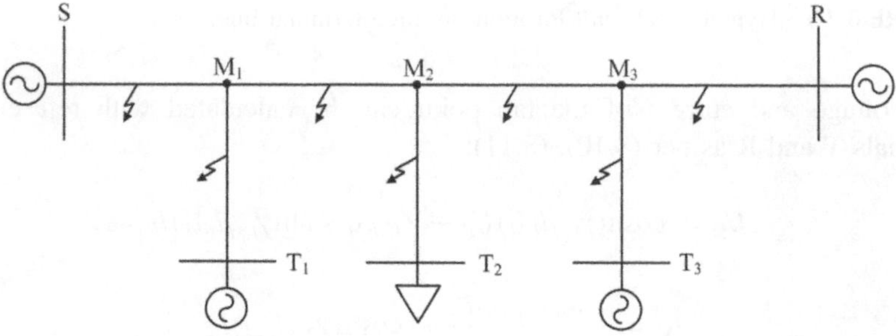

FIGURE 8.11 SLD of a five-terminal transmission line.

lines (T_1M_1, T_2M_2, T_3M_3) and the main line. For the five-terminal line, there may be four hypothetical fault locations on the path of the main line (SM_1, M_1M_2, M_2M_3, M_3R) and three on the tapped lines (T_1M_1, T_2M_2, T_3M_3). Indices corresponding to each hypothetical fault location would be calculated to identify the faulty line segment and correct fault location.

8.4.2.2 Transient Frequency-Based Schemes

Most of the transient frequency-based algorithms utilize the traveling wave principle. Various researchers have presented the different traveling wave-based algorithms for two-terminal transmission lines [11–14]. The techniques mentioned above are further extended by several researchers for faulty section identification and location on a multi-terminal line [15–18]. The high-frequency-based techniques described in [15], [17] depend on the comparison of the measured TT and real TT for traveling along the individual transmission line, whereas reference [18] proposed the FDRM that is used to identify the faulty section in a three-terminal line. For a hypothetical fault at point F_R in Figure 8.10, the term l_{SdR}, which is the fault distance between the local terminal S and fault point F_R, is given by (8.14):

$$l_{SdR} = \frac{l_{SR} + v(t_S - t_R)}{2} = l_{SM} + l_{MdR} \tag{8.14}$$

Hence, the l_{SdR} is the correct fault distance. Here, the term Δt_{dS} is the time taken by the wave to travel through the line F_S-S from the fault point F_S. Terms Δt_{dT}, Δt_{dM}, Δt_{MS}, and Δt_{MR} have similar meanings like Δt_{dS}. Then, the time difference of wave in reaching terminals S and T is given by (8.15):

$$\begin{aligned} t_S - t_T &= \Delta t_{dS} - \Delta t_{dT} = (\Delta t_{dM} + \Delta t_{MS}) - (\Delta t_{dM} + \Delta t_{MT}) \\ &= \Delta t_{MS} - \Delta t_{MT} \end{aligned} \tag{8.15}$$

Hence, the fault distance of line ST when the fault occurs in line section MR is given by (8.16):

$$\begin{aligned} l_{SdT} &= \frac{l_{ST} + v(t_S - t_T)}{2} = \frac{l_{ST} + v(\Delta t_{MS} - \Delta t_{MT})}{2} \\ &= l_{SM} \end{aligned} \tag{8.16}$$

Hence, the tap point M can be called a pseudo-fault point, while terminals S and T are the local and remote terminals, respectively. In this case, the traveling wave from F_R refracts towards the tap point. Other fault distances, i.e. l_{RdS}, l_{RdT}, l_{TdS}, and l_{TdR}, are derived in the same manner to l_{SdR} and l_{SdT}.

The FDRM is used in [18] to identify the faulty line section. The elements of FDRM for the three-terminal transmission lines are expressed in (8.17), where the symbol "x" denotes the element that does not exist.

$$\begin{bmatrix} & S & T & R \\ S & \times & l_{SdT}/l_{SM} & l_{SdR}/l_{SM} \\ T & l_{TdS}/l_{TM} & \times & l_{TdR}/l_{TM} \\ R & l_{RdS}/l_{RM} & l_{RdT}/l_{RM} & \times \end{bmatrix} \tag{8.17}$$

According to the (8.14) and (8.16), the following relationship is obtained for fault in the line section MR:

$$l_{SdR}/l_{SM} > 1$$

$$l_{SdT}/l_{SM} = 1$$

In this way, the elements of FDRM are obtained as (8.18):

$$\begin{bmatrix} & S & T & R \\ S & \times & =1 & >1 \\ T & =1 & \times & >1 \\ R & <1 & <1 & \times \end{bmatrix} \tag{8.18}$$

In (8.18), except for the diagonal elements, all the elements of row R are less than 1, whereas the elements of column R are greater than 1. Here, it is to be noted that this is the case for a fault in the line section MR. If the fault is in another line section (MS or MT) instead of MR, the same FDRM matrix is calculated to identify the faulty section. After the calculation of FDRM, the following rules are utilized to identify the faulty section. If all the elements of one row are less than one and greater than one for the elements of a column belonging to the same terminal, except the diagonal one, then the fault is identified in the branch connected to the corresponding terminal. For a fault at the tap point, all the elements of FDRM are one except the diagonal.

This method can be extended for a multi-terminal line having more than three terminals. For example, the FDRM for a five-terminal line can be derived and shown in (8.19):

$$\begin{bmatrix} & S & T_1 & T_2 & T_3 & R \\ S & \times & l_{SdT_1}/l_{SM_1} & l_{SdT_2}/l_{SM_1} & l_{SdT_3}/l_{SM_1} & l_{SdR}/l_{SM_1} \\ T_1 & l_{T_1dS}/l_{T_1M_1} & \times & l_{T_1dT_2}/l_{T_1M_1} & l_{T_1dT_3}/l_{T_1M_1} & l_{T_1dR}/l_{T_1M_1} \\ T_2 & l_{T_2dS}/l_{T_2M_2} & l_{T_2dT_1}/l_{T_2M_2} & \times & l_{T_2dT3}/l_{T_1M_1} & l_{T_2dR}/l_{T_2M_2} \\ T_3 & l_{T_3dS}/l_{T_3M_3} & l_{T_3dT_1}/l_{T_3M_3} & l_{T_3dT_2}/l_{T_3M_3} & \times & l_{T_3dR}/l_{T_3M_3} \\ R & l_{RdS}/l_{RM_3} & l_{RdT_1}/l_{RM_3} & l_{RdT_2}/l_{RM_3} & l_{RdT_3}/l_{RM_3} & \times \end{bmatrix} \tag{8.19}$$

8.5 AI OR KNOWLEDGE-BASED SCHEMES

In traditional hard computing protection schemes, the accuracy is limited due to signal processing, measurement, and uncertainty. Such protection schemes are derived under some assumptions that are far from facts in certain critical conditions. The intelligent system recognizes the incorporation of logic, reasoning, and decision-making as main associates joined to establish a trade-off basis between uncertainty and precision. Improvement in technology brings definite advancements in the automation of substation. It is possible to achieve faster computation of the criteria signals and precise input signals after proper filtering. Standardized hardware with communication capability made it easy to utilize the synchronized measurements. From a wide perspective, the intelligent system underlies as soft computing.

AI explores how machines can mimic human beings' thoughts and actions in the numeric, non-numeric, and symbolic calculations. The artificial intelligence comprises making rational decisions and dealing with absent/lost data, adjusting to present situations, and improving itself on a long-time scale by utilizing the gained experience.

There have been numerous research activities going on in educational universities, industries, and research institutes on AI-based protection algorithms for different transmission lines configurations. However, very few research articles are available for a multi-terminal line.

8.5.1 ANN-Based Schemes

ANNs are biologically inspired and contain organized elements related to the human brain's anatomy. These organized elements are called neurons, and they have the capability of realizing fundamental logic functions. Various models are developed by using ANNs for applications in different fields, including fault identification in transmission lines. In a multi-terminal transmission line, ANN can be used for fault-type classification and faulty section identification.

8.5.2 Fuzzy Interference Systems

FISs use fuzzy logic to perform inference operations. In fuzzy logic, the truth's degree is indicated by any value in the range [0, 1]. Here, the value zero "0" represents the absolute falsehood and one "1" meaning truth. The FIS has three stages: the fuzzification stage, the inference stage, and the defuzzification stage [19]. In the fuzzification stage, various existing membership functions are used to assign membership degrees to the inputs. Further, in the inference stage, several if-then rules are applied to the first stage's fuzzified inputs. The last and third stage is defuzzification that offers the final decision based on second (inference) stage results, such as the input classification decision.

8.5.3 Support Vector Machine-Based Schemes

SVM is a popular supervised learning algorithm that analyzes the data used for classification and regression analysis [20], [21]. The SVM algorithm consists of a decision boundary or line that produces several classes segregated from an n-dimensional space. These lines or boundaries are called a hyperplane. The SVM parameter optimization process is a convex optimization challenge due to which falling into local optima can be avoided. Due to these properties, SVM is used for fault detection on a transmission line, classifying the fault type, and fault location identification.

Many researchers have proposed SVM-based hybrid algorithms for the protection of the two-terminal line, whereas very little work has been done related to the protection of three or multi-terminal lines using SVM. Ref [22] presents an SVM classifier-based algorithm for fault detection, identification of fault type, faulty line section, and identification of faulty half side of the section. In this SVM classifier-based algorithm, the wavelet energy of voltage transients is calculated, and its normalized form is used as an input to the SVM classifier.

Fault-type classification is performed using four binary SVM_i ($i = 1, \ldots, 4$). The SVMs are trained to identify the faulty phase. SVM_1, SVM_2, and SVM_3 are trained to detect the fault in phase a, b, and c, respectively, whereas SVM_4 is trained to distinguish ground or phase fault. The SVM_i ($i = 1, \ldots, 4$) value remains either +1 or −1. For example, in case of a grounded fault on phase a, the value of SVM_1 and SVM_4 is +1, while the values of SVM_2 and SVM_3 are −1.

Firstly, a range of fault conditions is used to train the SVM classifiers for a given topology and the efficacy of the classifiers is checked for separate fault scenarios. The input of the SVM classifiers is the normalized wavelet energy of postfault three-phase and zero-mode transient voltages. The suitability of different mother wavelets, i.e. Daubechies-4 (db-4), db-8, and Meyer, are checked for the SVM classification. As the classification accuracy of db-4 mother wavelet is the highest among other mother wavelets, it is used as the mother wavelet [22]. Further, it is commonly accepted wavelets in the literature. This algorithm follows the following steps.

1. The first step is to calculate the aerial and zero-mode voltages using Clarke's modal transformation.
2. DWT is applied to phase voltages (V_a, V_b, V_c) and zero-mode voltage (V_0) to calculate the scale-2 *WTC*s. The squared value of *WTC*s are then used to detect the wave instants and designated as WTC^2s.
3. The wavelet energy of voltages i.e. E_{Vk} ($k \in \{a, b, c,$ and $0\}$) is computed by summation of WTC^2 over one cycle after the detection of fault as per (8.20).

$$E_{N_{Vk}} = \sum_{m=0}^{M-1} WTC_k^2(m) \text{ for } k \in \{a, b, c \text{ and } 0\} \tag{8.20}$$

where M is the number of samples per cycle.

4. The wavelet energies are normalized according to (8.21):

$$E_{N_{Vk}} = \frac{E_{Vk}}{E_{Va} + E_{Vb} + E_{Vc} + E_{V0}} \text{for } k \in \{a, b, c \text{ and } 0\} \tag{8.21}$$

In order to achieve optimal decision functions, the training of the classifiers has to be carried out. The input features $E_{N_{Vk}}$ are calculated using terminals S and R voltages. The training matrix "$N \times 8$" stores the processed values in which each column represents a feature and each row represents a training sample. Here, "N" is the maximum possible fault conditions. These fault conditions have been created by simulating faults with the variation in fault location, fault type, FIAs, and R_F. Figure 8.12 shows a flowchart of this algorithm. Similarly, Figure 8.13 shows the optimal separating hyperplane of the SVM algorithm in a two-dimensional (2-D) feature space. Similar to this 2-D featured space, the training of the SVM classifiers is carried out with the help of a training matrix that corresponds to an 8-D feature space. After completion of this training process for each phase, the required decision-making features for the two-class separation are identified. Therefore, the SVMs are trained to classify any new set of incoming data.

After the fault type has been classified, two SVM classifiers are utilized to determine the faulty line. Then, the task of identification of faulty-half line section is carried out using SVM. Aerial mode component in scale-2 is eventually observed to determine the traveling-wave arrival time corresponding to the peaks of WTC^2s. Then, the fault location is determined using time differences between consecutive waves.

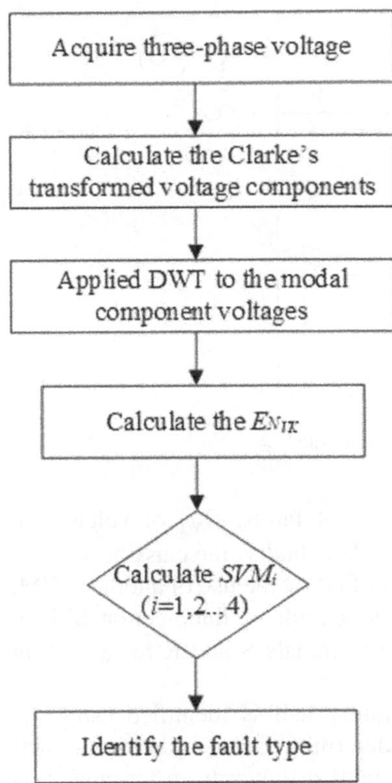

FIGURE 8.12 Flowchart of SVM-based algorithm for multi-terminal line.

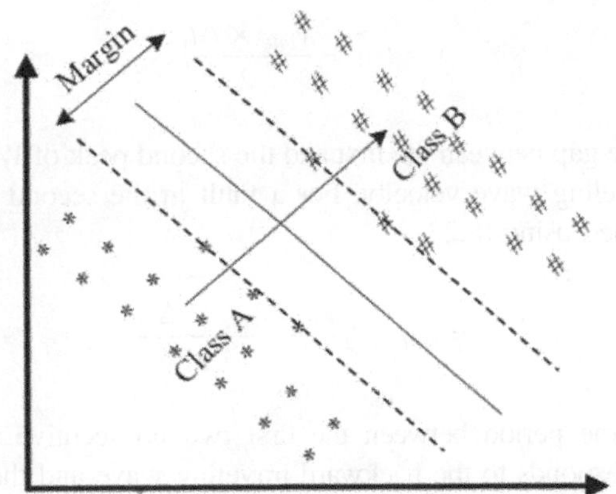

FIGURE 8.13 2D feature space with the optimal separating hyperplane.

Faulty line identification involves two steps. The two binary SVMs (SVM_n, $n = 1, 2$) are trained to identify the line section. SVM_1 is trained to detect the faults in line section SM, and SVM_2 for line section RM. If the value of both SVMs is –1, then the fault is in line section TM.

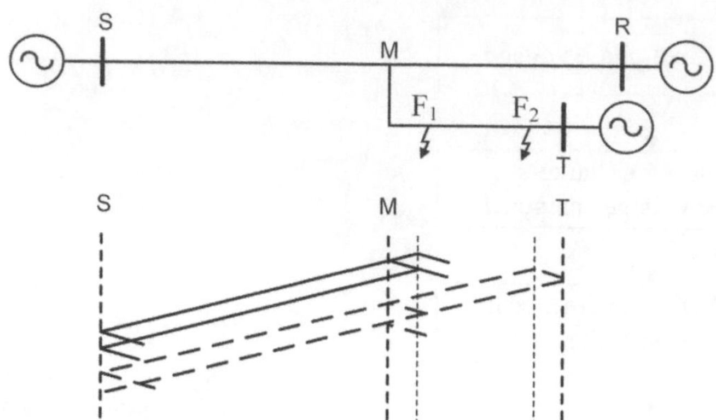

FIGURE 8.14 Bewley lattice diagram for three-terminal line.

Individual SVM is designed for different types of faults. $E_{N_{Vk}}$ of voltages at terminals S and R are processed as inputs of both SVM faulty line classifiers. After the faulty line is determined, the faulty half is identified by the use of another SVM, which is trained by using $E_{N_{Vk}}$ at terminal S for a fault in line section SM, at terminal R for a fault in line section RM, and at terminals S and R for a fault in line TM.

For the faults in line SM or RM, first the faulty half is identified using the corresponding SVMs, and then fault-location is determined. For a fault in the first half of the line TM, i.e. F_1 in Figure 8.14, the fault distance from terminal S is calculated using (8.22).

$$x = l_{SM} + \frac{v_{line} \times \Delta t_1}{2} \tag{8.22}$$

where Δt_1 is the time gap between the first and the second peak of WTC^2s at terminal S. The v_{line} is the traveling wave velocity. For a fault in the second half (F_2), the fault distance is determined using (8.23):

$$x = l_{SM} + l_{TM} - \frac{v_{line} \times \Delta t_2}{2} \tag{8.23}$$

where Δt_2 is the time period between the first two consecutive peaks of WTC^2 at terminal S that corresponds to the backward traveling wave and the reflected forward traveling wave from terminal T, respectively.

Here, the Gaussian RBF is considered for training and testing of SVM classifiers. The kernel function parameter (γ) is adjusted to obtain the high accuracy for each SVM. Generally, the value of γ is tuned by varying in range of 0.1 to 10 in steps of 0.1 and the optimized value is obtained using heuristic method. However, the value of γ can be different for different fault types and other conditions.

8.6 ADAPTIVE PROTECTION SCHEMES

To avoid problems such as overreaching and underreaching due to multiple infeed and high-fault resistance in a multi-terminal line, the use of adaptive relaying is increasing day by day. Adaptive relaying offers the following benefits:

1. Reduced numbers of mal-operations.
2. Enhances the relaying reliability.
3. Offers greater sensitivity during high resistance faults.

An example of adaptive DUTT scheme is discussed here for a three-terminal line that adaptively modifies the zone-1 setting of the distance relay installed at all the three terminals, as shown in Figure 8.15.

In Figure 8.15, the relays R_S, R_T, and R_R are located at terminals S, T, and R, respectively. The zone-1 of the R_S and R_T relay is set up to 80% of the impedance of the line segment connected between terminals S and T. Thus, the zone-1 of R_S relay covers up to point A for line segment MT and up to point B for line segment MR, respectively. Conversely, the zone-1 of the R_T relay covers up to point C for line segment MS and up to point D for line segment MR, respectively. Similarly, the zone-1 reach of the relay R_R is set to 80% of the line section's impedance connected between terminals R and T (up to point E for line section MR). The overlapping of zone-1 of three relays is also shown in Figure 8.15. The impedance corresponding to overlapping of reach between T and R relays (Z_{OT}), and between S and R relays (Z_{OS}) are given by (8.24) and (8.25), respectively:

$$Z_{OT} = kZ_{ST} - (Z_{TM} + Z_{ME}) \tag{8.24}$$

$$Z_{OS} = kZ_{ST} - (Z_{SM} + Z_{ME}) \tag{8.25}$$

where $k = 0.8$ or 80%.

For perfect DUTT operation, zone-1 reach of all the relays should be overlapped to each other. Here, the infeed current is the major cause of reduction of zone-1 reach of relays and hence, reduces the effective overlapping in between the zone-1

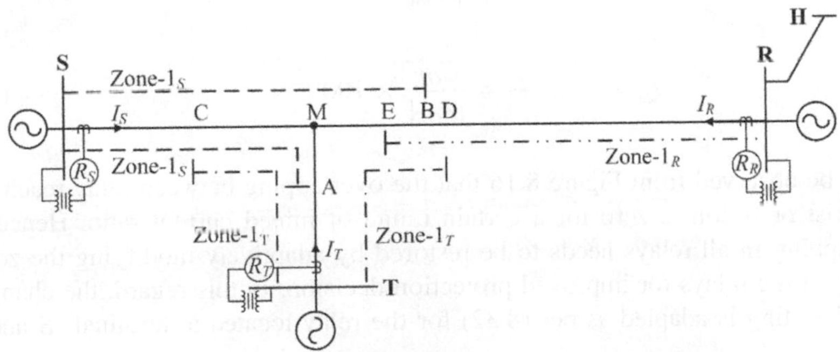

FIGURE 8.15 SLD for three-terminal line indicating reach distance relay.

reach. The higher the value of infeed current, the more under-reach problem in the relay. The three-phase fault draws maximum current in comparison to any other type of fault at the same location. Hence, maintaining the overlap of zone-1 reach for a three-phase fault would automatically ensure for other types of faults. The impedance measured by the relay at terminal S for a three-phase fault at point E is given by (8.26):

$$Z_{seen} = Z_{SM} + \left(1 + \frac{I_T}{I_S}\right) Z_{ME} \tag{8.26}$$

Thus, the error caused by the infeed current is given by (8.27):

$$Z_{error} = \frac{I_T}{I_S} Z_{ME} \tag{8.27}$$

The term Z_{error} is related to underreach or decrement in overlapping between the relay located at terminal S and R. Overlapping of relay setting ensures detection of any fault by at least one of the relay. The amount of effective overlapping present between the relay settings of S and R in the presence of infeed current can be formulated as (8.28) and (8.29), respectively:

$$C_{OS} = |Z_{OS}| - \left| \frac{|I_T|}{|I_S|} Z_{ME} \right| \tag{8.28}$$

$$C_{OT} = |Z_{OT}| - \left| \frac{|I_S|}{|I_T|} Z_{ME} \right| \tag{8.29}$$

Figure 8.16 shows the effective zone overlapping (in %) of relays at terminals S and T with reference to zone setting of a relay at terminal R, as per (8.30) and (8.31):

$$k_S = \frac{C_{OS}}{|Z_{OS}|} \times 100 \tag{8.30}$$

$$k_T = \frac{C_{OT}}{|Z_{OT}|} \times 100 \tag{8.31}$$

It can be observed from Figure 8.16 that the overlapping between relay reach does not exist or becomes zero for a certain range of infeed current ratio. Hence, the overlapping in all relays needs to be restored by adaptively modifying the zone-1 setting of the relays for improved protection decision. In this regard, the change in zone-1 setting is adapted as per (8.32) for the relay located at terminals S and T:

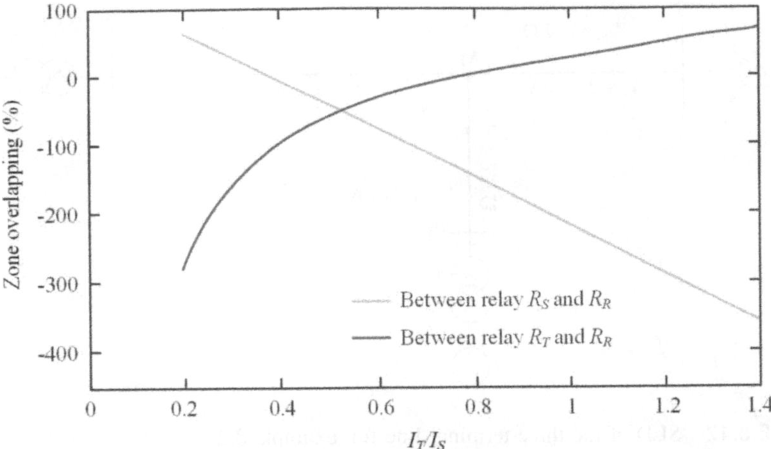

FIGURE 8.16 Variation in the overlapping of relay reach against ratio of infeed currents.

$$\Delta Z_S = \left|\frac{I_T}{I_S}\right| Z_{ME} \text{ and } \Delta Z_T = \left|\frac{I_S}{I_T}\right| Z_{ME} \tag{8.32}$$

Example 8.2: Figure 8.17 shows a SLD of a 100-kV three-terminal transmission line. The current distribution for a three-phase fault and the line impedances for each line section are mentioned in Figure 8.17.

a. Calculate the overlapping zone between the relays installed under DUTT scheme.
b. Determine the effective or actual overlapping of zone settings in the presence of infeed current for the fault case shown in Figure 8.17.

Solution:

a. The zone setting of the relays installed at terminals S, T, and R is given by:

$$\text{Zone}-1_S = k \times Z_{ST} = 0.8 \times (Z_{SM} + Z_{MT}) = 0.8 \times (5.7 + 2.6) = 6.64\ \Omega$$

$$\text{Zone}-1_T = 0.8 \times (Z_{MT} + Z_{SM}) = 0.8 \times (2.6 + 5.7) = 6.64\ \Omega$$

$$\text{Zone}-1_R = 0.8 \times (Z_{MR} + Z_{MT}) = 0.8 \times (12 + 2.6) = 11.68\ \Omega$$

The obtained zone setting boundaries are shown in Figure 8.18. It is observed from Figure 8.18 that the reach of the relays located at terminals S and T reaches (beyond the tap point M) up to points B and D, respectively, whereas the reach of the relay located at terminal R remains up to point E.

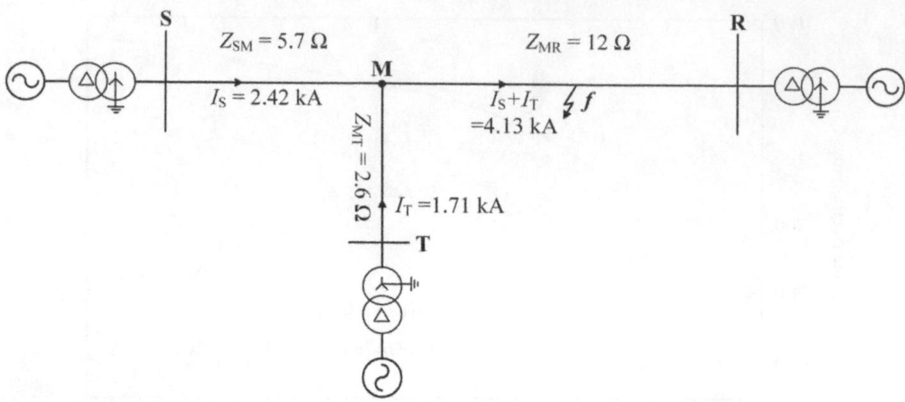

FIGURE 8.17 SLD of the three-terminal line for example 8.2.

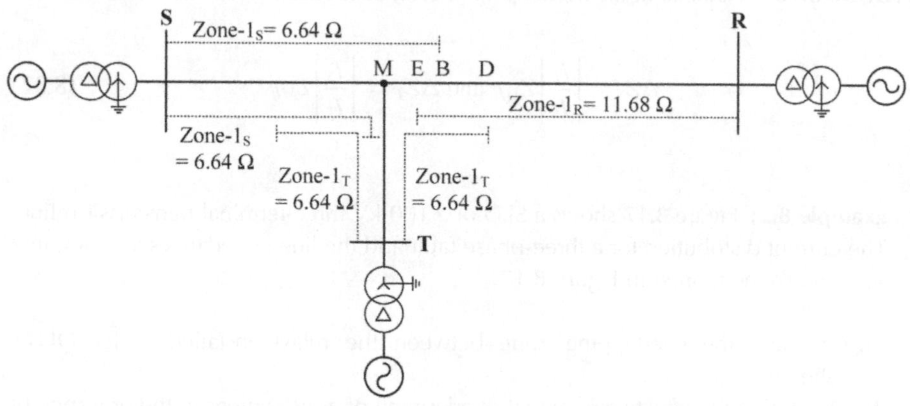

FIGURE 8.18 SLD of a three-terminal line showing overlapping.

The zone overlapping (Z_{OT}) between relay at terminals T and R is calculated as follows:

$$Z_{OT} = k \times Z_{ST} - (Z_{MT} + Z_{ME}) = 6.64 - (2.6 + 0.32) = 3.72\ \Omega$$

Similarly, $Z_{OS} = k \times Z_{ST} - (Z_{MS} + Z_{ME}) = 6.64 - (5.7 + 0.32) = 0.62\ \Omega$

b. $C_{OS} = |Z_{OS}| - \left|\frac{|I_T|}{|I_S|} Z_{ME}\right| = 0.62 - \frac{1.71}{2.42} \times 0.32 = 0.3938\ \Omega$

$$C_{OT} = |Z_{OT}| - \left|\frac{|I_S|}{|I_T|} Z_{ME}\right| = 3.72 - \frac{2.42}{1.71} \times 0.32 = 3.2671\ \Omega$$

Hence, the actual zone overlapping of the relay installed at terminals S and R is given by:

$$k_S = \frac{C_{OS}}{|Z_{OS}|} \times 100 = \frac{0.3938}{0.62} \times 100 = 63.51\%$$

$$k_T = \frac{C_{OT}}{|Z_{OT}|} \times 100 = \frac{3.2671}{3.72} \times 100 = 87.82\%$$

8.7 CONCLUSION

In this chapter, application of different AI techniques such as ANN, Fuzzy logic, SVM, and a combination of AI-based techniques with other transient tools like Wavelet for the protection of a multi-terminal transmission line is discussed. Primarily, these techniques develop a statistical model and train it with a database representing a wide variation of faults and other abnormalities. The major task of these techniques for three-terminal line protection includes fault detection, faulty section identification/classification, and fault location.

REFERENCES

[1] S. Sarangi and A. K. Pradhan, "Adaptive direct under-reaching transfer trip protection scheme for the three-terminal line," *IEEE Trans. Power Del.*, vol. 30, no. 6, pp. 2383–2391, Dec. 2015.

[2] R. K. Aggarwal, D. V. Coury, A. T. Johns, A. Kalam, "A practical approach to accurate fault location on extra high voltage teed feeders," *IEEE Trans. Power Del.*, vol. 8, no. 3, 874–883, 1993.

[3] A. A. Girgis, D. G. Hart, and W. L. Peterson, "A new fault location technique for two- and three-terminal lines," *IEEE Trans. Power Del.*, vol. 7, no. 1, 98–107, 1992.

[4] C. W. Liu, K. P. Lien, Ching-Shan Chen, and Joe-Air Jiang, "A universal fault location technique for n-terminal (N ≥ 3) transmission lines," *IEEE Trans. Power Del.*, vol. 23, no. 3, pp. 1366–1373, 2008.

[5] Y. H. Lin, C. W. Liu, and C. S. Yu, "A new fault locator for three-terminal transmission lines using two-terminal synchronized voltage and current phasors," *IEEE Trans. Power Del.*, vol. 17, no. 2, pp. 452–459, 2002.

[6] J. Izykowski, E. Rosolowski, M. M. Saha, M. Fulczyk, and Przemyslaw, "A fault location method for application with current differential relays of three-terminal lines," *IEEE Trans. Power Del.*, vol. 22, no. 4, pp. 2099–2107, 2007.

[7] V. Gaur, B. Bhalja, and M. Kezunovic, "Novel fault distance estimation method for three-terminal transmission line," *IEEE Trans. Power Del*, vol. 36, no. 1, pp. 406–417.

[8] P. Cao, Hongchun Shu, B. Yang, Y. Fang, Y. Han, J. Dong, and T. Yu, "Asynchronous fault location scheme based on voltage distribution for three-terminal transmission lines," *IEEE Trans. Power Del.*, vol. 35, no. 5, pp. 2530–2540, October 2020.

[9] M. A. Elsadd, and A. Y. Abdelaziz, "Unsynchronized fault-location technique for two- and three-terminal transmission lines," *Elect. Power Sys. Res.*, vol. 158, pp. 228–239, 2018.

[10] S. M. Brahma, "Fault location scheme for a multi-terminal transmission line using synchronized voltage measurements," *IEEE Trans. Power Del.*, vol. 20, no. 2, pp. 1325–1331, 2005.
[11] I. Niazy and J. Sadeh, "A new single ended fault location algorithm for combined transmission line considering fault clearing transients without using line parameters," *Elect. Power Energy Sys.*, vol. 44, no. 1, pp. 816–823, 2013.
[12] O. Naidu and A. K. Pradhan, "A traveling wave-based fault location method using unsynchronized current measurements," *IEEE Trans. Power Del.*, vol. 34, no. 2, pp. 505–513, April 2019.
[13] B. Sahoo and S. R. Samantaray, "An enhanced travelling wave-based fault detection and location estimation technique for series compensated transmission network," 7th International Conference on Power Systems (ICPS), Pune, India, 21–23 December 2017, pp. 61–68.
[14] B. Sahoo and S. R. Samantaray, "An enhanced fault detection and location estimation method for TCSC compensated line connecting wind farm," *Elect. Power Energy Sys.*, vol. 96, pp. 432–441, 2018.
[15] R. J. Hamidi and H. Livani, "Traveling-wave-based fault-location algorithm for hybrid multi-terminal circuits," *IEEE Trans. Power Del.*, vol. 32, no. 1, pp. 135–144, Feb. 2017.
[16] J. Ding, X. Wang, Y. Zheng, and L. Li, "Distributed traveling-wave based fault-location algorithm embedded in multi-terminal transmission lines," *IEEE Trans. Power Del.*, vol. 33, no. 6, pp. 3045–3054, Dec. 2018.
[17] B. K. Chaitanya, and Anamika Yadav, "Decision tree aided travelling wave based fault section identification and location scheme for multi-terminal transmission lines," *Measurement*, vol. 135, pp. 312–322, 2019.
[18] Z. Yongli and F. Xinqiao, "Fault location scheme for a multi-terminal transmission line based on current traveling waves," *Electr. Power Energy Sys.*, vol. 53, pp. 367–374, 2013.
[19] H.-J. Zimmermann, *Fuzzy sets, decision making, and expert systems*. Boston, MA, USA: Kluwer, 1987.
[20] V. Vapnik, *Statistical learning theory*. New York: Wiley, 1998.
[21] N. Cristianini and J. Shawe-Taylor, *An introduction to support vector machines and other kernel-based learning methods*. Cambridge, UK: Cambridge University Press, 2000.
[22] H. Livani, and C. Y. Evrenosoglu, "A fault classification and localisation method for three terminal circuits using machine learning," *IEEE Trans. Power Del.*, vol. 28, no. 4, pp. 2282–2290, 2013.

9 Data Mining-Based Protection Methodologies for Series Compensated Transmission Network

S. K. Singh[1], *D. N. Vishwakarma*[2], and *R. K. Saket*[2]

[1]Department of Electrical Engineering, Shri Ramswroop Memorial University, Lucknow, India

[2]Department of Electrical Engineering, Indian Institute of Technology, Varanasi, India

9.1 INTRODUCTION

The increasing dependency on electrical energy for daily life style and transportation in the current scenario consequently leads to ever growing requisition of electrical power supply. In order to preserve the balance between generation and increasing power demand, newer generating plants along with renewable energy sources are being gradually added in the modern power network. However, the addition of newer generating units has little significance if there is no competent transmission system for transporting the energy from the generating plants to the load centers. It creates a situation where there is a surplus power generation but still not able to fulfill the energy demands due to the power transfer limiting constraints of the network. It draws the attention of power engineers towards the amelioration and up-gradation of the extant power network for enhancing the power transfer competency. For reducing the intricacy of the grid network, instead of fabricating new lines more emphasis has been given to enhance the transfer capability of extant transmission system. The power-carrying competency of the line can be readily improved by reducing the overall line reactance as compared to other factors like voltage level variation which needs redesigning of network insulation; similarly, excessive variation of δ directly affects stability of the system. The line compensation technique has considerably changed the scenario, as it effectively helps in amplifying the power flow limits with better voltage control of the system. It is an excellent alternative over other reinforcement ways (like diffuseness of transmission network) for fulfilling the continuous growing requisition of power in the current scenario. Fixed series capacitor (FSC) or FACTS (flexible AC transmission systems) controller installation in the network offers an effectual way for adjusting

DOI: 10.1201/9780367552374-9

the line flow limit along with preferable network stability control. FACTS controller like TCSC (Thyristor controlled series capacitor) endowing variable level of line compensation and helps in slacking the inter-area oscillations with dynamic power flow control. There are many benefits of line compensation, on account of distance relaying the incorporation of series capacitor/FACTS device creates additional protection challenges. Over the past few decades, numerous articles have been presented for the protection of compensated power transmission network. However, the cases such as transforming faults, cross-country events, impact of CTs saturation, etc. are still almost untouched.

This chapter thoroughly addresses the critical protection challenges evolved due to incorporation of line compensation mechanism. In addition it describes multiple data mining–based protection methodologies for compensated transmission network that are well competent of dealing all kinds of shunt fault events and transforming events in the transmission system. These protective schemes are effectual in providing accurate output despite of changes in fault conditions such as varying fault inception angles, fault locations in the network, fault resistances, percentage of line compensation, and different topologies of transmission system.

9.2 RELAYING CHALLENGES IN SERIES COMPENSATED TRANSMISSION NETWORK

Albeit of significant benefits of compensating device (CD), its presence in the faulted network and the nonlinear behavior of associated safety circuit adversely influence the functioning of lineal distance relaying [1,2]. The erratic operating modes of line compensating units during abnormalities in the power network perverted the current and voltage signals. It creates additional critical challenges for the relaying system. Authors in [3–5] have discussed that the line compensation causes erratic tripping of the switchgear system. It has been suggested that the protection zone boundaries must be adopted according to the modified impedance due to the involvement of the compensating device in the faulted network. Authors in [6] have explained the variation of modified impedance with the connected compensated series reactance in the transmission network. Similarly, in [7,8] it has been reported that the relay zone settings are directly incumbent on the reactance and extinction angle of the compensating devices. Current inversion (CI), voltage inversion (VI), underreach, overreach, and additional transient harmonics are the prime unsuited phenomena that arise due to line compensation.

9.2.1 Under- and Overreaching of Relays

In case of compensated power network, the incorporation of the compensating unit comprehensively transformed the impedance observed by the relaying system [9]. Figure 9.1 clearly shows that due to the incorporation of compensating unit, the observed impedance got modified as represented by solid lines. It usually causes malfunctioning of the relaying system and hence creates undesirable tripping of the breakers.

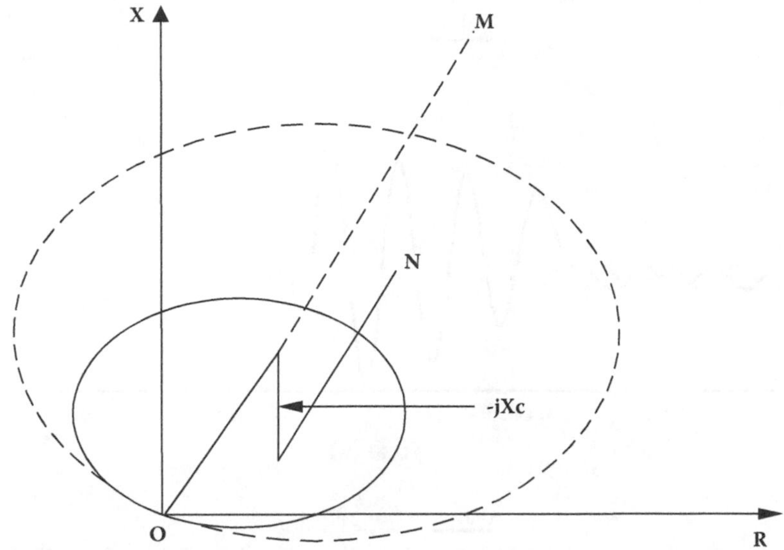

FIGURE 9.1 Modification in seen impedance due to CD in the transmission network.

9.2.2 Current and Voltage Inversion

Current inversion usually causes a significant variation in current phase angle (90° or more). It has been observed in a compensated network if a shunt fault is actuated just after the CD, then the equivalent circuit along one side of the fault is capacitive and at the other side of the fault circuit is inductive. However, it is uncommon during high-current events, as CD will be bypassed by the spark gap or Metal Oxide Varistor (MOV) under such events. It is commonly observed during high-resistance faults. The CI case shown in Figure 9.2 is observed during a line to ground fault event close to a compensating unit. Similarly, voltage inversion transformed the voltage phase angle by 90° or more. It has been observed if the net impedance from the supply source to the events location is inductive and the net impedance between relaying points to the location of fault is capacitive. CI and VI significantly enhance the complexity in directionality discrimination of the fault events [10–12]. In [13] the authors have explained the directionality issues that arise in a TCSC-compensated power transmission network.

9.2.3 Precarious Operation of MOV

The functioning of MOV unit during the fault occurrence in the network is erratic in nature which directly affects the distance relaying mechanism. In case of high instantaneous voltage, the capacitor is bypassed as the MOV starts conducting and as a result of which the net seen impedance will be the impedance of MOV. However, in case of low-current events, the seen impedance is the sum of the series capacitor and MOV as during such events MOV remains in

(a)

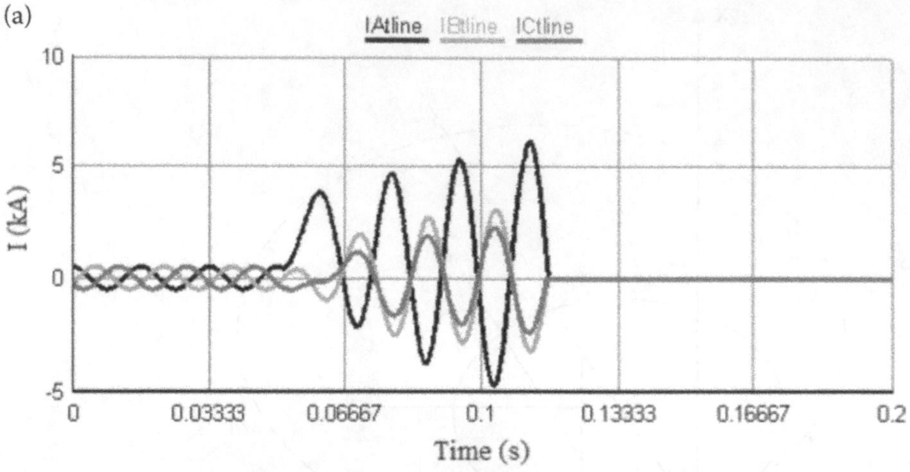

(b)

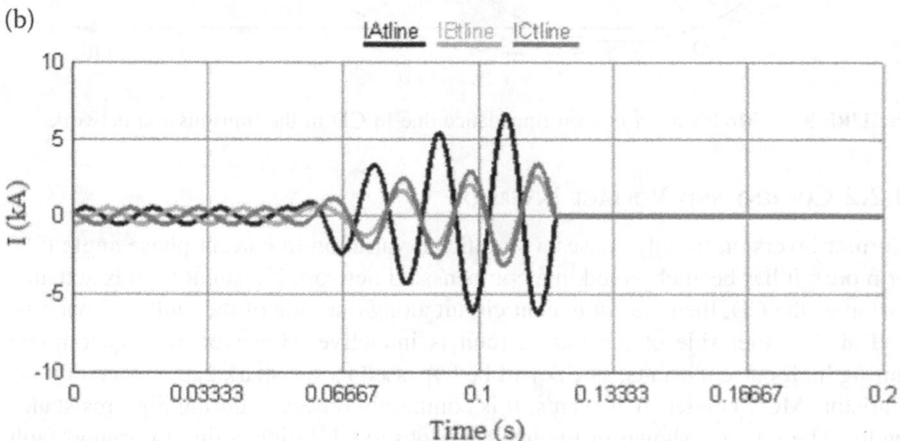

FIGURE 9.2 A case of current inversion due to CD in the transmission network; (a) A-G fault ahead of CD; (b) A-G fault immediately upon CD.

non-conducting mode. Hence, the relaying system faces overreaching and underreaching behaviors if the aforementioned situations are not considered in a relay setting. Authors in [14–19] have described the effects of modified impedance seen by the relaying circuit due to compensating devices in the faulted network.

9.2.4 Harmonics and Transients

The presence of CD in the power lines causes the addition of multiple transients and subharmonics in the grid. It comprehensively affects the exact assessment of voltage and current. Consequently, the seen current magnitude will be more than the actual. Hence overreaching issues arise.

9.3 DATA MINING–BASED PROTECTION MECHANISM

The data mining mechanism simply helps in identifying the patterns and relationship among the large data set. The acquired information or pattern from data mining mechanism can be efficiently used for making the decision regarding the type of particular fault events or its position in the transmission network. In the proposed protection methodologies, critical patterns in the post fault current samples has been identified after pre-processing the three-phase current samples and are utilized in machine learning (ML)–based models for ascertaining the categories and location of the faults in the network. Three different Non-parametric ML (K-nearest neighbor (KNN), Support Vector machine (SVM), and Probabilistic Neural Network (PNN))-based methodologies are explained for the protection of a series compensated power transmission network. Figure 9.3 shows the brief strategy of the proposed protection methodologies for compensated power network.

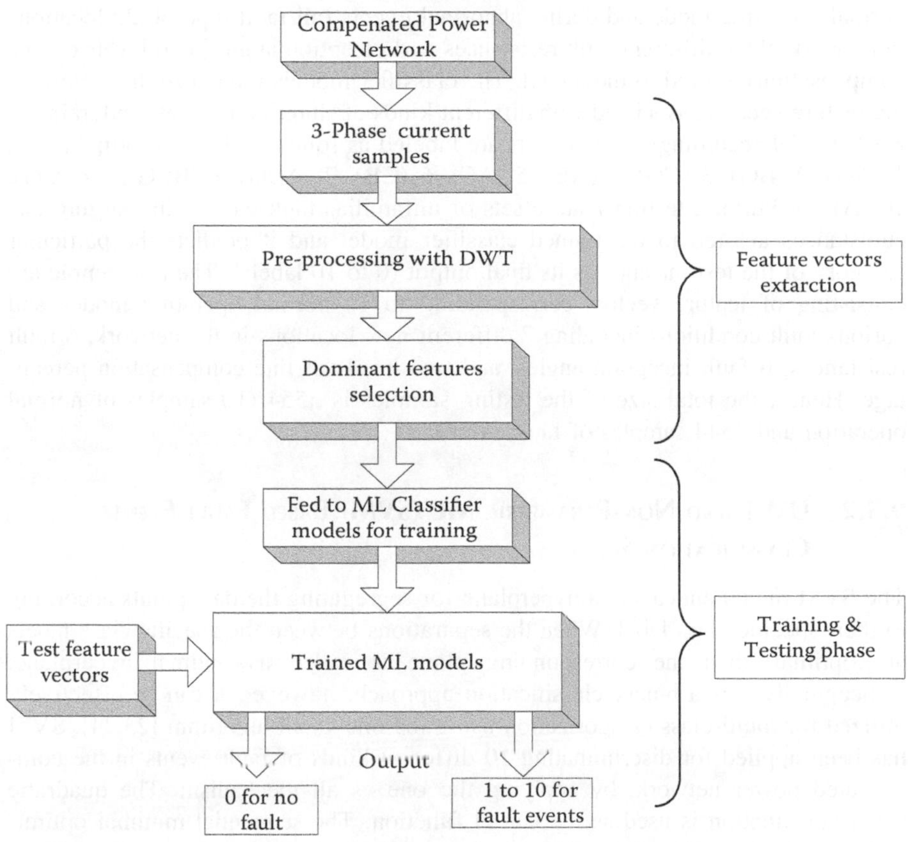

FIGURE 9.3 Flowchart of the proposed fault events classification scheme.

9.3.1 DWT and Non-Parametric ML (KNN) Based Fault Events Classification Scheme

The non-parametric algorithms are also termed instance-based learning algorithms, since they accumulate the training feature sets in a lookup table and interpolate from those. The KNN algorithm directly estimates the location of the unfamiliar test samples from the trained pattern set. The category of the test samples has been decided by the majority votes of its closest neighbors [20–22]. The data mining mechanism has been applied for identifying the feature pattern of the fault current signal. Afterward, the ascertained dominant features are applied to the KNN classifier models for categorizing the fault events in the network. First of all, the retrieved three-phase post-fault current samples are segmented into different levels using Db5 mother wavelet. Consequently, the effective fault signatures have been extracted as the feature vectors in terms of norm entropy of the DWT coefficients (e_a, e_b, and e_c). Thereafter, the computed entropy of each phase is applied as input to the non-parametric ML classifier models as training and testing data sets. During the training phase, the feature data set corresponding to various scenarios such as normal operating mode and during abnormal events (all fault types at 20 locations in network, three different fault resistances and inception angles, two levels of line compensation) are fed to the model. The classifier models recognize the pattern of the feature vectors associated with different kinds of shunt fault events and train the KNN model accordingly. The events are labeled as follows: class 0 (normal case), 1 (AG), 2 (BG), 3 (CG), 4 (AB), 5 (AC), 6 (CB), 7 (ABG), 8 (BCG), 9 (ACG), 10 (ABC). Further, testing feature sets of unfamiliar fault cases with varying circumstances are fed to the trained classifier model and it predicts the particular category of the test instance as its final output (0 to 10 labels). The test sample are consisting of feature vectors corresponding to 10 normal operating modes and various fault conditions including 7 different new locations in the network, 6 fault resistances, 6 fault inception angles, and two levels of line compensation percentage. Hence, the total size of the testing samples is 5554 (10 samples of normal operation and 5544 samples of fault cases).

9.3.2 DWT and Non-Parametric ML (SVM) Based Fault Events Classification Scheme

The SVM model endeavors a hyperplane for segregating the data points according to their specific class label. When the separations between the specific class labels are optimal, then the corresponding plane is called as optimal hyperplane. Conceptually it is a binary classification approach; however, it can be effectively utilized for multi-class categorization using the one vs all algorithm [23,24]. SVM has been applied for discriminating 10 different kinds of fault events in the compensated power network, by applying the one vs all mechanism. The quadratic Gaussian function is used as the kernel function. The sequential minimal optimization method is utilized for the separation of the hyperplane. During the training phase, here also the same extracted feature data set corresponding to various scenarios is fed to the SVM-based model. The classifier models recognize the pattern of the feature vectors associated with different cases using its computing algorithm and trained the model accordingly. The labeling of events is similar as aforementioned in the KNN mechanism. Later on, testing the same 5554 data samples has been utilized. The SVM-based model predicts the category of the test instance i.e. 0 to 10 as its final output.

9.3.3 DWT and Non-Parametric ML (PNN) Based Fault Events Classification Scheme

The PNN-based models are conceptually based on calculation of probability density function (PDF) of feature samples for individual class labels and the categorization is done in accordance with Bayes' rules [25]. The input layer neurons assess the nearness of the applied feature samples from the trained categories using equation (9.1):

$$f_A(x) = \frac{1}{(2\pi)^{k/2}\sigma^k} \exp\left[-\frac{(x - x_{Ai})^T (x - x_{Ai})}{2\sigma^2}\right] \tag{9.1}$$

where k represents the dimension; σ is smoothing parameter; i is the pattern number; and x_{Ai} is the ith training pattern from category A. Subsequently, summation layer sums up all received inputs from previous pattern layer for every category and provides the probability. Ultimately, once the sum of weighted votes is calculated, the Bayes' rule is applied for ascertaining the category of test cases. The category having the highest vote probability is called as the class label of the test case. During training and testing, the same feature data set has been utilized as aforementioned.

9.4 FEASIBILITY AND COMPETENCY ANALYSIS

All of the previous three methodologies have been extensively tested for 5554 testing samples (including normal operating mode, all types of events, five different fault resistance, six different fault inception angles, seven unknown locations in the network, and two levels of line compensation) for analyzing the performance in diagnosing the fault events in simulated compensated power network. The structure of the simulated network is shown in Figure 9.4. Four multiple compensation levels i.e. 30%, 35%, 40%, and 45% have been considered. Figure 9.5 shows the three-phase post-fault current samples retrieved during an AG fault event at 30 km in the network from the sending side on different inception angles.

The fault events classification accuracy of all the methodologies is computed by using the following equation:

$$Classification_{accuracy}(\%) = \frac{(\text{Total correct classified events})}{\text{Total number of test events}} \times 100 \tag{5.2}$$

Table 9.1 shows the fault events classification accuracy percentage acquired by the first (KNN)-based approach. The average accuracy obtained by the proposed KNN classifier model-based scheme is 99.523%.

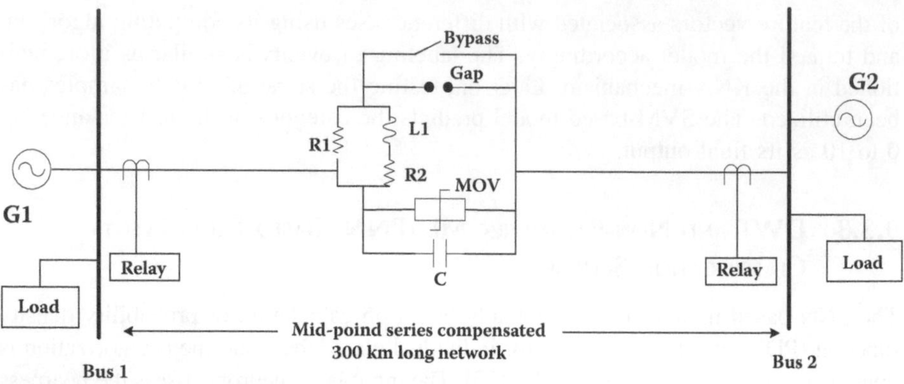

FIGURE 9.4 Simulated network.

Table 9.2 provides the fault events classification accuracy percentage obtained by an SVM-based scheme. The average accuracy obtained is 99.51%. Table 9.3 shows the corresponding confusion matrix obtained during testing of the SVM-based scheme.

Table 9.4 presents the fault events classification accuracy percentage obtained by a PNN-based scheme. However, during the validation of a PNN-based scheme the case of normal operation is neglected and only fault cases have been considered. The overall average fault events classification accuracy obtained by the proposed PNN classifier-based scheme is 99.65%. Figure 9.6 represents the associated confusion matrix obtained during testing.

The response time for predicting the precise fault events categories in the network by the proposed fault events classification scheme is demonstrated in Table 9.5. By seeing the obtained results for varying fault scenarios, it has been reaffirmed that all the three aforementioned approaches are well effective in ascertaining the fault events in series compensated power network irrespective of varying fault conditions. It has been seen that the PNN-based approach is having better accuracy as compared to the KNN- and SVM-based approaches. However, the time of response of a PNN-based scheme is slightly higher than the KNN- and SVM-based schemes.

9.4.1 Transforming Fault Events Identification

Transforming events are the abnormal events in the network which commence in one line and afterward within a short time span involve other lines also [26,27]. The transformation of involved power lines during short intervals makes these events more complicated than usual shunt events. Hence, these events cause additional challenges to the normal relaying system. The efficacy of aforementioned methodologies is also assessed for transforming events in the network. Multiple transforming events are simulated in a considered test network with an initial event commence in first line at 0.00166 seconds and within 5 ms involves other lines also.

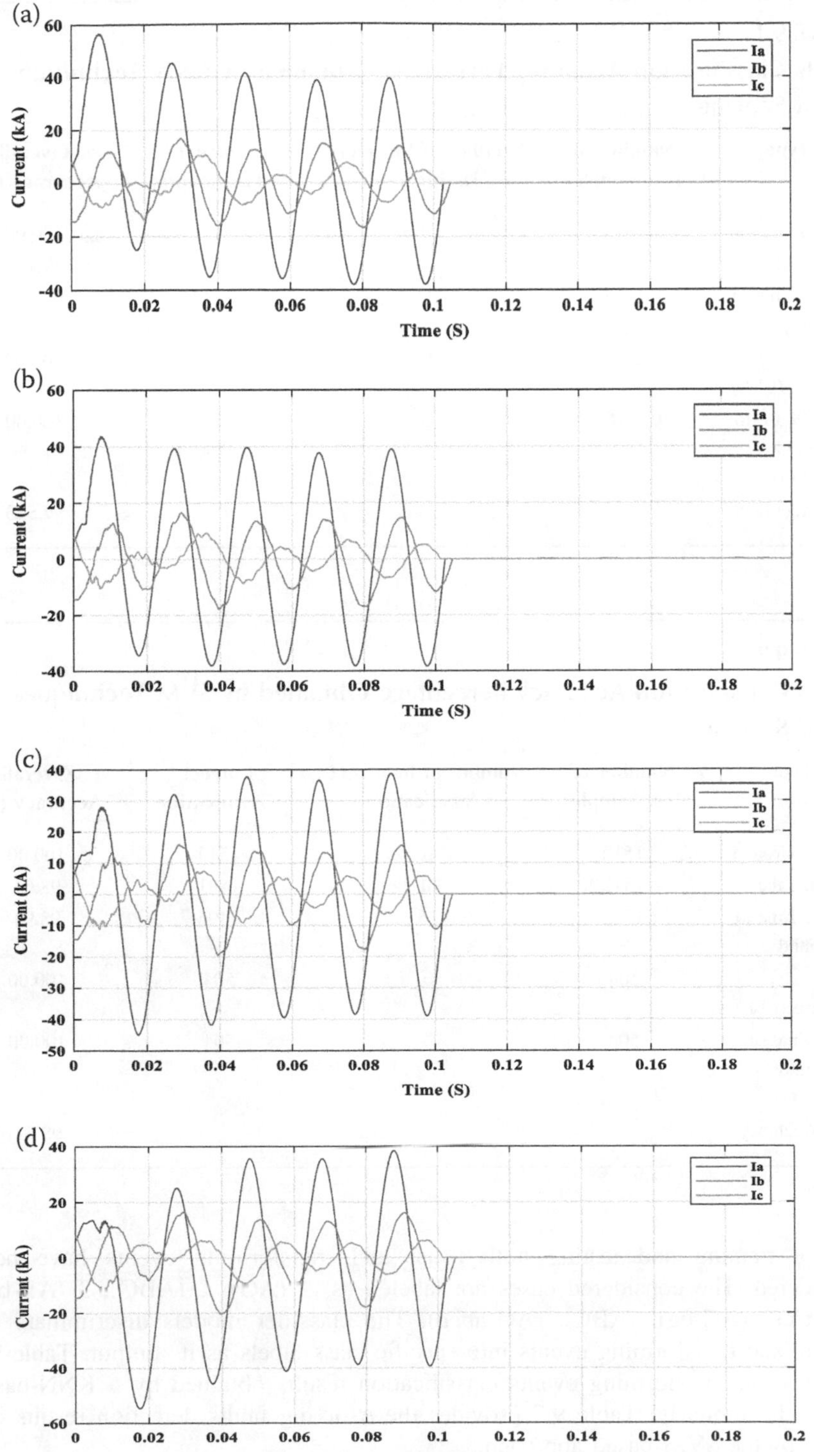

FIGURE 9.5 Current samples during AG event at different inception angles: (a). 30 degree, (b). 60 degree, (c). 90 degree, and (d). 120 degree.

TABLE 9.1
Faults Classification Accuracy Percentage Obtained by KNN Technique-Based Scheme

Fault Type	Number of Test Samples	Number of Incorrect Classifications	Correct Classification	Overall Accuracy (%)
Line to Ground	1512	0	1512	100.00
Line to Line	1512	13	1499	99.140
Double Line to Ground	1512	23	1489	98.478
Three Phase (LLL)	504	0	504	100.00
Three Phase to Ground (LLLG)	504	0	504	100.00
Avg. Accuracy				**99.523**

TABLE 9.2
Faults Classification Accuracy Percentage Obtained by SVM Technique-Based Scheme

Fault Type	Number of Test Samples	Number of Incorrect Classification	Correct Classification	Overall Accuracy (%)
Line to Ground	1512	0	1512	100.00
Line to Line	1512	21	1491	98.61
Double Line to Ground	1512	16	1496	98.94
Three Phase (LLL)	504	0	504	100.00
Three Phase to Ground (LLLG)	504	0	504	100.00
Avg. Accuracy				**99.51**

During training and testing both usual and transforming events have been considered. The considered cases are labeled as: 1 (AG), 2 (ABG), 3 (AB-bg), 4 (ACG), (AG-cg), (ABC), (AG-abcg). The classifier models discriminate the normal and transforming events into specific class labels as its output. Table 9.6 presents the transforming events classification results obtained by a KNN-based approach. Similarly, Table 9.7 provides the evolving faults detection results obtained by the SVM-based approach.

TABLE 9.3
Confusion Matrix for the SVM-Based Scheme

Actual Fault Events	Sample Size	Predicted Fault Events												
		AG	BG	CG	AB	AC	BC	ABG	BCG	ACG	ABC	ABCG	No Fault	Accuracy (%)
AG	**504**	**504**	0	0	0	0	0	0	0	0	0	0	0	**100**
BG	**504**	0	**504**	0	0	0	0	0	0	0	0	0	0	**100**
CG	**504**	0	0	**504**	0	0	0	0	0	0	0	0	0	**100**
AB	**504**	0	0	0	**493**	0	0	**11**	0	0	0	0	0	**97.8**
AC	**504**	0	0	0	0	**498**	0	0	0	**6**	0	0	0	**98.8**
BC	**504**	0	0	0	0	0	**500**	0	**4**	0	0	0	0	**99.2**
ABG	**504**	0	0	0	**5**	0	0	**499**	0	0	0	0	0	**99.0**
BCG	**504**	0	0	0	0	0	**3**	0	**501**	0	0	0	0	**99.4**
ACG	**504**	0	0	0	0	**8**	0	0	0	**496**	0	0	0	**98.4**
ABC	**504**	0	0	0	0	0	0	0	0	0	**504**	0	0	**100**
ABCG	**504**	0	0	0	0	0	0	0	0	0	0	**504**	0	**100**
No fault	**10**	0	0	0	0	0	0	0	0	0	0	0	**10**	**100**

TABLE 9.4
Faults Classification Accuracy Percentage Obtained by PNN Technique-based Scheme

Fault Type	Number of Test Samples	Number of Incorrect Classification	Correct Classification	Overall Accuracy (%)
Line to Ground	1512	0	1512	100.00
Line to Line	1512	7	1505	99.53
Double Line to Ground	1512	14	1498	99.07
Three Phase (LLL)	504	0	504	100.00
Avg. Accuracy				99.65

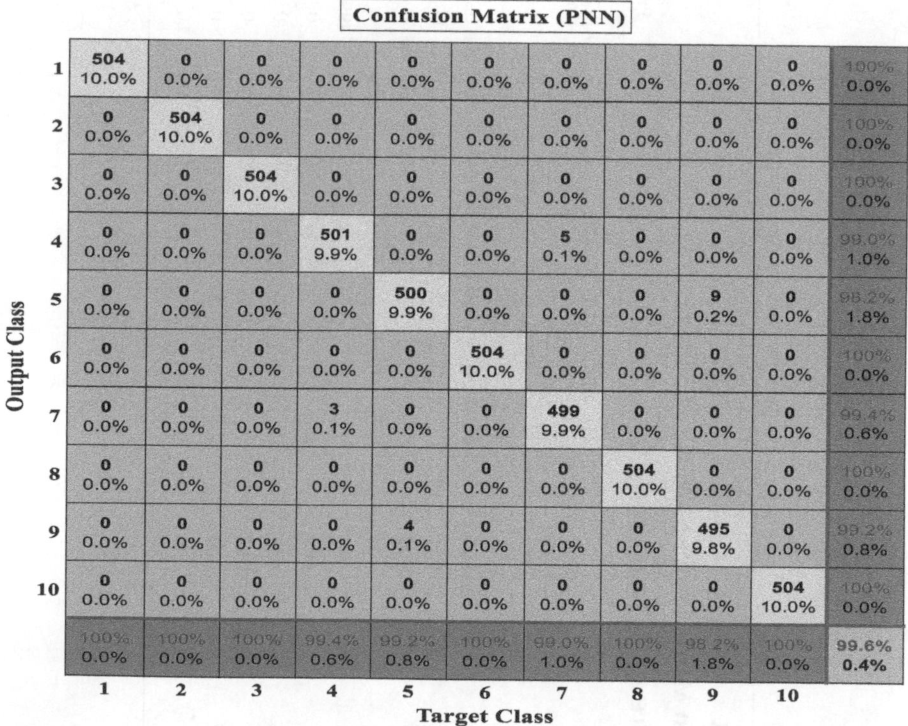

FIGURE 9.6 Confusion matrix during events classification using PNN classifier model.

TABLE 9.5
Time of Response of Different Non-Parametric ML Scheme

S. No	Classifier Model Utilized	Time of Response
i.	K-NN	6.05e-02 s
ii.	SVM	4.14e-02 s
iii.	PNN	1.59e-01 s

TABLE 9.6
Identification of Evolving Events Using KNN Classifier-Based Scheme

S. No	Primary Fault Type	Secondary Fault Type	Time Interval (ms)	Evolving Fault or Not	Classifier Output
1.	AG	–	–	No	1
2.	ABG	–	–	No	2
3.	AG	bg	5	Yes	3
4.	ACG	–	–	No	4
5.	AG	cg	5	Yes	5
6.	ABC	–	–	No	6
7.	AG	abcg	5	Yes	7
8.	AG	bg	10	Yes	3
9.	AG	cg	10	Yes	5
10.	AG	abcg	10	Yes	7
11.	AG	bg	15	Yes	3
12.	AG	cg	15	Yes	5
13.	AG	abcg	15	Yes	7

It had been observed that both KNN- and SVM-based approaches are also effective in ascertaining evolving events in the network.

9.5 SUMMARY

This chapter thoroughly addresses the additional critical protection issues observed in compensated transmission networks and describes multiple data mining–based protection methodologies for the same. The described methodologies are well capable of detecting the normal shunt and transforming fault events irrespective of varying network topologies and fault conditions. Recently, wide-area measurement devices like PMUs are inevitably implemented over the grid, which generates a huge measurement data set. Therefore, as compared with mathematical network

TABLE 9.7
Identification of Evolving Fault Events Using SVM Classifier-Based Scheme

S. No	Primary Fault Type	Secondary Fault Type	Time Interval (ms)	Evolving Fault or Not	Classifier Output
1.	AG	–	–	No	1
2.	ABG	–	–	No	2
3.	AG	bg	5	Yes	3
4.	ACG	–	–	No	4
5.	AG	cg	5	Yes	5
6.	ABC	–	–	No	6
7.	AG	abcg	5	Yes	7
8.	AG	bg	10	Yes	3
9.	AG	cg	10	Yes	5
10.	AG	abcg	10	Yes	7
11.	AG	bg	15	Yes	3
12.	AG	cg	15	Yes	5
13.	AG	abcg	15	Yes	7

modeling-based protection methods, advanced data mining–based protection methodologies are more pertinent and competent in providing adaptive protection to the compensated transmission network. The faster response, lesser complexity, and better competency of handling large measurement data sets make the data mining–based protection schemes more superior and adaptive than network modeling–based methods.

APPENDIX

Simulated Test Network Data Set

S. no	R (Ω/km)	X (H/km)	C (F/km)
Zero Sequence	0.3864	4.1264e-3	7.751e-9
Positive Sequence	0.01273	0.9337e-3	12.74e-9
Negative Sequence	0.01273	0.9337e-3	12.74e-9
Degree of line compensation	X_C (Ω)	C_S (μf)	
30%	26.390	120.6	
35%	30.788	103.4	
40%	42.24	62.8	
45%	39.585	80.41	

REFERENCES

[1] Vyas B., Maheshwari R. P., Das B., "Protection of series compensated transmission line: Issues and state of art," *Electric Power Systems Research*, 2014; 107: pp. 93–108.

[2] Tanbhir hoq Md, Wang J., Taylor N., "Review of recent developments in distance protection of series capacitor compensated lines," *Electric Power System Research*, Jan. 2021; 190: p. 106831. https://doi.org/10.1016/j.epsr.2

[3] Khederzadeh M., Sidhu T. S., "Impact of TCSC on the protection of transmission lines," *IEEE Trans. Power Delivery*, 2006; 21: pp. 80–87.

[4] Sidhu T. S., Khederzadeh M., "TCSC impact on communication aided distance-protection schemes and its mitigation," *IEE Proceedings - Generation, Transmission and Distribution*, 2005; 152: pp. 714–728.

[5] Sidhu T. S., Khederzadeh M., "Series compensated line protection enhancement by modified pilot relaying schemes," *IEEE Transactions on Power Delivery*, 2006; 21(3): 1191–1198.

[6] Zellagui M., Chaghi A., "Impact of GCSC on measured impedance by distance relay in the presence of single phase to earth fault," *International Journal of Electrical, Electronic Science and Engineering*, 2012; 6(10): 1112–1117.

[7] Zellagui M., Chaghi A., "A comparative study of impact series FACTS devices on distance relay setting in 400 kV transmission line," *Journal of Electrical and Electronics Engineering*, 2012; 5(2): pp. 111–116.

[8] Gilani N. S. et. al., "Data-mining for fault zone detection of distance relay in FACTS –Based transmission," IEEE Texas Power and Energy Conference, 2020.

[9] Shojaei A., Madani S. M., "Analysis of measured impedance by distance relay in presence of SSSC," 5th IET International Conference on Power Electronics, Machines and Drives, Brighton, UK, 2010.

[10] Jena P., Pradhan A. K., "A positive-sequence directional relaying algorithm for series-compensated line," *IEEE Transactions on Power Delivery*, 2010; 25(4): pp. 2288–2298.

[11] Jena P., Pradhan A. K., "Directional relaying in the presence of a thyristor-controlled series capacitor," *IEEE Transactions on Power Delivery*, 2013; 28(2): pp. 628–636.

[12] Biswal M., Pati B. B., Pradhan A. K., "Directional relaying for double circuit line with series compensation," *IET Generation, Transmission and Distribution*, 2013; 7(4): pp. 405–413.

[13] Enrique R.-A. et. al., "TCSC impact on communication aided distance-protection schemes and its mitigation," *IEEE Latin America Transactions*, 2020; 19(1): 147–154.

[14] Jamali S., Kazemi A., "Distance relay over-reaching due to installation of TCSC on next line," *IEEE International Symposium on Industrial Electronics*, 2006. DOI: 10.1109/ISIE.2006.295

[15] Dash P. K., Pradhan A. K., Panda G., "Apparent impedance calculations for distance-protected transmission lines employing series-connected FACTS devices," *Electric Power Components and Systems*, 2010; 29(7): pp. 577–595.

[16] Manori A., Tripathy M., Gupta H. O., "Advance compensated Mho relay algorithm for a transmission system with shunt flexible AC transmission system device," *Electric Power Components and Systems*, 2014; 42(16): pp. 1802–1810.

[17] Ibrahim A. M. et al., "An artificial neural network based protection approach using total least square estimation of signal parameters via the rotational invariance technique for flexible AC transmission system compensated transmission lines," *Electric Power Components and System*, 2011; 39(1): pp. 64–79.

[18] Jamali S., Shateri H., "Locus of apparent impedance of distance protection in the presence of SSSC," *European Transaction on Electrical Power*, 2011; 21: pp. 398–412.

[19] Biswas S. et al., "The effect of kernels in SVM on the fault classification accuracy of a transmission line compensated with TCSC," International Conference on Advances in Electronics, Electrical & Computational Intelligence (ICAEEC), 2020. DOI: 10.2139/ssrn.3577488.

[20] Friedman, J. H., Bentley, J. L., Finkel, R. A., "An algorithm for finding best matches in logarithmic expected time," *ACM Transactions on Mathematical Software*, 1977; 3(3): pp. 209–226

[21] Bay S. D., "Nearest neighbor classification from multiple feature subsets," *Intelligent Data Analysis*, 1999; 3(3): pp. 191–209

[22] Singh S. K., Vishwakarma D. N., Saket R. K., "An intelligent scheme for categorizing fault events in compensated power network using k-nearest neighbor technique," *International Journal of Power and Energy Conversion*, 2020; 11(4): pp. 352–368

[23] Cortes C., Vapnik V., "Support-vector networks," *Machine Learning*, 1995; 20(3): pp. 273–297.

[24] Singh S., Vishwakarma D. N., "An approach for discriminating abnormalities in compensated power transmission circuit," IEEE International Conference on Innovative Smart Grid Technologies (IEEE ISGT Asia, 2018. DOI: 10.1109/ISGT-Asia.2018.8467868.

[25] Singh S. K., Vishwakarma D. N., Saket R. K., "An intelligent scheme for categorization and tracing of shunt abnormalities in compensated power transmission network," *Journal of Electrical Systems*, 2019; 15(1): pp. 68–80.

[26] Singh S. K., Vishwakarma D. N., Saket R. K., "Intelligent Computing Based Scheme for Evolving Fault Events Location in Series Compensated Power Networks," *Journal of Electrical Systems*, 2019; 15(2): pp. 303–313.

[27] Kumar A. N. et al., "Adaptive neuro-fuzzy inference system based evolving fault locator for double circuit transmission lines," *IAES International Journal of Artificial Intelligence; Yogyakarta*, 2020; 9(3): pp. 448–455.

10 AI-Based Protective Relaying Schemes for Transmission Line Compensated with FACTS Devices

Bhupendra Kumar[1], *Anamika Yadav*[2], *and Almoataz Y. Abdelaziz*[3]

[1]Department of Electrical Engineering, VEC, India

[2]Department of Electrical Engineering, National Institute of Technology, India

[3]Faculty of Engineering and Technology, Future University in Egypt, Egypt

10.1 INTRODUCTION

In the modern world, an energy requirement has been fulfilled by adding new power generating stations or by enhancing the power transferal capability of the prevailing transmission line by using the emerging Flexible AC Transmission System (FACTS) technology. Thyristor Controlled Series Compensator (TCSC), Static Var Compensator (SVC), Static Synchronous Compensator (STATCOM), Static Synchronous Series Compensator (SSSC), and Unified Power Flow Controller (UPFC) hold explicit characteristics over the different FACTS devices. These FACTS controllers provide control of the real and reactive power flows among the transmission network by providing a variable voltage injection with any phase relationship concerning the current [1]. However, FACTS devices' operation during a fault condition disturbs the relaying signals results in the conventional relaying output [2–7]. A variety of AI techniques have been utilized for fault analysis in FACTS-compensated transmission lines in the literature. However, presented work in this chapter is restricted to high adherence regular reported journals to the said subject. Analysis of different fault diagnosis in a power system using an intelligent system application has been discussed in [8]. AI-based relaying schemes are found to be most suitable to solve problems associated with FACTS-compensated transmission line protection. Presently used digital protection embraces three modules in one i.e. detection, classification, and location of the fault. However, system operators receive a huge amount of operational data with

DOI: 10.1201/9780367552374-10

the means of Intelligent Electronics Devices (IED) and Phasor Measurement Units (PMU), which can be further used for fault diagnosis purposes. Availability of real-time data from fault recorders and virtually simulated fault data form the different electromagnetic simulation tools motivated the use of AI-based algorithms for fault diagnosis purposes. In this book chapter, a comprehensive review is covered for the recent development in an AI-based algorithm for FACTS compensated line. Comparisons between different AI approaches in terms of fault detection classification and location are presented. This study is useful for academia and practice engineers to gain insight into the AI application in the field of protection of transmission line incorporating FACTS devices.

This book chapter is organized as follows, Section 10.1 introduces this chapter. Section 10.2 discusses the FACTS technology devices and their advantages, and Section 10.3 deals with the protection issues in the FACTS-compensated transmission line. Section 10.4 discusses the overview of AI. Section 10.5 deals with the different AI approaches to the protection of FACTS-compensated transmission lines and comparisons concerning fault detection classification and location. Section 10.6 concludes this chapter with a future perspective.

10.2 FACTS TECHNOLOGY

FACTS devices are defined as "Alternating Current Transmission Systems incorporating power electronic-based and other static controllers to enhance controllability and increase power transfer capability" [1]. Based on their controller characteristics, FACTS devices are divided into different types as shown in Table 10.1. Based on the construction features FACTS controller are broadly classified as thyristor-based or VSC-based FACTS controller. Furthermore, based on the operating principle FACTS controllers are termed Series (TCSC/SSSC), Shunt (SVC/STATCOM), combined Series-Shunt (UPFC), and combined Series-Series (IPFC) FACTS controllers. FACTS controllers prove many advantages to today's large interconnected complex power system [1].

TABLE 10.1
Types of FACTS Devices

Kind of Connection	Controlled Parameter		
	P	**Q**	**P and Q**
Shunt-Connected		SVC and STATCOM	
Series-Connected	TSSC, TCSC, and SSSC		
Combined Shunt and Series-Connected	TCPST and TCPAR	TCVR	UPFC and IPFC

10.3 PROTECTION ISSUES WITH FACTS TECHNOLOGY INTEGRATION

Although many advantages of FACTS device in the transmission line operation and control, it will create issues for installed conventional protection system. Whenever FACTS device come into the fault loop it will change the apparent impedance measured value at the relaying location, which is responsible for nuisance tripping. However, it will not affect the measured impedance value if FACTS device is not considered in the fault loop. The presence of FACTS controllers in fault loop modifies the transmission line impedance seen by the distance relay. Hence the possibility here of underreach or overreach of the distance relays, depends totally on the type of the FACTS controller and the application for which it is applied, and also the location of the FACTS device is a very important consideration in the power system [8–13]. Integration of different FACTS devices in the system may create different protection issues, such as the inclusion of TCSC in series with the transmission line reactance create complication in form of overreaching and underreaching. However, system disturbance and random switching of capacitor bank will result in the harmonic injection in the system, which in turn changes the relaying terminal voltage and current signal, voltage inversion, current inversion, sub-harmonic oscillations, and additional transient may result in mal-operation of distance relaying. Moreover, Voltage Source Converter (VSC)–based FACTS devices results in switching transient due to high switching and control operation may result in malfunctioning of the conventional distance relaying. In the other side, the compensated injected reactance by different VSC-based converters affects the measured impedance at the relaying point. However, the compensation level and the different reference setting of different FACTS devices, makes the measured impedance unpredictable, henceforth it is not practically feasible to set each individual reach setting for different operating modes of FACTS devices. Moreover, equal compensation provided for any unsymmetrical fault may result in the nuisance fault classification. Such a complex condition brought the need for more reliable protection schemes for a FACTS-compensated transmission line.

Various solutions have been proposed to solve the problem associated with the transmission line protection in the presence of FACTS devices. One of the most common solutions reported by many authors is adaptive relay setting for the transmission line in the presence of FACTS devices [14–17]. In this method, the adaptive trip boundary has been formulated based on the apparent impedance calculation, by varying fault resistance, while the fault location changes from the relaying bus to load end of the transmission line. The adaptive protection schemes were dependent on the knowledge regarding power system condition and FACTS parameters. In the absence of desired information, the adaptive procedure couldn't be performed. In such a situation, the distance relay will follow the predefined setting. The adaptive relaying technique can only identify faults within its reach but is unable to classify the fault type and fault phase. To show the impact of UPFC integration in conventional relaying a 500 kV, 60 Hz power system network has been considered [14]. A 100 MVA GTO based UPFC is considered at the sending end bus and considered as in an automatic power flow control mode. In automatic control (APFC) mode, a shunt converter works as a STATCOM and series converter as an SSSC to control both power flow and the voltage level at the connecting bus simultaneously. The measured signals at the relaying bus changes dynamically due to automatic action and number of switching action in the UPFC control. However, the switching dynamics change the conventional relaying output may result in a malfunction of the relaying scheme.

In a conventional distance and alpha plane line current differential (APLCD), the protection scheme is tested using simulated fault events obtained using MATLAB®/Simulink [13].

Double line to a ground fault created at 70% of the line, and apparent impedance trajectory is drawn with the zone-1 setting of the distance relay for three different conditions, such as when UPFC is not considered in the fault loop, when UPFC is operated in the voltage regulation mode, and when UPFC is operated in the APFC mode, as shown in Figure 10.1(a). From Figure 10.1(a) it is observed that the distance relay impedance trajectory does not enter zone 1 even the fault has been simulated in the first zone of the relay. Moreover, UPFC is considered in the sending end bus to show its impact on a complex differential plane [13]. To observe the UPFC impact on a relaying scheme, different variations have been done with the UPFC control. Different variations in the control setting of real power, reactive power, and bus terminal voltage have been done in the UPFC control. Figure 10.1(b) depicts the effect in the APLCD protection for a line to ground fault. Figure 10.1(b) shows that APLCD protection performance during different configuration of UPFC whether it is in the fault loop of not. When UPFC is in a fault loop it gives nuisance tripping. However, with a variation in the different parameter of UPFC controller, not much affects the complex current ratio trajectory, as shown in Figure 10.1(b). From Figure 10.1 it is observed that the conventional relaying scheme is affected with the inclusion of a FACTS device in the fault loop. It is not practically feasible to consider all the dynamics of a FACTS controller into the conventional relaying scheme; therefore modern signal processing and advanced computational intelligence technique is required in the dependable operation of the protective relaying algorithm.

10.4 OVERVIEW OF AI

Artificial intelligence (AI) involved modeling the neurons in the brain, which resulted in the field of neural networks. An artificial neural network consists of a large collection of neural units (artificial neurons), whose behavior is roughly based on how real neurons communicate with each other in the brain. Each neural unit is connected with many other neural units, and links can enhance or inhibit the activation state of adjoining units. The network architecture consists of multiple layers of neural units. A signal initiates at the input layer, traverses through hidden layers, and finally culminates at the output layer [18]. AI has been stocked with the uniqueness of parallel processing, nonlinear mapping, associative memory, and offline and online learning abilities. The wide uses of AI with its overcoming outcomes make it an efficient problem-solving means in electrical power systems.

A deep learning neural network differs from older neural networks in that they often have more hidden layers. Furthermore, deep learning networks can be trained using both unsupervised and supervised learning.

AI can be broadly classified into five groups as training-based methods (machine learning [ML], learning probabilistic methods), statistical models, exploration/search

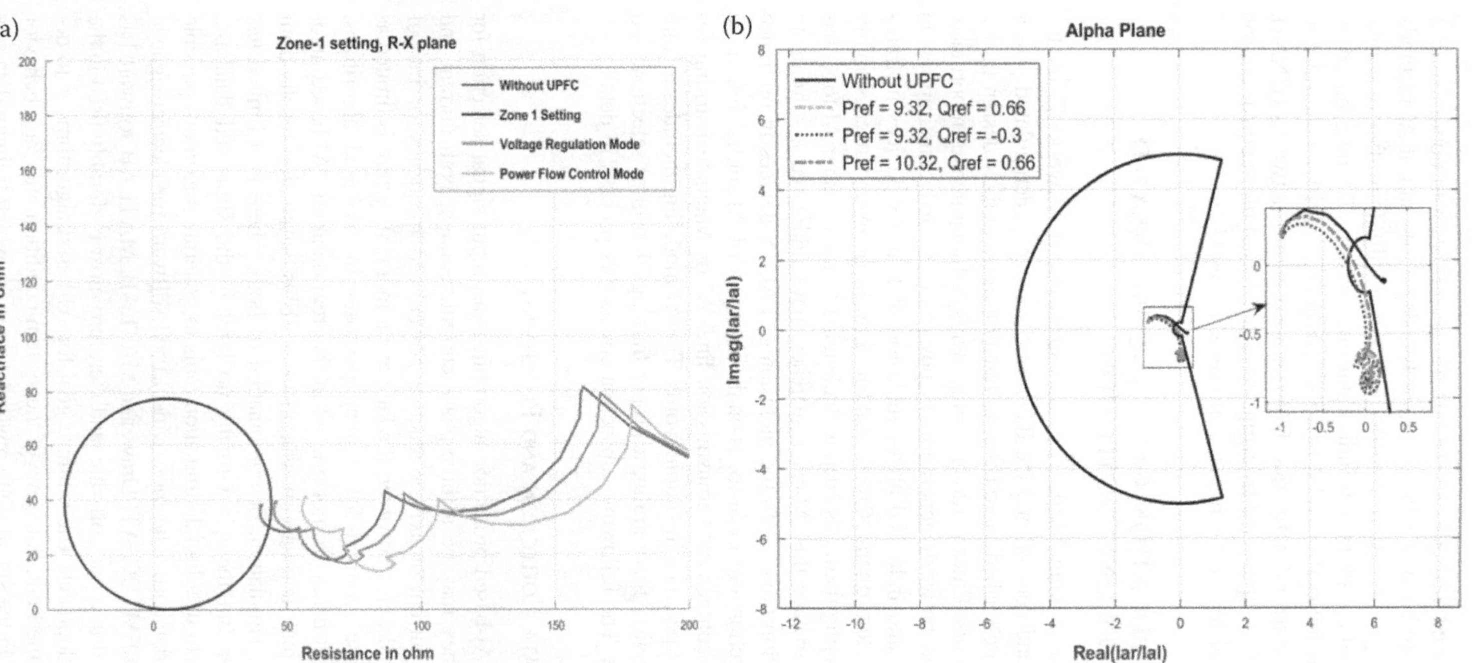

FIGURE 10.1 Ccnventional relaying, protecting a UPFC compensated transmission line (a) distance relay; (b) alpha plane current differential relay.

methods, optimization-based methods (genetic algorithm, particle swarm optimization), game theory, and decision-making methods. Figure 10.2 depicts a chart of AI methods.

In AI, ML provides systems with the ability to automatically learn from gigantic historical or synthetic data without any individual intervention. Training-based algorithms such as neural networks (NNs) and its variants e.g. artificial neural networks (ANNs), recurrent neural networks (RNNs), deep neural networks (DNNs), supervised learning, and unsupervised learning have established more consideration and have been functional to several fields of power systems [19].

10.5 AI-BASED APPLICATION IN FACTS-COMPENSATED TRANSMISSION LINE PROTECTION

Electric power systems are among the most complex structures in the world. Given the continually increasing load demand and the dawn of the recently deregulated power market, power systems are pushed more often to function close to their design limits. In this context, Transmission Lines (TLs) are components of fundamental importance, providing the connection between generation in power plants and consumption in residential, commercial, and industrial areas. Inclusion of FACTS devices in transmission lines helps a system operator to operate it close to or in enhanced stability limits. As a result, protecting them has been a challenging task and using AI tools has become more common to face this. Many researchers have addressed a solution to transmission line protection issues with the inclusion of FACTS devices in reference [15]. However, in this section, our main focus on the overview of AI application in the protection of a FACTS-compensated transmission line. AI or Machine learning focuses on learning from experience in simple words. The AI modeling includes texture identification, data collection for training, a proper training algorithm selection, and testing of unknown data. The following subsection covers this process in detail.

10.5.1 Training Data Collection and Processing

An important aspect in AI-based protection algorithm is an input signal selection for preparing a database. They may be a three-phase current, voltage, both current and voltage, zero sequence current, active or reactive power, and frequency or rate of change of frequency signals, etc. However, the fault event in a real system is limited as compared to the database required for training purposes in AI-based algorithms. Besides, some of the fault cases are not present in this real database. AI-based algorithms are based on the real field data measurement by IEDs, simulation results from different electromagnetic simulation, or combination of both. There is a limited test case in the real database but there is virtually no limit in the simulated database. Therefore it is necessary to use digital simulation tools to generate maximum possible fault or abnormal event database. The most widely used software for electromagnetic simulation is the PSCAD-EMTDC, ATPDraw, and MATLAB. MATLAB software has several libraries integrated into one software, and therefore most preferably used in the AI-based algorithm development. The ultimate goal for AI-based algorithms is to detect, classify, and locate the fault, but some of the system condition and fault behavior changed the information in the signal. Different conditions for which training data sets

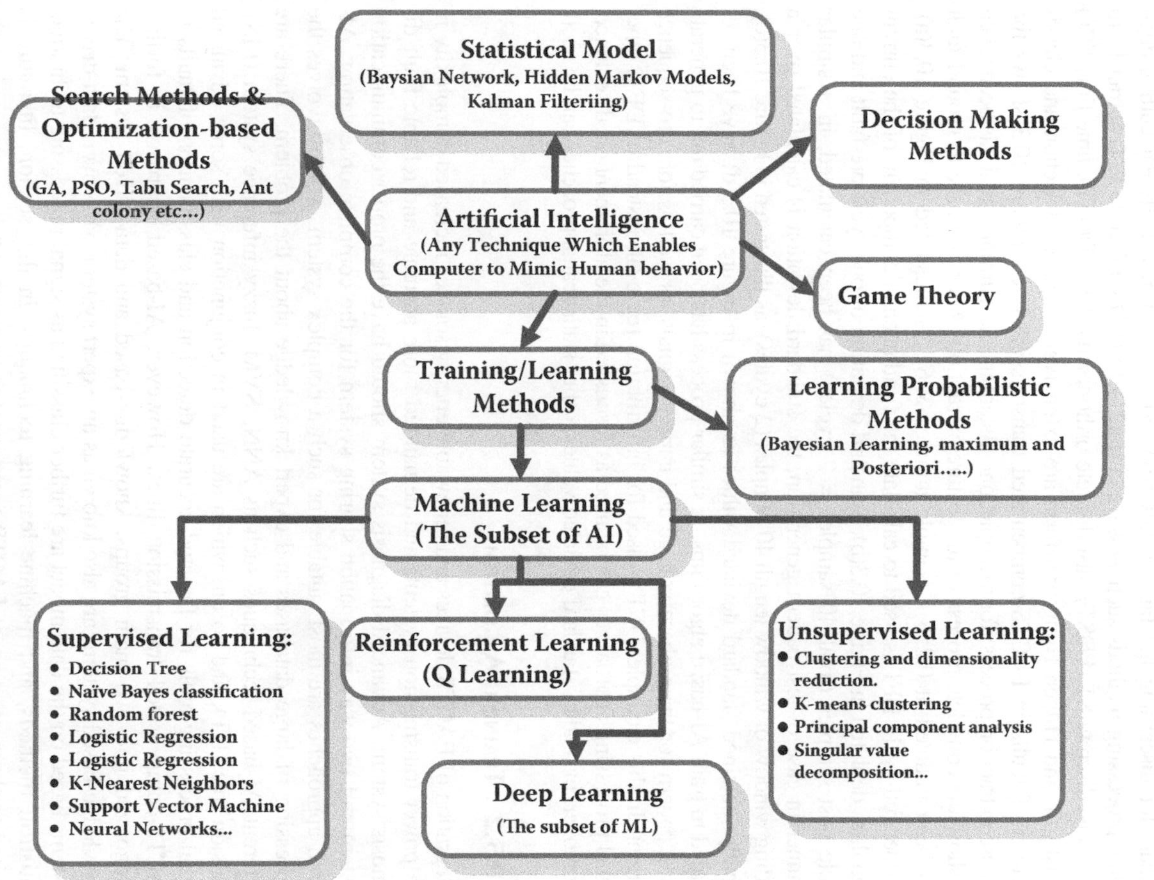

FIGURE 10.2 AI technique.

have been generated is fault type, fault inception angle, fault resistance, fault location, and compensation level of the FACTS devices. Once the input signal is collected, it will be processed for different signal processing tools to retrieve the valuable information. Signal processing is a method applied to extract useful features from measured raw signals for discerning the fault events from non-fault events. Different cutting-edge signal processing methods such as wavelet transform (WT), Discrete WT, S-transform (ST), fast discrete ST (FDST), fast discrete orthonormal ST (FDOST), time-time (TT) transform, and Hilbert-Hung transform are extensively used for detection and classification of faults in FACTS-compensated transmission lines. Figure 10.3 shows the feature extraction process; for example, one instantaneous current signal has been taken to show the complete process of the feature extraction. A single phase to ground fault has been initiated at 0.255 s (sample number 306), as illustrated in Figure 10.3(a). Further full-cycle DFT is used to calculate the fundamental component of the current signal, as depicted in Figure 10.3(b). Standard deviation of one cycle pre-fault and one cycle post samples (total 40 samples = 2 cycles) has been calculated, in a similar manner in case of test pattern generation, the standard deviation is calculated over a sliding window of window length 40 samples (2 cycles), as illustrated in Figure 10.3(c). Finally, obtained standard deviated values as shown in Figure 10.3(d) have been recorded to train AI-based algorithms. A similar process has been carried out to generate a test pattern for three-phase voltage and current signals, as well as for zero-sequence current. In this example, DFT is used for information retrieval; instead of DFT, other signal processing approaches can be used as discussed in the aforementioned sentence. Moreover, instead of standard deviated values, other statistical approaches can be used.

10.5.2 Training Algorithms

Integration of FACTS devices and renewable energy sources increased complexity in the power transmission systems and demanded more accurate and reliable fault diagnosis system. Advanced diagnosis system should have the proper communication system and two-way information sharing system for the complete observation. AI-based approaches are most suitable for such a complex system, which involves the processing of large databases and expert knowledge about the problem. There are different AI-based techniques such as ANN, SVM, fuzzy inference system (FIS), decision tree (DT), and so on, which are used in conjunction with some advanced signal processing tools for fast and accurate detection and classification of faults in FACTS-compensated transmission lines. However, AI-based approach is further categorized into two main groups, knowledge-based and data-driven system. The knowledge-based systems are also known as an expert system. Moreover, data-driven systems based on the data model are further classified as signal processing technique, statistical methods, and machine learning techniques. In this section, literature is limited to the protection of FACTS compensated transmission line presented in reputed journals and conferences having high adherence. Table 10.2 summarized some of the important findings for the series compensated transmission line from the literature. From reference [20–39] of Table 10.2 incorporated AI-based approaches in terms of fault detection, classification, and location. A complete protection scheme comprises these three modules in one.

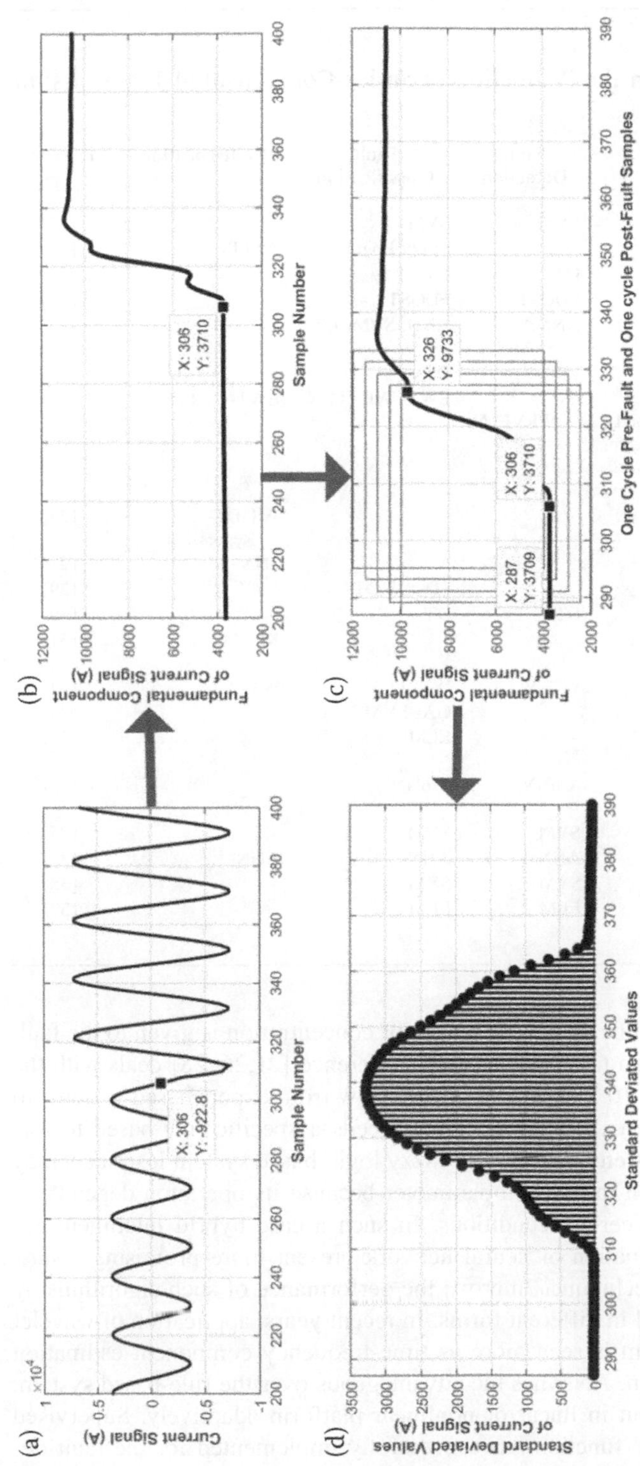

FIGURE 10.3 Feature extraction.

TABLE 10.2
AI-Based Application in the Protection of Series-Compensated Transmission Line and References

FACTS Device	Technique Used	Fault Detection	Fault Classification	Fault Location	References
TCSC	Data Mining, SVM	SVM	SVM	–	[20]
TCSC	Fuzzy Logic, ANFIS	–	Fuzzy Logic	ANFIS	[21]
TCSC	Decision Tree (DT)	DT	DT	–	[22]
TCSC	FDOST	FDOST	FDOST	–	[23]
FSC	ANN, SVM, DT	ANN, SVM, DT	ANN, SVM, DT	–	[24]
TCSC	DT	DT	DT`	–	[25]
FSC	Empirical Wavelet Transform (EWT) and weighted random vector functional link network (WRVFLN)	EWT-WRVFLN	EWT- WRVFLN	EWT- WRVFLN	[26]
FSC	WT-DT-Regression	–	–	WT-DT-Regression	[27]
FSC	kNN	–	–	kNN	[28]
FSC	FDOST DT		FDOST DT	–	[29]
FSC	DT	DT	DT	–	[30]
TCSC	Kernel extreme learning machine (KELM)	–	–	KELM	[31]
TCSC	WT-FL	WT- FL	WT- FL	WT- FL	[32]
TCSC	GA-SVM	–	GA- SVM	–	[33]
FSC	Extreme learning machine ELM	–	ELM	–	[34]
TCSC	Chebyshev Neural Network (ChNN)	(ChNN)	(ChNN)	–	[35]
TCSC	SVM	SVM	SVM	–	[36]
TCSC	ANN	ANN	ANN	ANN	[37]
FSC	SVM	SVM	SVM	–	[38]
FSC	Extreme learning machine (ELM)	ELM	ELM	–	[39]

From Table 10.2 it can be observed that major concentration is given to the fault detection and classification task, whereas only reference [21,26–28] deals with the fault location task. Table 10.2 gives a brief overview from expert-based systems to data-driven systems. The expert-based system needs a specific rule-based to formulate fault diagnosis system [21,32]. The fuzzy logic-based system loses accuracy with the large variation in the system parameter because its operation depends on the specific rule base in certain conditions. In such a case hybrid intelligent approaches with the combination of neural network present more promising results [21]. Signal processing techniques improve the performance of such algorithms by analyzing the input signal in different forms. In recent years, application of wavelet transform and S-transform is seen more as time-frequency component estimation contains more information. ANN has the advantageous over the rule-based system, as it can map input-output in linear or nonlinear platform adaptively. Supervised learning with radial basic function has successfully implemented for the fault detection, classification, and location mode in reference [37].

Supervised learning of input data sets are trained with the desired known output set of the variables. However, for the set of an unknown data set (not used in the training data set), it exhibits excessive data fit or over-fitting. Another approach for fault classification problem is a support vector machine (SVM). SVM is a statistical learning algorithm and requires fewer data sets as compared to ANN. Moreover, it does not require prior knowledge of the problem [20,24,33,36,38]. However, DT-based fault classification approaches are taking less time for training the network as compared to ANN- and SVM-based approaches for the same data set [22,24,29,30]. Moving forward to signal processing technique for feature extraction from voltage and current signal, WT is not only restricted for feature extraction, but it also used for fault detection and classification [40]. However, S-transform and FDOST-initiated SVM-based fault detection and classification scheme are presented in [41]. In addition to the aforementioned approaches, numerical optimization-based GA-SVM [33] is presented for fault detection and classification task.

Table 10.3 shows different AI-based approaches for the FACTS-compensated transmission line. In this table different approaches are categorized for the three functions i.e. fault detection, classification and location in FACTS compensated transmission line. From Table 10.3 it can be seen that AI-based application in transmission line fault location estimation is limited in presence of a FACTS device [40–51]. Signal processing technique WT, intrinsic time decomposition, DWT, and

TABLE 10.3
AI-based Application in the Protection of FACTS-Compensated Transmission Line and References

FACTS Device	Technique Used	Fault Detection	Fault Classification	Fault Location	References
UPFC	Intinsic Time-Decomposition (ITD)	–	–	ITD	[42]
UPFC	Wavelet Transform	WT	WT	–	[40]
SSSC	Wavelet Transform	WT	WT	–	[43]
STATCOM	DWT-ANN	DWT-ANN	DWT-ANN	–	[44]
S	ANN	ANN	ANN	–	[45]
UPFC	ANN	ANN	ANN	–	[46]
SVC	Differential Energy Spectrum (DES)	DES	DES	–	[47]
UPFC	DT-Fuzzy Logic (FL)	DT- FL	DT- FL	–	[48]
UPFC	FDOST-SVM	FDOST-SVM	FDOST- SVM	–	[41]
UPFC	WT	WT	WT	–	[49]
TCSC/UPFC	DT	DT	DT	–	[50]
TCPST/TCSC/UPFC	ANN	ANN	–	–	[51]

FDOST imitated, and data-driven AI approaches have been used for fault detection and classification task. However, ITD-based fault location estimation for UPFC compensated transmission line is presented in reference [42]. In reference [40,43] WT is used for fault detection and classification task. ANN-based approaches have been presented in reference [44–46,51]. The hybrid method with the combination of DT and fuzzy logic has been presented in reference [48]. Reference [40,41,43–51] deals with the only fault detection and classification task and has not discussed fault location estimation in the FACTS-compensated transmission line. It has been observed that ANN-based algorithms are most commonly used to solve a complex problem and proven its reliability in terms of fault detection and classification.

For an example, system studies in reference [46,49], have been taken to show its reliability in fault detection, and classification for FACTS-compensated transmission line. A general block diagram of a commonly used ANN-based algorithm is shown in Figure 10.4. In the studies system, UPFC is considered at the source end and voltage and current signal are measured at source bus. In any AI-based algorithm, the training data collection is a very important part of the algorithm. However, different signal processing tools have been utilized to prepare the training data, as discussed in Section 10.5.1. In Figure 10.4, two ANN modules are designed, first ANN module for fault detection and it has been trained such that the output of the fault detector is 1 for fault situations, else 0 for the no-fault situation. Moreover, the second ANN module has been taken for all 10 types of shunt fault classification. Classifier module gives four outputs, from which three outputs represent the three individual phases involved and one output represents ground involvement. Based on the type of fault in the protected section of the transmission line, a classifier shows any one of the fault types out of 10 types of shunt faults, as shown in Figure 10.4.

The graphical and simulated output of the ANN-based algorithm is shown in Figure 10.5.

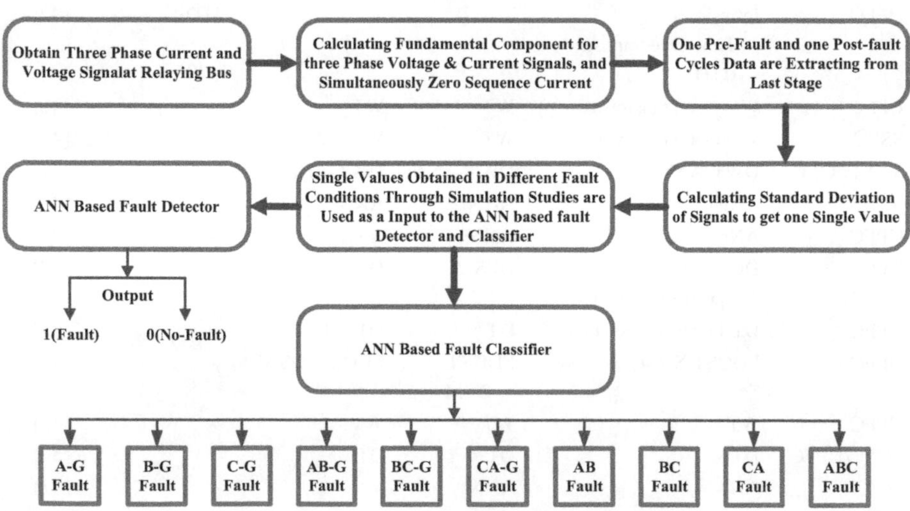

FIGURE 10.4 Block diagram of the ANN-based algorithm.

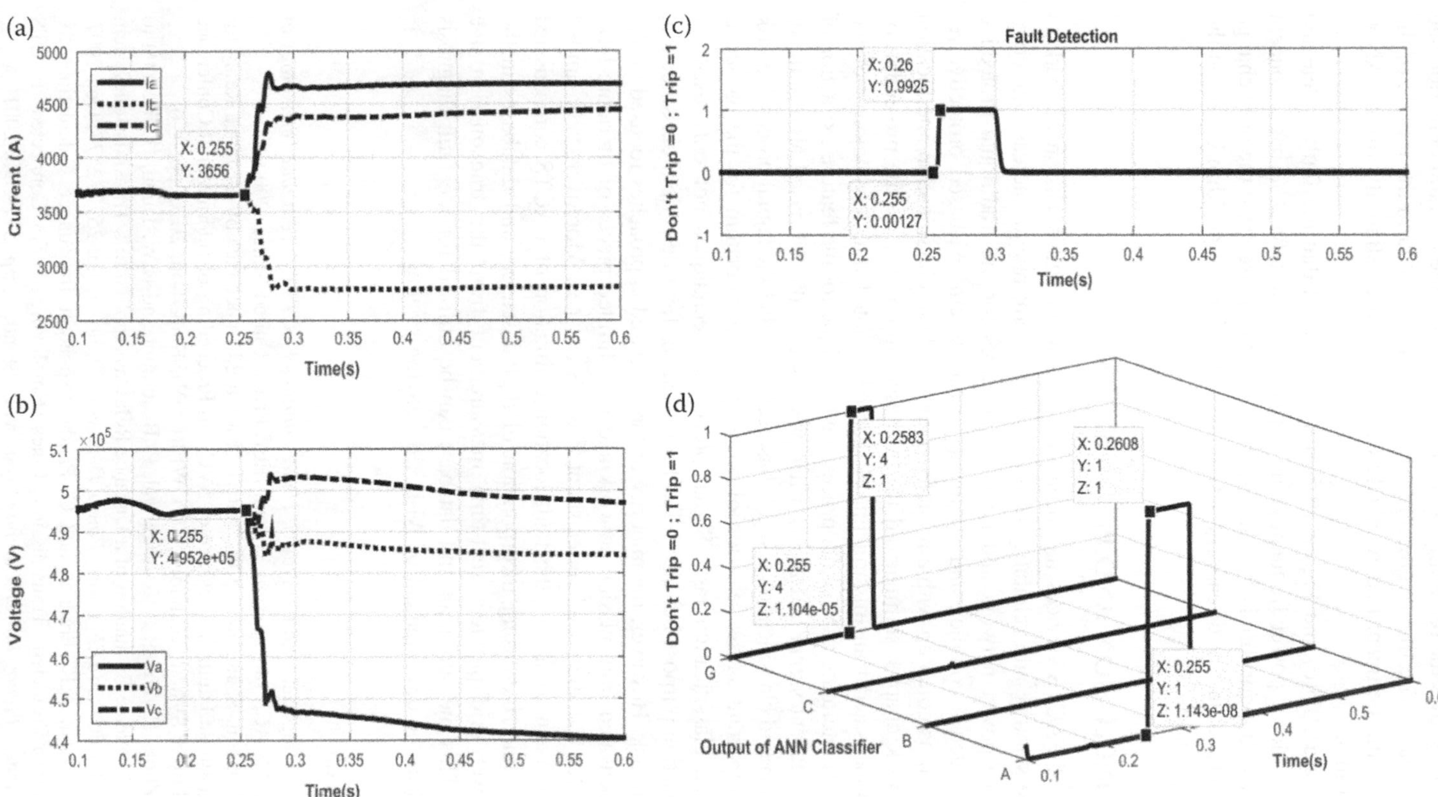

FIGURE 10.5 (a) Current signals; (b) voltage signals; (c) fault detection; (d) fault classification.

Figure 10.5 shows the reliability of the both ANN module for fault detection and classification task. The line to ground fault has been simulated in the middle of the transmission line compensated with UPFC to check the reliability of the ANN-based algorithm.

From Figure 10.5(c) and Figure 10.5(d) it is observed that the fault is detected within one cycle of power frequency and classified accurately. The basic concept for any AI-based algorithm is the same as discussed in Figure 10.4; the only change is in the precise data identification for training and training algorithm based on the different application.

10.6 CONCLUSION AND PERSPECTIVES

Integration of FACTS devices has been increased in the transmission network to improve the stability and reliability of the power system; results increase the complexity of the power network and impose challenges for the traditional protective relaying. In such a complex and nonlinear power system, AI-based protective relaying algorithms come up with a reliable solution in fault classification and location task. This book chapter summarized the AI-based approaches to the protection of FACTS-compensated transmission line. It has been noted that the majority of the proposal in the literature uses signal processing technique for the feature extraction of voltage and current signal and then further processed in the AI-based algorithm for fault detection and classification tasks. It has been seen that the signal processing task is equally important as the selection of proper training algorithm for the particular detection and classification task. Signal processing techniques not only used for feature extraction purpose but also it has been used for fault detection, classification, and location task. However, the majority of the AI-based approaches focused on the series-compensated transmission network and very limited pieces of literature have been found for the shunt and series-shunt FACTS devices. Moreover, very limited literature has been found for the fault location estimation of FACTS-compensated transmission line. Increasing incorporation of digital sensors and databases in the modern power grid provides real-time measurement from the transmission line equipment. This measurement and database can be utilized for the fault diagnosis process and to the development of AI-based relaying systems.

REFERENCES

[1] N. G. Hingorani and L. Gyugyi, *Understanding FACTS Concept and Technology of Flexible AC Transmission System*. IEEE Press, United States, 2000.

[2] A. Kazemi, S. Jamali and H. Shateri, "Measured Impedance by Distance Relay for Inter Phase Faults in Presence of SVC," in Proceeding of International Conference on Power System Technology (POWERCON), Hangzhou, 2010, pp. 1–6.

[3] X. Zhou, H. Wang, R.K. Aggarwal, P. Beaumont and R.W. Dunn, "Correction to "Performance Evaluation of a Distance Relay as Applied to a Transmission System with UPFC"," *IEEE Trans. Power Del.*, vol. 22, no. 4, pp. 2577–2577, Oct. 2007.

[4] P.K. Dash, A.K. Pradhan and G. Panda "Apparent Impedance Calculations for Distance-Protected Transmission Lines Employing Series-Connected FACTS Devices," *Electr. Power Compon. Syst.*, vol. 29, no. 7, pp. 577–595, 2001.

[5] S. Jamali, A. Kazemi and H. Shateri, "Distance Relay Over-Reaching Due to Installation of TCSC on Next Line," in Proceeding of IEEE International Symposium on Industrial Electronics, 2006, vol. 3, pp. 1954–1959.
[6] P.K. Dash, A.K. Pradhan, G. Panda and A.C. Liew "Digital Protection of Power Transmission Lines in the Presence of Series Connected FACTS Devices," in Proceeding of IEEE Power Engineering Society Winter Meeting. Conference Proceedings, 2000, pp. 1967–1972.
[7] A. Ghorbani, M. Khederzadeh and B. Mozafari, "Impact of SVC on the Protection of Transmission Lines," *Int. J. Electr. Power Energy Syst.*, vol. 42, no. 1, pp. 702–709, 2012.
[8] V.H. Ferreira, R. Zanghi, M.Z. Fortes, G.G. Sotelo, R.B.M. Silva, J.C.S. Souza, C.H.C. Guimarães and S. Gomes, "A Survey on Intelligent System Application to Fault Diagnosis in Electric Power System Transmission Lines," *Electr. Power Syst. Res.*, vol. 136, pp. 135–153, 2016.
[9] P.K. Dash, A.K. Pradhan and G. Panda, "Apparent Impedance Calculations for Distance-Protected Transmission Lines Employing Series-Connected FACTS Devices," *Electr. Power Compon. Syst.*, vol. 29, no. 7, pp. 577–595, 2001.
[10] S. Jamali, A. Kazemi and H. Shateri, "Distance Relay Tripping Characteristic in Presence of UPFC," in Proceeding of International Conference on Power Electronics, Drives and Energy Systems, 2006. PEDES '06, 12–15 Dec. 2006, pp. 1–6.
[11] A. Kazemi, S. Jamali and H. Shateri, "Adaptive Distance Protection in Presence of UPFC on a Transmission Line," in Proceeding of 10th IET International Conference on Developments in Power System Protection (DPSP 2010). Managing the Change, 29 Mar. 2010–1 Apr. 2010, pp. 1–6.
[12] R. Dubey, S.R. Samantaray, B.K. Panigrahi and G.V. Venkoparao, "Adaptive Distance Relay Setting for Parallel Transmission Network Connecting Wind Farms and UPFC," *Int. J. Electr. Power Energy Syst.*, vol. 65, pp. 113–123, Feb. 2015, ISSN 0142-0615.
[13] B. Kumar and A. Yadav, "Impact of UPFC on Alpha Plane Line Current Differential Protection," in *Proceeding of, Pune, India,* 2017, pp. 67–73.7th International Conference on Power Systems.
[14] B. Kumar, A. Yadav and M. Pazoki, "Impedance Differential Plane for Fault Detection and Faulty Phase Identification of Fact Compensated Transmission Line," *Int. Trans. Elect. Energy Syst.*, vol. 19, no. 4, pp. 1–28, Dec. 2018.
[15] S. Biswas and P.K. Nayak, "State-of-the-Art on the Protection of FACTS Compensated High-Voltage Transmission Lines: A Review," *High Voltage*, vol. 3, no. 1, pp. 21–30, Mar. 2018, doi: 10.1049/hve.2017.0131.
[16] P.K. Dash, A.K. Pradhan and G. Panda, "Distance Protection in the Presence of Unified Power Flow Controller," *Electr. Power Syst. Res.*, vol. 54, no. 3, pp. 189–198, 2000.
[17] M.C.R. Paz, R.C. Leborgne and A.S. Bretas, "Adaptive Ground Distance Protection for UPFC Compensated Transmission Lines: A Formulation Considering the Fault Resistance Effect," *Int. J. Electr. Power Energy Syst.*, vol. 73, pp. 124–131, 2015.
[18] R. Neapolitan and X Jiang. *Artificial Intelligence*. New York: Chapman and Hall/CRC, 2018. https://doi.org/10.1201/b22400
[19] A.K. Ozcanli, F. Yaprakdal and M. Baysal, "Deep Learning Methods and Applications for Electrical Power Systems: A Comprehensive Review," *Int. J. Energy Res.*, vol. 44, pp. 7136–7157, 2020.
[20] N.S. Gilani, M.T. Bina, F. Rahmani and M.H. Imani, "Data-mining for Fault-Zone Detection of Distance Relay in FACTS-Based Transmission," in Proceeding of 2020 IEEE Texas Power and Energy Conference (TPEC), College Station, TX, USA, 2020, pp. 1–6.

[21] M. Zand, M.A. Nasab, O. Neghabi, M. Khalili and A. Goli, "Fault Locating Transmission Lines with Thyristor-controlled Series Capacitors by Fuzzy Logic Method," in Proceeding of 2020 14th International Conference on Protection and Automation of Power Systems (IPAPS), Tehran, Iran, 2019, pp. 62–70.
[22] S.K. Mohanty, A. Karn and S. Banerjee, "Decision Tree Supported Distance relay for Fault Detection and Classification in a Series Compensated Line," in Proceeding of 2020 IEEE International Conference on Power Electronics, Smart Grid and Renewable Energy (PESGRE2020), Cochin, India, 2020, pp. 1–6.
[23] P.K. Mishra, A. Yadav and M. Pazoki, "Resilience-Oriented Protection Scheme for TCSC-Compensated Line," *Int. J. Elect. Power Energy Syst.*, vol. 121, pp. 1–11, 2020.
[24] A. Swetapadma, A. Yadav and A.Y. Abdelaziz, "Intelligent Schemes for Fault Classification in Mutually Coupled Series-Compensated Parallel Transmission Lines," *Neural Comput. Appl.*, vol. 32, pp. 6939–6956, 2020.
[25] S.K. Mohanty, A. Karn and S. Banerjee, "Decision Tree Supported Distance Relay for Fault Detection and Classification in a Series Compensated Line," in Proceeding of 2020 IEEE International Conference on Power Electronics, Smart Grid and Renewable Energy (PESGRE2020), Cochin, India, 2020, pp. 1–6.
[26] M. Sahani and P.K. Dash "Fault Location Estimation for Series-Compensated Double-Circuit Transmission Line Using EWT and Weighted RVFLN," *Eng. Appl. Artif. Intell.*, vol. 88, pp. 1–11, 2020.
[27] A. Swetapadma and A. Yadav, "A Hybrid Method for Fault Location Estimation in a Fixed Series Compensated Lines," *Measurement*, vol. 123, pp. 8–18, 2018.
[28] A. Swetapadma, P. Mishra, A. Yadav and A.Y. Abdelaziz, "A Non-Unit Protection Scheme for Double Circuit Series Capacitor Compensated Transmission Lines," *Electr. Power Syst. Res.*, vol. 148, pp. 311–325, 2017.
[29] P.K. Mishra, A. Yadav and M. Pazoki, "FDOST-Based Fault Classification Scheme for Fixed Series Compensated Transmission System," *IEEE Syst. J.*, vol. 13, no. 3, pp. 3316–3325, Sep. 2019.
[30] M. Mohammad Taheri, H. Seyedi, M. Nojavan, M. Khoshbouy and B. Mohammadi Ivatloo, "High-Speed Decision Tree Based Series-Compensated Transmission Lines Protection Using Differential Phase Angle of Superimposed Current," *IEEE Trans. Power Del.*, vol. 33, no. 6, pp. 3130–3138, Dec. 2018.
[31] P. Tripathi, G. Pillai and H.O. Gupta, "Kernel-Extreme Learning Machine-Based Fault Location in Advanced Series-Compensated Transmission Line," *Electr. Power Compon. Syst.*, vol. 44, no. 20, pp. 2243–2255, 2016.
[32] G.R. Rajeswary, G.R. Kumar, G.J.S. Lakshmi and G. Anusha, "Fuzzy-Wavelet Based Transmission Line Protection Scheme in the Presence of TCSC," in Proceeding of 2016 International Conference on Electrical, Electronics, and Optimization Techniques (ICEEOT), Chennai, 2016, pp. 4086–4091.
[33] P. Tripathi, G.N. Pillai and H.O. Gupta, "New Method for Fault Classification in TCSC Compensated Transmission Line Using GA Tuned SVM," in Proceeding of 2012 IEEE International Conference on Power System Technology (POWERCON), Auckland, 2012, pp. 1–6.
[34] P. Ray, B.K. Panigrahi and N. Senroy, "Extreme Learning Machine Based Fault Classification in a Series Compensated Transmission Line," in Proceeding of 2012 IEEE International Conference on Power Electronics, Drives and Energy Systems (PEDES), Bengaluru, 2012, pp. 1–6.
[35] B.Y. Vyas, R.P. Maheshwari and B. Das, "Improved Fault Analysis Technique for Protection of Thyristor Controlled Series Compensated Transmission Line," *Int. J. Electr. Power Energy Syst.*, vol. 55, pp. 321–330, 2014.
[36] B. Yashvantrai Vyas, R.P. Maheshwari and B. Das, "Pattern Recognition Application of Support Vector Machine for Fault Classification of Thyristor

Controlled Series Compensated Transmission Lines," *J. Inst. Eng. India Ser. B*, vol. 97, pp. 175–183, 2016.

[37] A. Hosny and M. Safiuddin, "ANN-Based Protection System for Controllable Series-Compensated Transmission Lines," in Proceeding of 2009 IEEE/PES Power Systems Conference and Exposition, Seattle, WA, 2009, pp. 1–6.

[38] U. B. Parikh, B. Das and R. Maheshwari, "Fault Classification Technique for Series Compensated Transmission Line Using Support Vector Machine," *Int. J. Electr. Power Energy Syst.*, vol. 32, no. 6, pp. 629–636, 2010.

[39] V. Malathi, N.S. Marimuthu, S. Baskar and K. Ramar, "Application of Extreme Learning Machine for Series Compensated Transmission Line Protection," *Eng. Appl. Artif. Intell.*, vol. 24, no. 5, pp. 880–887, 2011.

[40] J. Pardha Saradhi, R. Srinivasarao and V. Ganesh, "Wavelet-Based Algorithm for Fault Detection and Discrimination in UPFC-Compensated Multiterminal Transmission Network," in *Innovations in Electrical and Electronic Engineering. Lecture Notes in Electrical Engineering*, vol. 661. Springer, United Nation, 2021.

[41] Z. Moravej, M. Pazoki and M. Khederzadeh, "New Pattern-Recognition Method for Fault Analysis in Transmission Line With UPFC," *IEEE Trans. Power Del.*, vol. 30, no. 3, pp. 1231–1242, Jun. 2015.

[42] S. Mishra, S. Gupta, A. Yadav, "Intrinsic Time Decomposition Based Fault Location Scheme for Unified Power Flow Controller Compensated Transmission Line," *Int. Trans. Electr. Energy Syst.*, vol. 30, 2020, Art. no. e12585.

[43] H.V. Gururaja Rao, N. Prabhu and R.C. Mala, "Wavelet Transform-Based Protection of Transmission Line Incorporating SSSC with Energy Storage Device," *Electr. Eng.*, vol. 102, pp. 1593–1604, 2020.

[44] S.K. Mishra and L.N. Tripathy, "A Novel Relaying Approach for Performance Enhancement in a STATCOM Integrated Wind-Fed Transmission Line Using Single-Terminal Measurement," *Iran J. Sci. Technol. Trans. Electr. Eng.*, vol. 44, pp. 897–910, 2020.

[45] F. Fayaz and G.L. Pahuja, "ANN-Based Relaying Algorithm for Protection of SVC-Compensated AC Transmission Line and Criticality Analysis of a Digital Relay," *Recent Adv. Comput. Sci. Commun.*, vol. 13, pp. 381–393, 2020.

[46] B. Kumar and A. Yadav, "Statistical and Machine Learning Technique to Detect and Classify Shunt Fault in a UPFC Compensated Transmission Line," *Majles J. Electr. Eng.*, vol. 13, no. 3, pp. 37–48, Sep. 2019.

[47] S. K. Mishra, L. N. Tripathy and S. C. Swain, "Fault Classification and Detection of SVC Integrated Double Circuit Line," in Proceeding of 2017 Innovations in Power and Advanced Computing Technologies (i-PACT), Vellore, 2017, pp. 1–6.

[48] M. K. Jena, S. R. Samantaray and L. Tripathy, "Decision Tree-Induced Fuzzy Rule-based Differential Relaying for Transmission Line Including Unified Power Flow Controller and Wind-farms," *IET Gener., Transm. Distrib.*, vol. 8, no. 12, pp. 2144–2152, Dec. 2014.

[49] B. Kumar and A. Yadav, "Wavelet Singular Entropy Approach for Fault Detection and Classification of Transmission Line Compensated With UPFC," in Proceeding of 2016 International Conference on Information Communication and Embedded Systems (ICICES), Chennai, 2016, pp. 1–6.

[50] S. R. Samantaray, "Decision Tree-based Fault Zone Identification and Fault Classification in Flexible AC Transmissions-based Transmission Line," in *IET Gener., Transm. Distrib.*, vol. 3, no. 5, pp. 425–436, May 2009.

[51] P.K. Dash, A.K. Pradhan, G. Panda and A.C. Liew, "Digital Protection of Power Transmission Lines in the Presence of Series Connected FACTS Devices," in Proceeding of 2000 IEEE Power Engineering Society Winter Meeting. Conference Proceedings (Cat. No.00CH37077), Singapore, 2000, pp. 1967–1972.

11 AI-Based PMUs Allocation for Protecting Transmission Lines

Abdelazeem A. Abdelsalam and Karim M. Hassanin
Electrical Engineering Department, Suez Canal University, Egypt

11.1 INTRODUCTION

Transmission networks are granted with solutions more advanced than distribution networks. Improved monitoring tools for transmission and distribution networks is conceived through the use of modern measurement techniques and the creation of new data collection algorithms. Phasor Measurement Units (PMUs) represent the modern measurement devices that are used to enhance the monitoring of transmission networks. PMUs provide measurements for voltage magnitude and phase angle with a high sampling frequency and these measurements are synchronized with other measurements using time stamps created by the Global Positioning System (GPS). Connecting the PMUs together constructs a WAMS that represents the modern structure to migrate the power grid from conventional to smart. The best situation to monitor all power system buses and branches is to allocate a PMU at each bus but the cost of PMUs and their communication infrastructure will be very high. So the number of allocated PMUs should be minimum and with a high observability of system buses. There are many conventional mathematical and artificial intelligence techniques worked in this area to solve this problem of minimizing the number of PMUs with full system monitoring.

This chapter introduces the principles of PMUs and wide area monitoring systems (WAMS) that are used to protect the transmission networks. The PMUs based WAMS perform three jobs; data collection and acquisition, data transportation and monitoring, control and protection. The first job is performed by PMUs and their communication infrastructure provides the second job. While the last job is provided by the energy management system that includes control and protection actions. The conventional mathematical methods and advanced AI which are used to optimal allocate the PMUs and WAMS are presented and a comparison is conducted. A case study of finding the minimum number of PMUs with full system monitoring using an advanced AI technique is introduced. Finally, the modern applications of PMUs and WAMS in protecting the power systems is covered in this chapter.

DOI: 10.1201/9780367552374-11

11.2 BASICS OF PMUS AND WAMS

A Phasor Measurement Unit (PMU) is a measurement tool that measures the waveforms in an electrical power system using time sources to monitor the state of this system. A PMU provides the data at any system bus in phasor forms that are complex numbers. It adds an improvement of the reliability of electrical power system infrastructure [1]. Synchrophasor devices are PMUs that provide simultaneous phasor measurements at different system buses. These phasor measurements are synchronized using time stamping. This time stamping is provided by the global positioning system (GPS) [2].

PMUs help in monitoring, controlling, analyzing, and protecting the power systems. Where they enable the utilities to use the collected data to assess the status of the power networks and take a quick action to any disturbance in the power system. Thus, applying PMUs in the power transmission systems and substations is necessary to achieve smart power grids [3].

PMUs can measure phase voltages and phase currents, positive sequence components, and frequency [4]. Online applications of PMUs involve monitoring and observation of big area for the voltage and frequency stability, while offline applications include analysis of post events, data compilation, and validating model [2,3]. This analysis of post events helps in correcting the measurement in the future [5].

11.2.1 Basic PMU Structure

A PMU helps in observing the power system by obtaining synchronized measurements for voltage phasors, current phasors, and frequency at a particular node [6,7].

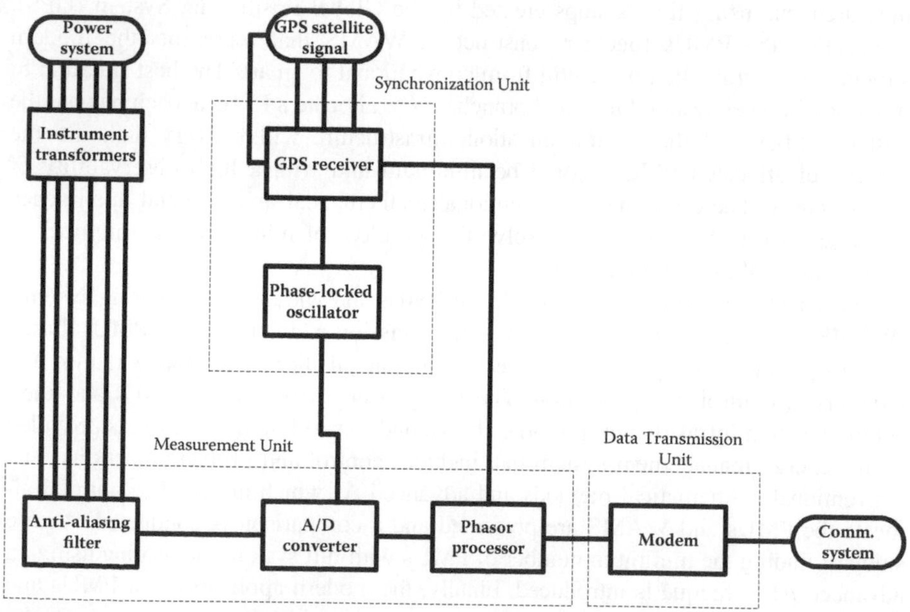

FIGURE 11.1 Basic PMU structure.

Figure 11.1illustrates the main parts that are essential to compute voltage and current phasors by taking the same time reference using GPS signals. The basic structure of a PMU consists of measurement unit, synchronization unit, and data transmission unit [8].

A time synchronization unit comprises a GPS receiver and phase-locked oscillator to provide the sampling stamp to the measuring unit. A GPS time stamping can improve the synchronization accuracy to be better than 1 μs [7,9]. A control system is equipped to generate output signals to compare them with the phases of input signals. The control signal is produced by an electronic circuit that consist of frequency generator and a phase detector. The frequency generator is an oscillator that provides the matching of phases between input and output signals, while the phase detector compares the input periodic signals with the generated signals.

The measurement unit consists of three parts:

- **Anti-aliasing filter:** assures that all the analog signals have the same attenuation and phase shift. Therefore, ensuring that the magnitudes and relative phase angle differences of the different signals are unchanged.
- **Analog to digital converter:** converts the signals from continuous time form into digital time form with sampling rate.
- **Phasor microprocessors:** they are used to integrate different PMUs in a large-scale system by modern technology [9].

Finally, the Data Transmission Unit transmits the received data from the measurement devices using a MODEM (modulator-demodulator). A MODEM modulates and demodulates a signal. A MODEM can transmit more than one analog signal at the same time [7,9].

The most important purpose of optimal PMUs in placement in electrical power systems is the full observability of the power system, which means that voltage phasors of all system buses and current phasors of all system lines are known. After placing a PMU, whatever the method used, observability of the system must be checked. If the system becomes observable, stop the PMUs placement process, else the placement process must be continued.

There are two types of power system observability analysis: topology observability and numerical observability [10]:

- **Topology observability:** In this method, the decision is based on a logical operation. Thus, it needs information about network connectivity and measurement types with their locations. The power system is observable when the full rank spanning tree can be constructed by currents measurements set. It uses a graph theory and decoupled measurement model.
- **Numerical observability:** This method is based on the numerical factorization of the measurement Jacobean or gain matrix of measurements information. The system is observable when any of these matrices is full rank. It uses either decoupled or fully coupled measurement model [10].

$$Z = Hx + V \tag{11.1}$$

where:

Z: is the metrical vector of m dimensions

H: is the Jacobian matrix of m× (2N – 1) dimensions.

x: is the voltage phasor of (2N – 1) dimensions.

V: is the vector of metrical noise of m dimensions.

N: is the number of system buses.

Numerical observability methods are not suitable for large systems because they are involved with huge matrix complexity and manipulation [10].

11.2.2 PMU Placement Rules

A power system is said to be fully observed if all its buses are observable. The observable term means that the voltage at this bus can be measured using direct or indirect methods [11].

The direct method means that the voltage at any bus and the currents in all lines connected to this bus are measured using the installed PMU. While the indirect measurement means that there is no installed PMU at this bus and the voltage and currents are calculated using the adjacent buses known voltages and currents by applying Ohm's law and Kirchhoff's Current Law (KCL). The detailed explanations of direct and indirect measurements are provided in the following rules:

- **Rule one**: If the PMU is located at a bus, the voltage and currents of connected lines to this bus are measured directly using the PMU.
- **Rule two**: By using Ohm's law, the voltage of the bus that is adjacent to the bus with PMU is calculated. This is called pseudo measurements.
- **Rule three**: If the voltages at the sending and receiving ends a line are known, the current of that branch is calculated using Ohm's law.

A simple system shown in Figure 11.2(a) is introduced to clarify the application of the PMU direct and indirect measurements. In Figure 11.2(a), suppose a PMU existed at bus 1, rule one is applied. This means voltage at bus 1 and the currents of the connected lines, lines 1–2 and line 1–3, are measured using the PMU. Since lines 1–2 and 1–3 are observed, rule two is applied, the voltage of buses 2 and 3 are evaluated using Ohm's law. Since the voltages of buses 2 and 3 are known, rule three is applied, the current of line 2–3 is calculated using Ohm's law.

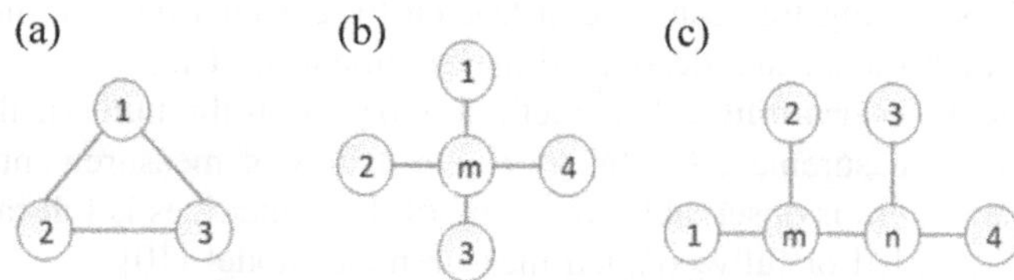

FIGURE 11.2 Modeling PMUs placement rules.

A zero injection bus (ZIB) is an efficient factor in reducing the number of PMUs for fulfilling the complete power system observability. The ZIB is defined as a bus has no injected or absorbed power. The sum of currents phasors of all incident branches of that ZIB equals zero according to KCL. The presence of ZIB can assess the power system observability based on the following three rules [12,13]:

- **Rule four**: If the ZIB is observed and all its adjacent buses are observed except a single bus, this bus will be observed using KCL at the ZIB.
- **Rule five**: If the ZIB is not observed and all adjacent buses are observed, the ZIB will be observed using the nodal voltage method.
- **Rule six**: When all buses adjacent to a group of unobservable ZIBs are observable, the voltage phasors of these unobservable ZIBs buses can be obtained through the node equation and they will be identified as observable buses.

Figure 11.2(b) and 11.2(c) are used to illustrate the explanation of these rules. In Figure 11.2(b), the ZIB is bus m. If the buses ZIB, 1, 2, and 3 are observed and bus 4 is not observed, bus 4 will be observed using KCL by calculating the current in the line m-4 and consequently the voltage can be calculated (rule four). Due to rule five, the ZIB m will be observable by using nodal voltage method at m if the buses 1, 2, 3, and 4 are observed (rule five), Figure 11.2(b). In Figure 11.2(c), using rule six, if buses 1, 2, 3, and 4 are observed, and ZIB buses m and n are not observed, the voltage of buses m and n are estimated using the nodal voltage method.

11.2.3 PMU Placement Problem Formulation

The main objective of this problem is to determine the minimum number of required PMUs and their locations to achieve a full observable power system. In this chapter, the full observability of the power system is checked for different test cases that in turn make changes in the OPP problem formulation. Following are the test cases and problem formulation in each case.

11.2.3.1 Case #1: Base case

The objective in this case is to find out the minimum number of PMUs and their locations needed for full observable power system without considering zero-injection buses or any contingencies. The objective function can be formulated by:

$$Minimize \sum_{i=1}^{N} x_i \tag{11.2}$$

$$Subject\ to\ A\,(X) \geq b \tag{11.3}$$

where

N is the number of power system buses

$A(X)$ is the observability constraint, which is non-zero if the power system is full observable with respect to the given sets of measurements according to the rule mentioned above and zero otherwise.

X is the binary decision variable vector, its elements x_i that decide feasibility of PMUs on the ith bus and these decision variables are defined as:

$$x_i = \begin{cases} 1 & if\ \ PMU\ \ is\ \ installed\ \ at\ \ i^{th}\ \ bus \\ 0 & otherwise \end{cases} \tag{11.4}$$

And A is the bus binary connectivity matrix, and its elements can be identified using the following equation:

$$A_{i,j} = \begin{cases} 1 & if\ \ i = j \\ 1 & if\ \ bus\ \ i\ \ is\ \ connected\ \ to\ \ bus\ \ j \\ 0 & otherwise \end{cases} \tag{11.5}$$

b is a column vector and all elements are one to insure the full observability of the system as follows:

$$b = [1\ 1\ 1 \ldots 1]^T_{N\times 1} \tag{11.6}$$

Measurement redundancy is the number of all measured data obtained by PMUs.

To explain how these equations are formulated, take the IEEE 14-bus test system shown in Figure 11.3 as an example:

By using Eq. (11.5), the structure of the connectivity matrix A is built for the IEEE 14-bus system as follows:

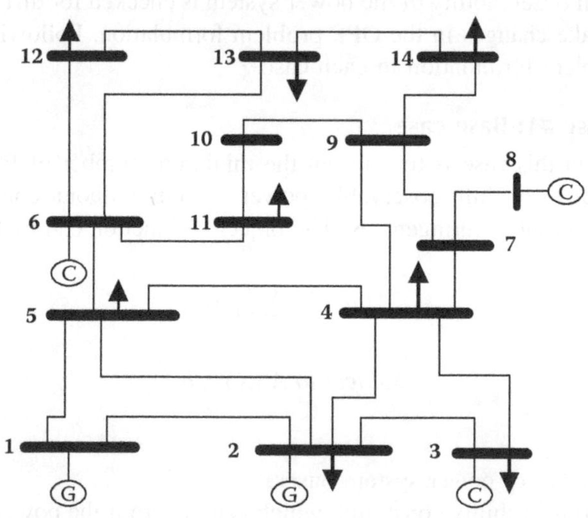

FIGURE 11.3 IEEE 14-bus test system.

$$[A] = \begin{bmatrix} 1 & 1 & 0 & 0 & 1 & 0 & 0 & 0 & 0 & 0 & 0 & 0 & 0 & 0 \\ 1 & 1 & 1 & 1 & 1 & 0 & 0 & 0 & 0 & 0 & 0 & 0 & 0 & 0 \\ 0 & 1 & 1 & 1 & 0 & 0 & 0 & 0 & 0 & 0 & 0 & 0 & 0 & 0 \\ 0 & 1 & 1 & 1 & 1 & 0 & 1 & 0 & 1 & 0 & 0 & 0 & 0 & 0 \\ 1 & 1 & 0 & 1 & 1 & 1 & 0 & 0 & 0 & 0 & 0 & 0 & 0 & 0 \\ 0 & 0 & 0 & 0 & 1 & 1 & 0 & 0 & 0 & 0 & 1 & 1 & 1 & 0 \\ 0 & 0 & 0 & 1 & 0 & 0 & 1 & 1 & 1 & 0 & 0 & 0 & 0 & 0 \\ 0 & 0 & 0 & 0 & 0 & 0 & 1 & 1 & 0 & 0 & 0 & 0 & 0 & 0 \\ 0 & 0 & 0 & 1 & 0 & 0 & 1 & 0 & 1 & 1 & 0 & 0 & 0 & 1 \\ 0 & 0 & 0 & 0 & 0 & 0 & 0 & 0 & 1 & 1 & 1 & 0 & 0 & 0 \\ 0 & 0 & 0 & 0 & 0 & 1 & 0 & 0 & 0 & 1 & 1 & 0 & 0 & 0 \\ 0 & 0 & 0 & 0 & 0 & 1 & 0 & 0 & 0 & 0 & 0 & 1 & 1 & 0 \\ 0 & 0 & 0 & 0 & 0 & 1 & 0 & 0 & 0 & 0 & 0 & 1 & 1 & 1 \\ 0 & 0 & 0 & 0 & 0 & 0 & 0 & 0 & 1 & 0 & 0 & 0 & 1 & 1 \end{bmatrix} \tag{11.7}$$

The inequality constraints of the A-matrix are structured as follows:

$$f(x) = \begin{cases} f_1 = x_1 + x_2 + x_5 \quad \geq 1 \quad (a) \\ f_2 = x_1 + x_2 + x_3 + x_4 + x_5 \geq 1 \quad (b) \\ f_3 = x_2 + x_3 + x_4 \quad \geq 1 \quad (c) \\ f_4 = x_2 + x_3 + x_4 + x_5 + x_7 + x_9 \quad \geq 1 \quad (d) \\ f_5 = x_1 + x_2 + x_4 + x_5 + x_6 \quad \geq 1 \quad (e) \\ f_6 = x_5 + x_6 + x_{11} + x_{12} + x_{13} \quad \geq 1 \quad (f) \\ f_7 = x_4 + x_7 + x_8 + x_9 \quad \geq 1 \quad (g) \\ f_8 = x_7 + x_8 \quad \geq 1 \quad (h) \\ f_9 = x_4 + x_7 + x_9 + x_{10} + x_{14} \quad \geq 1 \quad (i) \\ f_{10} = x_9 + x_{10} + x_{11} \quad \geq 1 \quad (j) \\ f_{11} = x_6 + x_{10} + x_{11} \quad \geq 1 \quad (k) \\ f_{12} = x_6 + x_{12} + x_{13} \quad \geq 1 \quad (l) \\ f_{13} = x_6 + x_{12} + x_{13} + x_{14} \quad \geq 1 \quad (m) \\ f_{14} = x_9 + x_{13} + x_{14} \quad \geq 1 \quad (n) \end{cases} \tag{11.8}$$

The previous constraints mean that, for example, according to constraint (11.8b), if PMU is located at bus 2, buses 1, 2, 3, 4, and 5 will be observed.

11.2.3.2 Case #2: Considering ZIBs

As mentioned earlier, consideration of ZIBs helps in reducing the number of PMUs required to achieve full observability for the power system. Merging process is used to formulate the existence of ZIBs in the system, which includes merging every ZIB with one of its adjacent buses. Furthermore, this merging causes redefining for network equations to reflect the changes.

This proposed case considers the presence of ZIBs and radial buses and their effect on the optimal PMUs placement. The proposed case has three rules by considering ZIBs and radial buses as follows:

- **Rule one:** combine ZIB and next radial bus together.

In Figure 11.4, the ZIB is m and the radial bus is 2 so ZIB and bus 2 are combined to construct one bus 2 m. A PMU must be located at bus 1 if bus 3 is ZIB to fulfill the observability of bus 2 m.

- **Rule two:** merging ZIB with the adjacent bus of the maximum number of branches.

In Figure 11.5, the ZIB is m and bus 3 has the maximum number of connected branches so bus m is combined with bus 3 to construct bus 3 m.

- **Rule three:** combine ZIB and the adjacent radial bus together

In Figure 11.6, the ZIB bus m is combined with the radial bus 1 of the maximum number of connected branches to construct the bus 1 m.

Applying these rules in the IEEE 14-bus test system is shown in Figure 11.3, which has bus 7 as ZIB. Bus 7 has a radial bus of its adjacent buses, which is bus 8.

FIGURE 11.4 Rule one of modeling of ZIB merging process.

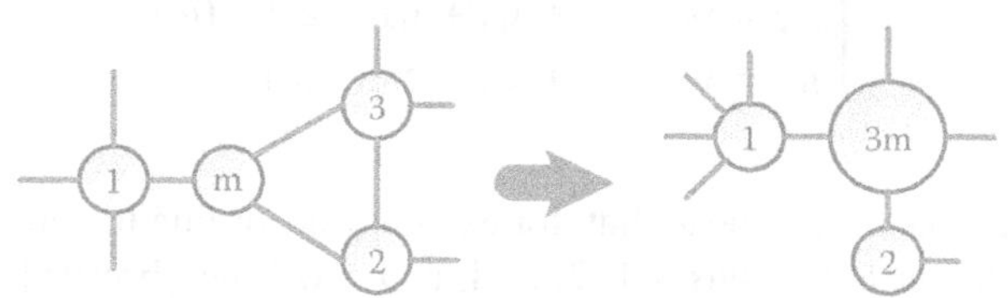

FIGURE 11.5 Rule two of modeling of ZIB merging process.

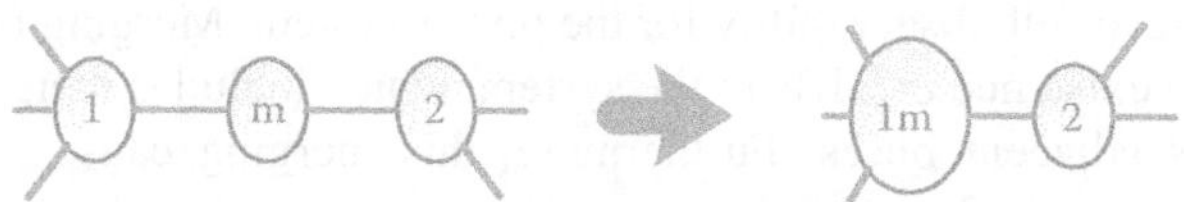

FIGURE 11.6 Rule three of modeling of ZIB merging process.

Thus, rule 1 of the merging process will be applied, which includes the ZIB must be merged with its adjacent radial bus, then bus 7 will be removed from the system, and bus 8 is now directly connected to buses 4 and 9. This means that the inequality constraints of buses 4, 7, 8, and 9 must be modified. Eq. 11.9 shows inequality constraints after the merging process. Figure 11.7 shows the bus connections before and after this merging process. Figure 11.8 shows a flowchart of the merging process rules application.

$$f(x) = \begin{cases} f_4 = x_2 + x_3 + x_4 + x_5 + x_{8'} + x_9 & \geq 1 \ (a) \\ f_{8m} = x_4 + x_{8'} + x_9 & \geq 1 \ (b) \\ f_9 = x_4 + x_{8'} + x_9 + x_{10} + x_{14} & \geq 1 \ (c) \end{cases} \tag{11.9}$$

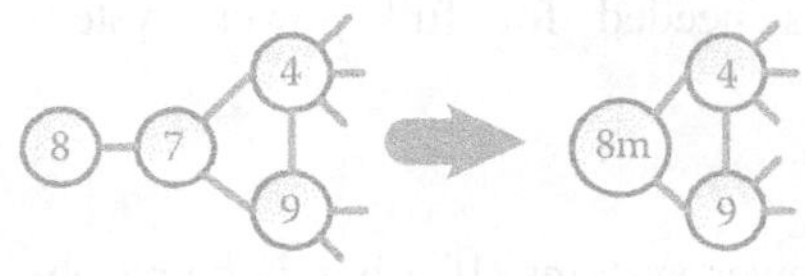

FIGURE 11.7 Modeling of ZIB for IEEE 14-bus system.

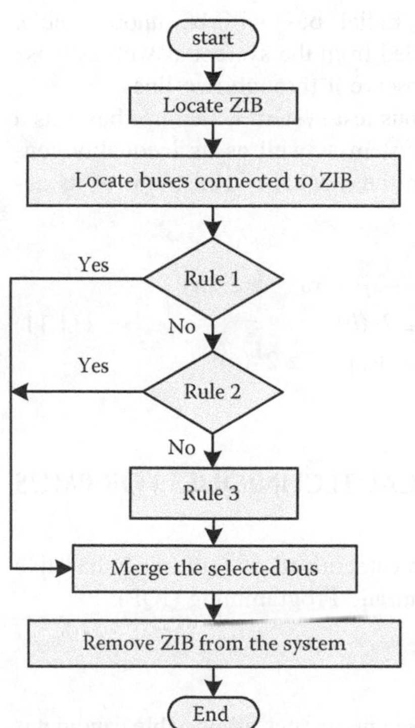

FIGURE 11.8 Flowchart of ZIBs merging process.

11.2.3.3 Case #3: Loss of a Single PMU

To enhance the reliability and fulfill the system full observability under contingencies such as a failure of PMU, each bus in the power system must be observed by at least two PMUs. This can be revealed in the constraints by multiplying the right-hand side of equation (11.6) by 2. Then:

$$b = [2\ 2\ 2\ldots\ldots.2]_{N\times 1}^{T} \tag{11.10}$$

The placement of PMUs is highly dependent on the number of radial buses existing in the system. The more radial buses in the system, the more PMUs are needed to ensure reliability and full observability of the system. As the radial has a connection with only one bus, PMUs should be located at that radial bus and the bus that it connects to it, to insure observability of the radial bus by at least two PMUs. So, radial buses increase the number of PMUs needed for full power system observability.

11.2.3.4 Case #4: Single Line Outage

Single line outage is a common fault in the power systems. If a bus is being observed by two PMUs, then a related line loss will not affect the complete observability of the power system. This can be modeled by modifying the constraints as given in Eq. (11.10). In this case, every radial bus has only one incident branch. So when losing that incident branch, that radial bus will be unobservable. Wherefore, these radial buses should be excluded from the system as well as those inequality constraints, as it is impossible to observe it through two lines.

By applying this condition on an IEEE 14-bus test system which has bus 8 as a radial bus, bus 8 will be excluded from the system as well as its inequality constraint will be removed from Eq. (11.8). Then the inequality constraint of its adjacent bus will be modified as follows:

$$f(x) = \begin{cases} f_4 = x_2 + x_3 + x_4 + x_5 + x_{7'} + x_9 & \geq 2\ (a) \\ f_{7'} = x_4 + x_{7'} + x_9 & \geq 2\ (b) \\ f_9 = x_4 + x_{7'} + x_9 + x_{10} + x_{14} & \geq 2\ (c) \end{cases} \tag{11.11}$$

11.3 CONVENTIONAL MATHEMATICAL TECHNIQUES FOR PMUS ALLOCATION

These conventional mathematical methods are categorized as follows: Exhaustive search, Integer Programming, and Integer Quadratic Programming (IQP).

11.3.1 Exhaustive Search

ES method is a general optimization method that enumerates all possible candidates found for the solution and then selects the proper candidate that satisfies the problem constraints at the most optimum value of the introduced objective function. The main advantage of this method is guaranteeing the global optimum solution. However, it is not appropriate for large-scale power systems with a large search space [14]. An exhaustive binary search methodology is introduced in [15] to obtain the minimum number of PMUs required to ensure full observability of the system. In case of many existing placement sets, which have the same minimum PMU number, a method is introduced to select the best one of them that results in maximum measurement redundancy. As a result of its exhaustive nature, the technique gives the global optimum solution, and hence reports the results obtained for different standard test systems for providing benchmark solutions for researchers. In [16], the authors developed an exhaustive binary search algorithm to solve OPP problem in addition to calculating the availability of channels, the measurement system availability of each PMU located, and the availability of measurement and maximizing the ability of identifying multiple line outages of the system in [17].

11.3.2 Integer Programming

Integer programming (IP) is a mathematical programming technique to solve optimization problems that has decision variables which are integer, in nature while the problem objective function and also the constraints are either linear or nonlinear as well as quadratic, thus results in an integer linear programming (ILP) method, integer nonlinear programming (INLP) method, and IQP method, respectively [14,18]. In [19], an IP was utilized to find optimal PMU locations for system observability and also to detect and eject bad data found in critical measurement pairs that can result in unsolvable state estimation problems or undetectable bad data. In [20], a new method is introduced for optimal PMUs placement in systems during contingencies. The ILP method is suitable to get an OPP problem solution which can ensure full observability of the whole system using minimum number of PMUs. A new algorithm is proposed in [21] to obtain the minimum required number of PMUs to achieve full observability of the electrical power system under any component outage. The proposed method runs in four stages: the conventional measurements for a known system state is found in first stage. In the second stage, the method selects redundant measurements sets during any contingency. In the third stage, the minimum redundant measurements set is picked out from these aforementioned conventional measurements using binary IP technique. Finally, these aforementioned measurements are arranged in such a way that minimizes PMU locations. The results of IEEE test systems demonstrate proposed method which achieves the system full observability under the given contingency. By using ILP in [22], the contributions are: a new OPP model is designed in such a way that models dc lines for observability analyses. The new OPP problem model aims to minimizing the cost of PMU placement while the PMU cost is a function of its measurement communication channels. This practical assumption assures locating more than one PMU at each single node in the model because of limited measurement channels of PMUs due to system's technical limitations. The performance of the proposed model is checked on IEEE test systems and a practical power system. An ILP is utilized to obtain the optimal number and locations of PMUs to ensure the system observability and then it can be used in system state estimation. When locating PMUs in a power system that already includes SCADA measurement units, it is preferable to find out the minimum number of PMUs firstly and then install them in phases in such a way that each phase has a certain number of installed PMUs in locations ensuring the highest possible level of observability [23,24].

In [25], the OPP problem is solved utilizing ILP, taking sensitivity analysis into account. As the PMU is very expensive, so it is important to allocate minimum number of PMUs as well as guaranteeing safe and secure operation of the power system. Thus, the most sensitive system buses based on load changes are determined and PMUs are optimally located so that all the sensitive buses are certainly observed. A proposed technique in [26] for PMUs allocation, in multiple stages during a given time period, achieves full observability of the electrical power system even under contingencies such as a single line outage or a single PMU failure. The introduced method is based on ILP and a multi-criterion decision-making approach. An ILP algorithm is utilized in the first stage to find the minimum number of PMUs that ensures the complete full observability of the power system. This PMU allocating includes the contingency cases such as a single-line outage and a single-PMU failure. A MCDM method is subsequently developed to preferring the optimal PMUs sites in the system. This criterion includes three indices, namely a voltage control area observability index (VOI), a tie-line oscillations observability index (TOI), and a bus voltage observability index (BOI). Finally, selecting the PMU locations in each stage by classifying the optimal places using the previously mentioned indices in the MCDM process. The number of PMUs are further minimized and hence decreasing the total system cost.

The authors in [27] introduce the varied aspects of OPP problem employing binary integer linear programming (BILP) technique. They also introduce cases with and without considering the existence of ZIBs. OPP problem may have various optimal solutions; therefore, two indices, namely BOI and SORI, have been introduced to classify these solutions. An ILP model is introduced in [28] to obtain the minimum number of PMUs and their optimal sites in the system according to the feasibility of the power system proposed backup protection strategy.

In [29], the authors presented an optimization model to obtain the minimum PMUs in the electrical power network. That model contains observability requirements which are generally based on a set of probabilistic criteria. The non-linear mathematical term related with the probability criteria of an observability index is altered to a linear model by introducing a useful linearization technique. The previously mentioned linear model is described as a Mixed Integer Programming (MIP) problem. Thus, that proposed model is suitable with the Mixed Integer Programming solving tools and is tested also on large-scale electrical power systems. Different formulations that are numerical and use integer programming permit easy analysis of network full observability [28]. An effective and inclusive formulation, based on a mixed integer linear programming (MILP) technique, is used for solving the OPP problem and while considering the channel availability and different contingencies associated with the power systems, i.e. a single-line outage and a loss of PMU were also considered in [30].

In [31], a MILP framework was introduced for the proposed model which simultaneously optimizes two objective functions: (i) first function is minimizing the number of PMUs needed and (ii) second function is maximizing the system's measurement redundancy. Incorporating the impact of existing ZIBs, the proposed formulation is utilized to properly evaluate the probability of unobservability of system buses that result from a single line outage and a PMU loss. An ILP-based technique for PMU placement has been introduced to achieve full observability of the power system under contingencies. During optimal placement difficulties, there are two proposed indices in the proposed model to obtain the optimal solutions; namely BOI, which indicates the total number of synchro-phasors that observe a given system bus and to obtain the sum of all BOI for the whole system, the second index is SORI. An optimal PMU placement scheme is proposed considering different PMU communications channel capacities. This has been specified to be more useful to the power system planner in extending from the point of view of application-cost to PMUs-PDCs installation in the WAMS-based power system at the system planning stage. Not only the cost factor is considered, but also different criteria are suggested for optimally allocating of PMUs in a given power system, such as network observability, system state estimation accuracy, in addition to robustness [32–38].

In [39], the authors introduced ILP with a fast and new practical model to solve the OPP problem. It contains power system contingencies such as a single-line outage and a PMU loss. The proposed model is more flexible because of the ability of handling each situation separately or altogether at the same time. Also, the proposed model contains measurement limitations such as communication constraints of electrical power systems. The main advantage of this method is that it takes a little time to find the optimal solution and it also gives a local optimal solution for a large-scale power system.

The OPP problem is formulated and solved, in [40], as an ILP problem. The results indicate the minimum number of PMUs that are strategically located to eliminate measurement shortages in the entire system. This denotes that any bad data appears on any single measurement will be accurately detected.

A new multi-stage PMU placement technique was proposed to maximize the power system observability during different time horizons. An ILP is widely used to solve the optimal PMU placement problem. An optimal placement of PMUs to ensure full power system observability during the existence of conventional measurements as well as considering the existence of zero-injection buses is also presented in [41–43].

A Binary Integer Programming technique is proposed in [44] for instantaneous PMU placement as well as conventional electrical power flow measurements for observing fault in electrical power systems. The formulation at the beginning is unsolvable because it is based on a nonlinear programming problem. Thus, initial formulation produces a binary non-linear integer programming problem that is changed later to a corresponding ILP problem using Boolean implications. The change provides a very high scale of scalability. In [45], the Binary Integer Linear Programming method is introduced to solve the OPP problem. The authors considered a local redundancy of the existing conventional measurements in the PMU allocation strategy. Besides, considering limiting the number of PMUs that are implemented yearly, a multistage PMU allocation strategy is introduced according to the betterment on local redundancy. A new method is proposed in [46] to solve the optimal PMU placement problem to assure that all WAMPAC applications attain full measurements observability. The authors categorize the problem according to the type of measurement shared. The first category combines the applications that need phasor measurement related to a network's stability. The second category requires a bus's normal voltage and a third category requires measurements under other faulted conditions.

In [47], authors introduce a new PMU placement algorithm, that minimizes the number of PMUs, considering network connectivity, existence of ZIBs, system reliability, N-1-line outage, and the level of voltage stability. By including the concept of system reliability, the proposed algorithm can efficiently reduce the number of PMUs while guaranteeing the full observability during major system contingencies. The proposed method utilizes the Voltage Stability Index (VSI) also to weigh the priority level of system buses, which is used to show the buses during various system contingencies.

11.3.3 Integer Quadratic Programming

Integer quadratic programming (IQP) is used to solve the optimization problems that have quadratic objective functions and linear constraints. In this method, all the design variables take only integer values [48]. An IQP-based method is utilized in [49] to obtain the minimum number of PMUs and their optimal sites in the power system to achieve full topological observability of the power system. In [50], an IQP algorithm is utilized to obtain the optimal locations of minimum PMUs required. The optimization process attempts to achieve two objectives: (1) the first is to minimize the number of PMUs required to ensure full observability of the system during normal operating conditions as well as contingency conditions such as single line outage and a PMU failure, and (2) to maximize the system measurement redundancy of all buses in the system. The proposed technique can be used to find PMUs locations in the case if existing conventional measurements, such as line power flows and power injections measurements. While the authors in [51] introduce a new method for solving the OPP problem while ensuring complete power system observability. They formulate the problem as a quadratic minimization problem combining continuous decision variables that are subjected to nonlinear system's observability constraints. The optimal solution is determined by an unconstrained weighted least square (WLS) approach with a nonlinear nature.

11.4 AI APPLICATION TO PMUS ALLOCATION

A proper methodology is needed to solve the OPP problem and obtain the minimum number of the PMUs ensures full observability of the power networks. In order to solve the OPP problem, various mathematical methods have been utilized. Issues such as difficulties of find out local minimum and dealing with different constraints in conventional mathematical techniques are overcome by other optimization methodologies like heuristic techniques, such as simulated annealing, genetic algorithms, differential evaluation, tabu search, ant colony optimization and particle swarm optimization, etc. Table 11.1 shows objective functions and contribution of heuristic methods that are utilized in solving OPP problem. Table 11.2 shows a comparison between conventional mathematical optimization techniques and Artificial Intelligence (AI) techniques.

11.5 CASE STUDY

The flower pollination algorithm (FPA) is utilized to obtain a minimum number of Phasor Measurement Units (PMUs) needed to ensure full observability for the power system as well as their optimal locations to achieve maximum measurement redundancy [88,89]. The MATLAB® programming environment is used to get these results by applying FPA on several test systems. To ensure the efficiency and robustness of the proposed method in solving the problem, it should be applied on different test systems as well as comparing the obtained results with those of other methods. This chapter presents the application of the proposed optimization algorithm to determine the minimum number of PMUs and their optimal locations for different IEEE standard test systems at different test cases.

11.5.1 IEEE 14-Bus System

It has five synchronous machines, three of which are synchronous condensers used for supporting reactive power, and the other two are generators. The single line diagram is shown in Figure 11.9. The ZIB locations and radial buses are illustrated in Table 11.3.

The results of applying FPA on this system under different cases are discussed in the following subsections.

11.5.1.1 Case #1: Base Case

This case aims to achieve full power system observability with minimum number of PMUs as well as maximizing measurement redundancy without considering the existence of ZIBs and without considering any contingencies. The constraint of this case is to observe each bus in the system at least one time. Table 11.4 shows the optimum number and locations of PMUs needed as well as their corresponding redundancy index for IEEE 14-bus test system. From the results shown in Table 4.2, the proposed method obtained five location sets with the four PMUs at each set. However, the best solution is the location set (2, 6, 7, 9) that achieved maximum measurement redundancy that equals 19. This means that many buses in the system are observed more than once.

11.5.1.2 Case #2: Considering ZIBs

From the problem formulation mentioned previously, it is clear that the existence of ZIBs greatly affects the minimum number of PMUs needed to obtain a full observable power system. A ZIB in the system helps in excluding one bus from the system buses, which are needed to be directly observed by PMUs, and this in turn

TABLE 11.1
Objective Function and Contribution of Heuristic Methods

Method	Objective Function and Contributions	Main Consideration	Advantage	Ref./ Year
Simulated annealing (SA)	Sensitivity constrained PMUs allocation to achieve a fully observed power system has been investigated using SA method. Obtained locations of PMUs ensure full observability of the power system and provide more worthy dynamic data of the electrical power systems.	ZIBs- Placing PMUs on system buses with higher sensitivity	Considering parameters sensitivity	[52]/ 2005
	Minimizing the number of the PMUs. Modeling observability depth, using spanning trees of the power system graph	An incomplete system observability based on the depth of unobservability	Communication-constrained PMUs placement	[53]/ 2005
	Optimal allocation of PMUs to enhance the ability of bad data detection. SA methodology with stochastic new solution generating is used	Critical measurements recognition	Power system is full observable with critical measurements free	[54]/ 2007
Stochastic simulated annealing (SSA)	Minimizing PMUs number using the proposed SSA algorithm	Including critical measurements recognition is as a penalty function.	Including critical measurements recognition.	[55]/ 2008
Genetic Algorithm (GA)	Minimizing number of PMUs and locating them optimally, ensuring a fully observed power system. A GA-based technique is used and equipping PMUs with	Include PMUs locations and their relation with current phasors that must be measured	Considering current measurements channels required in the optimization problem	[56]/ 2003

(Continued)

TABLE 11.1 (Continued)
Objective Function and Contribution of Heuristic Methods

Method	Objective Function and Contributions	Main Consideration	Advantage	Ref./ Year
	current phasors measurements as the maximum number of concurrent lines in system buses is highlighted			
	Multi objective proposed model, minimizing PMUs number and maximizing measurements redundancy.	A single line outage and a single PMU failure are considered	The population size is very small and less iterations	[57]/ 2011
Genetic algorithm (GA) and Branch and bound	Minimizing the number of PMUs, and formulate a topological based observability.	Considering the existence of zero-injection buses	The problem is formulated as mixed integer linear and nonlinear programming	[58]/ 2009
Immune genetic algorithm (IGA)	Minimizing number of PMUs, introducing improved IGA that is based on usage of the local and prior knowledge that related to the existed problem	Considering the existence of zero-injection buses, and adding three new impactful vaccines	The process speed is increased remarkably, the algorithm is applied to a large scale system and prevents the familial reproduction	[59]/ 2009
Non-dominated sorting genetic algorithm (NSGA)	Minimizing the number of PMUs as well as maximizing system's measurement redundancy. Pareto-optimal solutions are determined instead of a single optimal solution	Measurement redundancy, ZIBs	Providing optimal solution for conflicting objectives as well as correction of infeasible solutions	[60]/ 2003
Hybrid Genetic algorithm and simulated annealing (HGS)	The number of PMUs and RTUs are minimized with the existence of free critical measurement	Conventional measurements and remote terminal units (RTUs), bad data detection, and current measurement loss	Applicable to current electrical power networks that are monitored using RTUs.	[61]/ 2009

(*Continued*)

TABLE 11.1 (Continued)
Objective Function and Contribution of Heuristic Methods

Method	Objective Function and Contributions	Main Consideration	Advantage	Ref./Year
Particle swarm optimization (PSO)	Minimizing PMUs installation costs and modeling a non-uniform cost of PMUs for different system buses	Non-uniform cost of PMUs placements	Considering realistic installation costs of PMUs and minimizing the total cost instead of minimizing the number of PMUs.	[62]/ 2010
Binary particle swarm optimization (BPSO)	Minimizing PMUs number for full system observability. Introducing a hybrid algorithm based on BPSO algorithm and immune mechanism.	Maximum measurement redundancy is achieved, and considering a single PMU and multi PMU fault	The process speed is increased and the function is simplified	[63]/ 2008
	Minimizing PMUs number with presence of conventional measurements such as injections and flows	Zero-injection buses are considered as well as conventional measurement, and achieving maximum measurement redundancy	Modeling various types of system conventional measurements in OPP problem formulation.	[64]/ 2011
	Minimizing the number of PMUs and fault tolerant formulation of PMUs placement.	Control system reconfigurability criterion as well as modeling data loss at a given PMUs number.	Robustness of the obtained solutions against data loss at a given number of PMUs	[65]/ 2013
Modified binary particle swarm optimization (BPSO)	Minimizing the number of required PMUs as well as introducing a novel rule of topological observability of power systems.	Zero-injection buses-PMU/line outage	Introducing a new rule of topological observability for reducing PMUs number.	[66]/ 2011
Differential evolution (DE)	Minimizing the mean square error (MSE) by finding minimum PMUs number, with or without presence of	Conventional measurements are present and obtaining Minimum square error (MSE) of	Accurate, high speed and simple process, applicable for multi-objective problems	[67]/ 2010

(Continued)

TABLE 11.1 (Continued)
Objective Function and Contribution of Heuristic Methods

Method	Objective Function and Contributions	Main Consideration	Advantage	Ref./ Year
	conventional measurements	system state estimation.		
	Minimizing the mean square error (MSE) of system state estimation by minimizing the PMUs number.	Considering persistent changes in the topology of power system	Incorporating OPP into state estimation problem	[68]/ 2011
	Minimizing number of PMUs to ensure both normal and fault observability of the system.	Fault observability and considering zero- injection buses	Using minimum number of PMUs for power system observability during normal and faulty condition	[69]/ 2013
Non-dominated sorting differential evolution (NSDE)	Minimizing the number of required PMUs and maximizing reliability of measurements and it is a multi-objective optimization algorithm	ZIBs and Maximum measurement reliability.	Accurate and complete achievement as well as flexibility.	[70]/ 2010
Tabu Search (TS)	Two competing objectives are considered including minimum number of needed PMUs and maximum redundancy	Maximum measurement redundancy and Zero-injection buses	Solutions with high accuracy as well as less computational complexity	[71]/ 2006
A new parallel Tabu search (TS)	Minimizing the number of needed PMUs as well as Proposing a new parallel TS algorithm	Zero-injection buses are considered, communication constraint and State estimation matrix condition	Less computational time.	[72]/ 2008
Recursive tabu search (RTS)	Minimizing number of PMUs utilizing a recursive Tabu search method.	Zero-injection buses and maximum measurement redundancy	Applied to large scale power systems.	[73]/ 2013
An adaptive clonal algorithm (CLONALG)	Two objectives include minimum PMUs and maximum system's measurement redundancy.	Maximum measurement redundancy	High speed of process as well as Obtaining feasible schemes	[74]/ 2006

(Continued)

TABLE 11.1 (Continued)
Objective Function and Contribution of Heuristic Methods

Method	Objective Function and Contributions	Main Consideration	Advantage	Ref./ Year
Branch and bound (B and B)	Minimizing PMUs number and Monitoring pilot buses that are required for the control of secondary voltage.	Zero-injection buses, improving the performance of secondary voltage control	Monitoring pilot buses to increase speed of voltage control scheme	[75]/ 2008
Matrix reduction	Minimum PMUs number, Using preprocessing technique and solving the problem utilizing mathematical based methods	reduction of Virtual data elimination preprocessing method and algorithm of matrix reduction,	Reduction of the placement model's size as well as computational effort.The method is applied to large scale power system.	[76]/ 2008
Improved ant colony optimization	Minimizing the number of required PMUs as well as introducing an improved ant colony algorithm	Maximum system's measurement redundancy,	High speed process	[77]/ 2009
Bacterial Foraging algorithm (BFA)	Minimizing the number of needed PMUs and maximizing measurement redundancy, considering existence of conventional measurements	Zero-injection buses and Maximum measurement redundancy	Suitable for real world current power system due to the ability to model conventional measurements	[78]/ 2010
Iterated local search (ILS)	Minimizing PMUs number using a page rank placement algorithm (PPA) and ILS.	PMU failure	Considering contingencies, easy to understand and implement.	[79]/ 2010
Biogeography based optimization (BBO)	Minimizing the number of meters and PMUs needed considering outage of single branch/meter and outage of single branch/PMU.	Zero-injection buses, line outage, PMU failure, PMU/line outage and SCADA meter outage	Robustness against the outages, Utilizing virtual bus reduction method for reducing the scale of the power system,	[80]/ 2011
Multi objective biogeography	Two objectives: minimum number of	Zero-injection buses are considered as	Producing distributed Pareto optimal	[81]/ 2012

(*Continued*)

TABLE 11.1 (Continued)
Objective Function and Contribution of Heuristic Methods

Method	Objective Function and Contributions	Main Consideration	Advantage	Ref./ Year
based optimization (MO-BBO)	required PMUs and maximum system's measurement redundancy.	well as measurement redundancy	solutions better than NSGA-II and NSDE	
Mutual information (MI)	Minimum number of required PMUs using an information theoretic concept and uncertainty modeling	Conventional measurement as well as considering PMU failure	Modeling the uncertainties in the power system states	[82]/ 2013
Binary imperialistic completion algorithm (BICA)	Minimizing PMUs number required for complete system observability and maximizing measurement redundancy. A novel topological observability rule of ZIBs is introduced	Zero-injection buses, Line outage, PMU failure, PMU/line outage, Measurement redundancy	Fast convergence, very small deviation, as well as zero standard deviation	[83]/ 2013
Chemical reaction optimization (CRO)and simplified version of CRO (SCRO)	Minimum number of needed PMUs is obtained to reach full power system observability	Zero-injection buses, large scale power systems	Efficiency, simple structure, adaptability and less computational time	[84]/ 2013
Cellular learning automata (CLA)	A multi-objective optimal placement of Minimum PMUs number and maximum measurement redundancy. PMUs placement is also done considering the existence of conventional non-synchronous measurements by using a generalized observability function	PMU failure, Line outage, maximum measurement redundancy, PMU/ line outage and conventional measurements	Efficient in large scale power systems	[85]/ 2013

(Continued)

TABLE 11.1 (Continued)
Objective Function and Contribution of Heuristic Methods

Method	Objective Function and Contributions	Main Consideration	Advantage	Ref./ Year
Combination of Minimum spanning tree algorithm with improved genetic algorithm (MST-GA)	Minimizing PMUs number using a hybrid method	Maximum measurement redundancy	This method is capable of repairing infeasible solutions and balancing between reparation efficiency and solutions quality.	[86]/ 2013
Artificial bee colony(ABC)	ABC concept is applied to obtain PMUs minimum number to achieve a full observable power system and satisfying maximum measurement redundancy	Zero-injection buses are considered as well as Single line outage	Feasibility and performance of the proposed method are demonstrated through comparing the simulation obtained results with the earlier works	[87]/ 2014
Flower Pollination Algorithm (FPA)	Obtaining minimum number of PMUs that achieves full power system observability and maximizes measurements redundancy	Minimum PMUs number, full observability and maximum redundancy.	High speed method, applicable for large power systems and less computational procedures.	[88]/ 2017
	Obtaining minimum number of PMUs, at normal and contingency operation, that achieves full power system observability and maximizes measurement redundancy	Minimum PMUs number, full observability, PMU failure, single line outage and maximum redundancy.	High speed method, applicable for large power systems and less computational procedures.	[89]/ 2020

helps in reducing the number of PMUs needed to observe all system buses. The constraint of this test case is to make every bus in the system to be observed at least once. Table 11.5 shows the optimum number and locations of PMUs needed for the IEEE 14-bus system to be full observable when considering the existence of the ZIBs. From the results shown in Table 11.5, the existence of ZIBs helps in reducing

TABLE 11.2
Comparative Analysis of Mathematical

Properties	Mathematical Technique	Artificial Intelligence Techniques
Problem size	Mathematical methods are typically used when the system is not very complicated. But, some complicated problems are solved using conventional mathematical techniques.	AI are computational methods that attempt to simply imitate the human knowledge capability so as to resolve optimal PMU placement problem.
Assumptions	Incorporating several assumptions into the models, is always used to supplement the lack of physical understanding.	There are less simplifications and assumptions incorporated in the problem.
Data requirement	Conventional technique first uses the principles (viz. physical laws) to deduce the relationships between different parameters of the system, which is usually justifiably simplified using various assumptions and need former knowing about the relationships nature.	AI models do not depend on the data that count on the data alone to find out the parameters and structure that solve a system problem, but use lesser assumptions about the physical behavior of the power system than mathematical models.
Problem solving approach	Mathematical models depend on assuming the model's structure in advance. Hence many mathematical models be unable to simulate the more complex problems.	In AI techniques, the model is developed based on training of data input-output pairs to obtain the parameters and structure of the model.
Upgrade of results	Conventional mathematical techniques are not able to upgrade the past results.	AI models are always updated to obtain the best results by introducing new training examples every time new data become available.
Efficiency	Conventional mathematical techniques need more time to resolve the complex model problem.	Artificial intelligence techniques can solve complex problems efficiently and during less time.
Precision & accuracy	As compared to AI techniques, mathematical models provide a lower level of accuracy due to the assumptions that are incorporated in the modeling.	As compared to mathematical techniques, AI techniques provide a higher level of precision and accuracy due to less assumptions as well as less chances of error.
Transparency & model uncertainty	In mathematical models, such as MIP, The existence of integer decision variables in the problem model makes the problem hard to be resolved using a conventional log barrier.	AI techniques are still facing classical contraposition due to some inveterate inadequacy that need more attention in the future including the lack of transparency, model uncertainty, and knowledge extraction.

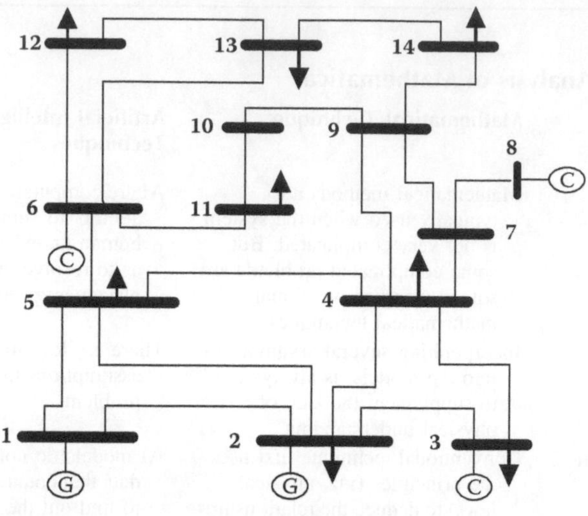

FIGURE 11.9 The single line diagram of IEEE 14-bus system.

TABLE 11.3
Locations of ZIBs and Radial Buses for IEEE 14-Bus Test System

Zero Injection Buses		Radial Buses	
No.	*Locations*	*No.*	*Locations*
1	7	1	8

TABLE 11.4
OPP Results for IEEE 14-Bus System at Base Case

No. of PMUs	PMUs Locations Sets	R. Index
4	2, 6, 7, 9	19
	2, 7, 11, 13	16
	2, 7, 10, 13	16
	2, 6, 8, 9	17
	2, 8, 10, 13	14

TABLE 11.5
OPP Results for IEEE 14-Bus System with Considering ZIBs

No. of PMUs	PMUs Locations Sets	R. Index
3	2, 6, 9	15

TABLE 11.6
OPP Results for IEEE 14-Bus system with Losing a Single PMU

No. of PMUs	PMUs Locations Sets	R. Index
9	2, 4, 5, 6, 7, 8, 9, 11, 13	39
	1, 2, 4, 6, 7, 8, 9, 11, 13	37
	2, 3, 5, 6, 7, 8, 9, 10, 13	36

the number of PMUs needed by 1 PMU. This reduction proves the importance of ZIBs in solving the OPP problem.

11.5.1.3 Case #3: Loss of a Single PMU

To increase the reliability of monitoring power systems, it is important to guarantee the full observability of the power system under contingencies such as loss of PMU or single line outage. This case aims to keep the power system fully observed even if PMU is lost. Losing PMU means losing observability for buses. Therefore, the constraint of this case is to make every bus in the system to be observed by two PMUs at least. Therefore, if PMU, which observed a bus, is lost, the other PMU will still observing that bus. Table 11.6 shows the optimum number of PMUs and their locations. From the results shown in Table 11.6, to increase the reliability of the power system the number of PMUs should be increased to guarantee power system full observability while losing PMU. This increase of PMUs depends on the number of system buses, the connection between them as well as the number of radial buses. As for each radial bus, a PMU should be pre-allocated at that radial bus and another PMU should be pre-allocated at the bus incident to that radial bus to insure its observability while losing a PMU. The minimum number of PMUs is 9 witha variety of location sets, the best of these sets is (2, 4, 5, 6, 7, 8, 9, 11, 13) that results in maximum measurement redundancy, which equals 39.

11.5.1.4 Case #4: Single Line Outage

To increase the reliability of the power system, the system full observability should be guaranteed even if a line is lost. This case has the same condition of the previous case which is every bus should be observed at least twice through two lines whose ends are buses with installed PMUs. As if a line is lost, the bus will be still observed through the other line. Table 11.7 shows the optimum number and locations of PMUs needed to solve OPP problem at this case. From the results shown in Table 11.7, it is noticed that the number of PMUs is less than those of the previous case although they have the same objective function and the same constraint. This is because of radial buses, which each one of them has a connection with only one incident bus. Therefore, it is impossible to be observed through two lines. In this case, all radial buses will be excluded from the systems and this is called radial branch outage. This helps to make the number of PMUs needed less than those of the previous case. The number of PMUs reduced by 2 for IEEE 14-bus system. The minimum number of PMUs is 7 with

TABLE 11.7
OPP Results for IEEE 14-Bus System with a Single Line Outage

No. of PMUs	PMUs Locations Sets	R. Index
7	2, 4, 5, 6, 9, 10, 13	33
	2, 4, 5, 6, 9, 11, 13	33
	1, 2, 4, 6, 9, 10, 13	31
	1, 2, 4, 6, 9, 11, 13	31

variety of location sets; the best of these sets is (2, 4, 5, 6, 9, 10, 13) that results in maximum measurement redundancy which equals 33.

11.5.2 IEEE 30-Bus System

It has six synchronous machines; four of these machines are synchronous condensers used for reactive power supporting, and the other two machines are generating units used for generation. The single line diagram is shown in Figure 11.10. The ZIB locations and radial buses are illustrated in Table 11.8. The results of applying FPA on this system under different cases are discussed in the following subsections.

11.5.2.1 Case #1: Base Case

Table 11.9 shows the results of optimal PMUs placement problem for IEEE 30-bus system at base case without considering the existence of ZIBs or any contingencies. To ensure full observability of IEEE 30-bus system, 10 PMUs are needed at least. As noticed from Table 11.9, the proposed method helps in obtaining a lot of PMU location sets with the same minimum number of PMUs, but the best location set is (2, 4, 6, 9, 10, 12, 15, 18, 25, 27) that achieves maximum measurement redundancy, which equals 52.

11.5.2.2 Case #2: Considering ZIBs

Table 11.10 shows the results of optimal PMUs placement problem for IEEE 30-bus system with considering the existence of ZIBs. The existence of 7 ZIBs helps in reducing the number of PMUs needed by 3. To ensure full observability of IEEE 30-bus system, 7 PMUs are needed at least. As noticed from Table 11.10, the proposed method helps in obtaining a lot of PMU location sets with the same minimum number of PMUs and the same measurement redundancy. The most important factor that controls the number of PMUs needed, their locations, and measurement redundancy is the number of system buses and the connection between these buses.

11.5.2.3 Case #3: Loss of a Single PMU

Table 11.11 shows the results of optimal PMUs placement problem for IEEE 30-bus system with considering the loss of a single PMU. Considering the loss of a single PMU increases the number of PMUs needed for full power system

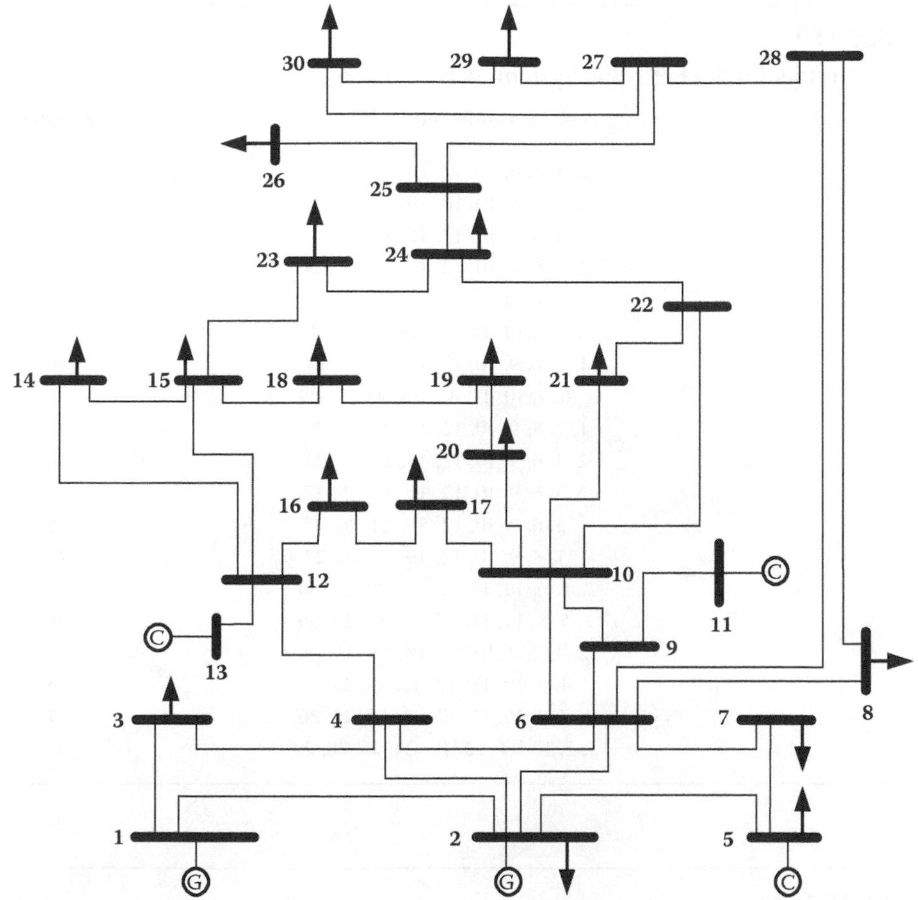

FIGURE 11.10 The single line diagram of IEEE 30-bus system.

TABLE 11.8
Locations of ZIBs and Radial Buses for IEEE 30-Bus Test System

Zero Injection Buses		Radial Buses	
No.	*Locations*	*No.*	*Locations*
7	6, 9, 22, 25, 27, 28	3	11, 13, 26

observability to be almost twice the PMUs needed at base case. To ensure full observability of IEEE 30-bus system, 21 PMUs are needed at least. As noticed from Table 11.11, the proposed method helps in obtaining a lot of PMU location sets with the same minimum number of PMUs, but the best location set is (1, 2, 3, 5, 6, 9, 10, 11, 12, 13, 15, 17, 18, 20, 22, 24, 25, 26, 27, 28, 29) that achieves maximum measurement redundancy, which equals 83.

TABLE 11.9
OPP Results for IEEE 30-Bus System at Base Case

No. of PMUs	PMUs Location Sets	R. Index
10	2, 4, 6, 9, 10, 12, 15, 18, 25, 27	52
	2, 4, 6, 9, 10, 12, 19, 24, 25, 27	51
	2, 4, 6, 9, 10, 12, 15, 19, 25, 29	50
	2, 3, 6, 9, 10, 12, 19, 24, 25, 29	47
	3, 5, 6, 10, 11, 12, 18, 23, 26, 29	40
	1, 5, 6, 10, 11, 12, 15, 20, 25, 30	44
	1, 2, 6, 9, 10, 12, 15, 20, 25, 27	50
	1, 6, 7, 10, 11, 12, 18, 23, 26, 29	40
	1, 7, 8, 9, 10, 12, 18, 24, 25, 27	42
	2, 3, 6, 9, 10, 12, 15, 20, 25, 27	50
	3, 5, 6, 9, 10, 12, 15, 19, 25, 27	48
	1, 5, 6, 9, 10, 12, 18, 24, 26, 27	45
	1, 7, 8, 9, 10, 12, 19, 24, 26, 27	40
	2, 4, 6, 10, 11, 12, 18, 23, 25, 30	46
	1, 5, 8, 10, 11, 12, 15, 18, 25, 30	39
	2, 4, 6, 9, 10, 12, 18, 24, 25, 29	49
	2, 4, 6, 10, 11, 12, 15, 18, 25, 30	48
	1, 7, 8, 10, 11, 12, 18, 23, 25, 30	37
	3, 5, 10, 11, 12, 19, 24, 25, 28, 29	39

TABLE 11.10
OPP Results for IEEE 30-Bus System with Considering ZIBs

No. of PMUs	PMUs Location Sets	R. Index
7	1, 7, 10, 12, 18, 24, 29	29
	3, 5, 10, 12, 18, 24, 29	29
	1, 5, 10, 12, 19, 24, 29	29
	1, 5, 10, 12, 18, 24, 29	29
	1, 7, 10, 12, 19, 24, 29	29

11.5.2.4 Case #4: Single Line Outage

Table 11.12 shows the results of optimal PMUs placement problem for IEEE 30-bus system with considering a single line outage. In this case, exclusion of radial buses helps to make the number of PMUs needed less than those of the previous case; the number of PMUs is 5. To ensure full observability of IEEE 30-bus system,

TABLE 11.11
OPP Results for IEEE 30-Bus System with Losing a Single PMU

No. of PMUs	PMUs Location Sets	R. Index
21	1, 2, 3, 5, 6, 9, 10, 11, 12, 13, 15, 17, 18, 20, 22, 24, 25, 26, 27, 28, 29	83
	2, 3, 4, 6, 7, 8, 9, 10, 11, 12, 13, 15, 17, 19, 20, 21, 23, 25, 26, 27, 30	82
	1, 2, 3, 6, 7, 9, 10, 11, 12, 13, 15, 17, 18, 20, 22, 23, 25, 26, 27, 28, 29	82
	1, 2, 4, 6, 7, 9, 10, 11, 12, 13, 15, 17, 19, 20, 21, 24, 25, 26, 28, 29, 30	82

TABLE 11.12
OPP Results for IEEE 30-Bus System with a Single Line Outage

No. of PMUs	PMUs Location Sets	R. Index
16	1, 2, 4, 5, 6, 10, 12, 15, 16, 18, 19, 22, 24, 27, 28, 30	71
	2, 3, 4, 5, 6, 10, 12, 15, 16, 19, 20, 22, 24, 27, 28, 30	71
	2, 3, 4, 5, 6, 10, 12, 15, 17, 18, 20, 22, 24, 27, 28, 29	71
	1, 2, 4, 5, 6, 10, 12, 15, 17, 18, 20, 22, 24, 27, 28, 29	71
	2, 3, 4, 6, 7, 8, 10, 12, 15, 17, 19, 20, 22, 24, 27, 29	70
	2, 3, 4, 6, 7, 8, 10, 12, 15, 17, 18, 20, 22, 24, 27, 29	70
	1, 2, 3, 5, 6, 8, 10, 12, 15, 16, 18, 19, 22, 24, 27, 29	68
	1, 2, 3, 6, 7, 8, 10, 12, 15, 16, 18, 19, 22, 24, 27, 29	68
	1, 3, 5, 6, 7, 10, 12, 15, 17, 18, 19, 22, 24, 27, 28, 30	67

16 PMUs are needed at least. As noticed from Table 11.12, the proposed method helps in obtaining a lot of PMUs location sets with the same minimum number of PMUs, but the best location set is (1, 2, 4, 5, 6, 10, 12, 15, 16, 18, 19, 22, 24, 27, 28, 30) that achieves maximum measurement redundancy, which equals 71.

11.6 APPLICATION OF PMUS IN PROTECTING TRANSMISSION LINES

Improvements in protection schemes due to presence of PMUs are somewhat obvious. More precise voltage and current measurements result in more accurate power system protection. Adaptive relaying improves system protection by allowing the relay to receive data from remote locations to change its settings according to different system configurations [5,20]. It is obvious that these settings will be more precise if they are based on actual measurements rather than approximated measurements. Often, high impedance faults have low fault currents that may not exceed detection thresholds of protective devices. PMUs can provide precise phasor representations to help in the detection of a high impedance faults.

This detection allows proper isolation for high impedance faults with the aid of communication devices. On a secondary level, the exact impedance of a line can be calculated if there are two PMUs installed on that line; one PMU at each end. Many protective devices settings depend on line impedances, as these line impedances are used in determining fault currents.

The PMUs can actually provide 60-Hz waveforms and magnitudes data in faulted condition using individual channels [15]. This feature helps to provide the operator with a real-time representation of the power system during a fault. The PMUs measurements can be used for more accurate determination of the time, location, and consequences of any fault that is difficult to simulate, proving PMUs valuable applications in post-disturbance analysis.

REFERENCES

[1] Kezunovic M., "Accurate Fault Location in Transmission Networks Using Modeling, Simulation and Limited Field Recorded Data", PSERC publication, report, November 2002.

[2] America Recovery and Reinvestment Act of 2009. "Synchrophasor Technologies and Their Deployment in the Recovery Act Smart Grid Program", US Department of Energy, August 2013.

[3] Rahman W. U., Ali M., Ullah A., Rahman H. U., Iqbal M., Ahmad H., Zeb A., Ali Z., Shahzad M. A., Taj B., "Advancement in Wide Area Monitoring Protection and Control Using PMU's Model in MATLAB/SIMULINK", *Smart Grid Renew Energy*, November 2012, vol. 3, no. 4, pp. 294–307.

[4] De La Ree J., Centeno V., Thorp J. S., Phadke A. G., "Synchronized Phasor Measurement Applications in Power Systems", *IEEE Trans Smart Grid*, June 2010, vol. 1, no. 1, pp. 20–27.

[5] Novosel D., Vu K., "Benefit of PMU Technology for Various Applications", CIGRE Croatian National Committee, 4-0C, 7th Symposium on Power System Management Cavtat, November 2006.

[6] Dotta D., Chow J. H., Vanfretti L., Almas M. S., Agostini M. N., "A MATLAB-based PMU Simulator", IEEE Power and Energy Society General Meeting (PES), 2013.

[7] Hart D. G., Gharpure V., Novosel D., Karlsson D., Kaba M., "PMUs – A New Approach to Power Network Monitoring", Report No. 1-en, ABB, 2001.

[8] Pinte B., Quinlan M., Yoon A., Reinhard K., Sauer P. W., "A One-phase, Distribution-level Phasor Measurement Unit for Post-event analysis", Power and Energy Conference at Illinois (PECI), Champaign, IL, 2014, pp. 1–7.

[9] Singh B., Sharma N. K., Tiwari A. N., Verma K. S., Singh S. N., "Applications of Phasor measurement units (PMUs) in electric power system networks incorporated with FACTS controllers", *Int J Eng Sci Technol*, 2011, vol. 3, no. 3, pp. 64–82.

[10] Mohammadi-Ivatloo B., "Optimal Placement of PMUs for Power System Observability using Topology based Formulated Algorithms", *J Appl Sci* 2009, vol. 9, pp. 2463–2468.

[11] Su C., Chen Z., "Optimal Placement of Phasor Measurement Units with New Considerations", 2010 Asia-Pacific Power and Energy Engineering Conference (APPEEC), IEEE, Chengdu, China, 28-31 March 2010, pp. 1–4.

[12] Aminifar F., Khodaei A., Fotuhi-Firuzabad M., Shahidehpour M., "Contingency-Constrained PMU Placement in Power Networks", *IEEE Trans Power Syst*, 2010, vol. 25, no. 1, pp. 516–523.

[13] Esmaili M., "Inclusive Multi-Objective PMU Placement in Power Systems Considering Conventional Measurements and Contingencies", *Int Trans Electr Energy Syst*, 2016, vol. 26, no. 3, pp. 609–626.
[14] Nazari-Heris M., Mohammadi-Ivatloo B., "Application of Heuristic Algorithms to Optimal PMU Placement in Electric Power Systems: An Updated Review", *Renew Sustain Energ Rev*, 2015, vol. 50, pp. 214–228.
[15] Chakrabarti S., Kyriakides E., "Optimal Placement of Phasor Measurement Units for Power System Observability", *IEEE Trans Power Syst*, 2008, vol. 23, pp. 1433–1440.
[16] Albuquerque R. J., Paucar V. L., "Evaluation of the PMUs Measurement Channels Availability for Observability Analysis", *IEEE Trans Power Syst*, 2013, vol. 28, pp. 2536–2544.
[17] Wu J., Xiong J., Shil P., Shi Y., "Optimal PMU Placement for Identification of Multiple Power Line Outages in Smart Grids 2014 IEEE 57th International Midwest Symposium on Circuits and Systems (MWSCAS), College Station, TX, 2014, pp. 354–357.
[18] Thorp J. S., Phadke A. G., Karimi K. J., "Real Time Voltage-Phasor Measurements for Static State Estimation", *IEEE Power Eng Rev*, 1985, vol. PER-5, no. 11, pp. 32–33.
[19] Yuill W., Edwards A., Chowdhury S., Chowdhury S. P., "Optimal PMU Placement: A Comprehensive Literature Review", 2011 IEEE Power and Energy Society General Meeting, San Diego, CA, 2011, pp. 1–8.
[20] Tai X., Marelli D., Rohr E., Fu M., "Optimal PMU Placement for Power System State Estimation with Random Component Outages", *Electr Power Energy Syst*, 2013, vol. 51, pp. 35–42.
[21] Rakpenthai C., Premrudeepreechacharn S., Uatrongjit S., Watson N. R., "An Optimal PMU Placement Method against Measurement Loss and Branch Outage", 2007 Power Engineering Society General Meeting, Tampa, FL, IEEE, 2007, pp. 101–107.
[22] Aminifar F., Fotuhi-Firuzabad M., Safdarian A., Shahidehpour M., "Observability of Hybrid AC/DC Power Systems With Variable-Cost PMUs", *IEEE Trans Power Del*, 2014, vol. 29, pp. 345–352.
[23] Gomathi V., Ramachandran V., "Optimal Location of PMUs for Complete Observability of Power System Network", 2011 1st International Conference on Electrical Energy Systems (ICEES), Newport Beach, CA, 2011, pp. 314–317.
[24] Fish A. A., Chowdhury S., Chowdhury S. P., "Optimal PMU Placement in a Power Network for Full System Observability", 2011 IEEE Power and Energy Society General Meeting, San Diego, CA, 2011, pp. 1–8.
[25] Yammani C., Narsi R. K., Maheswarapu S., "Optimal Placement of PMU's Considering Sensitivity Analysis", 2015 International Conference on Technological Advancements in Power and Energy (TAP Energy), Kollam, 2015, pp. 40–44.
[26] Sodhi R., Srivastava S. C., Singh S. N., "Multi-Criteria Decisionmaking Approach for Multistage Optimal Placement of Phasor Measurement Units", *IET Gener, Transm Distrib*, 2011, vol. 5, pp. 181–190.
[27] Biswal A., Mathur H. D., "Identification of Optimal Locations of PMUs forWAMPAC in Smart Grid Environment", 2015 International Conference on Technological Advancements in Power and Energy (TAP Energy), Kollam, 2015, pp. 369–374.
[28] Zare J., Aminifar F., Sanaye-Pasand M., "Synchrophasor- Based Wide-Area Backup Protection Scheme with Data Requirement Analysis", *IEEE Trans Power Del*, 2015, vol. 30, pp. 1410–1419.
[29] Xu B., Abur A., "Observability analysis and measurement placement for systems with PMUs", IEEE PES Power Systems Conference and Exposition, 2004, vol. 2, pp. 943–946.

[30] Abiri E., Rashidi F., Niknam T., Salehi M. R., "Optimal PMU Placement Method for Complete Topological Observabilit of Power System under Various Contingencies", *Electr Power Energy Syst*, 2014, vol. 61, pp. 585–593.

[31] Aghaei J., Baharvandi A., Rabiee A., Akbari M. A., "Probabilistic PMU Placement in Electric Power Networks: An MILP-Based Multiobjective Model", *IEEE Trans Ind Inform*, 2015, vol. 11, pp. 332–341.

[32] Sodhi R., Srivastava S. C., Singh S. N., "Optimal PMU Placement to Ensure System Observability under Contingencies", 2009 IEEE Power & Energy Society General Meeting, Calgary, AB, 2009, pp. 1–6.

[33] Dua D., Dambhare S., Gajbhiye R. K., Soman S. A., "Optimal Multistage Scheduling of PMU Placement An ILP Approach", *IEEE Trans Power Del*, 2008, vol. 23, no. 4, pp. 1812–1820.

[34] Kavasseri R., Srinivasan S. K., "Joint Placement of Phasor and Power Flow Measurements for Observability of Power Systems", *IEEE Trans Power Syst*, 2011, vol. 26, no. 4, pp. 1929–1936.

[35] Rihan M., Ahmad M., Beg M. S., Anees M. A., "Robust and economical placement of phasor measurement units in Indian Smart Grid", 2013 IEEE Innovative Smart Grid Technologies-Asia (ISGT Asia), Bangalore, 2013, pp. 1–6.

[36] Bhonsle J. S., Junghare A. S., "An Optimal PMU-PDC Placement Technique in Wide Area Measurement System", 2015 International Conference on Smart Technologies and Management for Computing, Communication, Controls, Energy and Materials (ICSTM), India, 6–8 May 2015, pp. 401–405.

[37] Abbasy N. H., Ismail H. M., "A Unified Approach for the Optimal PMU Location for Power System State Estimation", *IEEE Trans Power Syst*, 2009, vol. 24, no. 2, pp. 806–813.

[38] Gupta N., Goyal M., Tripathy P., "A Novel Approach for Optimal Placement of PMUs with Minimum Measurement Channels", 2012 IEEE International Conference on Power and Energy (PECon), Kota Kinabalu, 2012, pp. 505–509.

[39] Dua D., Dambhare S., Gajbhiye R. K., Soman S. A., "Optimal Multistage Scheduling of PMU Placement: An ILP Approach", *IEEE Trans Power Del*, 2008, vol. 23, no. 4, pp. 1812–1820.

[40] Chen J., Abur A., "Placement of PMUs to Enable Bad Data Detection in State Estimation", *IEEE Trans Power Syst*, 2006, vol. 21, no. 4, pp. 1608–1615.

[41] Razavi S. E., Falaghi H., Ramezani M., "A New Integer Linear Programming Approach for Multi-Stage PMU Placement", Smart Grid Conference (SGC), 2013, Tehran, 2013, pp. 119–124.

[42] Wang F., Zhang W., Li P., "Optimal Incremental Placement of PMUs for Power System Observability, 2012 IEEE Power and Energy Society General Meeting, San Diego, CA, 2012, pp. 1–7.

[43] Korres G. N., Manousakis N. M., Xygkis T. C., Löfberg J., "Optimal Phasor Measurement Unit Placement for Numerical Observability in the Presence of Conventional Measurements Using Semi-Definite Programming", *IET Gener, Transm Distrib*, 2015, vol. 9, no. 15, pp. 2427–2436.

[44] Kavasseri R., Srinivasa S. K., "Joint Placement of Phasor and Conventional Power Flow Measurements for Fault Observability of Power Systems", *IET Gene Trans Distrib*, 2011, vol. 5, no. 10, pp. 1019–1024.

[45] Nan X., Beng G. H., "Optimal PMU Placement with Local Redundancy of Conventional Measurements", 2013 IEEE Symposium on Computational Intelligence Applications in Smart Grid (CIASG), Singapore, 2013, pp. 1–5.

[46] Fadiran J. I., Chowdhury S., Chowdhury S. P., "A Multi-Criteria Optimal Phasor Measurement Unit Placement for Multiple Applications", 2013 IEEE Power & Energy Society General Meeting, Vancouver, BC, 2013, pp. 1–5.

[47] Putranto L. M., Hara R., Kita H., Tanaka E., "Voltage Stability Based PMU Placement Considering N – 1 Line Contingency and Power System Reliability", 2014 International Conference on Bali Power Engineering and Renewable Energy (ICPERE), 2014, pp. 120–125.
[48] More K. K., Jadhav H. T. A., "Literature Review on Optimal Placement of Phasor Measurement Units", 2013 International Conference on Power, Energy and Control (ICPEC), 2013, pp. 220–224.
[49] Chakrabarti S., Kyriakides E., Albu M., "Uncertainty in Power System State Variables Obtained Through Synchronized Measurements", *IEEE Trans Instrum Meas*, 2009, vol. 58, no. 8, pp. 2452–2458.
[50] Chakrabarti S., Kyriakides E., Eliades D. G., "Placement of Synchronized Measurements for Power System Observability", *IEEE Trans Power Deliv*, 2009, vol. 24, no. 1, pp. 12–19.
[51] Manousakis N. M., Korres G. N., "A Weighted Least Squares Algorithm for Optimal PMU Placement", *IEEE Trans Power Syst*, 2013, vol. 28, no. 3.
[52] Hong-Shan Z., Ying L., Zeng-Qiang M., Lei Y., "Sensitivity Constrained PMU Placement for Complete Observability of Power Systems", 2005 IEEE/PES Transmission and Distribution Conference and Exhibition: Asia and Pacific, IEEE, 2005, pp. 1–5.
[53] Nuqui R. F., Phadke A. G., "Phasor Measurement Unit Placement Techniques for Complete and Incomplete Observability", *IEEE Trans Power Del*, 2005, vol. 20, no. 2381–2388.
[54] Kerdchuen T., Ongsakul W., "Optimal PMU Placement for Reliable Power System State Estimation", Second GMSARN International Conference, Pattaya, Thailand, 2007, pp. 1–6.
[55] Kerdchuen T., Ongsakul W., "Optimal PMU Placement by Stochastic Simulated Annealing for Power System State Estimation", *GMSARN Int J*, 2008, vol. 2, pp. 61–66.
[56] Marin F., Garcia-Lagos F., Joya G., Sandoval F., "Genetic Algorithms for Optimal Placement of Phasor Measurement Units in Electrical Networks", *Electron Lett*, 2003, vol. 39, no. 1403–1405.
[57] Bedekar P. P., Bhide S. R., Kale V. S., "Optimum PMU Placement Considering One Line/One PMU Outage and Maximum Redundancy Using Genetic Algorithm", Eighth International Conference on Electrical Engineering/Electronics, Computer, Telecommunications and Information Technology 2011(ECTI-CON), IEEE, 2011, pp. 688–691.
[58] Mohammadi-Ivatloo B., "Optimal Placement of PMUs for Power System Observability Using Topology Based Formulated Algorithms", *J Appl Sci*, 2009, vol. 9, pp. 2463–2468.
[59] Aminifar F., Lucas C., Khodaei A., Fotuhi-Firuzabad M., "Optimal Placement of Phasor Measurement Units Using Immunity Genetic Algorithm", *IEEE Trans Power Del*, 2009, vol. 24, pp. 1014–1020.
[60] Milosevic B., Begovic M., "Nondominated Sorting Genetic Algorithm for Optimal Phasor Measurement Placement", *IEEE Trans Power Syst*, 2003, vol. 18, pp. 69–75.
[61] Kerdchuen T., Ongsakul W., "Optimal Placement of PMU and RTU by Hybrid Genetic Algorithm and Simulated Annealing for Multiarea Power System State Estimation", *Greater Mekong Subregion Acad Res Network*, 2009, vol. 7, pp. 7–12.
[62] Su C, Chen Z., "Optimal Placement of Phasor Measurement Units with New Considerations", 2010 Asia-Pacific Power and Energy Engineering Conference (APPEEC), IEEE, 2010, pp. 1–4.

[63] Peng C., Xu X., "A Hybrid Algorithm Based on BPSO and Immune Mechanism for PMU Optimization Placement", 2008 Seventh World Congress on Intelligent Control and Automation (WCICA2008), IEEE, 2008, pp. 7036–7040.

[64] Ahmadi A., Alinejad-Beromi Y., Moradi M., "Optimal PMU Placement for Power System Observability Using Binary Particle Swarm Optimization and Considering Measurement Redundancy", *Expert Syst Appl*, 2011, vol. 38, pp. 7263–7269.

[65] Huang J., Wu N. E., "Fault-Tolerant Placement of Phasor Measurement Units Based on Control Reconfigurability", *Control Eng Pract*, 2013, vol. 21, pp. 1–11.

[66] Hajian M., Ranjbar A. M., Amraee T., Mozafari B., "Optimal Placement of PMUs to Maintain Network Observability Using a Modified BPSO Algorithm", *Int J Electr Power Energy Syst*, 2011, vol. 33, pp. 28–34.

[67] Ketabi A., Nosratabadi S., Sheibani M., "Optimal PMU Placement Based on Mean Square Error Using Differential Evolution Algorithm", 2010 First Power quality conference (PQC), IEEE, 2010, pp. 1–6.

[68] Al-Mohammed A., Abido M. A., Mansour M., "Optimal PMU Placement for Power System Observability Using Differential Evolution", 2011 11th International Conference on Intelligent Systems Design and Applications (ISDA), IEEE, 2011, pp. 277–282.

[69] Rajasekhar B., Chandel A. K., Vedik B., "Differential Evolution Based Optimal PMU Placement for Fault Observability of Power System", 2013 Students Conference on Engineering and Systems (SCES), IEEE, 2013, pp. 1–5.

[70] Peng C., Sun H., Guo J., "Multi-Objective Optimal PMU Placement Using a Non-Dominated Sorting Differential Evolution Algorithm", *Int J Electr Power Energy Syst*, 2010, vol. 32, pp. 886–892.

[71] Peng J., Sun Y., Wang H., "Optimal PMU Placement for Full Network Observability Using Tabu Search Algorithm", *Int J Electr Power Energy Syst*, 2006, vol. 28, pp. 223–231.

[72] Mesgarnejad H. S., Shahrtash, S. M., "Multi-Objective Measurement Placement with New Parallel Tabu Search Method", *IEEE Canada Electric Power Conference,* Vancouver, BC, Canada, 2008, pp. 1–6.

[73] Koutsoukis N. C., Manousakis N. M., Georgilakis P. S., Korres G. N., "Numerical Observability Method for Optimal Phasor Measurement Units Placement Using Recursive Tabu Search Method", *IET Gener Transm Distrib*, 2013, vol. 7, pp. 347–356.

[74] Bian X., Qiu J., "Adaptive Clonal Algorithm and Its Application for Optimal PMU Placement", 2006 International Conference on Communications, Circuits and Systems Proceedings, IEEE, 2006, pp. 2102–2106.

[75] Mohammadi-Ivatloo B., Hosseini S., "Optimal PMU Placement for Power System Observability Considering Secondary Voltage Control", 2008 Canadian Conference on Electrical and Computer Engineering (CCECE2008), IEEE, 2008, pp. 000365–000368.

[76] Zhou M., Centeno V. A., Phadke A. G., Hu Y., Novosel D., Volskis H. A., "A Preprocessing Method for Effective PMU Placement Studies", 2008 Third International Conference on Electric Utility Deregulation and Restructuring and Power Technologies (DRPT2008), IEEE, 2008, pp. 2862–2867.

[77] Wang B., Liu D., Xiong L., "An Improved Ant Colony System in Optimizing Power System PMU Placement Problem", 2009 Asia-Pacific Power and Energy Engineering Conference (APPEEC2009), IEEE, 2009, pp. 1–3.

[78] Mazlumi K., Vahedi H., "Optimal Placement of PMUs in Power Systems Based on Bacterial for Aging Algorithm", 2010 18th Iranian Conference on Electrical Engineering (ICEE), IEEE, 2010, pp. 885–888.

[79] Hurtgen M., Maun J.-C., "Optimal PMU Placement Using Iterated Local Search", *Int J Electr Power Energy Syst*, 2010, vol. 32, pp. 857–860.

[80] Jamuna K., Swarup K., "Power System Observability Using Biogeography Based Optimization", International Conference on Sustainable Energy and Intelligent Systems (SEISCON2011), IET, 2011, pp. 384–389.
[81] Jamuna K., Swarup K., "Multi-Objective Biogeography Based Optimization for Optimal PMU Placement", *Appl Soft Comput*, 2012, vol. 12, pp. 1503–1510.
[82] Li Q., Cui T., Weng Y., Negi R., Franchetti F., Ilic M. D., "An Information-theoretic Approach to PMU Placement in Electric Power Systems", *IEEE Trans Smart Grid*, 2013, vol. 4, pp. 446–456.
[83] Mahari A., Seyedi H., "Optimal PMU Placement for Power System Observability Using BICA, Considering Measurement Redundancy", *Electr Power Syst Res*, 2013, vol. 103, pp. 78–85.
[84] Xu J., Wen M. H., Li V. O., Leung K.-C., "Optimal PMU Placement for Wide-Area Monitoring Using Chemical Reaction Optimization", 2013 IEEE PES Innovative Smart Grid Technologies (ISGT), IEEE, 2013, pp. 1–6.
[85] Mazhari S. M., Monsef H., Lesani H., Fereidunian A., "A Multi-Objective PMU Placement Method Considering Measurement Redundancy and Observability Value under Contingencies", *IEEE Trans Power Syst*, 2013, vol. 28, pp. 2136–2146.
[86] Hui-Ling Z., Yuan-Xiu D., Xiao-Pan Z., Huan Q., Cheng-Xun H., "Hybrid of MST and Genetic Algorithm on Minimizing PMU Placement". Proceedings of the 2013 Third International Conference on Intelligent System Design and Engineering Applications, IEEEComputerSociety, 2013, pp. 820–823.
[87] Kulanthaisamy A., Vairamani R., Karunamurthi N. K., Koodalsamy C., "A multi-objective PMU placement method considering observability and measurement redundancy using abc algorithm", *Adv Electr Comput Eng* 2014, vol. 14, pp. 117–128.
[88] Hassanin K. M., Abdelsalam A. A., Abdelaziz A. Y., "Optimal PMUs Placement for Full Observability of Electrical Power Systems Using Flower Pollination Algorithm", 2017 the 5th IEEE International Conference on Smart Energy Grid Engineering, 2017, pp. 20–25.
[89] Abdelsalam A. A., Hassanin K. M., Abdelaziz A. Y., Haes Alhelou H., "Optimal PMUs Placement Considering ZIBs and Single Line and PMUs Outages", *AIMS Energy J*, 2020, vol. 8, no. 1, pp. 122–141.

12 An Expert System for Optimal Coordination of Directional Overcurrent Relays in Meshed Networks

Hajjar A. Ammar
Department of Electrical Engineering, Tishreen University, Syria

12.1 INTRODUCTION

Directional overcurrent relays (DOCR), which are simple and economic, are commonly used in power system protection, as primary protection in distribution and sub-transmission systems, and as secondary protection in transmission systems. The main problem that arises with this type of relays is the difficulty in performing relays coordination (relays setting), especially in meshed networks [1].

Therefore, since 1960 a great effort has been devoted to developing approaches for DOCR coordination. Some of these approaches are procedural [1–10], and the others are non-procedural [11–22]. In this respect, a trial-and-error approach is introduced in [2]; however, it is required a large number of iterations to be converged (if any). To make the coordination process faster, a systematic approach based on the topological analysis of the network is introduced in [3]. However, it is required a large number of matrixes manipulation and there is no guarantee that it will give optimal coordination.

In [4], a nonlinear programming (NLP) technique is used to solve the problem of DOCR coordination. However, due to the complexity of this technique, the optimal DOCR coordination is commonly carried out using the linear programming (LP) techniques, including the active set strategy two-phase method [1], the simplex method [5], Interior point method [6], etc. In these techniques, the coordination time interval (CTI) between the backup and primary (B/P) relays is considered as the coordination constraints, and the optimal coordination problem is solved based on the objective function (OF), relay characteristics curves, limits on the relays settings, and the coordination constraints.

In [7], optimal coordination of DOCR is obtained by solving the constraints only. However, the main disadvantages of this method are it gives contentious TDS and needs manipulation of high-dimension matrixes.

DOI: 10.1201/9780367552374-12

The abovementioned optimization techniques give continuous TDS that need rounding off to be suitable for DOCR settings. However, rounding off the TDS values may results in miscoordination cases [8]. To get discrete values of the pickup current (Ip) and TDS for each relay Reference [9] applied the binary integer programming technique (BILP). However, since the coordination problem of DOCR is non-linear and non-convex optimization one, the above-mentioned optimization techniques may be trapped in a local minimum. To get rid of this problem, mixed-integer linear programming (MILP) is applied in [10], by considering the Ip setting as a discrete optimization variable and the TDS as a continuous optimization variable. However, the problem of continuous TDS values still exists with this method. In reality, applying the BILP and MILP techniques to optimal coordination of DOCR increases the problem complexity.

Subsequently, application of various metaheuristic optimization algorithms (non-procedural methods) such as particle swarm optimization (PSO) [11], evolutionary algorithm (EA) [12], enhanced differential evolution algorithm (EDEA) [13], seeker optimization algorithm (SOA) [14], nature-inspired whale optimization algorithm (NIWOA) [15], hybrid genetic algorithm (HGA) [16], calculations, evaporation rate water cycle algorithm (ER-WCA) [17], etc., have also been proposed in the literature. However, it is well known that the metaheuristic algorithms take relatively a large simulation time in giving a reasonably good solution [18], [19]. Consequently, these algorithms are not suitable for online optimal coordination of DOCR.

In the literature, some conventional methods based on expert systems for DOCR coordination are presented in [20–22]. In this respect, Protective Device Coordination Expert System (PDCES) software is presented in [20], to help the electric distribution engineers in correcting miscoordinated protective device pairs in radial networks. In [21], an expert system for protection coordination of distribution system under the presence of distributed generators is presented. In [22], an expert system for DOCR coordination in meshed networks is presented based on the break points method (the relays by which the coordination process will start). However, this expert system required a large number of rules and there is no guarantee that it will give optimal coordination.

In this chapter, a novel Expert System (ES) is introduced for optimal coordination of DOCR in meshed networks, in which minimization is innately included by setting the time dials to a minimum and increasing their values progressively. Accordingly, it has inherent immunity from trapping in a local minimum, and the problem of miscoordination, which may arise due to rounding off the TDS values, is avoided. Moreover, it does not need any initial solution or objective function, or auxiliary variables, and it is suitable for coordinating both electromechanical and digital DOCR, offline, and online, respectively. The introduced ES is structured using the CLIPS shell, which is a forward-chaining rule-based language. The performance of the introduced ES is verified by comparing its results with those obtained using the conventional optimal method.

12.2 IMPORTANCE OF THE ES AND ITS OBJECTIVES

The process of DOCR coordination in meshed networks involves setting relays one by one so that at each stage the relay being set should be coordinated with all its primary relays. Consequently, the innate multi-loop structure of the modern power systems requires a large number of relays setting calculations to be done iteratively around all loops of the system until the system-wide coordination is obtained. In this respect, a logic approach, based on simple IF-THEN-rules, appears to be more suitable for protection coordination than procedural approaches due to the following reasons:

- The coordination process is an art more than a science since it demands the appreciable knowledge and skill of a senior protection engineer.
- Major parts of the relays settings knowledge are available in a rule style.
- The coordination problem is generally non-procedural and it is solved using heuristics and human expertise.
- The setting of the protective relays in meshed networks is a search problem in which every backup/primary (B/P) relays pair should be coordinated.
- ES can give explanations about their decisions.

The objective of this chapter is to introduce a novel ES, using a simple IF-THEN-rules structure, for optimal coordination of DOCR in meshed networks, which is more suitable than other optimization techniques used in [1–19], [22].

Justifications of using ES, with respect to other optimization techniques are provided as follows:

1. Nonlinear programming techniques are complex. So, the coordination of DOCR is commonly carried out using linear programming techniques. However, in all linear programming techniques, the objective function and the auxiliary variables must be defined, therefore the total number of auxiliary variables becomes equal to the number of constraints. Thus, the introduction of auxiliary variables requires a complicated solution. Moreover, most of the abovementioned techniques give continuous TDS that need rounding off, in so doing miscoordination cases may appear. Thus, these problems can be avoided using the introduced ES.
2. Solving the problem of DOCR coordination using traditional optimization techniques requires an initial guess, and results in multi-optimum points, thus the solution may be trapped in a local minimum. However, the introduced ES has inherent immunity from trapping at a local minimum.
3. Metaheuristic algorithms take relatively large simulation time in giving a reasonably good solution. As a result, these algorithms are not suitable for the online optimal coordination of DOCR.

12.3 PROBLEM FORMULATION OF THE OPTIMAL COORDINATION OF DOCR

DOCR has two parameters to be set, Ip and TDS, where Ip is a threshold current value above which the DOCR should operate. In this context, Ip can be determined previously based on the maximum load current and the minimum fault current [1], [5], [6]. The TDS defines the operating time (T) of the relay for each current value and it is generally given as a curve, T versus M, where M is a multiple of Ip, i.e. M = I/Ip, and I is the relay current (fault current). The optimal TDS can be determined using one of the traditional optimization techniques [1], [4–10].

In this chapter, the DOCR characteristics are conformed to the following IEC standard [23]:

$$T = \frac{k_1 TDS}{[M^{k_2} - 1]} \tag{12.1}$$

where k_1 and k_2 are constants, their values depend on the relay characteristic type (normal inverse, very inverse, extremely inverse) [23].

The problem of optimal coordination of DOCR can be illustrated with the help of Figure 12.1 assuming that the network has n relays.

To find the TDS of the DOCR using the traditional optimal techniques, the objective function, and the constraints are defined as follows [1], [4–10]:

$$\text{Minimize } (Obj = \textstyle\sum_{i=1}^{n} T_{ii}) \tag{12.2}$$

where T_{ii} is the operating time of the primary relay R_i for a maximum near-end fault located at line i.

Subject to:

The operating times of the backup relay(s) should be greater than the operating time of its primary relay by at least the CTI. This constraint can be expressed by the following inequality:

$$T_{ji} \geq T_{ii} + CTI \tag{12.3}$$

where T_{ji} is the operating time of the backup relay R_j for the same maximum near-end fault of the primary relay R_i.

The TDS for each relay should satisfy the following constraint:

$$TDS_i^{max} \geq TDS_i \geq TDS_i^{min} \tag{12.4}$$

where TDS_i^{max} and TDS_i^{min} are the upper and lower limits of the TDS_i, respectively.

The operating time T_i for each relay should satisfy the following constraint:

$$T_i^{max} \geq T_i \geq T_i^{min} \tag{12.5}$$

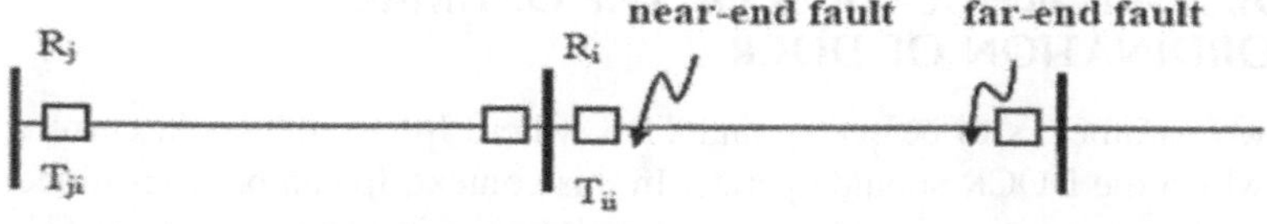

FIGURE 12.1 An illustrative diagram for basic definitions.

where T_i^{max} and T_i^{min} are the upper and lower limits of the relay operating time T_i, respectively.

Substituting (12.1) in (12.2) we get (12.6):

$$\text{Minimize } \left(Obj = \sum_{i=1}^{n} a_i \ TDS_i\right) \tag{12.6}$$

In reality, a_i are positive real numbers and have no effect on the optimal solution [1], [6], [24–26]. Thus, solving the constraints is optimal when all coefficients of the decision variables, of the objective function, are positive [24–26], i.e. the objective function value will be minimum when the decision variables TDS_1, TDS_2, …, TDS_n have the smallest values that satisfy the constraints.

Accordingly, the optimal coordination of DOCR can be found by solving the constraints only, using the introduced ES. In this context, the constraints of the introduced ES are constructed from (12.1), (12.3), (12.4), and (12.5). In this chapter, (12.3) is formed considering maximum near-end faults without considering the change in the network configuration.

12.4 STRUCTURE OF THE INTRODUCED ES

The ES is an intelligent computer program that uses knowledge and inference engine to solve problems that are difficult enough to require significant human expertise for their solution [27]. Figure 12.2 depicts a block diagram of the introduced ES for optimal coordination of DOCR. It consists of two parts: the first one is the ES that works under the CLIPS shell, the second one includes the load flow and fault analysis programs that work under the MATLAB® environment.

A brief description of the main components of the ES is given below:

- **User Interface:** It is the mechanism by which the user and the ES communicate.
- **Facts:** Facts are the data on which inferences are derived. Facts are changeable with respect to the network configuration.

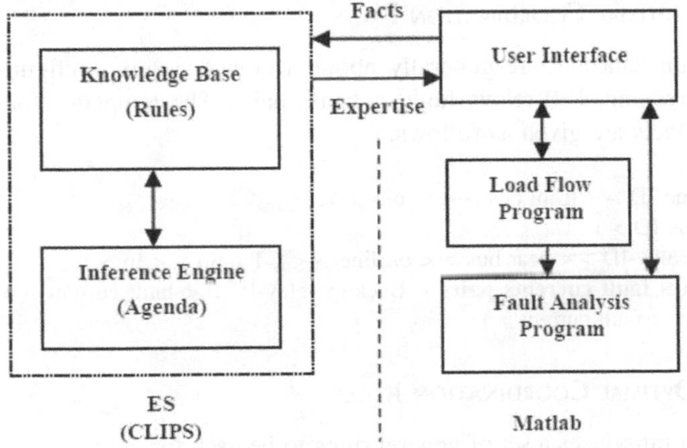

FIGURE 12.2 A block diagram of the ES for optimal coordination of DOCR.

- **Knowledge Base:** It contains all the user-defined rules in the form of (IF condition(s) THEN conclusion(s)). These rules are needed to allow proper inferences during the coordination process. Rules are permanent knowledge.
- **Inference Engine:** It is the main component in the ES; it is a built-in component in a CLIPS Shell, which processes knowledge to draw suitable conclusions. It provides methods to control the direction of reasoning, forward or backward chaining, as well as strategies like Depth, Random, Complexity, etc., to guide the search for a solution [27].
- **Agenda:** It is a prioritized list of rules created by the inference engine, whose rules are satisfied by the facts.

12.4.1 The Mechanism by Which the Introduced ES Work

- The user (protection engineer) triggers the load flow and fault analysis programs, through the user interface, to prepare some required facts for the ES.
- The ES compares these facts with the existing rules, and then the inference engine collects the satisfied rules on the agenda according to their priorities.
- The inference engine then executes the agenda according to the priorities. In so doing, the user obtains the optimal solution of the DOCR coordination problem in the form of expertise (the TDS for each relay that satisfies the CTI between this relay and its backup relay(s)).
- If any change occurs in the protected network, operational, and/or topological, the user repeats the previous steps.

One of the main tasks of the introduced ES is to find the B/P relay pairs, which are essential for forming the coordination pairs and to find the TDS for all relays, and consequently to calculate the B/P operating times of all relays that satisfy the coordination constraint.

12.5 AN ES FOR OPTIMAL COORDINATION OF DOCR

12.5.1 Optimal Coordination Facts

Facts contain data that are generally about a power system configuration (lines, buses); relays and, B/P relays fault currents pairs. The templates for the above-mentioned facts are given as follows:

(**line** < line-ID > < from bus > < to bus >)
(**bus** < bus-ID >)
(**relay** < relay-ID > < near bus > < on line > < CT ratio > < Ip >)
(**B/P relays fault currents pairs** < Backup relay-ID > < fault current > < Primary relay-ID > < fault current >)

12.5.2 Optimal Coordination Rules

This section introduces a set of general rules to be used for optimal coordination of DOCR in meshed networks.

Rule1 (Forming B/P relays pairs)
IF near-end relay fact is satisfied
AND the far-end relay fact, with different line-ID, is satisfied
THEN the near-end relay is back up to the far-end relay(s).
Rule2 (Initiating TDS with its lower bound value (LBV))
IF relay fact is satisfied
THEN let the initial value of the TDS of this relay = TDS_{min}.
Rule3 (Calculating the B/P times for this relay)
IF TDS, M_p, and M_b are satisfied
THEN calculates the values of POT and BOT for this relay.
where M_p and M_b refer to multiple I_P of the primary and backup relay, respectively, and POT, BOT refer to the primary and backup operating times of the relay, respectively.
Rule4 (Checking the lower time limit of the primary relay)
IF POT of the relay is less than the lower operating time limit of the relay T^{min}
THEN increment the TDS of the relay and recalculate POT, BOT for this relay.
Rule5 (Checking the coordination constraint)
IF the coordination time margin (CTM) between the B/P relay pairs is less than the CTI
THEN increment the TDS of the backup relay and recalculate POT, BOT for this relay.
Rule6 (Checking the upper bound value of the TDS)
IF the TDS for any relay is more than the TDS_{max}
THEN print no coordination could be found.

12.6 VERIFICATION OF THE INTRODUCED ES

To verify the introduced ES, it has been applied to three test systems: the IEEE 3-bus test system [4], the 8-bus test system [5], [6], and the IEEE 5-bus test system [24], [25]. In this chapter, all the DOCR are assumed identical, with normal inverse characteristics, and conformed to the IEC standard [23]. The characteristic constants of these relays are $k_1 = 0.14$, $k_2 = 0.02$, the TDS values range is [0.05–1.1], the CTI is taken as 0.2(S) for the first and third test systems, and 0.3 (S) for the second test system. Moreover, a three-phase near-end fault is simulated for each relay.

To apply the introduced ES, the I_P and current transformer (CT) for each relay are determined previously. The Facts for each test system are determined previously also. Hence, all data necessary for executing the introduced ES are ready now, and the ES is ready to perform the optimal coordination of DOCR (determining the B/P relays pairs, optimal TDS for each relay, and accordingly the B/P relay operating times that satisfy the coordination constraint). It is worth mentioning that, in the introduced ES one can choose two or three or four decimal places (DP) for the TDS value, hence there is no need to round off the TDS values as done in, [1], [4–10], [12]. In this context, the incremental step size of the TDS is taken as 0.01, 0.001, and 0.0001 for the two, three, and four DP, respectively.

12.6.1 IEEE 3-Bus Test System

Figure 12.3 depicts the one-line diagram of the IEEE 3-bus test system that consists of three buses, three lines, and three sources. A complete data of this system are given at [4]. The locations of the six DOCR are indicated in the Figure. Table 12.1 depicts the CT ratio and the I_P for each relay [4]. Table 12.2 depicts the B/P relays pairs as determined by the introduced ES.

Table 12.3 depicts the fault currents for each B/P relays pair [24], [25]. For a comparison purpose, Table 12.4 depicts the TDS value for each relay as determined by the introduced ES, and by the traditional LP technique based on the Interior Point Method LP (IPM). Given Table 12.4, it is clear that the difference between the

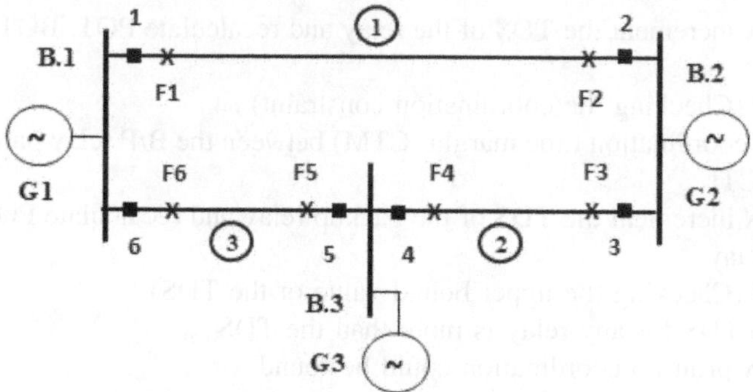

FIGURE 12.3 The one-line diagram of the IEEE 3-bus test system [4], [26].

TABLE 12.1
Pickup Current and CT Ratio for Each Relay of the IEEE 3-Bus Test System [4]

Relay No.	Ip (A)	CT Ratio
1	5	300/5
2	1.5	200/5
3	5	200/5
4	4	300/5
5	2	200/5
6	2.5	400/5

TABLE 12.2
The B/P Relays Pairs of the IEEE 3-Bus Test System

Backup Relay No.	Primary Relay No.	Backup Relay No.	Primary Relay No.
1	3	4	2
2	6	5	1
3	5	6	4

TABLE 12.3
The Fault Currents for Each B/P Relays Pairs of the IEEE 3-bus Test System [24], [25]

Backup Relay No.	Fault Current (A)	Primary Relay No.	Fault Current (A)
1	1163.7	3	2895.6
2	921.10	6	5016.9
3	1130.9	5	3439.9
4	1207.9	2	2939.8
5	846.50	1	5016.9
6	994.20	4	3303.2

TABLE 12.4
Optimal DOCR Setting Values for the IEEE 3-Bus Test System

Relay No.	TDS			
	LP (IPM)	ES (2DP)	ES (3DP)	ES (4DP)
1	0.0986	0.11	0.099	0.0987
2	0.1809	0.19	0.181	0.1805
3	0.1186	0.13	0.119	0.1187
4	0.1203	0.13	0.121	0.1202
5	0.1512	0.17	0.152	0.1513
6	0.1194	0.13	0.120	0.1194
$\Sigma_{i=1}^{n} (TDS)$	0.7892	0.860	0.792	0.7888
Mean TDS	0.1315	0.1433	0.132	0.1315
$\Sigma_{i=1}^{n} (T_i)$	1.6887	1.8419	1.6950	1.6881

TABLE 12.5
B/P DOCR Operating Times and the CTM for the IEEE 3-Bus Test System, Considering TDS with 2DP

Backup Relay No.	BOT (S)	Primary Relay No.	POT (S)	CTM (S)
1	0.5603	3	0.3314	0.2289
2	0.4737	6	0.2721	0.2016
3	0.5162	5	0.3046	0.2115
4	0.5540	2	0.3286	0.2254
5	0.4926	1	0.2657	0.2268
6	0.5584	4	0.3380	0.2203

TDS values is minor. Moreover, the difference between the summation of the TDS values, the mean value of the TDS, and the summation of the operating times (in seconds) of the primary relays of the ES and the LP (IPM) are minor also. In reality, this difference is due to the inherent processing in the MATLAB environment, which is based on a double precession (i.e. it has 15 DP) while the introduced ES uses 2, 3, and 4 DP only. Hence, the main advantages of the introduced ES are needles to round off the TDS values and its suitability for use with both electro-metrical and digital relays, in contrast to other optimization approaches [1], [4–10], [11–19]. The ES has other advantages such as needless to manipulate large dimensions matrixes and simplicity. Moreover, the introduced ES can be used online in contrast to meta-heuristic optimization algorithms (GA, PSO, etc.).

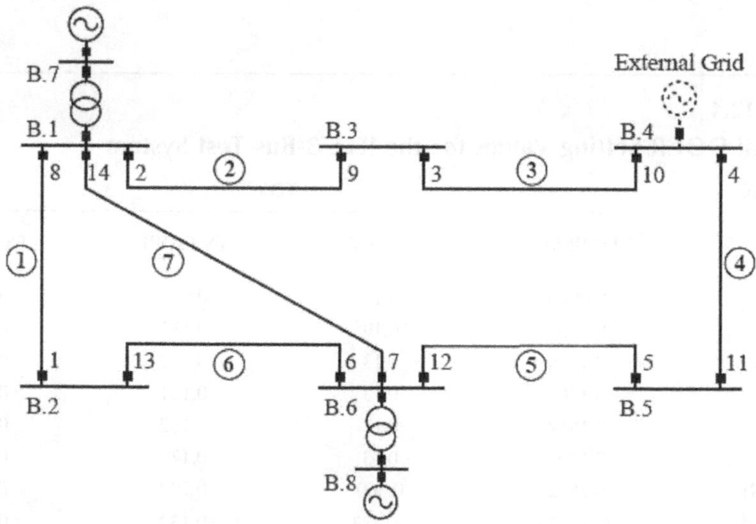

FIGURE 12.4 The one-line diagram of the 8-bus test system [5], [6].

Table 12.5 depicts the operating times of the B/P relays pairs and the coordination time margin (CTM) for the IEEE 3-bus test system, considering the TDS values with two DP. Given Table 12.5, it is clear that there is no any mis-coordination case between the B/P relays pairs.

TABLE 12.6
Pickup Current and CT Ratio for Each Relay of the 8-bus Test System [5], [6]

Relay No.	Ip (A)	CT Ratio
1	0.5	1200/5
2	2	1200/5
3	1.5	800/5
4	1.5	1200/5
5	1	1200/5
6	2	1200/5
7	1.5	800/5
8	1.5	1200/5
9	1.5	800/5
10	1.5	1200/5
11	1	1200/5
12	2	1200/5
13	0.5	1200/5
14	1.5	800/5

TABLE 12.7
The B/P Relays Pairs for the 8-Bus Test System

Backup Relay No.	Primary Relay No.	Backup Relay No.	Primary Relay No.
6	1	7	8
1	2	9	8
7	2	10	9
2	3	11	10
3	4	12	11
4	5	13	12
5	6	14	12
14	6	8	13
5	7	1	14
13	7	9	14

TABLE 12.8
The Fault Currents for Each B/P Relays Pair for the 8-Bus Test System [14]

Backup Relay No.	Fault Current (A)	Primary Relay No.	Fault Current (A)
6	3233	1	3233
1	996	2	5924
7	1889	2	5924
2	3556	3	3556
3	2243	4	3782
4	2401	5	2401
5	1198	6	6109
14	1873	6	6109
5	1198	7	5223
13	987	7	5223
7	1889	8	6092
9	1165	8	6092
10	2484	9	2484
11	2345	10	3884
12	3708	11	3708
13	987	12	5899
14	1873	12	5899
8	2990	13	2990
1	996	14	5199
9	1165	14	5199

12.6.2 The 8-Bus Test System

Figure 12.4 depicts the one-line diagram of the 8-bus test system that consists of eight buses, seven lines, two power transformers, and two generators. This system has a link to an external grid at bus 4, which is modeled by a short-circuit capacity of 400 MVA. A complete data of this system is given in [5], [6]. The locations of the 14 DOCR relays are indicated at the Figure. Table 12.6 depicts the CT ratio and I_P for each relay [5], [6]. Table 12.7 depicts the B/P relays pairs as determined by the introduced ES. Table 12.8 depicts the fault currents for each B/P relays pair, as given in [14].

Table 12.9 depicts the TDS values as determined by the LP(IPM) and the introduced ES. Given Table 12.9, it is clear that the difference between the TDS values is minor. Moreover, the difference between the summation of the TDS, the mean value of the TDS, and the summation of the operating times of the primary relays of the introduced ES and the LP(IPM) are minor.

Table 12.10 depicts the B/P relays pairs operating time and the CTM for the 8-bus test system, considering the TDS values with four DP. Given Table 12.10, it is clear that there is no any miscoordination case between the B/P relays pairs.

TABLE 12.9
Optimal DOCR Setting Values for the 8-Bus Test System

Relay No.	TDS			
	LP (IPM) [6]	ES (2DP)	ES (3DP)	ES (4DP)
1	0.4076	0.43	0.410	0.4077
2	0.3754	0.40	0.378	0.3755
3	0.3902	0.42	0.393	0.3903
4	0.3079	0.33	0.310	0.3080
5	0.2743	0.29	0.276	0.2743
6	0.3161	0.33	0.318	0.3162
7	0.3971	0.42	0.400	0.3972
8	0.3610	0.38	0.363	0.3612
9	0.2720	0.29	0.273	0.2722
10	0.3083	0.33	0.310	0.3085
11	0.3951	0.42	0.397	0.3953
12	0.3825	0.41	0.384	0.3826
13	0.4122	0.44	0.414	0.4123
14	0.4017	0.43	0.403	0.4019
$\Sigma_{i=1}^{n}(TDS)$	5.0015	5.3200	5.0290	5.0032
Mean TDS	0.3572	0.3800	0.3592	0.3574
$\Sigma_{i=1}^{n}(T_i)$	12.6056	13.4106	12.6751	12.6098

TABLE 12.10
B/P DOCR Operating Times and the CTM for the 8-Bus Test System, Considering TDS with Four DP

Backup Relay No.	BOT (S)	Primary Relay No.	POT (S)	CTM (S)
6	1.1384	1	0.8382	0.3002
1	1.3202	2	1.0199	0.3003
7	1.3200	2	1.0199	0.3001
2	1.2864	3	0.9864	0.3000
3	1.1953	4	0.8953	0.3000
4	1.1147	5	0.8146	0.3001
5	1.1751	6	0.8481	0.3269
14	1.3412	6	0.8481	0.4930
5	1.1751	7	0.8751	0.3000
13	1.3409	7	0.8751	0.4658
7	1.3200	8	0.8688	0.4512
9	1.1871	8	0.8688	0.3183

(Continued)

TABLE 12.10 (Continued)
B/P DOCR Operating Times and the CTM for the 8-Bus Test System, Considering TDS with Four DP

Backup Relay No.	BOT (S)	Primary Relay No.	POT (S)	CTM (S)
10	1.0965	9	0.7964	0.3001
11	1.1864	10	0.8864	0.3000
12	1.2833	11	0.9833	0.3000
13	1.3409	12	1.0409	0.3000
14	1.3412	12	1.0409	0.3003
8	1.1692	13	0.8689	0.3003
1	1.3202	14	0.8868	0.4333
9	1.1871	14	0.8868	0.3003

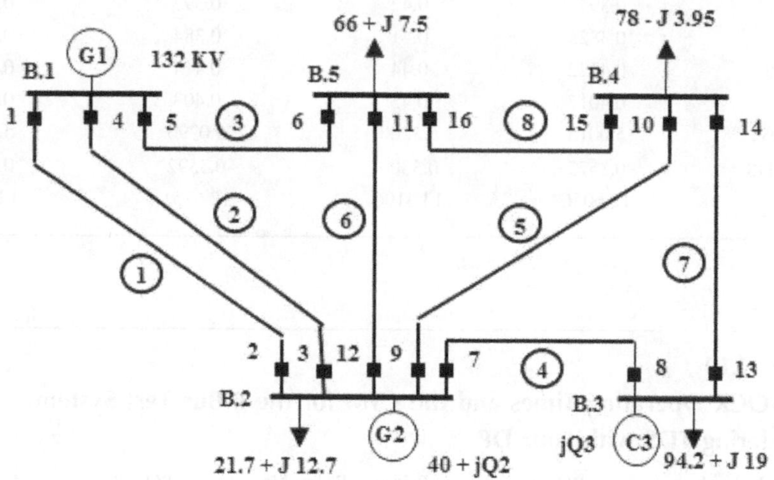

FIGURE 12.5 The one-line diagram of the IEEE 5-bus test system [24], [25].

TABLE 12.11
Pickup Current and CT Ratio for Each Relay of the IEEE 5-Bus Test System [24], [25]

Relay No.	Ip (A)	CT Ratio
1	6	400/5
2	1	200/5
3	1	200/5
4	6	400/5

(Continued)

TABLE 12.11 (Continued)
Pickup Current and CT Ratio for Each Relay of the IEEE 5-Bus Test System [24], [25]

Relay No.	Ip (A)	CT Ratio
5	8	300/5
6	1	200/5
7	8	300/5
8	1	100/5
9	6	300/5
10	1	200/5
11	1	200/5
12	7	200/5
13	1	100/5
14	4	200/5
15	1	100/5
16	7	300/5

TABLE 12.12
The B/P Relay Pairs for the IEEE 5-Bus Test System

Backup Relay No.	Primary Relay No.	Backup Relay No.	Primary Relay No.
1	3	8	12
1	7	9	14
1	9	9	15
1	12	10	2
2	4	10	3
2	5	10	7
3	1	10	12
3	5	11	2
4	2	11	3
4	7	11	7
4	9	11	9
4	12	12	6
5	11	12	16
5	16	13	10
6	1	13	15
6	4	14	8
7	13	15	6
8	2	15	11
8	3	16	10
8	9	16	14

TABLE 12.13
The Fault Currents for Each B/P Relays Pair for the IEEE 5-Bus Test System [24], [25]

Backup Relay No.	Fault Current (A)	Primary Relay No.	Fault Current (A)
1	981	3	2728.9
1	981	7	3328.1
1	981	9	3426.0
1	981	12	3403.6
2	605.7	4	3773.3
2	605.7	5	4093.1
3	605.7	1	3773.3
3	605.7	5	4093.1
4	981	2	2728.9
4	981	7	3328.1
4	981	9	3426.0
4	981	12	3403.6
5	915	11	1842.6
5	915	16	1636.5
6	375	1	3773.3
6	375	4	3773.3
7	1022.9	13	1571.1
8	446.8	2	2728.9
8	446.8	3	2728.9
8	446.8	9	3426.0
8	446.8	12	3403.6
9	850.7	14	1966.9
9	850.7	15	1436.2
10	301.7	2	2728.9
10	301.7	3	2728.9
10	301.7	7	3328.1
10	301.7	12	3403.6
11	282.4	2	2728.9
11	282.4	3	2728.9
11	282.4	7	3328.1
11	282.4	9	3426.0
12	840.3	6	1799.2
12	840.3	16	1636.5
13	638.3	10	1796.9
13	638.4	15	1436.2
14	788.6	8	1377.2
15	1043.3	6	1799.2
15	1043.3	11	1842.6
16	1188.8	10	1796.9
16	–1188.8	14	1966.9

12.6.3 The IEEE 5-Bus Test System

Figure 12.5 depicts the one-line diagram of the IEEE 5-bus test system, which is part of the IEEE 14-bus test system, which consists of five buses, eight lines, and three generators. A complete data of this system are given in [24], [25], all loads are given in MVA. The locations of the 16 DOCR relays are indicated in the Figure. Table 12.11 depicts the CT ratio and I_P for each relay, as given in [24,25]. Table 12.12 depicts the B/P relays pairs as determined by the introduced ES. Table 12.13 depicts the fault currents for each B/P relays pair [24], [25].

TABLE 12.14
Optimal DOCR Setting Values for the IEEE 5-Bus Test System

Relay No.	TDS			
	LP (IPM)	ES (2DP)	ES (3DP)	ES (4DP)
1	0.0500	0.06	0.050	0.0500
2	0.1521	0.17	0.154	0.1523
3	0.1521	0.17	0.154	0.1523
4	0.0500	0.06	0.050	0.0500
5	0.0567	0.07	0.058	0.0568
6	0.1199	0.14	0.120	0.1199
7	0.0728	0.08	0.073	0.0729
8	0.2129	0.23	0.215	0.2131
9	0.0739	0.08	0.074	0.0739
10	0.1369	0.15	0.138	0.1371
11	0.1305	0.14	0.131	0.1306
12	0.0969	0.11	0.098	0.0971
13	0.3053	0.32	0.306	0.3055
14	0.1244	0.14	0.126	0.1246
15	0.2524	0.27	0.253	0.2526
16	0.0809	0.09	0.082	0.0810
$\Sigma_{i=1}^{n}(TDS)$	2.0677	2.2800	2.0820	2.0697
Mean TDS	0.1292	0.1425	0.1301	0.1294
$\Sigma_{i=1}^{n}(T_i)$	4.3797	4.8670	4.4123	4.3842

TABLE 12.15
B/P DOCR Operating Times and the CTM for the IEEE 5-Bus Test System, Considering TDS with 3DP

Backup Relay No.	BOT (S)	Primary Relay No.	POT (S)	CTM (S)
1	0.4861	3	0.2446	0.2415
1	0.4861	7	0.2588	0.2273
1	0.4861	9	0.2247	0.2613

(Continued)

TABLE 12.15 (Continued)

B/P DOCR Operating Times and the CTM for the IEEE 5-Bus Test System, Considering TDS with 3DP

Backup Relay No.	BOT (S)	Primary Relay No.	POT (S)	CTM (S)
1	0.4861	12	0.2678	0.2183
2	0.3860	4	0.1662	0.2197
2	0.3860	5	0.1853	0.2006
3	0.3860	1	0.1662	0.2197
3	0.3860	5	0.1853	0.2006
4	0.4861	2	0.2446	0.2415
4	0.4861	7	0.2588	0.2273
4	0.4861	9	0.2247	0.2613
4	0.4861	12	0.2678	0.2183
5	0.6249	11	0.2303	0.3945
5	0.6249	16	0.4163	0.2086
6	0.3665	1	0.1662	0.2003
6	0.3665	4	0.1662	0.2003
7	0.6702	13	0.4697	0.2005
8	0.4695	2	0.2446	0.2249
8	0.4695	3	0.2446	0.2249
8	0.4695	9	0.2247	0.2448
8	0.4695	12	0.2678	0.2017
9	0.5971	14	0.3427	0.2544
9	0.5971	15	0.3969	0.2002
10	0.4684	2	0.2446	0.2238
10	0.4684	3	0.2446	0.2238
10	0.4684	7	0.2588	0.2096
10	0.4684	12	0.2678	0.2006
11	0.4600	2	0.2446	0.2154
11	0.4600	3	0.2446	0.2154
11	0.4600	7	0.2588	0.2012
11	0.4600	9	0.2247	0.2353
12	0.6173	6	0.2123	0.4049
12	0.6173	16	0.4163	0.2010
13	0.5973	10	0.2443	0.3529
13	0.5973	15	0.3969	0.2004
14	0.3427	8	0.3407	0.2033
15	0.4303	6	0.2123	0.2179
15	0.4303	11	0.2303	0.2000
16	0.5459	10	0.2443	0.3016
16	0.5459	14	0.3427	0.2031

Table 12.14 depicts the TDS values as determined by the LP(IPM) and the introduced ES. Given Table 12.14, it is clear that the difference between the TDS values is minor. Moreover, the difference between the summation of the TDS values, the mean value of the TDS, and the summation of the operating times of the primary relays of the introduced ES and the LP(IPM) are minor also.

Table 12.15 depicts the B/P relays pairs operating time and the CTM for the IEEE 5-bus test system, considering the TDS values with three DP. Given Table 12.15, it is clear that there is no any miscoordination case between the B/P relays pairs.

12.7 CONCLUSION

This chapter introduced a novel ES for optimal coordination of DOCR in meshed networks, in which the coordination knowledge is represented in a rule style. The introduced ES has inherent immunity from trapping in a local minimum. Moreover, it has got rid of the problem of miscoordination, which may arise due to rounding off the TDS values. Hence, it is suitable for coordinating both electromechanical and digital DOCR, off-line, and on-line, respectively. Extensive simulation of the introduced ES proved its efficiency in setting the DOCR optimally.

REFERENCES

[1] A. Y. Abdelaziz, H. E. A. Talaat, A. I. Nosseir, and Ammar A. Hajjar, "An adaptive protection scheme for optimal coordination of overcurrent relays," *Electr. Power Syst. Res.*, vol. 61, no. 1, pp. 1–9, Feb. 2002.

[2] R. E. Albrecht, M. J. Nisja, W. E. Feero, G. D. Rockefeller, and C. L. Wagner, "Digital computer protective device coordination program—I—General program description," *IEEE Trans. Power App. Syst.*, vol. PAS-83, no. 4, pp. 402–410, Apr. 1964.

[3] M. J. Damborg and S. S. Venkata, "Specification of computer-aided design of transmission protection systems," Department of Electrical Engineering, University of Washington, Tech. Rep. EPRI EL-3337 (Final Rep.), 1984.

[4] A. J. Urdaneta, R. Nadira, and L. G. Perez Jimenez, "Optimal coordination of directional overcurrent relays in interconnected power systems," *IEEE Trans. Power Del.*, vol. 3, no. 3, pp. 903–911, Jul. 1988.

[5] A. S. Braga and J. T. Saraiva, "Co-ordination of directional overcurrent relays in meshed networks using the simplex method," in Proceedings of 8th Mediterranean Electrotechnical Conference on Industrial Applications in Power Systems, Computer Science and Telecommunications (MELECON 96), vol. 3, pp. 1535–1538, 1996.

[6] A. A. Hajjar, S. M. Tarraf, and G. A. Isper, "Studying factors that affecting the optimal setting of numerical inverse overcurrent relays in meshed networks," *Tishreen Univ. J. Res. Sci. Stud.—Eng. Sci. Ser*, vol. 42, no. 2, pp. 250–272, 2020.

[7] H. A. Abyaneh, M. Al-Dabbagh, H. K. Karegar, S. H. H. Sadeghi, and R. A. H. Khan, "A new optimal approach for coordination of overcurrent relays in interconnected power systems," *IEEE Trans. Power Del.*, vol. 18, no. 2, pp. 430–435, Apr. 2003.

[8] W. El-Khattam, and T. S. Sidhu, "Restoration of directional overcurrent relay coordination in distributed generation systems utilizing fault current limiter," *IEEE Trans. Power Del.*, vol. 23, no. 2, pp. 576–585 Apr. 2008.

[9] R. Corrêa, G. Cardoso, O. C.B. de Araújo, and L. Mariotto, "Online coordination of directional overcurrent relays using binary integer programming," *Electr. Power Syst. Res.*, vol. 127, pp. 118–125, 2015.

[10] Y. Damchi, M. Dolatabadi H. R. Mashhadic, and Javad Sadeh, "MILP approach for optimal coordination of directional overcurrent relays in interconnected power systems," *Electr. Power Syst. Res.*, vol. 158, pp. 267–274, 2018.
[11] M. M. Mansour, S. Mekhamer, and N-S. El-Kharbawe, "A modified particle swarm optimizer for the coordination of directional overcurrent relays," *IEEE Trans. Power Del.*, vol. 22, no. 3, pp. 1400–1410, Jul. 2007.
[12] C. W. So, and K. K. Li, "Time coordination method for power system protection by evolutionary algorithm," *IEEE Trans. Ind. Appl.*, vol. 36, no. 5, pp. 1235–1240, Sep. 2000.
[13] M. Y. Shih, A. C. Enríquez, T. Y. Hsiaoc, and L. M. Torres Trevino, " Enhanced differential evolution algorithm for coordination of directional overcurrent relays," *Electr. Power Syst. Res.*, vol. 143, pp. 365–375, 2017.
[14] T. Amraee, "Coordination of directional overcurrent relays using seeker algorithm," *IEEE Trans. Power Del.*, vol. 27, no. 3, pp. 1415–1422, Jul. 2012.
[15] A. Wadood, T. Khurshaid, S. G. Farkoush, J. Yu, C. H. Kim, and S.-B. Rhee, ''Nature-inspired whale optimization algorithm for optimal coordination of directional over-current relays in power systems,'' *Energies*, vol. 12, no. 12, 2297, Jun. 2019.
[16] M. Javadi, A. Nezhad, A. Anvari-Moghadam, and J. Guerrero, "Hybrid mixed-integer non-linear programming approach for directional overcurrent relay co-ordination," *IET, J. Eng.*, vol. 2019, no. 18, pp. 4743–4747, Aug. 2019.
[17] N. El-Nailya, S.M. Saada, and F. A. Mohamedb, "Novel approach for optimum coordination of overcurrent relays to enhance microgrid earth fault protection scheme," *Sustain. Cities Soc.*, vol. 54,102006, Mar. 2020.
[18] M. N. Alam, B. Das, and V. Pant, "A comparative study of metaheuristic optimization approaches for directional overcurrent relays coordination," *Electr. Power Syst. Res.*, vol. 128, pp. 39–52, Nov. 2015.
[19] M. N. Alam, "Adaptive protection coordination scheme using numerical directional overcurrent relays," *IEEE Trans. Ind. Inform.*, vol. 15, no. 1, pp. 64–73, Jan. 2019.
[20] H. W. Hong, V. M. Mesa, and N. G. Steven, "Protective device coordination expert system," *IEEE Trans. Power Del.*, vol. 6, no. 1, pp. 359–365, Jan. 1991.
[21] K. Tuitemwong, and S. Premrudeepreechacharn, "Expert system for protection coordination of distribution system with distributed generators," *Electr. Power Energy Syst.*, vol. 33, pp. 466–471, 2011.
[22] K. Kawahara, H. Sasaki, and H. Sugihara, "An application of the rule-based system to the coordination of directional overcurrent relays," in Proceedings of 6th International Conference on Developments in Power Systems Protection, Nottingham, UK, 25–27 Mar. 1997.
[23] Measuring relays and protection equipment – Part 151: Functional requirements for over/under current protection, IEC Standard 60255- 151, 2009.
[24] A. A. Hajjar, "Adaptive coordination of overcurrent relays in power distribution networks," MSc. Thesis in Electrical Engineering, Ain Shams University, Jun. 1999.
[25] A. A. Hajjar, A. Y. Abdelaziz, H. E. A. Talaat, and A. I. Nosseir, "Optimal co-ordination of overcurrent relays by linear programming: An enhanced problem formulation," in Proceedings of 3rd CIGRE Regional Conference for Arab Countries, Doha, May 1999.
[26] S. S. Rao, *Engineering Optimization: Theory and Practice*. John Wiley & Sons;USA, 12 Nov. 2019, p. 832.
[27] J., Giarratano, *Expert Systems: Principles and Programming*. PWS Publishing Company Boston, USA, 1998, p. 602.

13 Optimal Overcurrent Relay Coordination Considering Standard and Non-Standard Characteristics

Ahmed Korashy, Salah Kamel, Loai Nasrat, and Francisco Jurado

[1]Department of Electrical Engineering, Faculty of Engineering, Aswan University, Egypt

[2]Department of Electrical Engineering, University of Jaén, Spain

NOMENCLATURE

Ant Colony Optimization ACO
Artificial ecosystem-based optimization AEO
Backup Relay Operating Time T_{backup}
Biogeography-Based Optimization BBO
Black Hole BH
Characteristics Relay Curve CRC
Constant Values for The Characteristic Of The Relay B
Constant Values for The Characteristic Of The Relay A
Convergence Characteristics CC
Coordination Time Interval CTI
Current Iteration value Iter
Directional Overcurrent Relays DOCRs
Decision Variable DV
DISTANCE Jaya Algorithm DJAYA
Enhanced Backtracking Search Algorithm EBSA
Fault Current If
Firefly Algorithm FFA
Genetic Algorithm GA
Gravitational Search Algorithm GSA
Group Search Optimization GSO

DOI: 10.1201/9780367552374-13

Harmony Search HS
Improved Moth-Flame Optimization IMFO
Improved Firefly Algorithm IFO
Institute of Electrical and Electronics Engineers IEEE
Inverse Definite Minimum Time IDMT
Linear Programming LP
Lower Ranges Of Decision Variables LoBo
Modified Version For Water Cycle Algorithm MWCA
Moth-Flame Optimization MFO
Modified Electromagnetic Field Optimization MEFO
Non-Linear Programming NLP
Number of Primary Relay P
Number of Maximum Iteration Maxi.Iter
Number of Decision Variables Nvars
Number Of Population Npop
Objective Function OF
Operating Time OT
Oppositional Jaya Algorithm OJAYA
Over-Current Relay OCR
Particle Swarm Optimization PSO
Pickup Current Ip
Primary Relay Operating Time T_{pri}
Seeker Algorithm SA
Seeker algorithm SA
Sequential Quadratic Programming SQP
Slim Mould Algorithm SMA
Teaching-Learning-Based Optimization TLBO
Time Dial Setting TDS
Upper Ranges Of Decision Variables UpBo
Water Cycle Algorithm WCA

13.1 INTRODUCTION

Present day, the stability and security of the electric power network are becoming more challenging over the past few decades because of the growing size of the power network. The electrical network is planned to be as possible faultless through careful planning, proper installation and regular maintenance of electrical equipment [1]. The fault causes instability and short- to long-term power outages. The main roles of a protective relay are to keep healthy parts in service, isolate only the fault part, and maintain the reliability of the electric network [2]. Due to low cost and simplicity, the DOCRs are usually used in protecting of sub-transmission and distribution system [3]. DOCRs operating time is based on two settings, which are considered as a DV. The correct chosen of relay settings (TDS and Ip) play a crucial role in solving the coordination problem [4].

13.1.1 Methods for Coordination of DOCRs

Different approaches have been suggested to find a solution to the problem of coordination. Several algorithms are implemented to deal with such problems such as graph theory-technique and manual methods. Where these methods successfully solved the coordination problem in case of a small system. However, such methods were very time-consuming in the large network [5]. Also, the trial-and-error technique was initiated to find the optimal relay setting using computers. This technique has a slow rate of convergence and the obtained TDS values of the relays using this approach are relatively high [6]. The linear programming was suggested to solve the problem of coordination in the 1980s, where the TDS is optimally calculated. However, the Ip in this method is assumed to be predefined [7]. Both relay setting (TDS and Ip) are optimally and simultaneously determined using nonlinear programming. The solution obtained from conventional methods is far away from a globally optimal solution. Recently, meta-heuristic and hybrid optimization methods have been developed and most widely used in the coordination of relay to get the globally optimal solution and able to escape from local minima solution [8]. Many optimization techniques such as TLBO [8], BBO [6], MEFO [9], and FFA [10] have been suggested to solve the problem of coordination. Many hybrid techniques have been suggested which utilize the features of nature-inspired and classical techniques, have been successfully proposed to solve the coordination problem of DOCRs [11].

The connecting of distributed generation (DG) to electric system network has numerous advantages and should be fully exploited to gain more profits. Even though the spread of DGs installing in electric power systems have many advantages, they have negative impacts on the protection system such as loss of coordination between relay pairs and increasing short-circuit levels [12]. The impact of DG on the performance of the protective relays must be considered carefully to maintain the security of electric power systems [13]. The performance DOCRs suffers degradation in the presence of DG. Despite there are different techniques that have been proposed as possible solutions to mitigate the negative impact of DGs penetration on the performance of the DOCRs coordination [14,15]. In this chapter, the DG impact has not been taken into consideration. In other words, the suggested algorithms have not considered dynamic changes in the electric power system topology.

13.2 DOCRS COORDINATION PROBLEM

The problem of DOCRs coordination was described as a non-linear problem of optimization with high boundaries. This problem dealt with calculating the optimal relay settings that minimize the total operating time for all primary relays at the same time maintain coordination margin between relay pairs. All protection relays in the electric system shall be chosen correctly to ensure good coordination among protective devices [4]. The problem coordination in this chapter is mathematically expressed as an optimization problem. The goal function for this problem can be expressed as [16]:

$$OF = Minimize \sum_{d=1}^{P} Tpri_d \tag{13.1}$$

where OF is the objective function and P is the number relays. T_{pri} is the primary relay operating time, which this time can be determined as [17]:

$$T_{pri} = \frac{A \times TDS^d}{\left(\frac{I_f^d}{I_p^d}\right)^B - 1} \tag{13.2}$$

where If is a fault current for relay d, TDS is time dial setting for relay d, and Ip is pickup current for relay d. The A and B are constant values [18]. These constant values represent the relay characteristics. These values are presented in Table 13.1 [18].

13.2.1 Boundaries of the Coordination Problem

The OF should be met under two categories of constraints. These categories include coordination constraints and relay characteristics.

13.2.1.1 Limits on Relay Characteristics

13.2.1.1.1 Limits on Pickup Current Setting

The constraint on Ip can be described as:

$$Ip_{M\ ini}^{d} \le Ip \le Ip_{M\ ax\ i}^{d} \tag{13.3}$$

where Ip_{Maxi} I and p_{Mini} are the maximum and minimum limits of Ip, respectively. These limits are depending on the maximum loading and minimum short circuit current to ensure that the DOCRs will not be initiated under normal current. Also, to ensure that at the smallest short-circuit current the relay will be sensitive [19], [20].

TABLE 13.1
Constants for Different Types of Standard DOCRs Characteristics

Curve type	A	β
Standard inverse	0.14	0.02
Very inverse	13.5	1
Extremely inverse	80	2
Long-time inverse	120	1

13.2.1.1.2 Limits on TDS

TDS boundaries can be described as:

$$TDS^{d}_{Mini} \le TDS^{d} \le TDS^{d}_{Maxi} \tag{13.4}$$

where TDS_{Maxi} and TDS_{Mini} are maximum and minimum values of TDS, respectively. These limits depend on the manufacturer of relay [21].

13.2.1.1.3 Parameters of Characteristics Relay Curve

In case of standard or conventional characteristic relay curve scenario, the normal inverse is chosen and the value of the relay parameter B and A in that scenario are set as 0.02 and 0.14, respectively [22]. In the case of non-standard or non-conventional characteristic curve scenario, the relay parameter *A* and *B* are set as a DV and considered as constraint, which these constraints can represent as follows:

$$A^{d}_{Mini} \le A^{d} \le A^{d}_{Maxi} \tag{13.5}$$

$$B^{d}_{Mini} \le B^{d} \le B^{d}_{Maxi} \tag{13.6}$$

The A_{Mini} is the minimum value of constant A for relay d and the A_{Maxi} is the maximum value of constant A for relay d. The B_{Mini} is the minimum value of constant B for relay d and the B_{Maxi} is the maximum value of constant B for relay d [18].

13.2.1.2 Boundaries on DOCRs Coordination

The avoidance of mal-operation of DOCRs is the main target of the boundaries on coordination. This goal could be achieved by the right sequence of operation between backup and primary relays. The main relays are considered as the first defences to isolate the faults. While the main and backup relays simultaneously sensed the fault. In order to prevent from mal-operation of DOCRs, a specified margin between relays is required, which this margin called the CTI [5]. This delay is a very important issue that guarantees the backup relays will be initiated after this delay in the event of failure of the main relay to initiated. The CTI can be calculated as [23,24]:

$$T_{backup} - T_{pri} \ge CTI \tag{13.7}$$

where T_{pri} is the main relay operating time. T_{backup} is backup relays operating times, respectively [24].

13.3 RECENT OPTIMIZATION TECHNIQUES

In mathematics and computer science, the optimization process is called mathematical programming as it is related to computer programming. The optimization process is used in many areas especially engineering systems. It is applying to get the better (minimum/maximum) solution for the objective function between numbers of variables under required constraints [25]. Different methods of optimization algorithm are suggested to solve the problem of coordination. Also, different enhanced and improvement are suggested to enhance the conventional technique performance. These techniques are:

- Water cycle algorithm (WCA) and its modified version (MWCA).
- Moth-flame optimization (MFO) and its improved version (IMFO).

13.3.1 WCA and MWCA

13.3.1.1 Conventional WCA

WCA is motivated by the observation of the water cycle in nature. In this method, water evaporates and transfer to the atmosphere then back to earth as raindrop [26]. The overall process for this algorithm can be described as:

1. *Raindrops Initialization:* The algorithm starts with drops. These drops initiated between minimum and maximum decision variables limits. The drops are classified as the sea, which is the best drop. Then better drop is set as river and the rest drops are selected as streams [26], [27]. The raindrops are randomly initiated, which this concept can be represented as:

$$Drop = [L_1, L_2, L_3, \ldots, L_{M\,\mathrm{var}}] \tag{13.8}$$

$$Raindrops\ Population = \begin{bmatrix} Drop_1 \\ Drop_2 \\ Drop_3 \\ \vdots \\ Drop_{M_{pop}} \end{bmatrix} = \begin{bmatrix} L_1^1 & L_2^1 & \ldots & \ldots & L_{N\,\mathrm{var}}^1 \\ L_1^2 & L_2^2 & \ldots & \ldots & L_{M\,\mathrm{var}}^2 \\ \ldots & \ldots & \ldots & \ldots & \ldots \\ L_1^{M_{pop}} & L_2^{M_{pop}} & \ldots & \ldots & L_{M\,\mathrm{var}}^{M_{pop}} \end{bmatrix} \tag{13.9}$$

where *Mpop is* the drops number. The Mvars is the number of decision variables. Based on the OF value, the cost of the drops can be expressed as [26]:

$$\mathrm{Cos}\ t_j = f(h_1^j, h_1^j, h_1^j, \ldots, h_1^j)\mathrm{j} = 1, 2, \ldots, \mathrm{M_{pop}} \tag{13.10}$$

Based on the value of (13.12), the drops are selected and classified as a sea, which has the least value, and the other as a river. The total number of one sea and rivers is calculated as in (13.11). The other drops that form the streams can be described as in (13.12).

$$M_{sr} = Rivers\ Numbe + 1 \tag{13.11}$$

$$M_{drops} = M_{pop} - M_{sr} \tag{13.12}$$

Based on flow intensity, the drops are classified as sea and rivers as:

$$MS_y = round\left\{\left|\frac{\text{Cos } t_y}{M_{\sum_{j=1}^{sr} \text{Cos } t_j}}\right| \times M_{Raindrops}\right\} \text{y} = 1, 2, \ldots . \text{M}_{\text{sr}} \tag{13.13}$$

2. *A streamflow*: The new location for and rivers and streams can be expressed as [26]:

$$H_{Stream}^{j+1} = H_{Stream}^{j} + rand \times G \times (H_{River}^{j} - H_{Stream}^{j}) \tag{13.14}$$

$$H_{River}^{j+1} = H_{River}^{j} + rand \times G \times (H_{Sea}^{j} - H_{River}^{j}) \tag{13.15}$$

where G is a number between 1 and 2. H_{river} is river location and H_{stream} is streams location. *H sea is* the sea *location* [26]. In case of stream successive to find a location better than a location found by the river, the exchange will occur between the stream and river. Also, the same case will occur between the sea and the river [27].

3. *Condition of Evaporation*: The evaporation condition protects the technique from stuck in local minima, which is applied to both steams and rivers during their movement toward the sea [27]. The evaporation condition is checked using the following pseudocode [26]:

If $\left|H_{Sea}^{j} - H_{River}^{j}\right| < X_{maxi}$ j = 1, 2, 3, …., $M_{sr} - 1$
Begin raining
End

where X_{maxi} is a small number. This value is reduced as follows:

$$X_{max}^{j+1} = X_{max}^{j} - \frac{X_{max}^{j}}{Maxi.\ Iter} \tag{13.16}$$

The evaporation condition is met then the raining will occur, when the gap between the river and sea is less than X_{maxi}. The condition of evaporation also check the streams that move toward the sea using the following pseudo-code [26]:

If $\left|H_{Sea}^{j} - H_{Streams}^{j}\right| < X_{maxi}$ i = 1, 2, 3, …., M_S
Begin raining
End

4. *Raining Cycle:* After meeting the condition of evaporation, the raining cycle begins and drops form streams in many positions. The position of the new stream can be expressed as [27]:

$$H_{Stream}^{new} = LoBo + rand \times (UpBo - LoBo) \tag{13.17}$$

where UpBo is maximum limits of DV and LoBo is minimum limits are of DV. For the new streams that move toward directly to the sea can be described as [26]:

$$H_{Stream}^{new} = H_{sea} + \sqrt{0.1} \times randn(1, M_{\text{var}}) \tag{13.18}$$

13.3.1.2 MWCA Algorithm

The performance of metaheuristic techniques can be improved through the right balance between the two conflicting elements. The first element that aims to search locally is known as exploitation. The second element that aims to search globally is known as exploration. Global minima can be guarantee and the search space is reduced by those elements and prevent the technique from stuck in local minima [28–30]. The balance between theses elements can be performed in WCA by the value of parameter G, which helps the streams to move towards the rivers in different positions [26]. In the onventional algorithm, the G is set to be equal two [26], [27]. In the MWCA, the G parameter is proposed to increase the exponential gradually throughout iterations from one to two using (13.19). This change helps the algorithm to enhance its capability to balance between search globally and search locally to find the best promising solution. Also, by replacing the G parameter from a fixed value to a variable throughout the cycle process helps to reduce search space. Therefore, the location for new rivers and streams can be expressed as [31]:

$$G^j = 2 - \left(1 - \frac{j}{Maxim.\ Iter}\right)^2 \tag{13.19}$$

$$H_{Stream}^{j+1} = H_{Stream}^{j} + rand \times G^j \times (H_{River}^{j} - H_{Stream}^{j}) \tag{13.20}$$

$$H_{River}^{j+1} = H_{River}^{j} + rand \times G^j \times (H_{Sea}^{j} - H_{River}^{j}) \tag{13.21}$$

For all new streams, the raining cycle can be determined using (13.20). This process enhances the algorithm capability to explore in a whole space for the global minimum. The process of raining for new streams can be determined using (13.20). In the WCA technique, (13.20) is only utilized for streams that move toward the sea [26]. The condition of evaporation for rivers and streams, which move toward the sea to begin the rain cycle in case the condition is true, can be accomplished through the following pseudocode:

If $\left|H^{j}_{Sea} - H^{j}_{Streams}\right| < X_{max}\ j = 1, 2, 3, \ldots., M_S$
Begin raining using (18)
End

If $\left|H^{j}_{Sea} - H^{j}_{River}\right| < X_{Maxi}\ j = 1, 2, 3, \ldots., N_S$
Begin raining using (18)
End

13.3.2 MFO and IMFO Algorithms

13.3.2.1 The MFO Algorithm

The MFO is a type of population-based algorithm, which is inspired from the movement of moths with respect to the moonlight. The moth exploits a mechanism called transverse orientation for movement. [32]. The moth in this technique is considered a nominated solution to the problem of coordination and the DV is considered the moth's location in space [33]. The randomness of the population for the MFO is initiated and the moths' set is described as [32]:

$$Z = \begin{bmatrix} zi_{1,1} & zi_{1,2} & \ldots & \ldots & zi_{1,q} \\ zi_{2,1} & zi_{2,2} & \ldots & \ldots. & zi_{2,q} \\ \ldots & \ldots & \ldots & \ldots & \ldots \\ zi_{u,1} & zi_{u,2} & \ldots. & \ldots. & zi_{u,l} \end{bmatrix} \tag{13.22}$$

where q is the number of DV and u is the number of moths. Regarding OF value, the moths are sorting. The storing moth can be represented as:

$$OZ = \begin{bmatrix} OZ_1 \\ OZ_2 \\ \ldots \\ OZ_u \end{bmatrix} \tag{13.23}$$

There are other components in the MFO technique that are called flames, and these components could be described [32]:

$$W = \begin{bmatrix} WI_{1,1} & WI_{1,2} & \ldots & \ldots & WI_{1,q} \\ WI_{2,1} & WI_{2,2} & \ldots & \ldots. & WI_{2,q} \\ \ldots & \ldots & \ldots & \ldots & \ldots \\ WI_{u,1} & WI_{u,2} & \ldots. & \ldots. & WI_{u,q} \end{bmatrix} \tag{13.24}$$

The flames are sorted regarding the OF value, which can be described as:

$$OW = \begin{bmatrix} OW_1 \\ OW_2 \\ \cdots \\ OW_q \end{bmatrix} \quad (13.25)$$

Both moths and flames are considered a candidate solution to the problem of optimization and the difference between them is the updating technique during the repetition cycle [33]. Flames in the MFO algorithm are considered flags. The moth explores the best solution around a flag. When a moth gets a better solution, it is updating its position regarding a flame [34]. This concept can mathematically be represented as [32]:

$$Z = S(Z_r, W_w) \quad (13.26)$$

$$O_r = |W_w - Z_r| \quad (13.27)$$

$$S(Z_r, W_w) = O_r.\ e^{nv}.\ \cos(2\pi v) + W_w \quad (13.28)$$

where Z_r refers to the rth moth and W_w indicates the wth flame. O_r refers to the difference between a moth and flame distance. The n is a fixed value. The v is a value that could be determined as [32]:

$$pp = -1 + It \times \left(\frac{-1}{Maxim.\ It} \right) \quad (13.29)$$

$$v = (pp - 1) \times rand + 1 \quad (13.30)$$

Generally, the MFO is accomplished using three functions that can be expressed as [32]:

$$MFO = (C, K, Y) \quad (13.31)$$

The C is a response to initializes a population randomly between boundaries of DV and determines the OF. K is the main function. This function is responsible in performing repetition of the process until the function Y becomes true. In case of conversions, criteria are met, the function Y will be true, and the best solution is obtained.

13.3.2.2 The IMFO Algorithm

The IMFO improves the performance of a conventional MFO. This improvement is performed in function K, where the IMFO utilizes the hierarchy of GWO leadership [35]. As known before, search agents in this method are the moths and flames are the best position for these search agents of moths. In the IMFO, there are three best flags. With respect to the position of these three best flags, the will moths update their position. This concept can be mathematically represented as follows [36]:

$$Z = S(Z_r, W_w) \tag{13.32}$$

$$O_{ALP} = \left| W_{ALP_w} - Z_{ALP_w} \right| \tag{13.33}$$

$$O_{BET} = \left| W_{BET_w} - Z_{BET_w} \right| \tag{13.34}$$

$$O_{DEL} = \left| W_{DEL_w} - Z_{DEL_w} \right| \tag{13.35}$$

$$H_1 = O_{ALP_r} \cdot e^{nv} . \ \cos(2\pi v) + W_{ALP_w} \tag{13.36}$$

$$H_2 = O_{BET_r} \cdot e^{nv} . \ \cos(2\pi v) + W_{BET_w} \tag{13.37}$$

$$H_3 = O_{DEL_r} \cdot e^{nv} . \ \cos(2\pi v) + W_{DEL_w} \tag{13.38}$$

$$S(Z_r, W_w) = \frac{H_1 + H_2 + H_3}{3} \tag{13.39}$$

where ALP is the first level and the BET is second in the hierarchy. The DEL is the third level in the hierarchy. The IMFO process can be concluded in Figure 13.1.

13.4 RESULTS AND DISCUSSION

The results from applying the suggested techniques in different systems to get a solution for the problem of coordination has been presented in this section, which can be summarized as below:

- Formulated the coordination problem using WCA and MWCA in case of conventional standard CRC.
- Formulated the coordination problem as using MFO and IMFO in case of conventional standard and non-conventional CRC.

The suggested methods are compared with recent and other optimization methods (EFO [9], MEFO [9], SMA [37], AEO [38], DE [39,40], (SA) [41], SFSA [22], HS [5], BBO [6], BH [8], GA-NLP [20], and GSA-SQP [42], GSA [42], PSO [39], GA

Input size of the population, number of DVs, and Maximum Iteration.

Function C (Phase #1)

Initialize solutions between a minimum and maximum boundaries of DVs and lower limits of DVs.
Determine OF value for each solution.

Function K (Phase #2)

```
for r = 1: u
for w = 1: q
    Update g.
    Determine O using (33),(34),and (35).
    Update Z(r,w) using (32) and (39).
end
End
```

Function Y (Phase #3)

```
If  Iter ≤ Maxi.Iter.
    Return to Function K( Phase #2).
else
end
```

FIGURE 13.1 IMFO steps.

[39], BBO-LP [6], SQP [39], FFA [41], IFA [41], BSA [43], EBSA [44], CSA [23], group search optimization (GSO) [45], DJAYA [46], and OJAYA [46])) to prove their superiority to solve the problem of DOCRs' coordination. The proposed techniques are implemented using the MATLAB® environment and the short circuit and load flow are accomplished using DIgSILENT Power Factory.

13.4.1 Description of Test Systems

In this section, nine-bus network and 15-bus network have been used to prove the efficiency of the suggested techniques. These systems are listed below:

- The nine-bus network system
- The 15-bus network system

13.4.1.1 The Nine-Bus Network

The nine-bus system is presented in Figure 13.2 [15]. There are 12 lines in this system. Each line has DOCRs in its end. There are 48 DVs for this system. The constraint limits for TDS in this system are 0.025 and 1.2, respectively. The co-ordination margin is equal to 0.2 s. Other data such as Ip and If are found in [14].

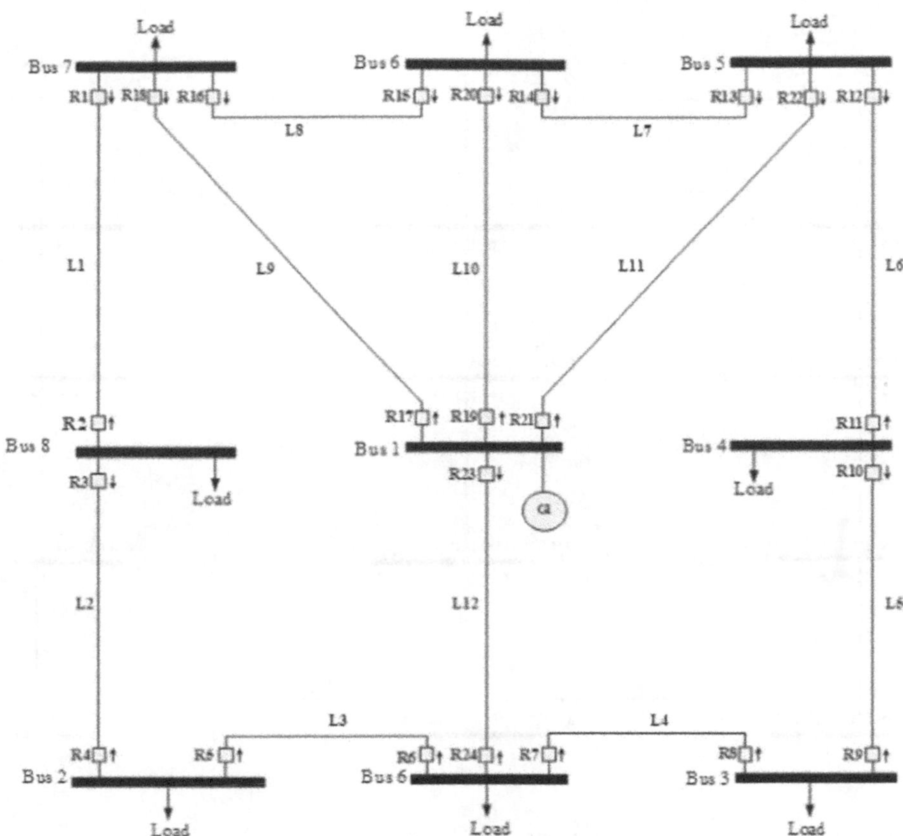

FIGURE 13.2 The network of the nine-bus network.

13.4.1.2 The 15-Bus Test System

The 15-bus system is presented in Figure 13.3 [47]. There are 21 lines in this system. Each line has DOCRs in its end. There are 84 DVs for this system. The constraint limits for TDS in this system are 0.1 and 1.1, respectively. The coordination margin is equal to 0.2 s. Other data such as the Ip and If are found in [9].

13.4.2 Formulated the Coordination Problem Using Standard-CRC

The performance of both techniques (WCA, MWCA) is assessed using the nine-bus system and 15-bus networks. The suggested methods are compared with other methods to show their effectiveness in solving the problem of coordination.

13.4.2.1 Using MWCA for Solving the Coordination Problem

13.4.2.1.1 Case 1: Nine-Bus Network

In this case, the WCA and MWCA are tested using the nine-bus, which is presented in Figure 13.2. The OT of the main relays and the OT of backup relays is shown in

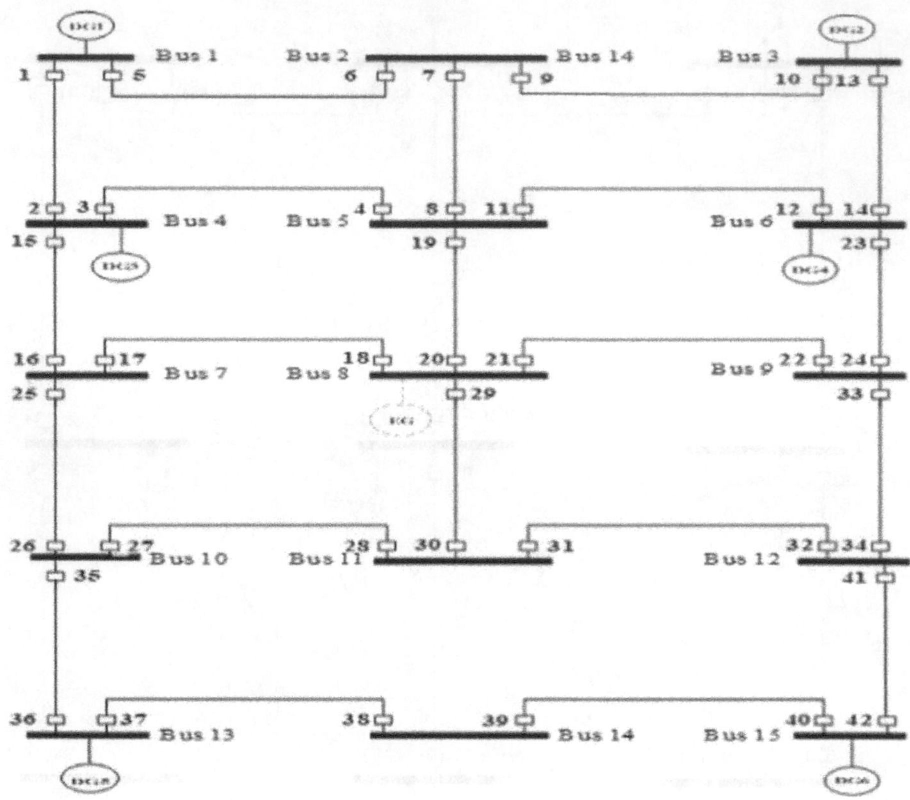

FIGURE 13.3 The network of the fifteen-bus network.

Figures 13.4 and 13.5. From these figures, it can be noticed that the backup relays will be initiated if the main relays fail to operate. It can be said that both techniques' (WCA and MWCA) successes find the optimal relay setting that maintains the sequential of operation between relay pairs.

The CC of WCA and MWCA techniques is presented in Figure 13.6. From Figure 13.6, it can be observed that the MWCA converge to the promising solution faster than WCA. Also, the OF value that is given from the MWCA is better than the OF value that is given from WCA, where the reduction on OF reaches 54.6%.

The comparison between well-known methods and WCA and MWCA is given in Table 13.2. The MWCA method presents the least OF, as noticed in this table. That proves the superiority and power of the MWCA for solving the coordination problem.

13.4.2.1.2 Case 2: 15-Bus Network

The WCA and MWCA are tested using the 15-bus network, which is shown in Figure 13.3. The main relays' OT and the backup relays' OT using WCA and

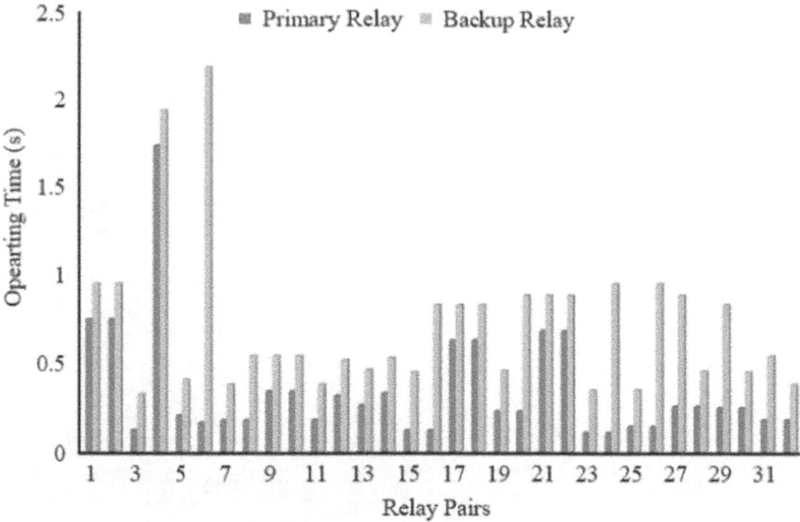

FIGURE 13.4 Relay pairs' OT of the nine-bus network using WCA.

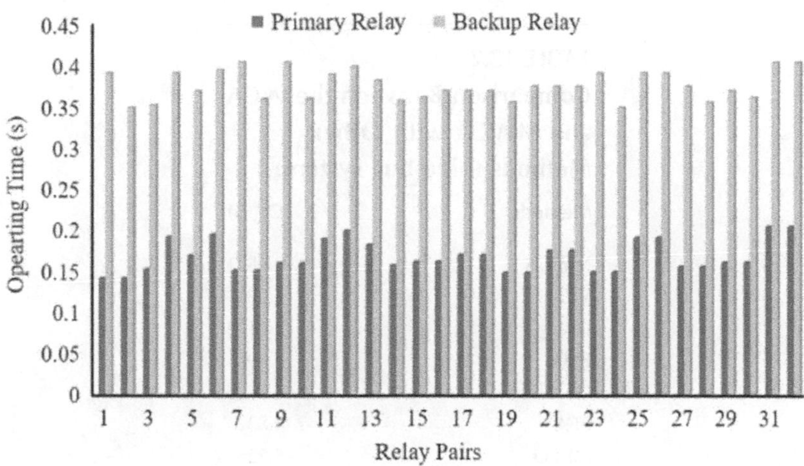

FIGURE 13.5 Relay pairs' OT of the nine-bus network using MWCA.

MWCA are shown in Figures 13.7 and 13.8. After delay time, the backup relays will be initiated if the main relays fail to operate as observed in Figures 13.7 and 13.8. It can be said that both techniques' successes get optimal settings that keep sequential operation between backup and primary relays.

The CC of WCA and MWCA methods is presented in Figure 13.9. From Figure 13.9, it can be noticed that the MWCA converge to the optimal solution faster than WCA. Also, the OF value given by the MWCA is better than the OF value that is given from WCA, where the reduction in OF reaches 26%.

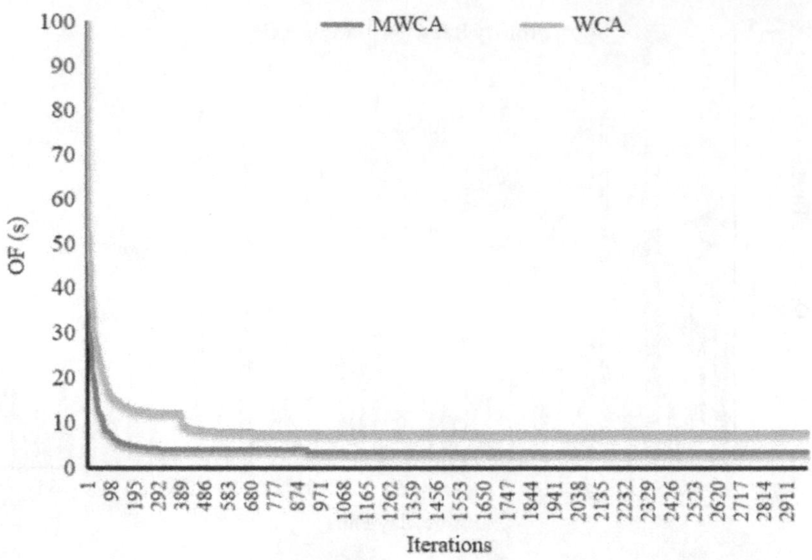

FIGURE 13.6 The MWCA and WCA fitness function for the nine-bus network.

TABLE 13.2
Comparison Between the WCA and MWCA with Other Methods (nine-bus system)

Methods	OF (s)
MWCA	3.6821
WCA	7.989
GA	14.542
PSO	6.895
GA-NLP	6.1786
FFA	4.832
MEFO	5.225
HS	4.9046
CSA	5.1836
DE	8.682
EFO	6.05

The comparison between well-known methods and WCA and MWCA is presented in Table 13.2. The MWCA method presents the least OF, as noticed in Table 13.3. That confirms the power of the MWCA to find the best solution for the problem of coordination.

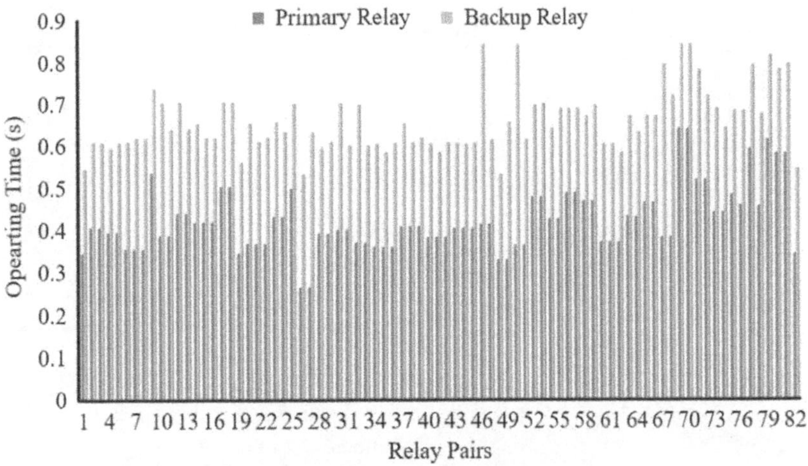

FIGURE 13.7 Relay pairs' OT of the 15-bus network using WCA.

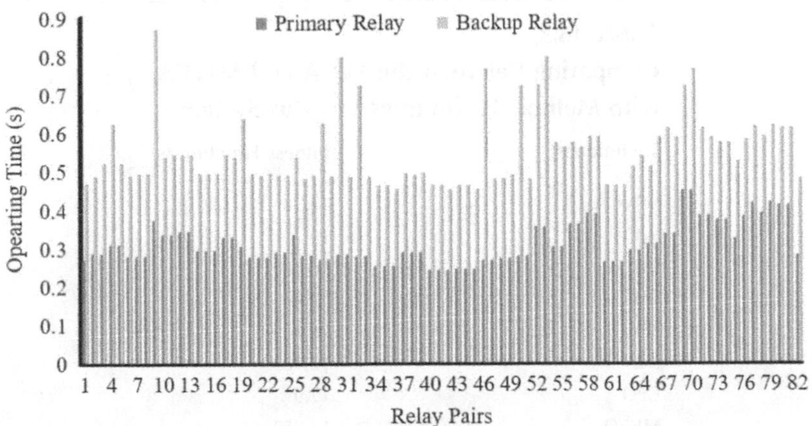

FIGURE 13.8 Relay pairs' OT of the 15-bus network using MWCA.

13.4.3 Solving the Problem of Coordination with Conventional CRC and Non- Conventional CRC

The performance of MFO and IMFO are assessed using nine-bus and 15-bus networks. The suggested methods are compared with other techniques to show their robustness in solving the problem of coordination. Both techniques are applied in two scenarios as:

- Scenario 1, using conventional CRC to find the best solution for the coordination problem.
- Scenario 2, using non-conventional CRC to find the best solution for the coordination problem.

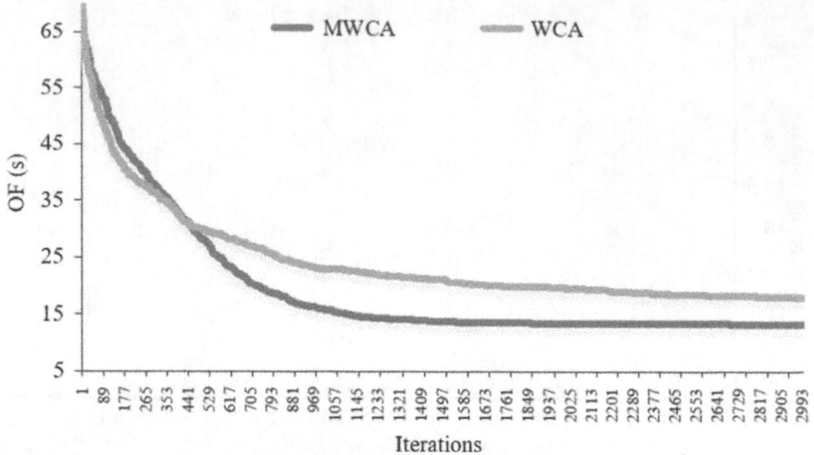

FIGURE 13.9 The MWCA and WCA fitness function for the 15-bus network.

TABLE 13.3
Comparing Between the WCA and MWCA with Method Techniques (15-Bus System)

Techniques	Fitness Function (s)
MWCA	13.308
WCA	18
BBO	16.58
BH	35.44
PSO	26.809
DE	18.9033
EFO	17.9
MEFO	13.953
CSA	19.552
GA	18.9033
FFA	22.71
BSA	16.293
GWO	15.37621

13.4.3.1 Scenario 1: Using Conventional CRC in Solving the Problem of Coordination

The suggested techniques (MFO, IMFO) in this scenario are tested using nine-bus and 15-bus networks. The standard characteristic is chosen in this scenario. So, the relay constants A and B are equal to 0.14 and 0.02, respectively.

13.4.3.1.1 Nine-Bus system

The MFO and IMFO are tested using the nine-bus system, which is presented in Figure 13.2. The main relays OT and the backup relay OT using MFO and IMFO are shown in Figures 13.10 and 13.11. The backup relays will be initiated if the main relays fail to operate, as noticed in these figures. It can be said that both techniques' successes get the best setting that keeps the operation sequential between DOCRs pairs.

The CC of MFO and IMFO techniques is given in Figure 13.12. From Figure 13.12, it can be noticed that the MWCA converge to the optimal solution faster than WCA. Also, the OF value given by the MWCA is better than the OF value that is given from WCA, where the reduction on OF reaches 70.27%.

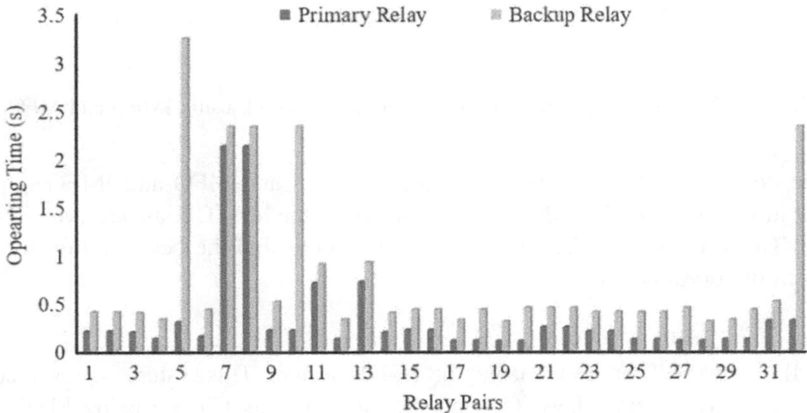

FIGURE 13.10 Relay pairs' OT of the nine-bus network using MFO.

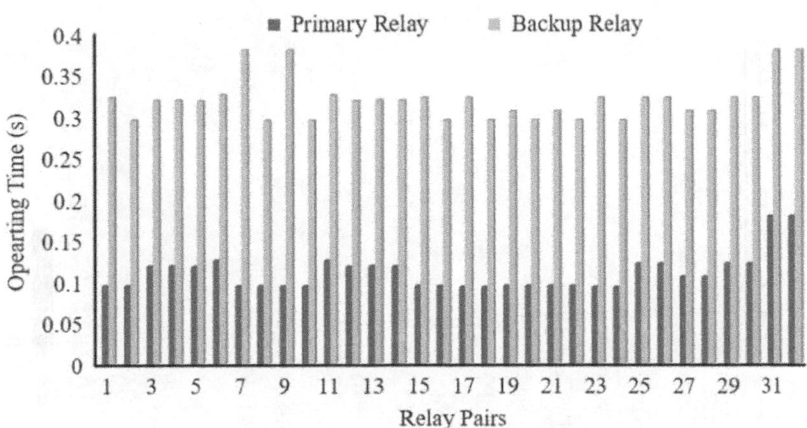

FIGURE 13.11 Relay pairs' OT of the nine-bus network using IMFO.

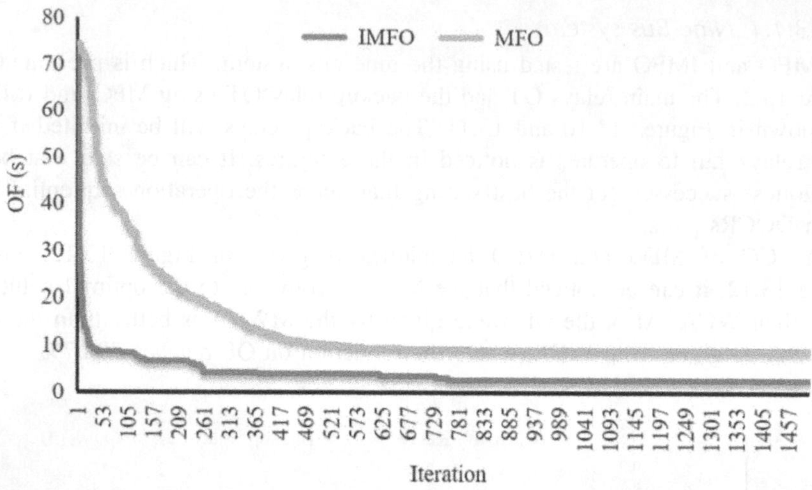

FIGURE 13.12 The fitness function for the nine-bus network using IMFO and MFO.

The comparison between well-known techniques and MFO and IMFO is presented in Figure 13.13. The IMFO technique gives the least OF, as observed in this figure. That proves the effectiveness of the IMFO to find the best solution for the problem of coordination.

13.4.3.1.2 15-Bus Network

The MFO and IMFO are tested using the 15-bus system. This system is presented in Figure 13.3. The main relays' OT and the backup relay OT are using MFO and IMFO as shown in Figures 13.14 and 13.15.

The backup relays will be initiated in case the main relays fail to operate after the specified margin, as noticed in these figures. It can be said that both

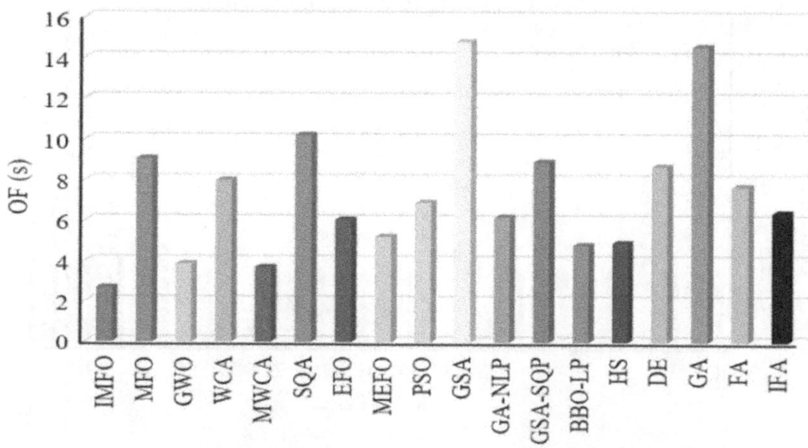

FIGURE 13.13 Comparing the IMFO with other optimization algorithms.

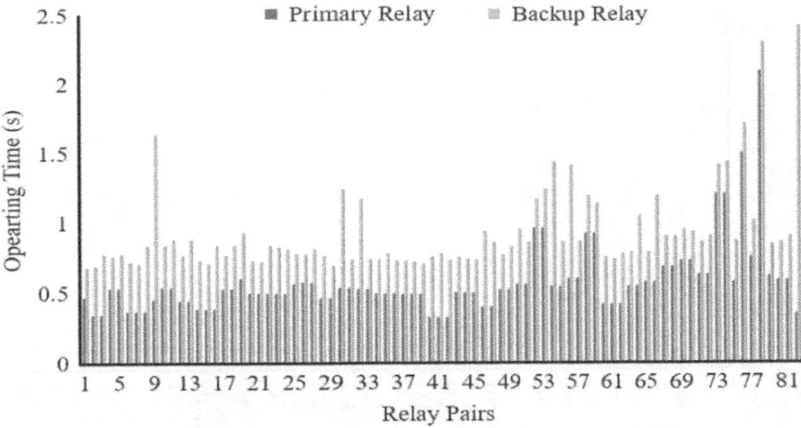

FIGURE 13.14 Relay pairs' OT of the 15-bus network using MFO.

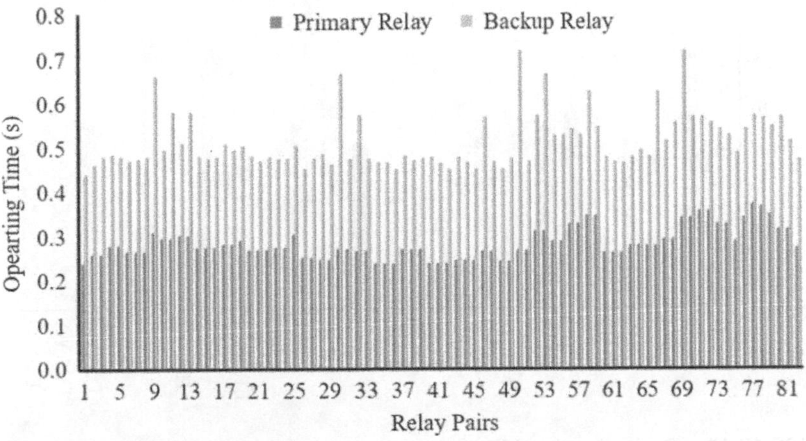

FIGURE 13.15 Relay pairs' OT of the 15-bus network using IMFO.

techniques' successes get the best setting that keeps the operation sequential between DOCRs pairs.

The CC of MFO and IMFO techniques is presented in Figure 13.16. The IMFO converge to the optimal solution faster than the MFO, as shown this figure. Also, the OF value given by the IMFO is better than the OF value that is given from MFO, where the reduction on OF reaches 52.94%.

The comparison between well-known methods and MFO and IMFO is presented graphically in Figure 13.17. The IMFO technique gives the least OF, as noticed in Figure 13.17. That proves the IMFO superiority to find the best solution for the problem of coordination.

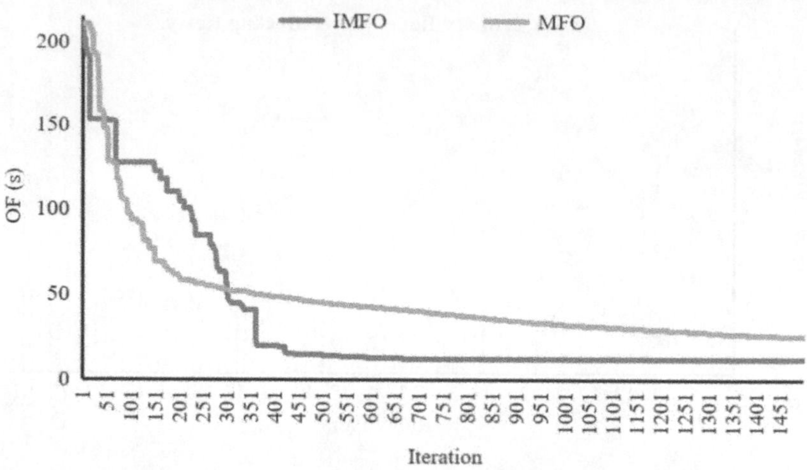

FIGURE 13.16 The fitness function for the 15-bus network using IMFO and MFO.

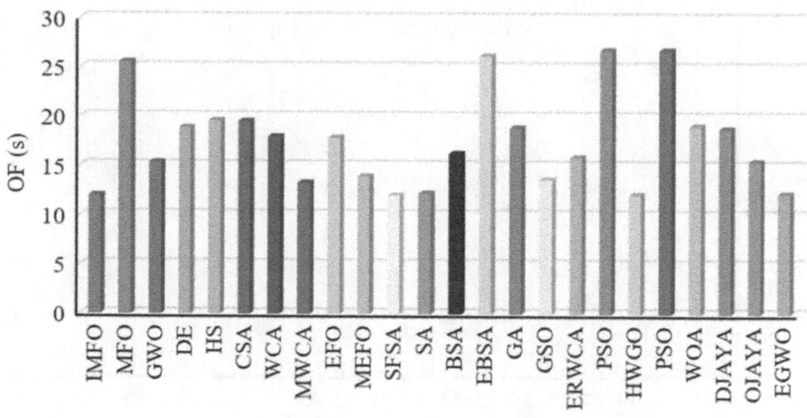

FIGURE 13.17 Comparing the IMFO with other optimization algorithms.

13.4.4 Scenario 2: Using Non-Conventional CRC in Solving the Problem of Coordination

The suggested techniques (MFO, IMFO) in this scenario are tested using a non-conventional CRC are selected in this scenario, where A and B are set as DVs and constraint for the relay constant (A, B) is set as follows:

- The upper and lower boundaries for A are 120 and 0.14, respectively.
- The upper and lower boundaries for B are 2 and 1, respectively.

13.4.4.1 Nine-Bus Network

The MFO and IMFO are tested using the nine-bus system. This network is presented in Figure 13.2. The comparison between the two scenarios is given in

Figure 13.20. Solving the problem of coordination using non-standard CRC gives less OF than scenario 1, as observed in this figure, whereas the reduction in OF in scenario 2 using MFO decreased more than 60% compared to the OF in scenario 1. The minimization in OF in scenario 2 using IMFO decreased more than 17% compared to OF in scenario 1. This result refers to the superiority and effectiveness of using the non-standard CRC in solving the coordination problem. Figure 13.18

13.4.4.2 15-Bus Network

The MFO and IMFO are tested using the 15-bus. This system is presented in Figure 13.3. The comparison between the two scenarios is presented in Figure 13.23. Solving the problem of coordination using non-standard CRC gives less OF than scenario 1, as observed in this figure. The reduction in OF in scenario 2 using MFO decreased more than 64% compared to the OF in scenario 1, while the reduction in OF in scenario 2 using IMFO decreased more than 59% compared to OF in scenario 1. This result indicates the superiority and effectiveness of using the non-standard CRC in solving the coordination problem. Figure 13.19

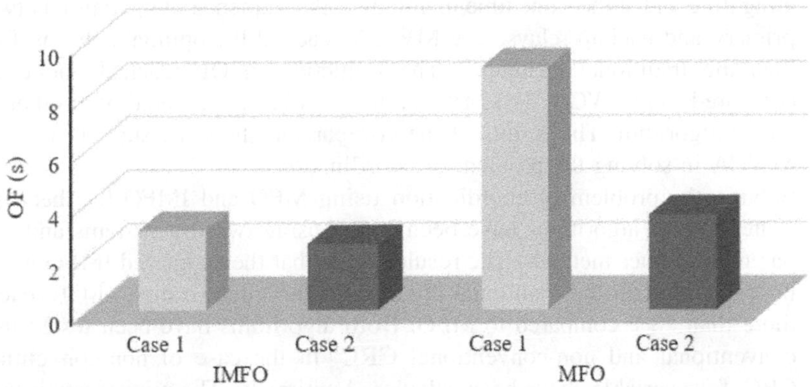

FIGURE 13.18 Conventional and non-conventional CRC comparison for the nine-bus network.

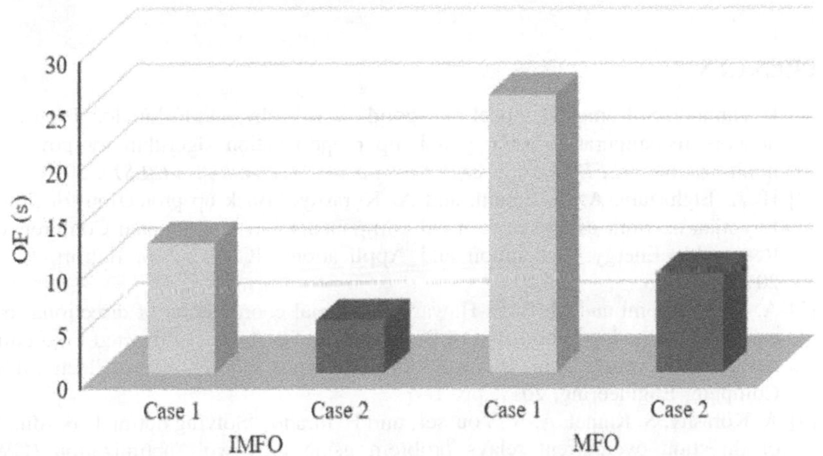

FIGURE 13.19 Conventional and non-conventional CRC comparison for the 15-bus network.

13.5 CONCLUSIONS

DOCRs' coordination is an important issue to preserve power system reliability. Finding the correct relay settings is the main target for solving the problem of coordination, which these settings reduce the DOCRs' operating time. In this chapter, different techniques have been proposed to find the optimal solution for this problem. Comparing well-known algorithms and suggested algorithms have been performed to show and prove the superiority of the suggested algorithms to deal with such a problem. The main achievements can be concluded as follows:

- Solving the problem of coordination using WCA and MWCA has been presented. The MWCA improved the performance of the traditional algorithm through the right balance between globally and locally searching that enhance the performance of the technique during the search to find the global minimum. This balance has been reached by increasing the G operator during the iterative process. The results from using MWCA prove that the suggested algorithm finds the optimal relay setting that reduced the summation operating time and at the same time maintained the sequence of operation between primary and backup relays. The MWCA reached the optimal solution faster than the traditional technique. The reduction in OF reached about 26% compared to the WCA. The MWCA also has been compared with other and recent algorithm. The results of the comparison show the superiority of the MWCA in solving the problem of coordination.
- Solving the problem of coordination using MFO and IMFO has been presented. These algorithms have been tested using two test systems and compared with other methods. The results show that the suggested IMFO gives a better result than the traditional algorithm. The minimization in OF reached more than 28% compared to MFO. Both algorithms have been tested using conventional and non-conventional CRC. In the case of non-conventional CRC, four variables have been calculated optimally. The minimization in OF using non-conventional CRC reached more than 59% compared to using conventional CRC.

REFERENCES

[1] H. Hamour, S. Kamel, H. Abdel-mawgoud, A. Korashy, and F. Jurado, "Distribution network reconfiguration using grasshopper optimization algorithm for power loss minimization," *IEEE Smart Energy Systems and Technologies (SEST)*, 2018.

[2] H. A. Elghazaly, A. M. Emam, and A. Korashy, "Back up protection for 500 kV Egyptian network using symmetrical components," in International Conference on Renewable Energy: Generation and Applications, ICREGA'16, Belfort, France, 2016, pp. 1–6.

[3] A. R. Al-Roomi and M. E. El-Hawary, "Optimal coordination of directional overcurrent relays using hybrid BBO-LP algorithm with the best extracted time-current characteristic curve," in IEEE 30th Canadian Conference on Electrical and Computer Engineering, 2017, pp. 1–6.

[4] A. Korashy, S. Kamel, A.-R. Youssef, and F. Jurado, "Solving optimal coordination of direction overcurrent relays problem using grey wolf optimization (GWO)

algorithm," in 2018 Twentieth International Middle East Power Systems Conference (MEPCON), 2018, pp. 621–625.

[5] V. N. Rajput and K. S. Pandya, "Coordination of directional overcurrent relays in the interconnected power systems using effective tuning of harmony search algorithm," *Sustainable Computing: Informatics and Systems*, 2017, 15, pp. 1–15.

[6] F. A. Albasri, A. R. Alroomi, and J. H. Talaq, "Optimal coordination of directional overcurrent relays using biogeography-based optimization algorithms," *IEEE Transactions on Power Delivery*, 2015, 30, pp. 1810–1820.

[7] A. J. Urdaneta, R. Nadira, and L. P. Jimenez, "Optimal coordination of directional overcurrent relays in interconnected power systems," *IEEE Transactions on Power Delivery*, 1988, 3, pp. 903–911.

[8] M. Singh, B. Panigrahi, and A. Abhyankar, "Optimal coordination of directional over-current relays using Teaching Learning-Based Optimization (TLBO) algorithm," *International Journal of Electrical Power & Energy Systems*, 2013, 50, pp. 33–34.

[9] H. Bouchekara, M. Zellagui, and M. A. Abido, "Optimal coordination of directional overcurrent relays using a modified electromagnetic field optimization algorithm," *Applied Soft Computing*, 2017, 54, pp. 267–283.

[10] A. Tjahjono et al., "Adaptive modified firefly algorithm for optimal coordination of overcurrent relays," *IET Generation, Transmission & Distribution*, 2017, 11, pp. 2575–2585.

[11] M. Ebeed, S. Kamel, and F. Jurado, "Optimal power flow using recent optimization techniques," in *Classical and recent aspects of power system optimization*. Elsevier, The Netherlands, 2018, pp. 157–183.

[12] D. K. Ibrahim, E. E. D. A. El Zahab, and S. A. E. A. Mostafa, "New coordination approach to minimize the number of re-adjusted relays when adding DGs in interconnected power systems with a minimum value of fault current limiter," *International Journal of Electrical Power & Energy Systems*, 2017, 85, pp. 32–41.

[13] V. Calderaro et al., "The impact of distributed synchronous generators on quality of electricity supply and transient stability of real distribution network," *Electric Power Systems Research*, 2009, 79, pp. 134–143.

[14] E. Purwar, D. N. Vishwakarma, and S. P. Singh, "A novel constraints reduction-based optimal relay coordination method considering variable operational status of distribution system with DGs," *IEEE Transactions on Smart Grid*, 2017, 1, pp. 889–898.

[15] A. S. Noghabi, J. Sadeh, and H. R. Mashhadi, "Considering different network topologies in optimal overcurrent relay coordination using a hybrid GA," *IEEE Transactions on Power Delivery*, 2009, pp. 1857–1863.

[16] A. Korashy, S. Kamel, A.-R. Youssef, and F. Jurado, "Evaporation Rate Water Cycle Algorithm for Optimal Coordination of Direction Overcurrent Relays," in 2018 Twentieth International Middle East Power Systems Conference (MEPCON), 2018, pp. 643–648.

[17] Rao, T.M., *Power System protection: Static relays*. 1989. Tata McGraw-Hill Education, New York.

[18] H. M. Sharaf, H. Zeineldin, D. K. Ibrahim, and E. Essam, "A proposed coordination strategy for meshed distribution systems with DG considering user-defined characteristics of directional inverse time overcurrent relays," *International Journal of Electrical Power & Energy Systems*, 2015, 65, pp. 49–58.

[19] P. P. Bedekar and S. R. Bhide, "Optimum coordination of directional overcurrent relays using the hybrid GA-NLP approach," *IEEE Transactions on Power Delivery*, 2011, 26, pp. 109–119.

[20] M. Singh, B. K. Panigrahi, A. R. Abhyankar, and S. Das, "Optimal coordination of directional over-current relays using informative differential evolution technique," *Journal of Computational Science*, 2014, 5, pp. 269–276.

[21] R. Mohammadi, H. A. Abyaneh, H. M. Rudsari, S. H. Fathi, and H. Rastegar, "Overcurrent relays coordination considering the priority of constraints," *IEEE Transactions on Power Delivery*, 2011, 26, pp. 1927–1938.

[22] A. El-Fergany and H. M. Hasanien, "Optimized settings of directional overcurrent relays in meshed power networks using stochastic fractal search algorithm," *International Transactions on Electrical Energy Systems*, 2017, 27, doi: https://doi.org/10.1002/etep.2395

[23] G. Darji, M. Patel, V. Rajput, and K. Pandya, "A tuned cuckoo search technique for optimal coordination of Directional Overcurrent Relays," in 2015 International Conference on Power and Advanced Control Engineering (ICPACE), 2015, pp. 162–167.

[24] V. N. Rajput, F. Adelnia, and K. S. Pandya, "Optimal coordination of directional overcurrent relays using improved mathematical formulation," *IET Generation, Transmission & Distribution*, 2018, 12, pp. 2086–2094.

[25] Kiranyaz, S., T. Ince, and M. Gabbouj, *Multidimensional particle swarm optimization for machine learning and pattern recognition*. 2014. Springer, New York.

[26] H. Eskandar, A. Sadollah, A. Bahreininejad, and M. Hamdi, "Water cycle algorithm–A novel metaheuristic optimization method for solving constrained engineering optimization problems," *Computers & Structures*, 2012, 110, pp. 151–166.

[27] A. A. Heidari, R. A. Abbaspour, and A. R. Jordehi, "An efficient chaotic water cycle algorithm for optimization tasks," *Neural Computing and Applications*, 2017, 28, pp. 57–85.

[28] W. Long et al. "Inspired grey wolf optimizer for solving large-scale function optimization problems." *Applied Mathematical Modelling*, 2018 60, pp. 112–126.

[29] W. Long and S. Xu, "A novel grey wolf optimizer for global optimization problems," in 2016 IEEE Advanced Information Management, Communicates, Electronic and Automation Control Conference (IMCEC), 2016, pp. 1266–1270.

[30] N. Mittal, U. Singh, and B. S. Sohi, "Modified grey wolf optimizer for global engineering optimization," *Applied Computational Intelligence and Soft Computing*, 2016, Art. ID.7950348.

[31] A. Korashy et al., "Modified water cycle algorithm for optimal direction overcurrent relays coordination," *Applied Soft Computing*, 2019, 74, pp. 10–25.

[32] S. Mirjalili, "Moth-flame optimization algorithm: A novel nature-inspired heuristic paradigm," *Knowledge-Based Systems*, 2015, 89, pp. 228–249.

[33] Y. Xu et al., "An efficient chaotic mutative moth-flame-inspired optimizer for global optimization tasks," *Expert Systems with Applications*, 2019, 129, pp. 135–155.

[34] H. Abdel-mawgoud et al. Optimal installation of multiple DG using chaotic moth-flame algorithm and real power loss sensitivity factor in distribution system," in 2018 International Conference on Smart Energy Systems and Technologies (SEST). 2018. IEEE.

[35] S. Mirjalili, S. M. Mirjalili, and A. Lewis, "Grey wolf optimizer," *Advances in Engineering Software*, 2014, 69, pp. 46–61.

[36] A. Korashy, S. Kamel, T. Alquthami, and F. Jurado, "Optimal coordination of standard and nonstandard direction overcurrent relays using an improved moth-flame optimization," *IEEE Access,* May 2020, 8(2): 87378–87392. doi: 10.1109/ACCESS.2020.2992566.

[37] S. Li, H. Chen, M. Wang, A. A. Heidari, and S. Mirjalili, "Slime mould algorithm: A new method for stochastic optimization," *Future Generation Computer Systems*, 2020, 111, 300–323.

[38] W. Zhao, L. Wang, and Z. Zhang, "Artificial ecosystem-based optimization: a novel nature-inspired meta-heuristic algorithm," *Neural Computing and Applications*, 2019, 32(4), pp. 1–43.
[39] M. N. Alam, B. Das, and V. Pant, "A comparative study of metaheuristic optimization approaches for directional overcurrent relays coordination," *Electric Power Systems Research*, 2015, 128, pp. 39–52.
[40] T. Amraee, "Coordination of directional overcurrent relays using seeker algorithm," *IEEE Transactions on Power Delivery*, 2012, 27, pp. 1415–1422.
[41] T. Khurshaid, A. Wadood, S. G. Farkoush, C.-H. Kim, J. Yu, and S.-B. Rhee, "Improved firefly algorithm for the optimal coordination of directional overcurrent relays," *IEEE Access*, 2019, 7, pp. 78503–78514.
[42] Radosavljević, J. and M. Jevtić, "Hybrid GSA-SQP algorithm for optimal coordination of directional overcurrent relays," *IET Generation, Transmission & Distribution*, 2016, 10(8): pp. 1928–1937.
[43] H. Bouchekara, M. Zellagui, and M. Abido, "Coordination of Directional Overcurret Relays Using the Backtracking Search Technique," *Journal of Electrical Systems*, 2016, 12(2): 387–405.
[44] A. Othman and A. Abdelaziz, "Enhanced backtracking search technique for optimal coordination of directional over-current relays including distributed generation," *Electric Power Components and Systems*, 2016, 2, pp. 278–290.
[45] M. Alipour, S. Teimourzadeh, and H. Seyedi, "Improved group search optimization technique for coordination of directional overcurrent relays," *Swarm and Evolutionary Computation*, 2015, 1, pp. 40–49.
[46] Yu, J., C.-H. Kim, and S.-B. Rhee, "Oppositional Jaya algorithm with distance-adaptive coefficient in solving directional over current relays coordination problem," *IEEE Access*, 2019, 7, pp. 150729–150742.
[47] V. A. Papaspiliotopoulos, G. N. Korres, and N. G. Maratos, "A novel quadratically constrained quadratic programming method for optimal coordination of directional overcurrent relays," *IEEE Transactions on Power Delivery*, 2015, 32(1), pp. 3–10.

14 Artificial Intelligence Applications in DC Microgrid Protection

Morteza Shamsoddini and Behrooz Vahidi
Amirkabir University of Technology

14.1 INTRODUCTION

DC microgrids are a promising solution for realizing the notional smart grids. They have salient features such as low loss, increased transmission capacity, and easy connection of different source types to the DC bus. Also, the widespread applications of DC microgrids are attracting more attention. However, implementing a DC microgrid requires critical considerations. For example, power electronic converters are vulnerable to the high current of DC faults (pole-to-pole and pole-to-ground). To avoid any damage to the grid components especially power electronic converters, DC microgrids should be equipped with a fast fault detection and isolation scheme. This chapter will address technical issues that should be considered in designing the DC microgrids protection system. Discovering the DC network behavior during fault incidents is the first step toward realizing a protection system. Hence, initially, fault current characteristics are thoroughly investigated. In order to address practical challenges, different types of grounding structures are described and compared. In terms of fault current characteristics, DC and AC networks have profound differences, caused by the incapability of AC circuit breakers to interrupt DC fault current. As a result, utilizing protective devices (PDs) compliant with DC fault current requirements is inevitable. Different types of invented and developed DC PD structures will be described, and compared in terms of their main features, including cost, losses, response time, and size.

To step toward future smart grids, it is an axiomatic fact that more intelligent and adaptive control and protection schemes should be designed. AI tends to play a vital role in the future of the power system. By investigating designed protection systems in the past studies, it is demonstrated that lack of high intelligence protection scheme which, in addition to protecting all possible operational situations, can provide a fast, reliable, and cost-efficient protection scheme, is sensed. In this chapter, the critical steps of AI-based protection system implementation are explained.

Prior to implementing a protection scheme based on the AI, technical considerations of DC microgrids should be addressed meticulously. In this regard, the

DOI: 10.1201/9780367552374-14

following section presents prerequisite knowledge for developing a protection scheme for DC microgrids.

14.2 TECHNICAL CONSIDERATIONS OF DC MICROGRID PROTECTION

For a proper design of the protection system and precise setting of the protective devices, the DC microgrid's comprehensive fault current analysis is essential. Hence, this section initially investigates the fault current behavior in a DC environment. For more investigation, other crucial challenges for designing a protection scheme for DC microgrids are presented.

14.2.1 DC Fault Current Characteristics

Protection system design considerations should be established based on two possible types of pole-to-pole (short-circuit) and pole-to-ground (earth-fault) faults (Figure 14.1a). Due to network characteristics, pole-to-ground fault occurrence is the possible type of fault, and it is just the opposite of the pole-to-pole fault. Also, a lower fault current for the pole-to-ground fault rather than the pole-to-pole fault is expected. As a result, in the settings of PDs of current-based protection schemes (see current-based protection schemes in Section 14.3), the minimum fault current which emanates from pole-to-ground fault is considered. So, PDs' setting, which is set based on the minimum fault current, will respond to pole-to-pole faults. Hence, in many studies [1–3], the pole-to-ground fault current equation is considered a key solution for designing a protection scheme.

In case of pole-to-ground faults or severe pole-to-pole faults in the DC network, the power electronic converters' controller will block the IGBTs for self-protection. It exposes the reverse diodes to the fault current [4]. Blocking the IGBTs and exposing the reverse diode to overcurrent causes the DC network to undergo two stages. In the first stage of the fault current, an RLC circuit, as shown in Figure 14.1b, including the equivalent of the fault path resistance (R), inductance (L), and DC link capacitor (C) is formed. In this stage, the natural response of the

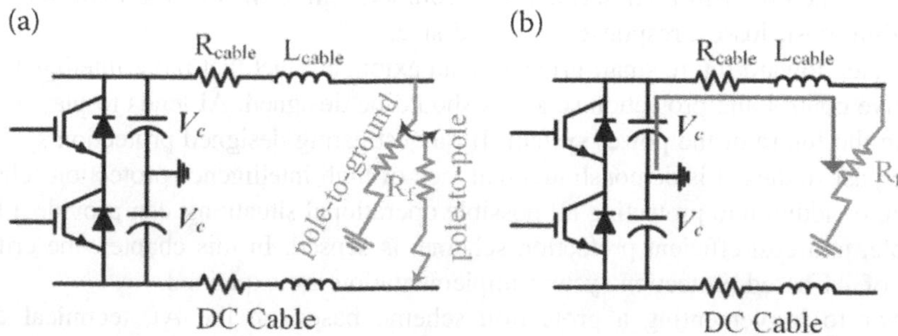

FIGURE 14.1 DC microgrid faults model a) two DC fault types; b) pole-to-ground fault first stage equivalent circuit [2].

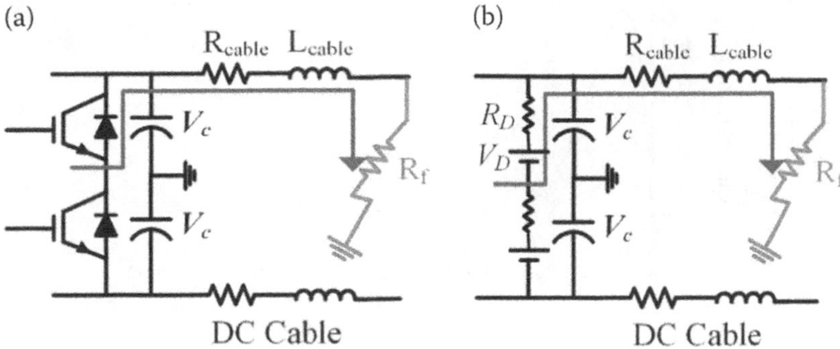

FIGURE 14.2 DC microgrid faults model a) pole-to-ground fault second stage; b) its equivalent circuit [2].

RLC circuit determines the fault current equation. When the fault current reaches its peak value, the second stage of the fault current will start; in this situation, the capacitor voltage drops to zero or becomes negative; thus, the reverse diode of the converter will be biased. In this stage, the participation of power sources to the fault current through converter interfaces to the fault current, as illustrated in Figure 14.2a and b, will be started, which exposes the reverse diodes to an extreme overcurrent up to 10 times greater than the rated value of the converter [5], [6]. Hence, if the second stage is not prevented, the converter would be damaged significantly. Both stages are analyzed in the following.

14.2.1.1 Analysis of the First Stage of the Fault Current

Once a pole-to-ground fault occurs, the DC-link capacitor discharges through the fault path impedance (cable and fault impedances), as shown in the equivalent circuit of Figure 14.1b. The fault current resulting from the DC-link capacitor's discharging through the fault path is illustrated in Figure 14.3. As can be observed, if none of the PDs operate, the peak magnitude of the fault current would go extremely higher than the rated current in a typical low-voltage DC microgrid. Assuming that the fault occurs at t_0, the natural response of the RLC circuit in the frequency domain can be written as [4]:

$$I(s) = \frac{v_C(t_0)/L_{eq} + s i_L(t_0)}{s^2 + \frac{R_{eq}}{L_{eq}} s + \frac{1}{L_{eq} C}} \tag{14.1}$$

where $i_L(t_0)$ and $v_C(t_0)$ are line current and DC voltage of the bus at instant t_0 before the fault. Also, R_{eq} and L_{eq} are fault path resistance and inductance, consisting of cable impedance and fault impedance. The time-domain of the above equation can be expressed as follows:

$$I(t) = \frac{v_{C0}}{L_{eq}(p_2 - p_1)}\left(e^{-p_1 t} - e^{-p_2 t}\right) + \frac{i_{L0}}{(p_2 - p_1)}\left(-p_1 e^{-p_1 t} + p_2 e^{-p_2 t}\right) \tag{14.2}$$

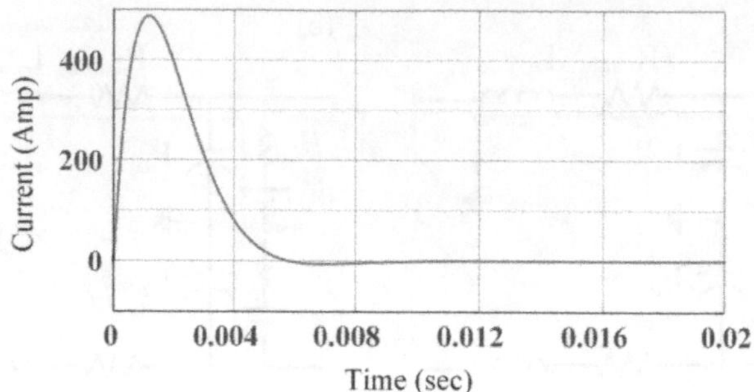

FIGURE 14.3 The fault current.

where p_1 and p_2 are equal to:

$$p_1, p_2 = \alpha \pm \omega \tag{14.3}$$

In (14.3), $\alpha = R_{eq}/2L_{eq}$, $\omega = \sqrt{\alpha^2 - \omega_0^2}$ and $\omega_0 = 1/\sqrt{L_{eq}C}$. If $\alpha^2 > \omega_0^2$, the current response would be over-damped; if $\alpha^2 < \omega_0^2$, the current response is under-damped, and if $\alpha^2 = \omega_0^2$, the current would be critically damped. The time it takes under-damped (t_{ud}^{pk}) and over-damped (t_{od}^{pk}) currents to reach their peak magnitudes can be formulated as [1]:

$$t_{ud}^{pk} = \frac{1}{\omega_0}\tan^{-1}\left(\frac{\omega_0}{\alpha}\right) \tag{14.4}$$

$$t_{od}^{pk} = \frac{\ln p_2/p_1}{p_1 - p_2} \tag{14.5}$$

The parameters calculated above are important and should be considered in the protection system design. Since after t_{ud}^{pk} or t_{od}^{pk} the DC-link capacitor voltage drops to zero, it forward biases the reverse diodes in voltage source converters (VSCs), causing the VSC to operate like a three-phase rectifier and grant the chance of contribution in the fault current to the AC side of the network which was prevented by IGBTs, blocking before t_{ud}^{pk} or t_{od}^{pk}. To prevent possible damages to the power electronic interfaces and DC load voltage collapse, the protection scheme should detect and disconnect the faulty section before these times.

14.2.1.2 Analysis of the Second Stage of the Fault Current

In the previous subsection, the reason for the start of the second fault current stage was explained in detail. The diode and IGBT shown in Figure 14.2a represent the equivalent of any conducting leg of a voltage source converter [7]. Due to this alternative path, the fault current response will change in this stage, regardless of the IGBTs' conducting state. The circulating current of the reverse polarity can be formulated as [8]:

$$I(t) = \frac{V_D}{R_D + R_{cable}} + A_1 e^{p_1 t} + A_2 e^{p_2 t} \tag{14.6}$$

in which A_1 and A_2 are coefficients determined based on the initial condition and fault impedance (see 14.2). As shown in Figure 14.2, R_{cable} is the cable resistance up to the fault point. The diode-capacitor parallel branch voltage can be written as $V(t) = I(t)R_{cable} + L_{cable}dI/dt$. By substituting 14.6, the voltage can be formulated as:

$$V(t) = \frac{V_D}{R_D + R_{cable}} R_{cable} + A_1(R_{cable} + p_1 L_{cable})e^{p_1 t} + A_2(R_{cable} + p_2 L_{cable})e^{p_2 t} \tag{14.7}$$

L_{cable} is the equivalent cable inductance up to the fault point. The fault current flowing in the cable (I_{cable}) and DC-link capacitor voltage (V_{cap}) during the second stage of the fault current is illustrated in Figure 14.4.

To thoroughly investigate the AC network's contribution through voltage source converter to the fault current, a three-phase short circuit analysis should be considered. During the second stage of the fault, the AC side starts to contribute to the fault current; for example, the phase A current during this stage of fault can be formulated as below [10]:

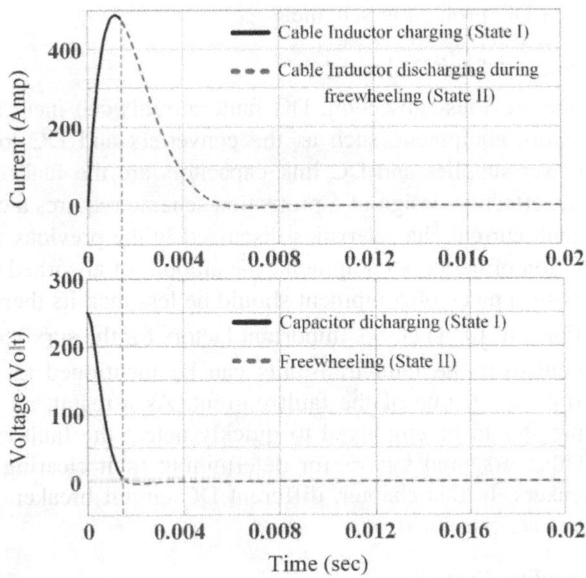

FIGURE 14.4 Flowing fault current in the cable (I_{cable}) and DC-link capacitor voltage (V_{cap}) [9].

$$I_{AC}^{A} = I_p sin(\omega_s t + \alpha - \varphi) + [I_{p_0} \sin(\alpha - \varphi_0) - I_p \sin(\alpha - \varphi)]e^{-t/\tau} \quad (14.8)$$

in which $\varphi = \tan^{-1}[\omega_0(L + L_{ac})/R]$, $\tau = (L_{ac} + L)/R$, φ_0 and I_{p_0} are initial phase angle and current, and L_{ac} is the AC side equivalent inductance. It is also possible to define I_{AC}^{B} and I_{AC}^{C} accordingly, which are flowing through the other legs of the converter. Consequently, the total current of all three phases ($I_{3\varphi}$) during the fault can be presented below:

$$I_{3\varphi} = I_{AC}^{A} + I_{AC}^{B} + I_{AC}^{C} \quad (14.9)$$

Up to now, the complete response of a typical DC network with voltage source converter, which can be mentioned as the most vulnerable component of a DC network, during fault is analyzed in two different stages. As a result, it is clarified that the protection system should operate and disconnect the faulty part within the peak reaching time (t_{ud}^{pk} or t_{od}^{pk}) to prevent converters and other components of the grid from being damaged. In addition to analyzing fault current for designing a protection scheme, practical considerations such as the grounding system, DC breakers, and other capabilities should be investigated. The following subsection addresses these considerations.

14.2.2 Technical Issues

As mentioned above, this section will also discuss and address a DC protection scheme's critical requirements. The text below will shed more light on other design considerations of a DC protection scheme.

14.2.2.1 Equipment Fault-Tolerant

As analyzed in the previous subsection, DC fault currents can incur significant damage to the network equipment, such as the converters and DC loads. In a DC microgrid, the power supplies and DC link capacitors are the fault current's main sources. Hence, an effective design of a protection scheme requires a comprehensive analysis of DC fault current characteristics discussed in the previous subsection.

For safe operation of network equipment, the amount of absorbed energy during the fault transient by a piece of equipment should be less than its thermal tolerance, which is proportional to $\int i^2$ [11]. So, important factors for the survivability of a DC network equipment over the fault transients can be mentioned as 1) amplitude, 2) shape, and 3) clearing time of the fault current. As a result, an advanced DC protection scheme should be employed to quickly detect the fault and isolate the faulty section. Other essential factors for determining fault clearing time also include circuit breakers. In this chapter, different DC circuit breakers and fault detection approaches are presented.

14.2.2.2 Grounding System

One of the most crucial aspects of designing a protection scheme in a DC microgrid is the grid grounding structure. The primary purposes of any grounding schemes can be mentioned as 1) facilitating the fault detection, 2) minimizing DC stray current, and 3) increasing safety of the personnel and equipment by decreasing common-mode voltage (CMV) [12].

Grounding resistance relates to the stray current with CMV. In fact, in high grounding resistance, the stray current would be very low, and CMV would be high. In contrast, low grounding resistance would lead to a low CMV and a high stray current [13]. IEC 60364-1 categorized the structure of DC grounding into five different types, including Isolated Terre (IT), Terre-Neutre (TN) with three different types of subclasses, and Terre-Terre (TT) [14]. Two letters of T and I stand for the direct connection and no connection of the source side to the earth, respectively. The two letters of T and N also stand for direct earthing and connection to the earth neutral of exposure side of the equipment, respectively. Figure 14.5a illustrates the TT grounding structure. In this grounding system, the source side's neutral conductor (power electronic interface) and the protective earth (PE) conductor of the load side are directly and separately connected to the earth point. Ease of implementation and intrinsic characteristics of preventing fault current from being transferred to the other sections of the grid are the main advantages of this scheme. Due to the available path, current circulation, which can result in voltage stress, is the main drawback of this grounding scheme [15].

The most widely used grounding topology is the TN scheme. This configuration's connections are direct grounding in the source side (like mid-point of the converter) and connecting the conductive exposed parts of the load side to the earth neutral. Ease of fault detection and adjustable grounding resistance, which provide the opportunity to control the amount of fault current, are the main advantages of this topology. However, in high grounding resistance, the touch voltage can increase higher than the permissible threshold in high voltage grids [16]. This topology grants this chance to connect load side conductive parts through neutral (N), PE, or even a combination of PE and Neutral (PEN) conductors. Hence, three subclasses of TN-S (S means **S**eparate PE from the neutral conductor, see Figure 14.5b), TN-C (C means **C**ombined PE and the neutral conductors, see Figure 14.5c), and TN-C-S (C-S defines the Combined and Separated conductors simultaneously, see Figure 14.5d) are provided in TN grounding scheme.

Since PE and N conductors are separated in the TN-S system, this scheme's electromagnetic compatibility (EMC) is higher than other TN grounding topologies. Also, it provides more reliability and personnel safety compared to other topologies because if neutral gets disconnected, the protective earth is still in service. This feature makes the TN-S grounding suitable for vital applications. Since the TN-C grounding scheme combines two conductors, it offers a cost-effective topology, making it suitable for widespread applications. Thanks to the combination of TN-S and TN-C topologies, the TN-C-S grounding structure benefits from both schemes. This structure suffers from difficult fault detection, if neutral conductor is disconnected.

The IT is another grounding topology in which the neutral conductor has no grounding point, and the conductive body of equipment is grounded separately (See Figure 14.5e). In the case of pole-to-ground, the fault current would be small in this

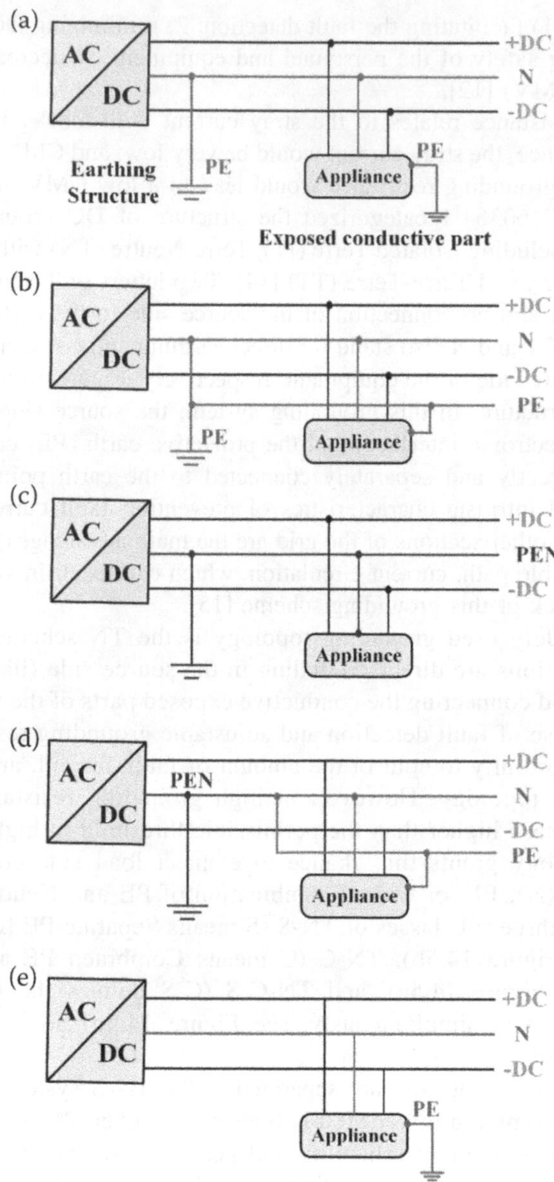

FIGURE 14.5 Different types of DC grounding structures a) TT; b) TN-S; c) TN-C; d) TN-C-S; e) IT [12].

configuration, granting the chance of continuous energy providing for loads. This characteristic makes this topology suitable for vital loads. However, difficult fault identification for the first pole-to-ground fault and severe fault current in the second pole-to-ground fault can be considered as the disadvantages of this topology [17]. Table 14.1 summarizes the comparison of these grounding topologies [18].

TABLE 14.1
Comparison of Different Grounding Structures [18]

	TT	TN-S	TN-C	IT
Safety of persons	• Good	• Good	• Good	• Good
Safety of property	• Good • Fault current less than a few dozen amperes	• Good • Poor • Fault current around a 1 kA	• Poor • Fault current a 1 kA	• Very good • Fault current less than a few dozen mA, but high for the second fault
Continuity of service	• Average	• Average	• Average	• Excellent
EMC	• Good • Risk of overvoltage/voltage imbalance • Equipotential problems • Require to manage devices which high leakage currents	• Excellent • Less equipotential problems • Require to manage devices with high leakage currents • High fault current (transient disturbance)	• Poor • High fault current (transient disturbances)	• Poor • Risk of overvoltage

It should be noted that different modes of ungrounded or floating, direct grounding, grounding with resistance or parallel resistance, and finally, semiconductor-based grounding (diode or thyristor) for the source side in DC microgrids are proposed [12]. To make it concise, only the advantages and disadvantages of these source side grounding approaches are provided in Table 14.2 [18].

14.2.2.3 DC Protective Devices

AC circuit breakers and DC protective devices (DCCBs) are two main PDs that are widely employed in DC environments. To offer a cost-effective PD for grids based on the VSCs, ACCBs are the first candidates; however, these PDs are not fast enough to cut the fault current before the second stage onset. So, without considering any solution for this issue, the fault current would harm the VSC's free-wheeling diodes severely. A viable solution to this issue is the to disconnect all power supplies during the fault and restore the grid after removing the faulty section [19]. Using this method might prevent possible damages to the grid equipment, but it also prevents continuous loads supplying and damaging the critical processes. DCCBs are other groups of PDs that are typically designed for DC environments

TABLE 14.2
Comparison of Different Grounding Strategies [18]

Grounding System	Advantages	Disadvantages
Ungrounded systems	• Continuous operation of DC microgrid during single line-to-ground fault • Low stray current • Simple and economical	• High CMV • Difficult fault detection • Severe fault situation in case of second fault in another pole
Solidly grounded system	• Low CMV • Require a low level of insulation • Easy fault detection	• High fault current • High stray current
Diode grounded system	• Low/Moderate CMV	• High level of corrosion • Moderate/High stray current
Thyristor grounded system	• Low/Moderate stray current	• Moderate/High stray current

where fault currents have no natural zero crossings. The PDs of this category are based on the solid-state components. In the following, the main PDs of the mentioned categories are briefly described.

- **Fuse:** Fuses, which are the essential part of every protection scheme, are known as analog and self-triggered PDs. It is applicable for a wide range of DC and AC, and its protection rating levels (current and voltage) have been developing. There are commercialized DC fuses for 36-kV applications [20–22]. These PDs with low inductance and high *di/di*, which can help minimize time to reach the melting point, can provide fast and effective protection. However, these PDs should be replaced after operating during faults. Hence, it is suggested to use fuses only as backup protection for another primary protection.
- **Mechanical circuit breaker (MCB):** The MCB comprises three main parts: a mechanical switch, an energy absorber circuit, and a commutation circuit. The operating mechanism of the MCBs is mainly based on hydraulic, pneumatic, spring, and magnetic. However, due to less moving parts, the magnetic-based MCB is more desired [12]. Proposing a simple structure and less operating time by Thomson coil MCB [23], [24], which its operating mechanism is based on the repulsion coil, attracted more attention in DC grids. Generally, when a fault occurs, to extinguish the fault current, mechanical switch contacts will separate; consequently, electric arcs will be created between two contacts of the MCB. Because of the alternative features of the current and the existence of natural zero-crossing in the AC environment, these arcs will be subsided and cleared completely. However, in the absence of natural zero-crossing in the DC environment, DC grids will face

several restrictions for employing MCBs. Several active and passive resonance circuits have been proposed [25]. The structure of these circuit breakers is shown in Figure 14.6. The operating principles of these structures consist of three phases: 1) disconnection of contacts, 2) interruption of the created arc due to contacts separation by resonance branch, and 3) dissipation of absorbed energy by resonance circuit through metal oxide varistor (MOV) branch.

- **Solid-state circuit breaker (SSCB):** In an attempt to achieve circuit breakers with faster operating time, SSCBs have been proposed. With increasing advancements in power electronic switches, it is possible to utilize these power switches as DCCBs. A typical structure of SSCB is illustrated in Figure 14.7. Among all of the available switches, thyristor-based switches have the lowest conduction loss. This characteristic results in a longer life cycle and lower cost. However, the lack of capability to actively interrupt the current has resulted in higher operation time than other SSCBs. This issue leads to higher fault current before fault current interruption. Insulated-gate bipolar transistor (IGBT), integrated gate commutated thyristor (IGCT), and gate-off thyristor (GTO) are active turn-off switches, which have naturally forced commutation. Since GTOs and IGCTs have much lower conduction loss than IGBTs, they are more applicable [24]. Up to now, the commercialized maximum

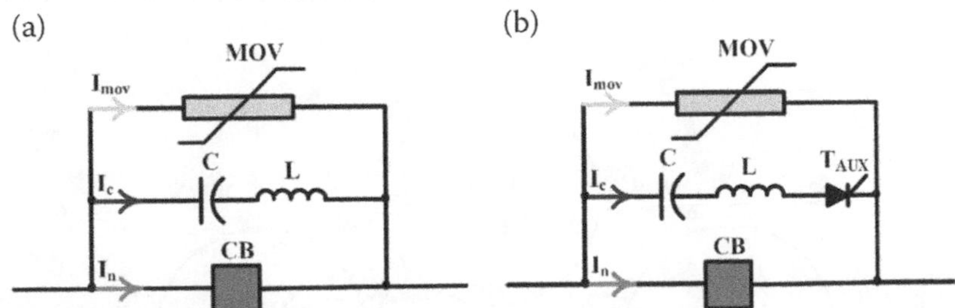

FIGURE 14.6 MCB with a) passive commutation structure and b) active commutation structure [12].

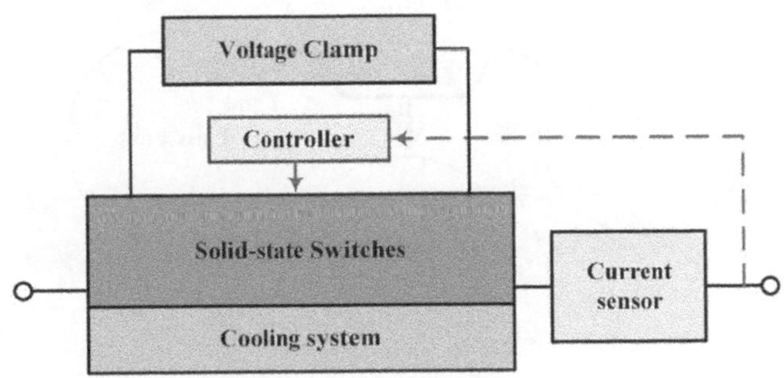

FIGURE 14.7 A typical SSCB structure [12].

voltage rating of DCCBs is 6.5 kV [26]. However, due to high conduction loss, MOSFET-based DDCB applications are limited to much lower voltage ratings.

Wide-band gap (WGB) materials are viable solutions to address technical issues and available limitations in the path of achieving ideal DCCBs (see Figure 14.8) using solid-state materials. Low conduction loss, high voltage, and high-temperature endurance are prominent features of WBG material. Hence, devices based on SiC or Gallium-Nitride (GaN) are the best choices for DCCBs. However, SiC-based or GaN-based offer more compact SSCBs with higher efficiency; several challenges such as the need for advanced gate-drive control and electromagnetic interference (EMI) filter should be addressed [27–30].

- **Hybrid circuit breaker (HCB):** To compromise between advantages and disadvantages of the MCBs and SSCBs, this class of circuit breakers has been designed. Hence, HCBs have a fast response time and low conduction loss. Also, the arc occurrence between mechanical contacts is prevented in this structure. Conventional HCBs structure consists of three main parallel branches, including mechanical switch branch, solid-state switch branch, and MOV (see Figure 14.9a) [31]. To limit the rate of fault current change, a current limiting reactor (CLR) is implemented in series with a residual circuit breaker (RCB). During the normal operation of the grid, the flowing current

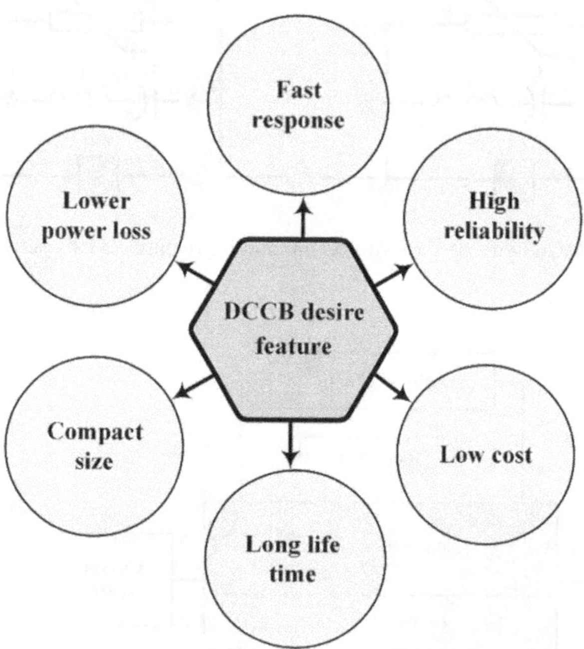

FIGURE 14.8 The key features of a DDCB [12].

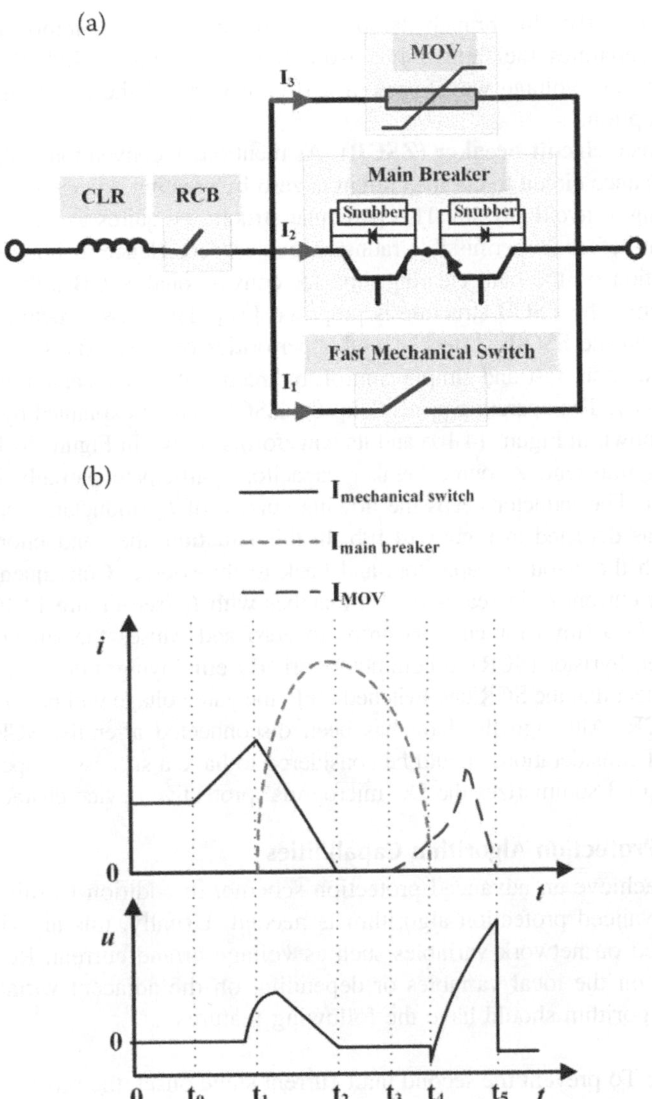

FIGURE 14.9 Conventional HCBs a) structure; b) current and voltage profile during faults.

passes through a mechanical switch. When a fault occurs, the mechanical switch's contacts will be separated, and an activation signal to the main breaker branch (solid-state breaker) will be sent. As a result, the fault current path will be transferred to this branch. The main breaker provides a path for the fault current to flow until the mechanical switch current completely subsides and reaches its full block voltage. In the next step of operation, the main breaker will be turned off. Due to inductors' presence, the voltage quickly increases until it reaches the MOV breakdown threshold.

Consequently, this branch damps the fault current completely. Finally, the RCB separates the faulted line from the grid. Figure 14.9b illustrates the current and voltage waveforms of different parts of the HCB during a fault interruption.

- **Z-source circuit breaker (ZSCB):** As mentioned, conventional SSCBs utilize a resonance circuit to cut the current at zero by creating voltage or current zero-crossing to avoid any arc. This particular structure requires to interrupt the fault current before its permissible ratings are exceeded. Hence, it imposes a critical restriction on the fault clearing time for conventional SSCBs. To solve SSCB problems, the ZSCB structure is proposed [32]. This new structure, developed based on the SSCBs, offers several superiorities over SSCBs, such as inherent commutation, fast and simple control, bidirectional power capability, and cost-efficiency. The operation principle of the ZSCB can be explained by the current path shown in Figure 14.10a and its waveforms shown in Figure 14.10b. In case of any transient, Z-source breaker capacitors participate partially in the fault current. The inductor keeps the flowing current of I_L (Inductance current) constant, as depicted in Figure 14.10b. In this situation, the conduction path goes through the z-source capacitors and back to the source. Consequently, the capacitor current I_C increases until it matches with I_L (see Figure 14.10b). At this point, I_{SCR} (thyristor current) drops to zero and causes the silicon-controlled rectifier/thyristor (SCR) to commutate off. By employing a detector circuit that can detect that the SCR has switched OFF, the gate voltage will be removed from the SCR. Although the fault has been disconnected after the SCR goes off, several considerations should be considered to have a successful operation [32]. Table 14.3 summarizes the DC microgrids' protective device characteristics.

14.2.2.4 Protection Algorithm Capabilities

In order to achieve an advanced protection scheme, in addition to suitable equipment, an advanced protection algorithm is needed. Usually, this algorithm is developed based on network variables such as voltage or/and current. Regardless of being based on the local variables or depending on the adjacent variables, every protection algorithm should have the following features:

- **Speed:** To prevent the second fault current stage onset, the protection scheme should detect and disconnect the faulty part within a few milliseconds.
- **Accuracy:** To avoid false tripping, the protection scheme should be accurately designed. It should be able to discriminate the faulty section from healthy sections.
- **Reliability:** The protection scheme should be designed to detect faults in the worst case of grid conditions. It should predict any unwanted event in which the protection system might face any malfunctions for safe operation.
- **Simplicity:** Fewer installed equipment means a more practical protection system. Avoiding communication infrastructure, less data acquisition, and less computational burden help achieve a more practical protection system.

Up to here, we have investigated the DC microgrid behavior during fault transients and basic practical requirements for designing an advanced protection system. Now,

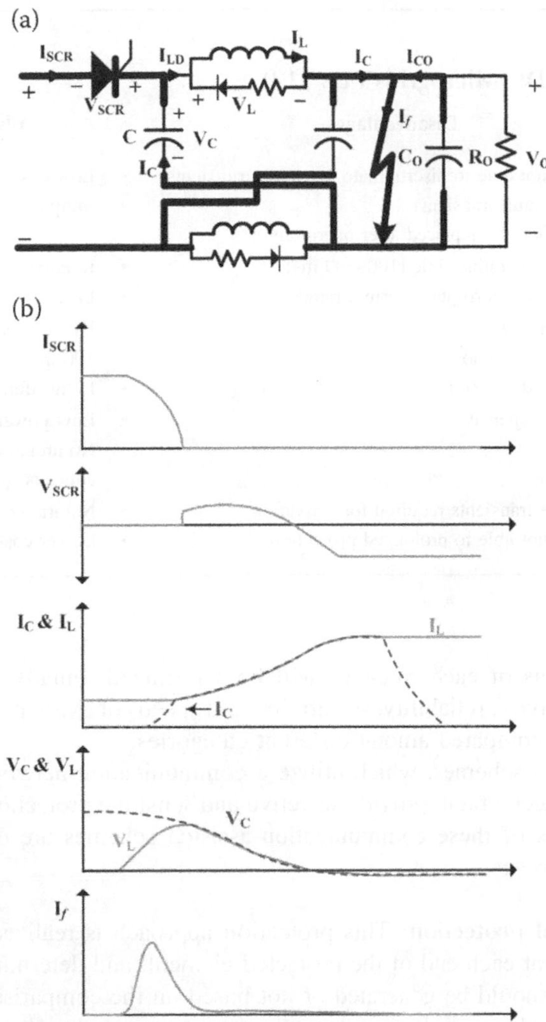

FIGURE 14.10 Z-source circuit breaker a) conduction path of fault current; b) fault-clearing waveforms [32].

it is better to investigate previous studies in this area to introduce the merits and demerits of each category and realize which solution would be the best fit for DC microgrid protection.

14.3 DC MICROGRID PROTECTION APPROACHES

To protect the DC microgrids, researchers have investigated different schemes. Generally, it is possible to categorize these schemes into two main categories: unit protection and non-unit protection. In the following section, operation principles,

TABLE 14.3
Comparison of DC Microgrids PDs [12]

PD type	Disadvantages	Advantages
Fuse	• It is not able to discriminate non-fault transients from fault transients • It should be replaced after operating	• Low cost • Simple structure
MCB	• Long operating time (100–300 ms) • Limited interruption current range	• Relatively low cost • Low power loss
SSCB	• Expensive • High power loss • Large dimensions	• Fast response time (t < 100 ms) • Long lifetime
HCB	• Very expensive	• Low power loss • No arcing on mechanical contacts
ZSCB	• Large transients required for activation • It is not able to prolonged protection	• Natural commutation • Lower cost rather than SSCB

the pros, and cons of each category will be scrutinized. Finally, the main characteristics (selectivity, reliability, sensitivity, and speed) of every protection scheme should have, are compared among different categories.

Unit protection schemes, which utilize a communication network to collect required data to detect a fault, provide selective and sensitive protection schemes. The prominent studies of these communication-assisted schemes are described in the following categories:

- **Differential protection:** This protection approach is realized by measuring the current at each end of the protected element, and determines whether the trip signal should be generated or not based on the comparison of both ends [33]. This scheme is highly dependent on time synchronization, which can cause false tripping in synchronization loss. To solve time synchronization, a discretized sequence of samples has been proposed in [34]. Furthermore, predefined thresholds might lose their validity when the DGs penetration level changes. Hence, adaptive differential protection to solve this issue has been proposed in [35], [36].
- **Centralized protection:** To provide integrated and definite protection, centralized schemes have been proposed in [37], [38]. All currents information will be acquired in a central processor, which decides about generating a trip signal.
- **Event-based protection:** Conventional differential protection requires a high bandwidth channel. Moreover, it requires accurate time synchronization. Researchers in [39], [40] have proposed a novel scheme that combines local and communication processes. In this scheme, the fault type is determined by

local measurement. In the next step, to issue the trip signal to circuit breakers, only fault type will be transferred to the adjacent protection units. Based on the processed data and the received data, each protection unit decides whether to generate a trip signal or not. Since in this scheme only fault types are communicated among protection units, the need for accurate, and high bandwidth channel is eliminated.

- **Permission-based protection:** These schemes are similar to AC permission-based protection schemes in which issuing a trip signal to the circuit breaker depends on two factors: 1) detecting a fault by protection units of both sides of the protected zone, 2) receiving permission signal by protection units of both sides of the protected zone. If the protection unit of any sides detects a fault, it generates a permission signal and issues it to the other side's protection unit. Due to this fact, this protection class takes advantage of the local-based protection approaches and communication-based protection approaches. In general, the local process of the reported studies for DC microgrids protection is based on the fault current magnitude and its direction toward the protection unit [41], estimation of the fault path inductance using the linear least square method [8], and transient parameters, exclusively oscillation frequency and power [9]. Finally, to improve the selectivity and coordinate the protection units in the mentioned studies, permissive signals are employed.

Although each of the schemes mentioned previously provides distinctive characteristics and proper protection for the DC microgrid, employing a communication channel adds complexity and imposes additional cost for DC microgrids' protection. Moreover, all of these schemes suffer from a broken communication channel, causing the protection system to fail. Therefore, utilizing backup protection is necessary.

Non-unit protection schemes utilize local sampling and processing to generate a trip signal. The categories of these communication-free schemes are presented below:

- **Pseudo-distance protection:** Implementing AC distance protection in DC microgrids is impossible due to lack of fundamental frequency and low cable impedance. Hence, for realizing distance protection, researchers rely on estimation methods, which result in pseudo-distance protection in [3], [42]. In these studies, if the desired parameter is estimated less than its threshold (the desired parameter in [42] is fault path resistance and in [3] is fault path inductance), a trip signal will be generated. Although these methods offer fast and effective fault detection and discrimination, their performance would face severe difficulties in faults with high impedance and faults on short cables adjacent to the long cables.
- **Overcurrent protection:** It is possible to modify and implement AC overcurrent protection for DC microgrids. Hence, some researchers have proposed modified overcurrent protection for DC microgrids [43–45]. Based on the reported studies, employing this protection class on DC microgrids with

complex configurations will face several challenges such as 1) arduous relays coordination due to complicated and time-consuming offline studies, 2) unnecessary disconnection of intact sections in case of some faults, and 3) low response time. To avoid the mentioned challenges, researchers in [43], [45] have utilized a communication channel with modified overcurrent protection.

- **Current derivatives protection:** Exorbitant increase of current in a very short time range after the fault incident leads to a dramatic change in the current derivative. Hence, this feature grants the chance of fault detection by current derivatives magnitude. Researchers in [1], [2], [46], [47] have employed feature of fault current derivative to develop a novel DC protection system. In [46], only the first derivative of the fault current is investigated for faults detection. To improve fault detection's selectivity by only the first derivative, the second-order derivative has been used in [1]. More advanced and selective current derivative-based fault detection using a sequential analyzing technique has been presented in [47]. Although this category approaches provide fast and effective protection schemes based on the local measurement, they require a high sampling rate for the precise calculation of the fault current derivatives. This issue causes noise amplification in the measurement equipment, which can result in protection system malfunction. To face this issue, noise cancellation methods and filtering techniques should be utilized. Defining definite thresholds for current derivative-based protection schemes is another issue because variable parameters such as microgrid cable characteristics and fault impedance type and magnitude are involved in the current derivative's magnitude.

The application of AI in power systems is now emerging, where it may revolutionize the protection of power systems. To step in this field, the following section develops a basic scheme of AI applications in designing an advanced protection system.

14.4 AI-BASED APPROACHES EFFECTIVENESS INVESTIGATION

Due to significant advancements in AI approaches in recent years, researchers tend to employ AI approaches, including neural network (NN), fuzzy logic, and expert system in power system applications such as control and protection schemes [48], [49]. Salient features of the artificial neural network (ANN) technique among all available AI-based techniques led to its widespread application for DC microgrid protection. ANN technique is more economical, less complex, and relatively fast; it is more suitable and practical rather than other AI techniques for real-time protection systems [50–52]. Variable parameters such as the fault current or/and voltage can be used as input for training the neural network [50], or more prevail technique is that initially signal features can be extracted and in the next step use them as the neural network input [51], [52]. Although utilizing the sampled fault current and/or voltage waveforms directly as input for ANN requires a time-consuming training process, long calculation time, and a complicated structure for precise operation, could be considered a practical solution for DC microgrid protection. Employing this method reduces the required data for training and facilitates online real-time implementation. Moreover, with feature vectors, the performance of the protection scheme is not affected by variations in electrical parameters such as fault resistance or loads. The purpose of employing ANN is to obtain a reliable and practical classifier that can recognize and classify the sources and fault causes.

Fast Fourier transform (FFT) and wavelet transform (WT) are the most well-known signal extraction methods used for the ANN-based protection schemes in frequency and time domains, respectively. It should be noted that the FFT method is disregarded as a method for developing a DC microgrid protection system due to its incapability to provide time-domain information. Signal extraction methods should be able to identify transients and sudden changes with short spikes [51]. On the other hand, the WT takes advantage of signal decomposition into its frequency components and provides multi-level frequency division, enabling the WT to provide signal decomposition with different levels of resolution. As a result, WT could be utilized in ANN-based protection approaches to provide an effective protection system by detecting sudden changes in the sampled data and identifying imposed transients by fault occurrence. Generally, in the WT-based ANN protection method, initially, the sampled signal features are extracted and sent as input for a trained NN. Based on the self-organized network, the NN investigates whether the extracted feature is for a normal situation or a faulty situation. Training an NN requires a time-consuming process, and also, the trained NN exclusively belongs to the considered microgrid. Hence, the mentioned drawbacks of ANN-based protection schemes should be alleviated by proposing novel AI-based protection schemes. There is an insufficient investigation in the AI-based protection schemes area, and presented protection schemes based on this method are broadly based on the conventional WT and discrete WT (DWT). The general flowchart of a feature extraction ANN-based protection scheme is presented in Figure 14.11. According to this flowchart, a concise description of the WT application in designing the protection system for DC microgrids is presented.

14.4.1 WT Principles

WT utilization for training a NN is based on the basic principles of WT theory, choosing efficient and appropriate WT function, and optimal decomposition level that are described in the following [51].

- **WT theory:** The WT can detect the transient of power signals with different WT coefficient scales; specifically, the transient signals with high-frequency components. Continuous WT (CWT) with $x(t)$ as input signal and $\psi(t)$ as mother WT function can be defined as below:

$$CWT(a, b) = \frac{1}{\sqrt{a}} \int_{-\infty}^{+\infty} x(t)\psi\left(\frac{t-b}{a}\right)dt \tag{14.10}$$

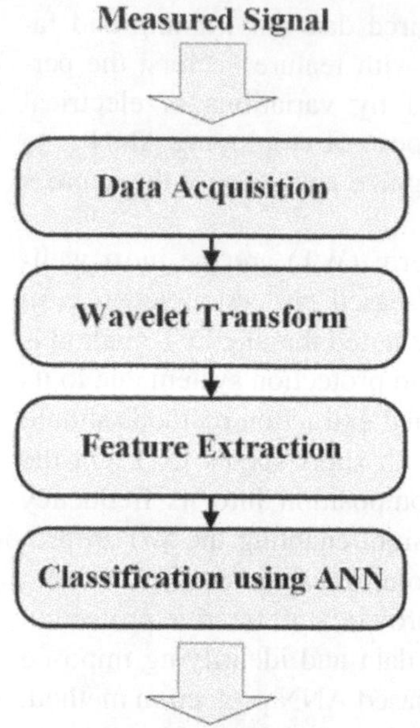

FIGURE 14.11 Flowchart of AI-based protection scheme using WT [51].

where a and b are scale and translation factors, respectively.

Digital systems require DWT in their process instead of CWT. Hence, continuous parameters of a, b, and t should be discretized. DWT can be written as:

$$DWT(m, n) = \frac{1}{\sqrt{a_0^m}} \sum_k x(k) \psi \left(\frac{k - nb_0 a_0^m}{a_0^m} \right) \tag{14.11}$$

The continuous parameters of a and b in (14.10) are changed to discrete functions based on m and n, which are integers. Also, k is the sample number in the input signal frame.

- **Mother WT selection:** Before applying WT to the sampled signal for its feature extraction, in order to achieve an accurate and time-efficient protection scheme, an appropriate WT function with efficient decomposition scales should be selected. General criteria for choosing the mother wavelet can be mentioned as follows [53]:
  1. Having enough numbers of vanishing moments
  2. Having sharp cutoff frequencies
  3. Obtaining the highest possible total wavelet energy
  4. Being based on an orthonormal basis

There are several wavelet mother families including the Symlets family (*symN*), Meyer family (*meyr*), Coiflets family (*coifN*), and Daubechies family (*dbN*) [54]; in the given abbreviations, N is the order of the wavelet function. Since Daubechies wavelets are faster in computation speed, they can be considered for real-time applications such as designing an AI-based protection scheme [52]. Due to the satisfactory performance of *db2–db10* in analyzing signals transient with sudden changes [54], [55], it is suggested that we choose them as mother wavelet candidates. Higher orders of *db10* are time-consuming functions that make them inappropriate for real-time protection applications.

High and low-pass filters response of mother wavelets of *db2* and *db10* are illustrated in Figure 14.12. Compared with *db2*, *db10* frequency response represents a sharp cutoff frequency and a drastic reduction in the overlapped area, culminating in low energy leakage during the decomposition process and high information remaining for feature extraction and classification [51].

- **Decomposition level selection:** To capture all signal features, it is preferred to employ higher decomposition levels. However, the decomposition level is directly related to computation time, implying that as the decomposition level increases, the computation time will impressively increase. Therefore, it is crucial to compromise between desired features and computation time for suitable decomposition level selection.

14.4.2 Feature Extraction

It is not efficient to use the obtained coefficients from WT as the classifier inputs directly. To reduce the computational burden, the feature extraction techniques such

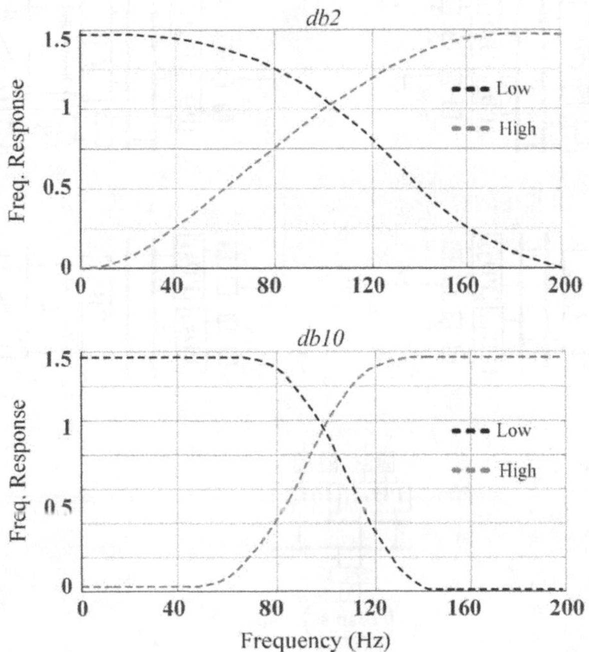

FIGURE 14.12 The frequency response of the high and low pass WT-based filters.

as mean, RMS, standard deviation, and Shannon-entropy are commonly employed to the obtain coefficients at every decomposition level. Since the mentioned extraction techniques are vulnerable to noise [56], researchers have proposed feature extraction based on the coefficients' energy through Parseval's theorem [51]. This section will investigate the extracted features of pole-to-pole and pole-to-ground faults using the energy-based feature extraction technique with *db10* mother wavelet.

14.4.3 Feature Extraction Results

To investigate the energy-based feature extraction technique, test DC microgrid of Figure 14.13 with given parameters of Table 14.4 is considered [57]. Features of pole-to-pole and pole-to-ground faults of bus1 are investigated and discussed in the following.

- **Pole-to-pole fault features:** As mentioned previously, by applying the energy-based feature extraction technique and *db10* mother wavelet to the current signal of the imposed short-circuit fault on bus1 (test microgrid of Figure 14.13) during normal and faulty operating conditions, the energy

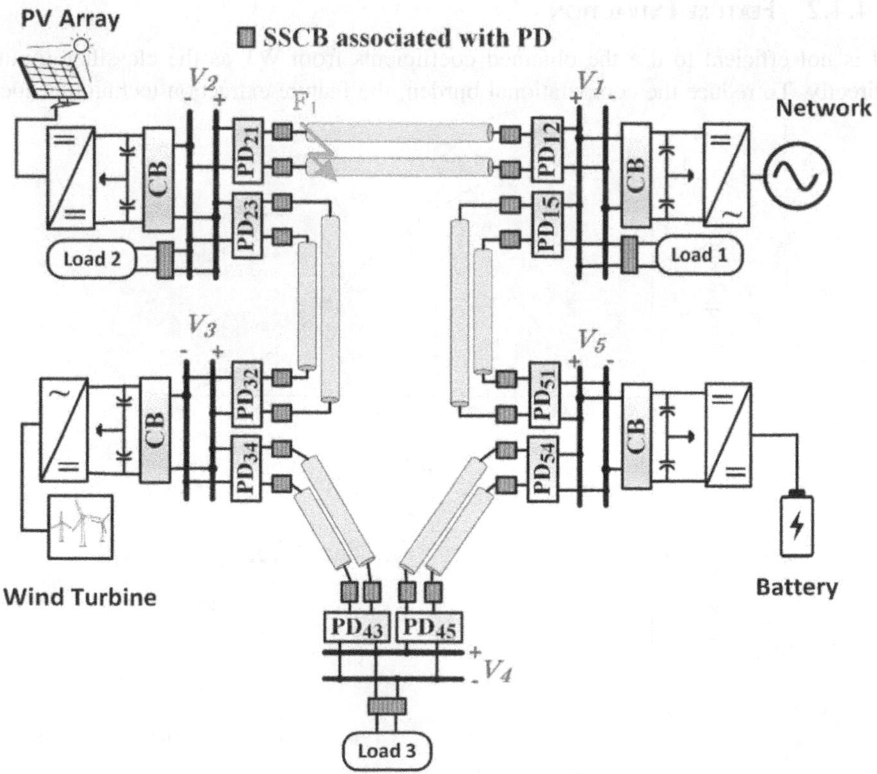

FIGURE 14.13 DC test microgrid [57].

TABLE 14.4
Test Microgrid Parameters [57]

Name	Advantages
Grid voltage	400 V DC
Grid VSC	250 kW
Battery	130, 0.4 kAh
Battery DC-DC Converter	125 kW
Solar panel	V_{mppt} = 29 V, I_{mppt} = 7.35 A
PV Converter	125 kW
Wind turbine	400 kW
Cable resistance	0.125 Ω/km
Cable inductance	0.232 mH/km
DC link capacitor	12.5 mF
Load	200 kW
Cable length (L_{ij}) m	L_{12} = 200, L_{23} = 500, L_{34} = 400, L_{45} = 700, L_{15} = 1000

distribution at different decomposition levels is illustrated in Figures 14.14 and 14.15. As can be seen from the results, comparing energy at different levels during normal and faulty situation, the faulty signal's energy level increases from decomposition levels of 5 to 8, due to the AC side fundamental frequency, and its second harmonic. By subtracting energy levels of the faulty situation from the normal situation, Figure 14.16 is obtained, which can be used to discriminate the faulty situation from the normal situation. Also, it is possible to compare x_6 vector with a predefined threshold to detect the short circuit fault. The PD of bus1 can issue a trip signal if x_6 exceeds the predefined threshold value.

- **Pole-to-ground fault features:** The extracted features of the pole-to-ground fault f_1, which is located at the end of line 1–2 with 2 Ω fault impedance, is

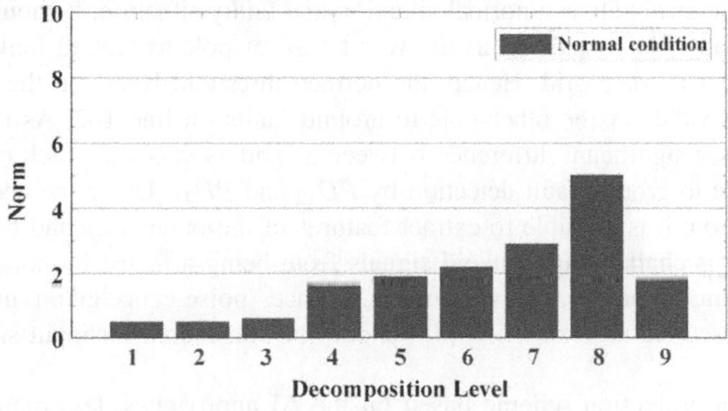

FIGURE 14.14 Energy distribution of each decomposition level of Line 1–2 under normal condition.

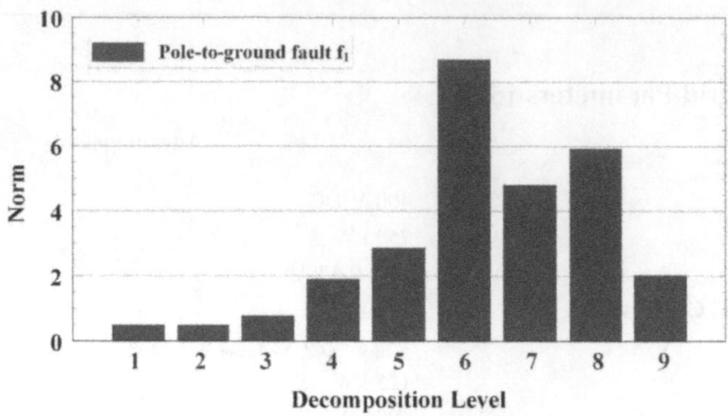

FIGURE 14.15 Energy distribution of each decomposition level of bus1 current under pole-to-pole fault condition.

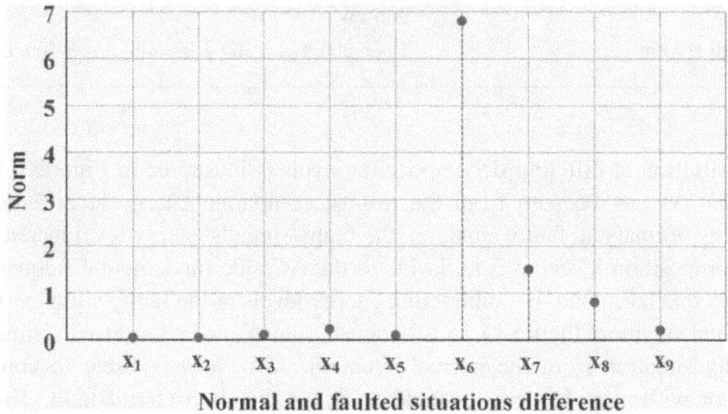

FIGURE 14.16 Extracted features of bus1 pole-to-pole fault with *db10* wavelet.

illustrated in Figure 14.17. Like the pole-to-pole fault, these features result from the difference between normal situation and faulty situation. It should be noted that fault f_1 is considered as the worst case of pole-to-ground fault in a low-voltage DC microgrid. Hence, the defined threshold based on the worst-case fault is valid for the other pole-to-ground faults on line 1–2. As it is shown, there is a significant difference between x_1 and x_2 vectors, which can be used for pole-to-ground fault detection by PD_{12} and PD_{21}. Using the above feature extraction, it is possible to extract features of the other lines and buses by the WT. It is challenging to avoid signals from being affected by noise, which is abundant in practical environments. Hence, noise-cancellation methods for WT-based feature extraction [58] should be considered for input signals.

To develop a protection scheme based on the AI approaches, less memory usage and low computation time are two essential factors that imply employing a feature extraction technique to reduce the size of input signals. After feature extraction, the

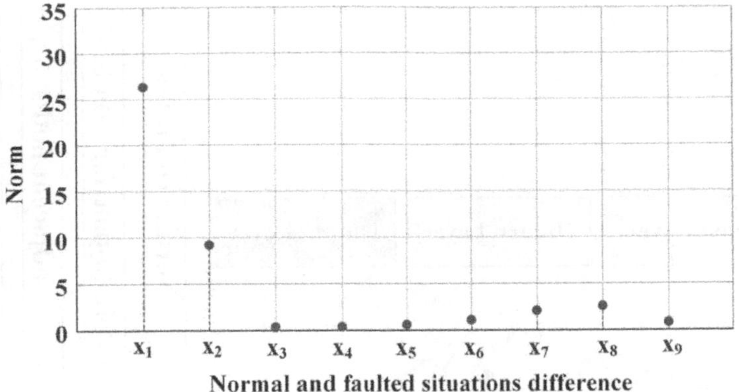

FIGURE 14.17 The extracted features of the pole-to-ground fault f_1.

obtained vectors are sent to ANN for pattern recognition. In the following, the ANN pattern recognition of the mentioned extracted features is described.

14.4.4 Pattern Recognition with ANN

The basic principle of ANN is based on recognizing the underlying patterns related to the output and input signals. In other words, without using any predefined thresholds, ANN can consider multiple features of the input signals at the same time and then compare the patterns based on the mutual similarities, which will be culminated in the generation of output signals. Among all different possible topologies for implementing ANN, the feed-forward network has been used broadly for fault diagnosis applications [59]. A multilayer feed-forward network consists of one input layer, one output layer, and at least one hidden layer between the output and input layers [60], [61]. The trial-and-error method is still widely used for the design of network configuration [59]. The structure of the implemented feed-forward ANN is illustrated in Figure 14.18.

Nine feature vectors extracted using WT are considered as inputs of the implemented neural network, leading to two output signals. The microgrid situation is determined based on the output results given in the following [51]:

$$\begin{aligned} &\varnothing: R^n \rightarrow \{0,\ 1\}^m \\ &\varnothing(\mathrm{x}) = (c_1, c_2, \ldots, c_m) \end{aligned} \tag{14.12}$$

with

$$c_i = \begin{cases} 1 \text{ fault} \\ 0 \text{ normal} \end{cases} \tag{14.13}$$

where $i = 1, 2, \ldots, \mathrm{m}$ denotes the number of output neurons, which is considered equal to 2 for each PD. Also, n shows the number of inputs and is considered equal to 9.

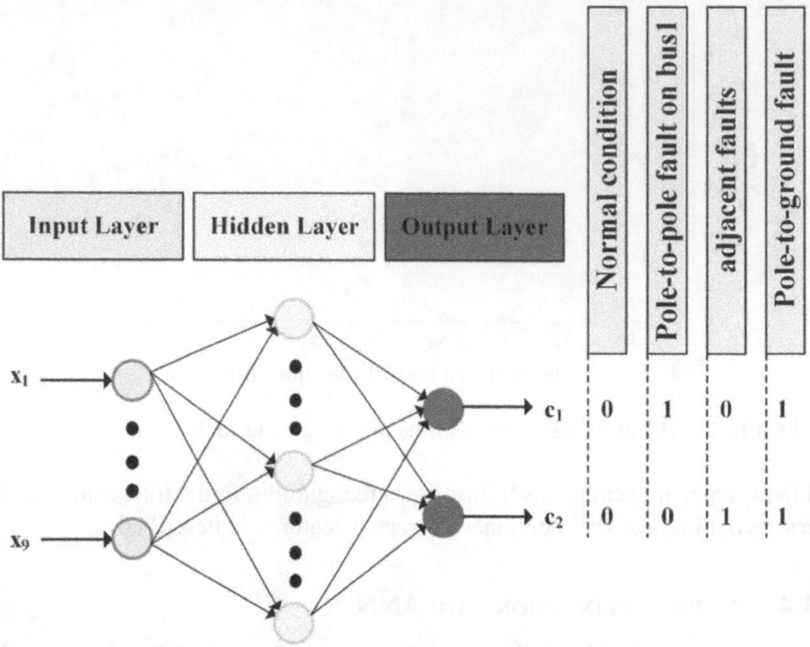

FIGURE 14.18 The structure of the implemented feed-forward ANN [51].

The number of neurons in the hidden layer has a significant impact on the network's convergence rate and stability. The efficient number will be attained via trial and error process. In the implemented network of Figure 14.17, the efficient number of hidden neurons is achieved as eight or more for this particular application. The error of the training process is set to less than 0.001. To avoid the high computational burden, eight hidden neurons are chosen to implement the ANN.

The log-sigmoid [62] function is employed for the hidden and output layers of implemented ANN. This activation function is suitable for binary output generation. Using a supervised manner and back propagation function, the ANN is trained [62], [63]. The training process continues until the error target is achieved. By updating the coefficients of the output and hidden layers based on the gradient descent with momentum, the average error is decreased until it reaches the predefined threshold.

14.4.5 Classification Results

The training process is accomplished in MATLAB® software package, and then the WT-based AI protection scheme is tested. Different operation conditions containing pole-to-pole fault, pole-to-ground fault, the disturbance caused by load changes, and normal operation are included in the training and testing cases. Details of the training and testing process are summarized in Table 14.5. The provided results demonstrate that the implemented AI-based protection scheme can correctly identify all operating

TABLE 14.5
ANN Classification Results

	Training	Testing	Result			
			Normal operation	P2P fault on bus1	Adjacent fault	P2G fault
Normal operation	40	60	60	0	0	0
Pole-to-pole fault on bus1	40	60	0	60	1	0
Adjacent faults	40	60	0	0	58	0
Pole-to-ground fault	40	60	0	0	1	60
Accuracy		100%	100%	96.7%	100%	
Overall accuracy	99.2%					

conditions with an accuracy of more than 99%, suggesting that AI techniques can develop a highly reliable protection system for DC microgrids.

14.4 CONCLUSION

Although DC microgrid offers numerous operational and potential benefits compared to AC microgrid, it cannot be simply protected by AC existing protective approaches and infrastructures. Implementing a DC microgrid cannot be realized without designing an effective protection system, which plays an integral part in the DC microgrid system design. To face associated challenges with DC microgrid protection, first, DC fault current in two main stages of capacitor discharge stage and freewheeling diodes stage is investigated. Second, technical issues that should be considered in designing a DC microgrid are described. Third, the grounding structure can also affect the safety of the microgrid. It has a significant impact on the protection schemes' ability to detect the grid's fault and survivability under a faulty situation. Hence, five grounding structure, including IT, TN-C, TN-S, T-N-C-S, and TT, and different grounding strategies (ungrounded, solidly grounded, thyristor grounded, and diode ground), are discussed and compared.

Due to the intrinsic difference between AC and DC fault current, it necessitates the utilization of PDs which comply with DC fault current requirements. The structure and operation principles of five classes of PDs are described and compared from different aspects, including cost, losses, response time, and size.

Two main categories of fault detection and isolation methods, including unit and non-unit protection with their subcategory studies, are discussed in detail. Although each of the investigated methods has specific advantages and disadvantages, most of them are evaluated directly on grounded radial or ring configurations under low fault impedance. Hence, identifying faults on the meshed type configurations with high impedance fault need more sophisticated schemes. As a result, to achieve such a sophisticated protection scheme, it should provide definite protection. AI-based protection schemes can satisfy all DC microgrids' protection requirements. To enhance the performance of AI-based protection, feature extraction techniques are the first candidates. A protection system based on the AI will be realized by extracting desired features and then training the ANN with efficient chosen parameters.

REFERENCES

[1] A. Meghwani, S. C. Srivastava, and S. Chakrabarti, "A non-unit protection scheme for DC microgrid based on local measurements," *IEEE Trans. Power Deliv.*, vol. 32, no. 1, pp. 172–181, 2017.

[2] M. Shamsoddini, B. Vahidi, R. Razani, and H. Nafisi, "Extending protection selectivity in low voltage DC microgrids using compensation gain and artificial line inductance," *Electr. Power Syst. Res.*, vol. 120, Art. no. 105992, 2020.

[3] M. Shamsoddini, B. Vahidi, R. Razani, and Y. A.-R. I. Mohamed, "A novel protection scheme for low voltage DC microgrid using inductance estimation," *Int. J. Electr. Power Energy Syst.*, vol. 188, Art. no. 106530, 2020.

[4] S. D. A. Fletcher, P. J. Norman, S. J. Galloway, and G. M. Burt, "Determination of protection system requirements for dc unmanned aerial vehicle electrical power networks for enhanced capability and survivability," *IET Electr. Syst. Transp.*, vol. 1, no. 4, pp. 137–147, 2011.

[5] R. M. Cuzner and G. Venkataramanan, "The status of DC micro-grid protection," in *2008 IEEE Industry Applications Society Annual Meeting*, 2008, pp. 1–8.

[6] M. E. Baran and N. R. Mahajan, "Overcurrent protection on voltage-source-converter-based multiterminal DC distribution systems," *IEEE Trans. power Deliv.*, vol. 22, no. 1, pp. 406–412, 2006.

[7] A. Berizzi, A. Silvestri, D. Zaninelli, and S. Massucco, "Short-circuit current calculations for DC systems," *IEEE Trans. Ind. Appl.*, vol. 32, no. 5, pp. 990–997, 1996.

[8] R. Mohanty and A. K. Pradhan, "Protection of smart DC microgrid with ring configuration using parameter estimation approach," *IEEE Trans. Smart Grid*, vol. 9, no. 6, pp. 6328–6337, 2017.

[9] R. Mohanty and A. K. Pradhan, "DC ring bus microgrid protection using the oscillation frequency and transient power,"*IEEE Systems J.*, vol. 13, no. 1, 875–884, 2018.

[10] J. Yang, J. E. Fletcher, and J. O'Reilly, "Short-circuit and ground fault analyses and location in VSC-based DC network cables," *IEEE Trans. Ind. Electron.*, vol. 59, no. 10, pp. 3827–3837, 2012.

[11] S. Barakat, M. B. Eteiba, and W. I. Wahba, "Fault location in underground cables using ANFIS nets and discrete wavelet transform," *J. Electr. Syst. Inf. Technol.*, vol. 1, no. 3, pp. 198–211, 2014.

[12] S. Beheshtaein, S. Member, R. M. Cuzner, and S. Member, "DC microgrid protection: A comprehensive review," *IEEE J. Emerg. Sel. Top. Power Electron.*, to be published, 2019.

[13] T. Dragičević, X. Lu, J. C. Vasquez, and J. M. Guerrero, "DC microgrids—Part I: A review of control strategies and stabilization techniques," *IEEE Trans. power Electron.*, vol. 31, no. 7, pp. 4876–4891, 2016.

[14] IEC 60364-1. "Low-voltage electrical installations–part 1: fundamental principles, assessment of general characteristics, definitions." 2005.

[15] J.-D. Park and J. Candelaria, "Fault detection and isolation in low-voltage DC-bus microgrid system," *IEEE Trans. power Deliv.*, vol. 28, no. 2, pp. 779–787, 2013.

[16] J. Mohammadi, F. B. Ajaei, and G. Stevens, "Grounding the dc microgrid," *IEEE Trans. Ind. Appl.*, vol. 55, no. 5, pp. 4490–4499, 2019.
[17] R. M. Cuzner, T. Sielicki, A. E. Archibald, and D. A. McFarlin, "Management of ground faults in an ungrounded multi-terminal zonal DC distribution system with auctioneered loads," in *2011 IEEE Electric Ship Technologies Symposium*, 2011, pp. 300–305.
[18] R. Talon, "Electrical installation guide: According to IEC International Standards." Schneider Electric, 2005.
[19] M. Farhadi and O. A. Mohammed, "Protection of multi-terminal and distributed DC systems: Design challenges and techniques," *Electr. Power Syst. Res.*, vol. 143, pp. 715–727, 2017.
[20] J. P. Brozek, "DC overcurrent protection-where we stand," *IEEE Trans. Ind. Appl.*, vol. 29, no. 5, pp. 1029–1032, 1993.
[21] M. Fang, L. Fu, R. Wang, and Z. Ye, "Coordination protection for DC distribution network in DC zonal shipboard power system," in *2011 International Conference on Advanced Power System Automation and Protection*, 2011, vol. 1, pp. 418–421.
[22] J. P. Vargas, B. Goss, and R. Gottschalg, "Large scale PV systems under non-uniform and fault conditions," *Sol. Energy*, vol. 116, no. 1, pp. 303–313, 2015.
[23] F. Wang, Z. Zhang, T. Ericsen, R. Raju, R. Burgos, and D. Boroyevich, "Advances in power conversion and drives for shipboard systems," *Proc. IEEE*, vol. 103, no. 12, pp. 2285–2311, 2015.
[24] C. Meyer, S. Schroder, and R. W. De Doncker, "Solid-state circuit breakers and current limiters for medium-voltage systems having distributed power systems," *IEEE Trans. power Electron.*, vol. 19, no. 5, pp. 1333–1340, 2004.
[25] K. Tahata, S. El Oukaili, K. Kamei, D. Yoshida, Y. Kono, R. Yamamoto, and H. Ito, "HVDC circuit breakers for HVDC grid applications," in 11th IET International Conference on AC and DC Power Transmission, Birmingham, U.K, 2015.
[26] Z. Chen *et al.*, "Analysis and experiments for IGBT, IEGT, and IGCT in hybrid DC circuit breaker," *IEEE Trans. Ind. Electron.*, vol. 65, no. 4, pp. 2883–2892, 2017.
[27] F. F. Wang and Z. Zhang, "Overview of silicon carbide technology: Device, converter, system, and application," *CPSS Trans. Power Electron. Appl.*, vol. 1, no. 1, pp. 13–32, 2016.
[28] R. J. Kaplar, J. C. Neely, D. L. Huber, and L. J. Rashkin, "Generation-After-Next Power Electronics: Ultrawide-bandgap devices, high-temperature packaging, and magnetic nanocomposite materials," *IEEE Power Electron. Mag.*, vol. 4, no. 1, pp. 36–42, 2017.
[29] X. She, A. Q. Huang, Ó. Lucía, and B. Ozpineci, "Review of silicon carbide power devices and their applications," *IEEE Trans. Ind. Electron.*, vol. 64, no. 10, pp. 8193–8205, 2017.
[30] T. Funaki *et al.*, "Power conversion with SiC devices at extremely high ambient temperatures," *IEEE Trans. Power Electron.*, vol. 22, no. 4, pp. 1321–1329, 2007.
[31] C. Peng, A. Q. Huang, and X. Song, "Current commutation in a medium voltage hybrid DC circuit breaker using 15 kV vacuum switch and SiC devices," in *2015 IEEE Applied Power Electronics Conference and Exposition (APEC)*, 2015, pp. 2244–2250.
[32] K. A. Corzine and R. W. Ashton, "A new Z-source DC circuit breaker," *IEEE Trans. Power Electron.*, vol. 27, no. 6, pp. 2796–2804, 2011.
[33] S. D. A. Fletcher, P. J. Norman, K. Fong, S. J. Galloway, and G. M. Burt, "High-speed differential protection for smart DC distribution systems," *IEEE Trans. Smart Grid*, vol. 5, no. 5, pp. 2610–2617, 2014.

[34] S. Dhar, R. K. Patnaik, and P. K. Dash, "Fault detection and location of photovoltaic based DC microgrid using differential protection strategy," *IEEE Trans. Smart Grid*, vol. 9, no. 5, pp. 4303–4312, 2017.
[35] S. Dhar and P. K. Dash, "Differential current-based fault protection with adaptive threshold for multiple PV-based DC microgrid," *IET Renew. Power Gener.*, vol. 11, no. 6, pp. 778–790, 2017.
[36] J. Naik, S. Dhar, and P. K. Dash, "Adaptive differential relay coordination for PV DC microgrid using a new kernel based time-frequency transform," *Int. J. Electr. Power Energy Syst.*, vol. 106, pp. 56–67, 2019.
[37] M. Monadi, C. Gavriluta, A. Luna, J. I. Candela, and P. Rodriguez, "Centralized protection strategy for medium voltage DC microgrids," *IEEE Trans. Power Deliv.*, vol. 32, no. 1, pp. 430–440, 2017.
[38] S. Augustine, M. J. Reno, S. M. Brahma, and O. Lavrova, "Fault current control and protection in a standalone DC microgrid using adaptive droop and current derivative," *IEEE J. Emerg. Sel. Top. Power Electron.*, 2020.
[39] M. Farhadi and O. A. Mohammed, "Event-based protection scheme for a multi-terminal hybrid DC power system," *IEEE Trans. Smart Grid*, vol. 6, no. 4, pp. 1658–1669, 2015.
[40] M. Farhadi and O. A. Mohammed, "A new protection scheme for multi-bus DC power systems using an event classification approach," *IEEE Trans. Ind. Appl.*, vol. 52, no. 4, pp. 2834–2842, 2016.
[41] A. A. S. Emhemed, K. Fong, S. Fletcher, and G. M. Burt, "Validation of fast and selective protection scheme for an LVDC distribution network," *IEEE Trans. Power Deliv.*, vol. 32, no. 3, pp. 1432–1440, 2017.
[42] P. Cairoli and R. A. Dougal, "Fault detection and isolation in medium-voltage DC microgrids: Coordination between supply power converters and bus contactors," *IEEE Trans. Power Electron.*, vol. 33, no. 5, pp. 4535–4546, 2018.
[43] A. A. S. Emhemed and G. M. Burt, "An advanced protection scheme for enabling an LVDC last mile distribution network," *IEEE Trans. Smart Grid*, vol. 5, no. 5, pp. 2602–2609, 2014.
[44] K. A. Saleh, A. Hooshyar, and E. F. El-Saadany, "Hybrid passive-overcurrent relay for detection of faults in low-voltage DC grids," *IEEE Trans. Smart Grid*, vol. 8, no. 3, pp. 1129–1138, 2017.
[45] A. Shabani and K. Mazlumi, "Evaluation of a Communication-Assisted Overcurrent Protection Scheme for Photovoltaic-Based DC Microgrid," *IEEE Trans. Smart Grid*, vol. 11, no. 1, pp. 429–439, 2019.
[46] A. Meghwani, S. C. Srivastava, and S. Chakrabarti, "A new protection scheme for DC microgrid using line current derivative," in *2015 IEEE Power & Energy Society General Meeting*, 2015, pp. 1–5.
[47] A. Meghwani, R. Gokaraju, S. C. Srivastava, and S. Chakrabarti, "Local measurements-based backup protection for DC microgrids using sequential analyzing technique," *IEEE Syst. J.*, vol. 14, no. 1, pp. 1159–1170, 2019.
[48] F. Filippetti, G. Franceschini, C. Tassoni, and P. Vas, "Recent developments of induction motor drives fault diagnosis using AI techniques," *IEEE Trans. Ind. Electron.*, vol. 47, no. 5, pp. 994–1004, 2000.
[49] B. K. Bose, "Expert system, fuzzy logic, and neural network applications in power electronics and motion control," *Proc. IEEE*, vol. 82, no. 8, pp. 1303–1323, 1994.
[50] N. K. Chanda and Y. Fu, "ANN-based fault classification and location in MVDC shipboard power systems," in *2011 North American Power Symposium (NAPS)*, 2011, pp. 1–7.

[51] W. Li, A. Monti, and F. Ponci, "Fault detection and classification in medium voltage DC shipboard power systems with wavelets and artificial neural networks," *IEEE Trans. Instrum. Meas.*, vol. 63, no. 11, pp. 2651–2665, 2014.
[52] C. S. Chang, S. Kumar, B. Liu, and A. Khambadkone, "Real-time detection using wavelet transform and neural network of short-circuit faults within a train in DC transit systems," *IEE Proceedings-Electric Power Appl.*, vol. 148, no. 3, pp. 251–256, 2001.
[53] V. Perrier, T. Philipovitch, and C. Basdevant, "Wavelet spectra compared to Fourier spectra," *J. Math. Phys.*, vol. 36, no. 3, pp. 1506–1519, 1995.
[54] W. Gao and J. Ning, "Wavelet-based disturbance analysis for power system wide-area monitoring," *IEEE Trans. Smart Grid*, vol. 2, no. 1, pp. 121–130, 2011.
[55] L. Zhang and P. Bao, "Edge detection by scale multiplication in wavelet domain," *Pattern Recognit. Lett.*, vol. 23, no. 14, pp. 1771–1784, 2002.
[56] A. M. Gaouda, E. F. El-Saadany, M. M. A. Salama, V. K. Sood, and A. Y. Chikhani, "Monitoring HVDC systems using wavelet multi-resolution analysis," *IEEE Trans. Power Syst.*, vol. 16, no. 4, pp. 662–670, 2001.
[57] M. Shamsoddini, G. B. Gharehpetian, R. Razani, and B. Vahidi, "Enhanced energy management system of hybrid DC microgrids with pulsed power load," in *Iranian Conference on Electrical Engineering (ICEE)*, 2018, pp. 1083–1088.
[58] H.-T. Yang and C.-C. Liao, "A de-noising scheme for enhancing wavelet-based power quality monitoring system," *IEEE Trans. Power Deliv.*, vol. 16, no. 3, pp. 353–360, 2001.
[59] M.-Y. Chow, R. N. Sharpe, and J. C. Hung, "On the application and design of artificial neural networks for motor fault detection. II," *IEEE Trans. Ind. Electron.*, vol. 40, no. 2, pp. 189–196, 1993.
[60] J. L. McClelland, D. E. Rumelhart, and P. D. P. R. Group, "Parallel distributed processing," *Explor. Microstruct. Cogn.*, vol. 2, pp. 216–271, 1986.
[61] R. Lippmann, "An introduction to computing with neural nets," *IEEE Assp Mag.*, vol. 4, no. 2, pp. 4–22, 1987.
[62] H. Haykin, *Neural Networks: A Comprehensive Foundation*. Piscataway, NJ, USA: IEEE Press, 1994.
[63] M. A. S. K. Khan and M. A. Rahman, "Development and implementation of a novel fault diagnostic and protection technique for IPM motor drives," *IEEE Trans. Ind. Electron.*, vol. 56, no. 1, pp. 85–92, 2008.

15 Soft Computing–Based DC-Link Voltage Control Technique for SAPF in Harmonic and Reactive Power Compensation

Pravat Kumar Ray[1] *and*
Sushree Diptimayee Swain[2]
[1]National Institute of Technology Rourkela, India
[2]O. P. Jindal University, India

15.1 INTRODUCTION

These days the degradation occurs in power quality due to the reinforce demand of power electronic converters in power system. The power quality issues results disturbances in current and voltage quality problems [1],[2]. The disturbance occurs in current quality because of harmonics. The harmonics in current flows through the line impedance in power system, produces disturbance in voltage. This disturbed voltage enforces the loads which are connected at the Point Of Common Coupling to withdraw current [3–5]. Existence of distortion in voltage and current rises losses in power system, reduce power factor and also the producer of malfunction in customer's equipment. The disturbances in power quality is produced because of non-linear loads such as uninterruptible power supplies (UPS), diode rectifier type load, moto drives, etc. In the middle of the 1940s the SAPF was developed. The selection of reference generation strategy plays a significant role in the SAPF system to get preference of harmonic cancellation. Different reference generation techniques like IRPT, Synchronous Reference Frame method. Perfect Harmonic Cancellation, are proposed next to the development of SAPF. The IRPT was first proposed by Akagi and with co-authors in 1984. This gives precise value of reference compensating voltage and also permits a clear distinction among reactive and active power. Even so, it results unsatisfiable under the condition of non-ideal supply. On that account, an advanced version of this algorithm was proposed in 2005 and was proved to be well enough with respect to the traditional one. Some researchers have announced that the hybrid control approach based synchronous reference frame method [6] is much more convenient than IRPT. Furthermore, articles facilitated the attention in incorporating the HSRF method in

DOI: 10.1201/9780367552374-15

research with contrast to traditional schemes for cancellation of harmonics in distribution systems. The aim of this chapter elaborates the described SAPF control strategies and to collate their performances with respect to DSVPWM for a non-linear load. The reference voltages [7],[8] are processed in a DSVPWM controller for generation of switching patterns in PWMVSI. The Fuzzy Logic Controller (FLC) [9],[10] currently attracts the attention of power system engineers regarding their implementation in SAPFs. The merits of FLC over traditional controllers are that the FLC does not require any mathematical model. It can also handle parameter mismatch and non-linearity in the system. The mamdani type of FLC has been employed as a controller in SAPF. It provides better results and also overcomes the drawbacks of the PI controller[11].

In recent times, FLCs have attracted large attention regarding their usefulness in SAPFs. The merits of FLCs over traditional controllers (PI) are that they don't need any exact mathematical model, can cope with non-linearity, can able to work with imprecise inputs, and are also more robust and reliable than PI controllers.

This chapter mainly emphasizes FLC with varieties of membership functions to examine the execution of the HSRF control approach for estimating reference currents of SAPF. DSVPWM has been employed for the switching pattern generation. In consequence, FLC-HSRF-DSVPWM [12],[13] gives a ultra-modern SAPF on applying distinct membership functions.

This chapter is organized as follows: the description of system topology of SAPF is described in Section 15.2, Section 15.3 depicts the reference generation techniques for SAPF system, Section 15.4 presents a proposed controller design technique for SAPF, Section 15.5 discusses the design principle of SAPF, Section 15.6 explains the simulation results for compensation of distortion using SAPF, Section 15.7 represents the experimental verification of proposed system, and Section 15.8 concludes the paper.

15.2 SYSTEM TOPOLOGY OF SAPF

SAPF is nothing but the combination of a small rated active filter and a series connected RL filter. Figure 15.1 presents the circuit diagram of SAPF. A three-phase full bridge PWMVSI, a DC-link capacitor, and three-phase high-frequency RL filters are the building blocks of SAPF. The functions of the DC-link capacitor in SAPF is to store reactive energy and also to minimize voltage fluctuation under load perturbation case. The high-frequency RL filters are useful in the suppression of switching surges of high frequency from the output voltage of PWMVSI. The SAPF is connected in end-to-end fashion with respect to source and non-linear load.

15.3 REFERENCE GENERATION TECHNIQUES FOR SAPF SYSTEM

15.3.1 Hybrid Control Approach Based Synchronous Reference Frame Method for Active Filter Design (HSRF)

In this technique, the estimated source currents (I_{sa}, I_{sb}, I_{sc}) and load voltages (V_{La}, V_{Lb}, V_{Lc}) are transfigured from a-b-c to rotating reference frame with fundamental frequency [14].

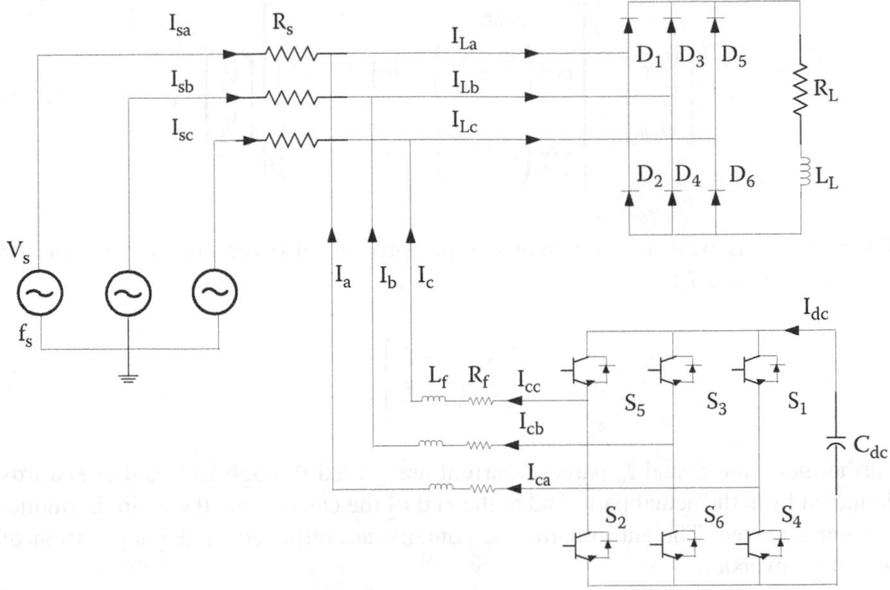

FIGURE 15.1 Shunt Active Power Filter (SAPF).

The phase angle of supply voltage has been traced by the PLL block (Phase Lock Loop); this one is mandatory for the conversion process. The procured voltages can be divisible into AC as well as DC components as follows:

$$\left.\begin{aligned} V_d &= \overline{V_d} + \widetilde{V_d} \\ V_q &= \overline{V_q} + \widetilde{V_q} \end{aligned}\right\} \tag{15.1}$$

where the nature of load voltages V_{La}, V_{Lb}, V_{Lc} is distorted. At first the components are passed through the Low Pass Filter (LPF) and then the output of LPFs are subtracted from the actual one. The DC portions are removed, since one can assume the total harmonic contents, are as follows:

$$\left.\begin{aligned} \widetilde{V_d} &= V_d - \overline{V_d} \\ \widetilde{V_q} &= V_q - \overline{V_q} \end{aligned}\right\} \tag{15.2}$$

By the inverted transformation one can procure the harmonic components of the load voltages as shown:

$$\begin{bmatrix} V_{Lah} \\ V_{Lbh} \\ V_{Lch} \end{bmatrix} = \begin{bmatrix} cos\theta & sin\theta \\ cos\left(\theta - \frac{2\pi}{3}\right) & sin\left(\theta - \frac{2\pi}{3}\right) \\ cos\left(\theta + \frac{2\pi}{3}\right) & sin\left(\theta + \frac{2\pi}{3}\right) \end{bmatrix} \begin{bmatrix} \widetilde{V}_d \\ \widetilde{V}_q \end{bmatrix} \tag{15.3}$$

The average as well as oscillating components are also present in the acquired currents of I_d and I_q:

$$\left.\begin{aligned} I_d = \overline{I_d} + \widetilde{I}_d \\ I_q = \overline{I_q} + \widetilde{I}_q \end{aligned}\right\} \tag{15.4}$$

Furthermore, the I_d and I_q parts of current are passed through LPF and afterwards deducted from the actual part. And at the end of the calculation, the entire harmonic content achieved. The entire harmonic contents are retrieved by the application of reverse conversion.

$$\begin{bmatrix} I_{sah} \\ I_{sbh} \\ I_{sch} \end{bmatrix} = \begin{bmatrix} cos\theta & sin\theta \\ cos\left(\theta - \frac{2\pi}{3}\right) & sin\left(\theta - \frac{2\pi}{3}\right) \\ cos\left(\theta + \frac{2\pi}{3}\right) & sin\left(\theta + \frac{2\pi}{3}\right) \end{bmatrix} \begin{bmatrix} \widetilde{I}_d \\ \widetilde{I}_q \end{bmatrix} \tag{15.5}$$

Figure 15.2 shows the combined source current and load voltage detection scheme for the generation of reference compensating signal. It is observed from Figure 15.2 that the inaccuracy between the reference and the actual value of DC-link voltage of PWMVSI has been through a PI/FL (Fuzzy Logic) controller and afterwards it is deducted from the alternating quantity on the d-axis. Later, the additional

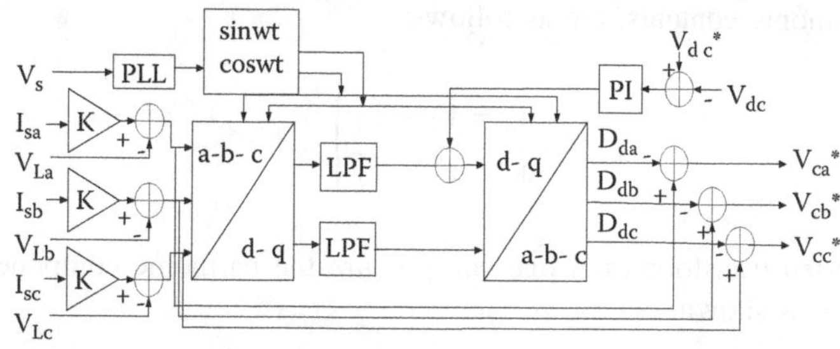

FIGURE 15.2 Hybrid Control Approach Based Synchronous Reference Frame (HSRF) Method.

fundamental units are included with harmonic units correspondingly. So far the expression for reference recompense voltages are illustrated below:

$$\left.\begin{aligned} V_{ca}^{*} &= KI_{sah} - V_{Lah} + \Delta V_{caf} \\ V_{cb}^{*} &= KI_{sbh} - V_{Lbh} + \Delta V_{cbf} \\ V_{cc}^{*} &= KI_{sch} - V_{Lbh} + \Delta V_{ccf} \end{aligned}\right\} \quad (15.6)$$

15.4 DESIGN OF PROPOSED FUZZY LOGIC CONTROLLER IN SAPF SYSTEM

FLC is independent of the system model. The design of FLC is completely based on framing of rules and processing of rules by the inference method. The rules are expressed in the English language with syntax like E is error, δE is change in error, y is control output, and the membership functions are denoted as: NB-negative big, NM-negative medium, NS-negative small, Z-zero, PS-positive small, PM-positive medium, PB-positive big [15]. These seven linguistic variables of membership function are similar for both input and output and characterized by employing triangular membership functions. The Takagi Sugeno fuzzy inference engine has been employed for designing a ultra-modern SAPF. The control rules for a fuzzy controller have been shown in Table 15.1

15.5 PROPOSED CONTROLLER DESIGN TECHNIQUE FOR SWITCHING PATTERN GENERATION IN SAPF SYSTEM

The discontinuous Space Vector Pulse Width Modulation (DSVPWM) is usually employed as a voltage regulater to regulate the switches of SAPF. The block diagram of a space vector for SAPF is shown in Figure 15.3a. The circuit topology shown in Figure 15.1 describes the presence of a two-level PWMVSI in SAPF. It comprises six switches (S_1–S_6) of IGBT. On the basis of combination of different switching patterns,

TABLE 15.1
Control Rule Table of a Fuzzy Controller

	Error (e)						
Change in Error (Δe)	NB	NB	NB	NB	NM	NS	ZE
	NB	NB	NB	NM	NS	ZE	PS
	NB	NB	NM	NS	ZE	PS	PM
	ND	NM	NS	ZE	PS	PM	PR
	NM	NS	ZE	PS	PM	PB	PB
	NS	ZE	PS	PM	PB	PB	PB
	ZE	PS	PM	PB	PB	PB	PB

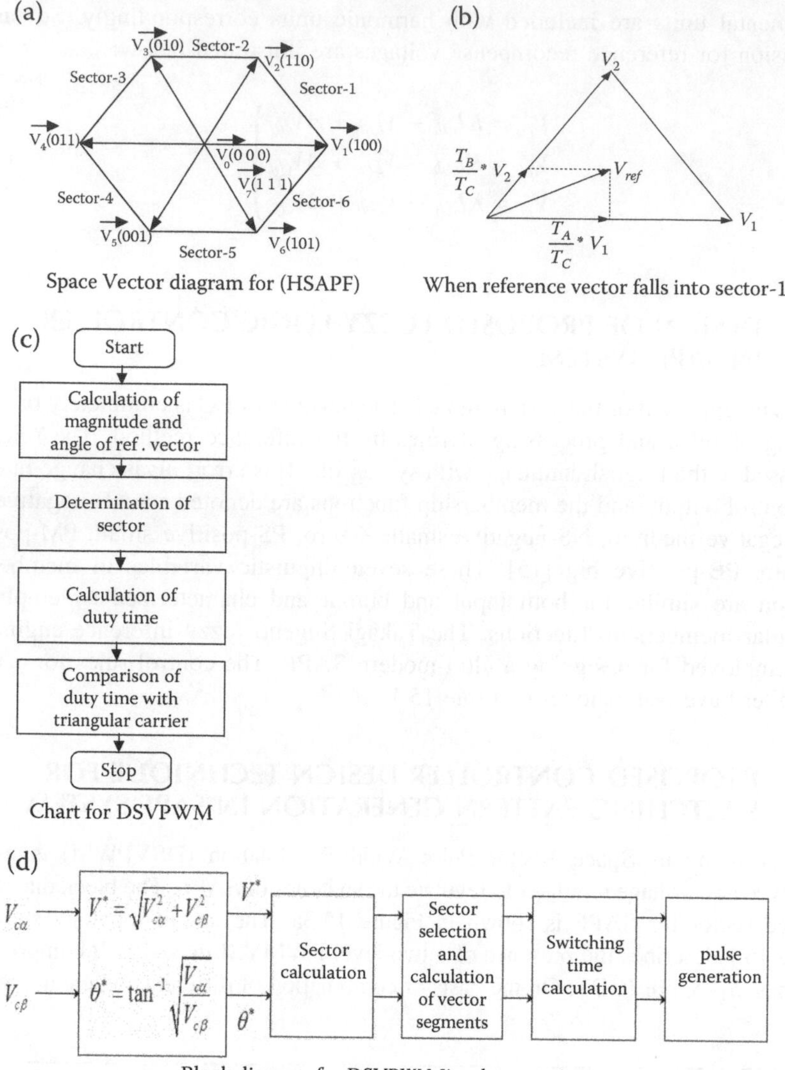

FIGURE 15.3 Diagrams for DSVPWM technique: (a). Space Vector diagram for (HSAPF) (b). When reference vector falls into sector-1 (c). Chart for DSVPWM (d). Block diagram for DSVPWM implementation.

PWMVSI generates distinct outputs. This DSVPWM controller is dissimilar from other techniques of PWM, because it generates a vector as a reference. Discontinuity occurs in SVPWM when one of the two zero vectors takes part in the execution of the SVPWM. One leg of the PWMVSI does not switch off throughout the switching period and remains tied up in either the positive or negative DC bus. This is well known as DSVPWM, since the switching is discontinuous. At the time of one switching interval,

TABLE 15.2
Switching Position of Three-Phase System

Switching State	a			b			c		
	S_1	S_2	V_{an}	S_3	S_4	V_{bn}	S_5	S_6	V_{cn}
1	ON	OFF	V_{dc}	ON	OFF	V_{dc}	ON	OFF	V_{dc}
0	OFF	ON	0	OFF	ON	0	OFF	ON	0

switching occurs in two branches: one branch is either tied up to the positive DC bus or -ve DC bus. When zero voltage (0 0 0) is removed, the leg voltage tied to the positive DC bus voltage is $0.5V_{dc}$ and when zero voltage (1 1 1) is eliminated, the leg voltage tied up to negative bus voltage is $-0.5V_{dc}$. Thus, the number of switching is reduced to 2/3. A reference vector is expressed as the reference compensating voltage on the $\alpha\beta$-axis. The ON-OFF position of the switch depends on the existence of the reference voltage vector on the $\alpha\beta$-axis. The odd number of switches like 1, 3, 5 are present at the top and even number of switches like 2, 4, 6 are present at the bottom of PWMVSI. Switches that are turned on mean 1 and switches turned off mean 0. If the top-most switches are turned on, then the top leg of VSI is turned on and a +ve value has been assigned to the terminal voltage. If overall the top-most switches are turned off i.e. zero, this is considered zero value for terminal voltage. Switching positions of three phase system are shown in Table 15.2.

The odd switches are correlative of even switches. Hence the expected combination of switching states are 001, 010, 011, 100, 101, 110. Total six switching stages are there, amongst the number of active stages are six i.e. (v_1–v_6) and remaining are zero switching. Control procedure of DSVPWM consists of four steps expressed as:

Step 1: Computation of reference angle and magnitude of voltage vector
Step 2: Calculation of time period T_a,T_b,T_c for each one
Step 3: Resolution of duty time for both even and odd switches
Step 4: Comparability of duty cycle with the saw-tooth to produce gate pulses

Figure 15.3d shows the steps for implementation of DSVPWM technique.

Step 1: Computation of magnitude and reference angle of voltage vector are depicted as:

$$|V^*| = \sqrt{V_{c\alpha}^{*2} + V_{c\beta}^{*2}} \tag{15.7}$$

$$\theta^* = tan^{-1}\left(\frac{V_{c\beta}^*}{V_{c\alpha}^*}\right) \tag{15.8}$$

where $V_{c\alpha}^*$ and $V_{c\beta}^*$ are the reference comp. voltage in $\alpha\beta$-axis, V^* is the reference vector of voltage, θ^* is the reference angle. Step 2: Determination of sector i.e finding the sector where reference outcome lie down, on computing switching time and switching sequence. The angle of reference voltage has been employed for sector selection process. Step 3: Time duration: V^* makes up two energetic and one inactive vector. For sector1 (0 to $\frac{\pi}{3}$): V^* can be constructed with V_0, V_1, and V_2. V^* can be obtained with respect to time duration as:

$$V_{ref}.\,T = V_1.\,T_1/T_2 + V_2.\,T_2/T + V_0.\,T_0/T \tag{15.9}$$

$$V^* = V_1T_1 + V_2T_2 + V_0T_0 \tag{15.10}$$

T is the total cycle:

$$T = T_1 + T_2 + T_0 \tag{15.11}$$

The location of V^*, V_1, V_2, V_0 is described in regards to magnitude and angle as follows:

$$V^* = V^* r^{j\theta} \tag{15.12}$$

$$V_1 = \frac{2}{3}V_{dc} \tag{15.13}$$

$$V_2 = \frac{2}{3}V_{dc}e^{j\frac{\pi}{3}} \tag{15.14}$$

$$V_0 = 0 \tag{15.15}$$

$$r^{j\theta} = cos\theta + jsin\theta \tag{15.16}$$

$$e^{j\frac{\pi}{3}} = cos\frac{\pi}{3} + jsin\frac{\pi}{3} \tag{15.17}$$

Place the values of V^*, V_1, V_2&V_0 in (20):

$$T * V^* * \begin{bmatrix} cos\theta \\ sin\theta \end{bmatrix} = T_1 * \frac{2}{3}V_{dc}\begin{bmatrix} 1 \\ 0 \end{bmatrix} + T_2 * \frac{2}{3}V_{dc}\begin{bmatrix} cos\frac{\pi}{3} \\ sin\frac{\pi}{3} \end{bmatrix} \tag{15.18}$$

Segregating the real and imaginary terms in (21), $T_1 and T_2$ can be computed as:

$$T_1 = T * \frac{\sqrt{3}\,V^*}{Vdc}sin\left(\frac{\pi}{3} - \theta\right) = T * a * sin\left(\frac{\pi}{3} - \theta\right) \tag{15.19}$$

$$T_2 = T * a * sin\theta, \tag{15.20}$$

$$0 < \theta < \frac{\pi}{3} \tag{15.21}$$

a is the modulation index. The imprecise computation of duty times for n number of sectors is enumerated as:

$$T_1 = T * a * sin\left(\frac{\pi}{3} - \theta + \frac{n-1}{3} * \pi\right) \tag{15.22}$$

$$T_1 = T * a\left[sin\left(\frac{n\pi}{3}\right) * cos\theta - cos\left(\frac{n\pi}{3}\right) * sin\theta\right] \tag{15.23}$$

$$T_2 = T * a\left[-cos\theta * sin\left(\frac{(n-1)\pi}{3}\right) + sin\theta * cos\left(\frac{(n-1)\pi}{3}\right)\right] \tag{15.24}$$

$$T_0 = T - T_1 - T_2 \tag{15.25}$$

Step 3: The estimation of duty time for each sector. Every sector comprises seven switching states for each cycle i.e. sector-1 passes through these switching states: 100 – 110 – 111 – 101 – 110 – 100, one round and again return back. This happens at time T_c and it is divisible among the seven switching states. Two vectors are zero vectors.

$$T = \frac{T_2}{2} + \frac{T_1}{2} + \frac{T_0}{2} + \frac{T_1}{2} + \frac{T_2}{2} \tag{15.26}$$

For sector-1, the switch is ON between $T_0 \& T - \frac{T_0}{2}$ in the first phase, between $\frac{T_0}{4} + \frac{T_1}{2}$ for the second phase, between $\frac{T_0}{2} + \frac{T_1}{2} + \frac{T_0}{2}$ and $T - \left(\frac{T_0}{2} + \frac{T_1}{2} + \frac{T_0}{2}\right)$ for the third phase, between $\frac{T_0}{2} + \frac{T_1}{2} + \frac{T_0}{2} + \frac{T_2}{2}$ and $T - \left(\frac{T_0}{2} + \frac{T_1}{2} + \frac{T_0}{2} + \frac{T_2}{2}\right)$ for the fourth phase and so on. Table 15.3 shows the duty tie for each sector. Step-4: Comparison among T_1, T_2, T_0 with the carrier to generate gate pulses for switches.

15.6 SIMULATION RESULTS FOR HARMONIC COMPENSATION USING SAPF

With regards to the potency of the proposed controller, an advance version of MATLAB® is employed for the simulation purpose of the SAPF system. The objective of the simulation is to analyze the result of existing controller with the proposed controller. In case of transient condition of load, minimum time was taken by the source current to attend steady state in the HSRF-FLC-DSVPWM method

TABLE 15.3
Switching Pulse Pattern for Each Sector of Three-Phase System

Sector	Upper Switches: S_1, S_3, S_5	Lower Switches: S_1, S_3, S_5
1	$S_1 = T_1 + T_2 + T_0$	$S_4 = T_0$
	$S_3 = T_2 + T_0$	$S_6 = T_1 + T_0$
	$S_5 = T_0$	$S_2 = T_1 + T_2 + T_0$
2	$S_1 = T_1 + T_0$	$S_4 = T_2 + T_0$
	$S_3 = T_1 + T_2 + T_0$	$S_6 = T_0$
	$S_5 = T_0$	$S_2 = T_1 + T_2 + T_0$
3	$S_1 = T_0$	$S_4 = T_1 + T_2 + T_0$
	$S_3 = T_1 + T_2 + T_0$	$S_6 = T_0$
	$S_5 = T_0$	$S_2 = T_1 + T_0$
4	$S_1 = T_0$	$S_4 = T_1 + T_2 + T_0$
	$S_3 = T_1 + T_0$	$S_6 = T_2 + T_0$
	$S_5 = T_1 + T_2 + T_0$	$S_2 = T_0$
5	$S_1 = T_b + T_0$	$S_4 = T_1 + T_0$
	$S_3 = T_0$	$S_6 = T_1 + T_2 + T_0$
	$S_5 = T_1 + T_2 + T_0$	$S_2 = T_0$
6	$S_1 = T_1 + T_2 + T_0$	$S_4 = T_0$
	$S_3 = T_0$	$S_6 = T_1 + T_2 + T_0$
	$S_5 = T_1 + T_0$	$S_2 = T_2 + T_0$

with respect to HSRF-PI-DSVPWM. The THD response of HSRF-FLC-DSVPWM based SAPF is very little in comparisons to the THD of SAPF with existing controller. Figures 15.4 and 15.5 correspondingly display the waveforms for supply voltage (Vs), load current (IL), source current (Is), and DC-link voltage for steady state condition of load. The source current without a filter looks like the load current. In transient condition, a sudden on/off of the load is taken into account. The creation of a transient state occurs by sudden switching on/off of load from t = 0.01 s to t = 0.05 s, respectively. The corresponding waveforms for both controllers are shown in Figures 15.6 and 15.7. It is observed from the results that the proposed controller is able to enforce the compensating voltage to track its reference very quickly with respect to change in load. But in the existing controller the process of reference tracking takes about one cycle for the DC-link voltage to track its reference value. Thus, it proves that the proposed controller-based SAPF system has more potency and also better efficiency for harmonic neutralization. Table 15.4 enlists the THD of different control strategies of the SAPF system i.e. HSRF-FLC-DSVPWM employing simulation results under both steady state as well as transient condition. The last column of Table 15.4 shows that the HCR factor maintained nearly equal but the deviation is less (0.28%) in the case of HSRF-FLC-DSVPWM in comparison to HSRF-PI-DSVPWM (0.92%). Values of different parameters used in experimental was is shown in Table 15.5.

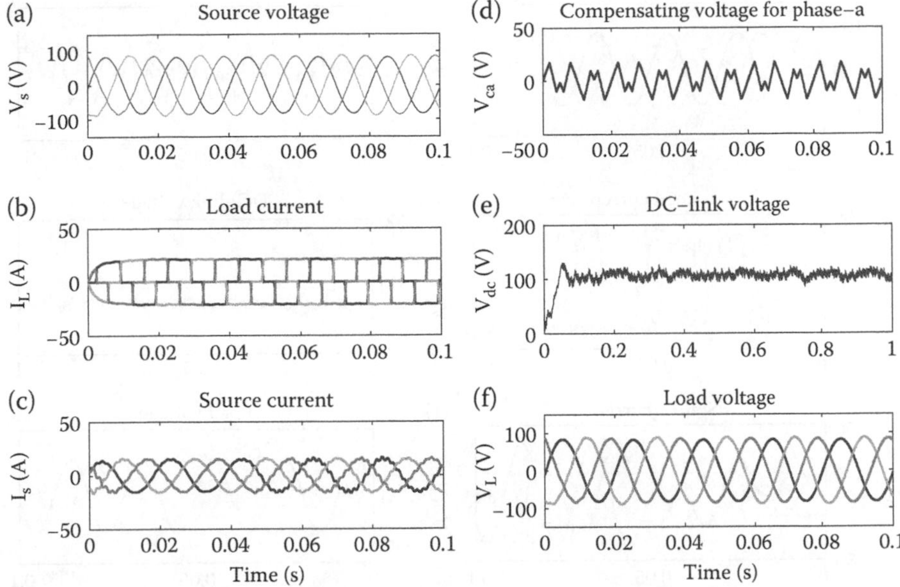

FIGURE 15.4 Steady-state response of HSRF-PI-DSVPWM control approach based SAPF system (a). Source Voltage (b). Load current (c). Source current (d). Compensating voltage for phase–a (e). DC-link voltage (f). Load voltage.

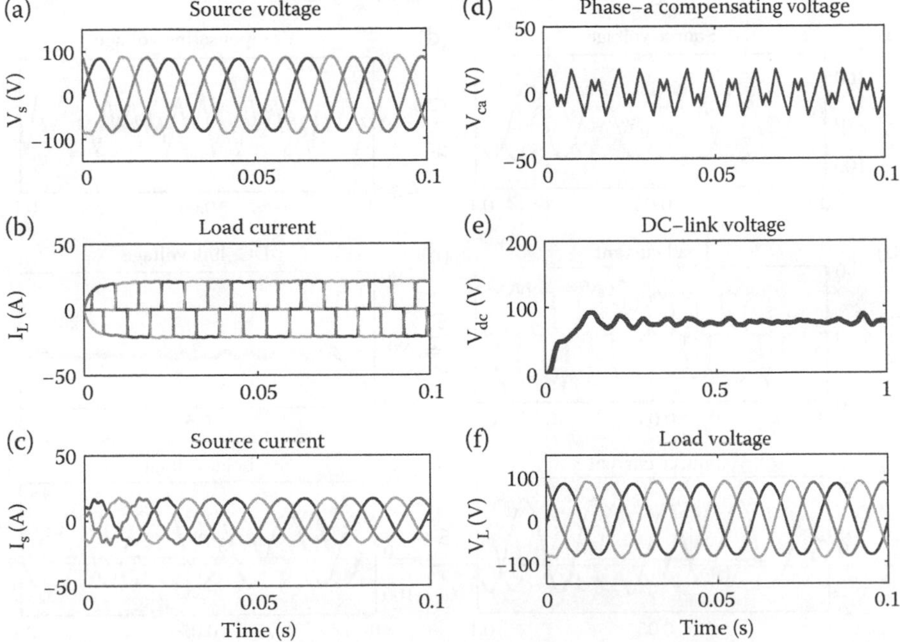

FIGURE 15.5 Steady-state response for SAPF system employing HSRF-FLC-DSVPWM control approach (a). Source Voltage (b). Load current (c). Source current (d). phase-a Compensating voltage (e). DC-link voltage (f). Load voltage.

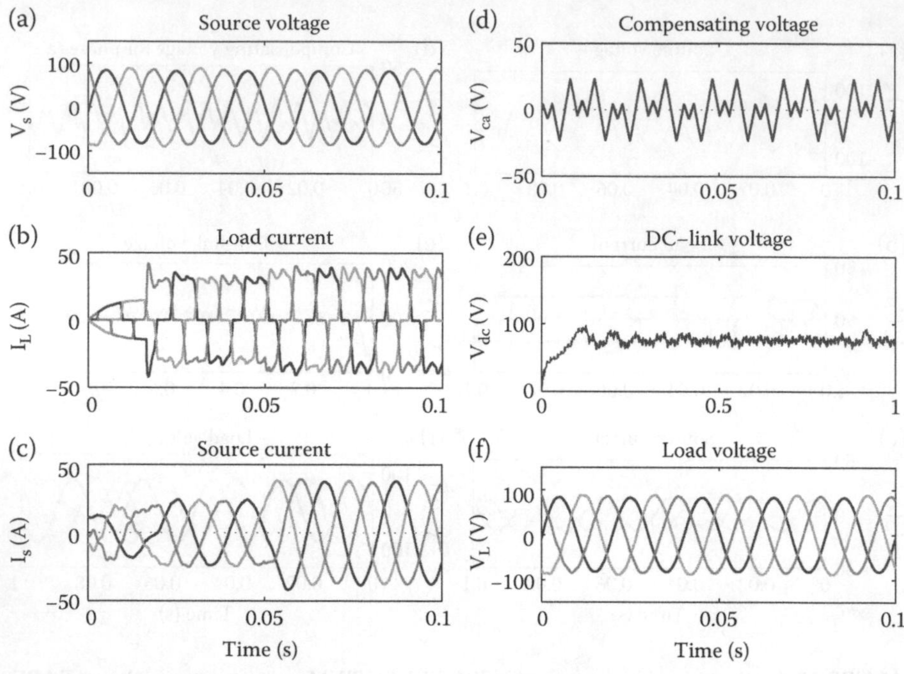

FIGURE 15.6 Response of SAPF system employing HSRF-PI-DSVPWM control approach under transient condition of load (a). Source Voltage (b). Load current (c). Source current (d). Compensating voltage (e). DC-link voltage (f). Load voltage.

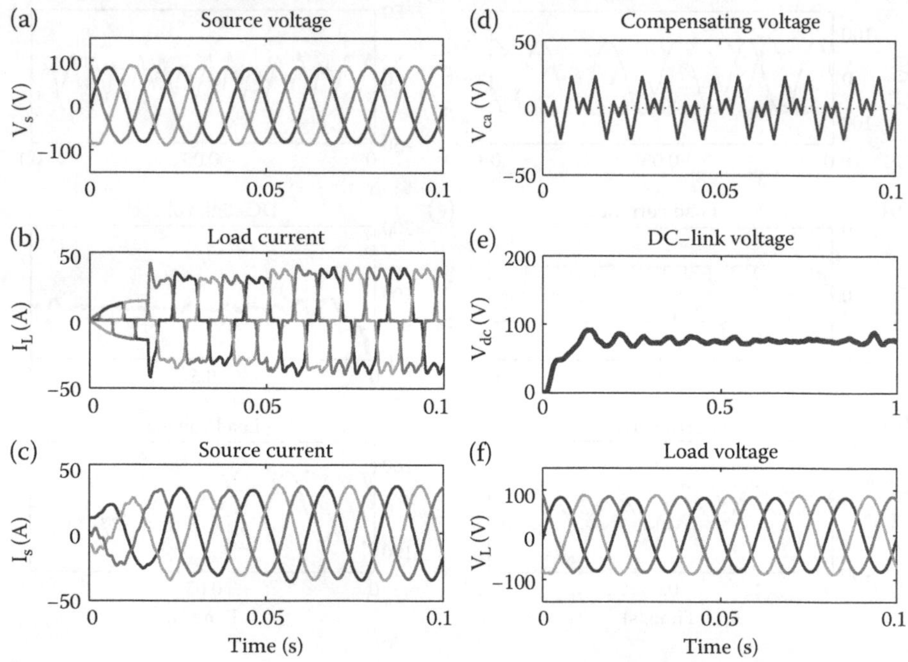

FIGURE 15.7 Response of SAPF system employing HSRF-FLC-DSVPWM control approach under transient condition of load (a). Source Voltage (b). Load current (c). Source current (d). Compensating voltage (e). DC-link voltage (f). Load voltage.

TABLE 15.4
THD of Control Strategies of Hsapf System (Simulation)

HSAPF Control Structure	Cases	Before Compensation THD (%)	After Compensation THD (%)	HCR (%)	HCR (SS)-HCR (TS)
HSRF-PI-DSVPWM	TS	25.7%	3.31	12.87	2*0.92
	SS	23.2%	3.2	13.79	
HSRF-FLC-DSVPWM	TS	25.7%	1.7	6.61	2*0.28
	SS	23.2%	1.6	6.89	

TABLE 15.5
Experimental Parameters

Line voltage and frequency	$V_s = 100V\,(RMS),\, f_s = 50Hz$
Line impedance	$L_s = 0.5mH,\, R_s = 0.1\Omega$
Current source type of nonlinear load impedance	$L_L = 20mH,\, R_L = 50\Omega$
Series active filter parameter	$C_f = 60\mu F,\, L_f = 1.35mH,\, C_{dc} = 2200\mu F$
Shunt passive filter parameter	$L_{pf} = 1.86mH,\, C_{pf} = 60\mu F$

It envisages from Table 15.4 that the HSRF-FLC-DSVPWM based SAPF is more preferable than the HSRF-PI-DSVPWM based SAPF for harmonic compensation. The HCR factor is calculated as follows:

$$HCR = \frac{THD\,(\%)\,AfterCompensation}{THD\,(\%)\,BeforeCompensation} * 100\% \tag{15.27}$$

15.7 EXPERIMENTAL RESULTS

The proposed algorithm of the controller has been executed through a laboratory prototype developed in our research lab, as shown in Figure 15.8(a). The effectiveness of the proposed algorithm is investigated with HSRF-FLC-DSVPWM. The HSRF with FLC is employed as the reference generation approach in the SAPF system. The experimental studies for steady-state condition of load are conducted and scrutinized as follows:

CASE 1 (Balanced supply voltage with steady state condition of load):

Figure 15.8 shows the hardware results of load current and source current after compensation for steady-state condition of the load. Figure 15.8 shows the behavior of the steady-state response of source currents for controlling strategy, by name, the HSRF approach in the SAPF system. But it is noticed from the result of the source current that

(a)

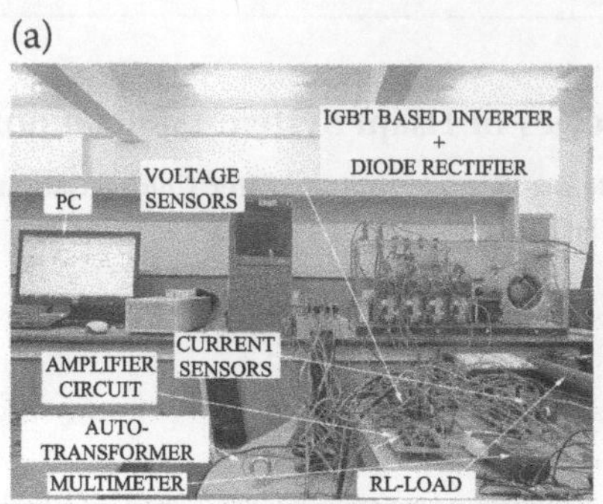

(b) (c)

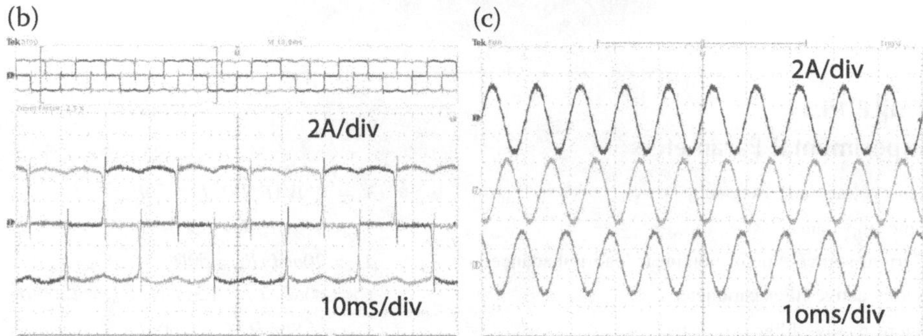

FIGURE 15.8 (a) Experimental setup of SAPF; (b) steady-state response of load current before compensation; (c) steady-state response of source current after compensation.

(a) (b)

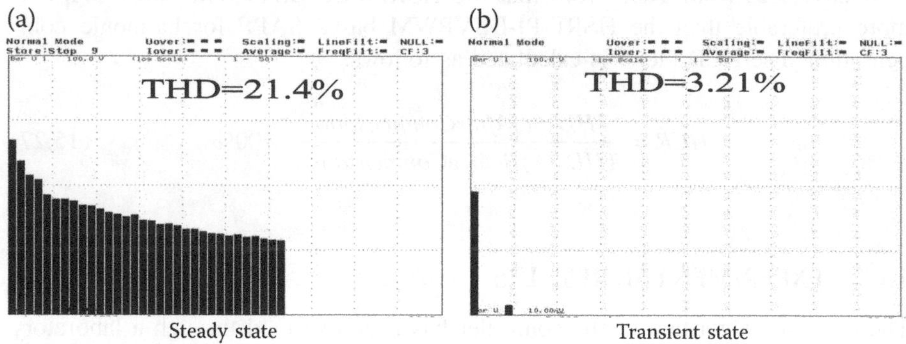

FIGURE 15.9 THD of source current (a). Under steady state condition before compensation; (b). Under transient state after compensation.

a Total Harmonic Distortion (THD) of 21.4% before compensation shown in Figure 15.9. However, it decreases to 3.21% after compensation with the HSRF, respectively, as depicted in Figure 15.9. The hardware realization proves that the HSRF-FLC-DSVPWM technique gives more attention on suppression of the voltage fluctuations across the dc capacitor, providing efficient harmonic elemnation performance.

It is justified from the study of laboratory prototype that the proposed HSRF-FLC-DSVPWM based SAPF has a superior steady-state response. It has been observed from the result of HSRF-FLC-DSVPWM based sysem that the result of the proposed control approach is mindblowing over the conventional one.

15.8 CONCLUSIONS

This chapter proposed a HSRF-FLC-DSVPWM based SAPF system. This SAPF minimizes harmonic components by incorporating proposed the HSRF with the FLC control algorithm. The HSRF-FLC-DSVPWM based SAPF has been elevated as very simple in construction and also attains better performance in comparison with different grid perturbations. But there is a time delay in the foregoing cases. Thus, SAPF gives a slow response. Hence, harmonic cancellation is fast and is not so absolute. The THD observed from simulation and experimental results is just touching the boundary line of IEEE-519 standard. Therefore, a HSRF-FLC-DSVPWM based SAPF system has been built. So it is noticed from the hardware as well as the simulation result that even if the load current is non-linear in nature, the HSRF-FLC-DSVPWM method has a superior performance in cancellation of harmonics than that of the HSRF-PI-DSVPWM. The proposed HSRF-FLC-DSVPWM strategy is very effective and has faster grid perturbations in comparison to the HSRF-PI-DSVPWM technique.

REFERENCES

[1] D. Rivas, L. Moran, J. W. Dixon, and Espinoza J. R., "Improving passive filter compensation performance with active techniques," *IEEE Transactions on Industrial Electronics*, vol. 50, no. 1, pp. 161–170, 2003.

[2] H.-L. Jou, J.-C. Wu, and K.-D. Wu, "Parallel operation of passive power filter and hybrid power filter for harmonic suppression," *IEE Proceedings-Generation, Transmission and Distribution*, vol. 148, no. 1, pp. 8–14, 2001.

[3] M. A. Mulla, C. Rajagopalan, and A. Chowdhury, "A novel control method for series hybrid active power filter working under unbalanced supply conditions," *International Journal of Electrical Power & Energy Systems*, vol. 64, pp. 328–339, 2015.

[4] P. Prabhakaran and V. Agarwal, "Novel boost-SEPIC type interleaved DC-DC converter for mitigation of voltage imbalance in a low-voltage bipolar DC microgrid," *IEEE Transactions on Industrial Electronics*, pp. 6494–6504, vol. 67, no. 8, 2020.

[5] S. Pirouzi, J. Aghaei, T. Niknam, M. H. Khooban, T. Dragicevic, H. Farahmand, M. Korpås, and F. Blaabjerg, "Power conditioning of distribution networks via single-phase electric vehicles equipped," *IEEE Systems Journal*, vol. 13, no. 3, pp. 3433–3442, 2019.

[6] R. Gregory, C. Azevedo, and I. Santos, "Study of harmonic distortion propagation from a wind park," *IEEE Latin America Transactions*, vol. 18, no. 6, pp. 1077–1084, 2020

[7] S. D. Swain, P. K. Ray, and K. B. Mohanty, "Improvement of power quality using a robust hybrid series active power filter," *IEEE Transactions on Power Electronics*, vol. 32, no. 5, pp. 3490–3498, 2016.

[8] A. P. Kumar, G. Siva Kumar, and D. Sreenivasarao, "Model predictive control with constant switching frequency for four-leg DSTATCOM using three-dimensional

space vector modulation," *IET Generation, Transmission Distribution*, vol. 14, no. 17, pp. 3571–3581, 2020.

[9] S. Saad and L. Zellouma, "Fuzzy logic controller for three-level shunt active filter compensating harmonics and reactive power," *Electric Power Systems Research, Elsevier*, vol. 79, no. 10, pp. 1337–1341, 2009.

[10] M. K. Rathi, and N. R. Prabha, "Interval type-2 fuzzy logic controller-based multilevel shunt active power line conditioner for harmonic mitigation," *International Journal of Fuzzy Systems, Springer*, vol. 21, no. 1 pp. 104–114, 2019.

[11] V. Narasimhulu, and Ch Sai, "Computational intelligence based control of cascaded H-bridge multilevel inverter for shunt active power filter application," *Journal of Ambient Intelligence and Humanized Computing, Springer*, pp. 1–9, 2020. doi: https://doi.org/10.1007/s12652-019-01660-0.

[12] P. K. Ray and S. D. Swain, "Performance enhancement of shunt active power filter with the application of an adaptive controller," *IET Generation, Transmission & Distribution*, vol. 14, no. 20, pp. 4444–4451, 2020.

[13] R. Kumar, H. O. Bansal, and H. P. Agrawal, "Development of fuzzy logic controller for photovoltaic integrated shunt active power filter," *Journal of Intelligent & Fuzzy Systems*, vol. 36, no. 6, pp. 6231–6243, 2019.

[14] H. Akagi, E. H. Watanabe, and M. Aredes, *Instantaneous Power Theory and Applications to Power Conditioning*. Wiley-IEEE Press, 2017.

[15] S. Hou, J. Fei, Y. Chu, and C. Chen, "Experimental investigation of adaptive fuzzy global sliding mode control of single-phase shunt active power filters," *IEEE Access*, pp. 64442–64449, 2019. doi: 10.1109/ACCESS.2019.2917020.

16 Artificial Intelligence Application for HVDC Protection

Zahra Moravej[1], *Amir Imani*[1], *and Mohammad Pazoki*[2]
[1]Faculty of Electrical and Computer Engineering, Semnan University, Iran
[2]School of Engineering, Damghan University, Iran

16.1 INTRODUCTION

Artificial intelligence (AI), or machine intelligence (MI), is one of the most popular current fields of study into which a large number of scientists carry on research. As a simple definition, AI refers to the simulation of human intelligence in machines that are programmed to think like humans and mimic their actions [1]. In recent years, AI has had numerous applications in extensive scientific studies such as economic forecasts, diagnosis in medical sciences, design of self-driving cars, and different engineering fields. Electrical power system studies with inherent complex nonlinear behavior made the experts utilize AI tools to solve different problems [2,3]. In recent decades, machine intelligence has been of use in many of these studies such as power systems' planning, analysis, monitoring, forecasting, operation, restoration, fault diagnosis, protection, etc. [4].

In recent decades smart power system protection deployment became facilitated thanks to signal processing and AI development. However, numerous challenges are emerging in this path. The first issue is about collecting raw online data of real-time grid status, which is done by intelligent electronic devices (IEDs) installed in different nodes [5,6]. Also, self-power non-intrusive sensors are being utilized in smart online monitoring and measurement. In addition, high-speed data transfer was necessary for communication-based protection methods carried out with the aid of a global positioning device (GPS) [7,8]. Furthermore, the computational capability of intelligent relays is a challenging aspect of smart protection to be investigated. Obviously, higher computation complexity leads to better fault identification performance e.g. appropriate discrimination and selectivity.

The main focus of this chapter is to investigate the application of AI approaches in the field of HVDC protection. In HVDC studies, protection issues are considered a more challenging and popular topic among the researchers in comparison with AC systems due to several reasons. For instance, the importance of fault identification

DOI: 10.1201/9780367552374-16

speed must be highlighted due to the inherent fast dynamics of HVDC systems (low inertia) [9]. In addition, high sensitivity of converter valves to the sudden variations of voltage and current is significant.

As the researchers study application of AI-based scheme protection in general, in-depth review of various methods for HVDC fault detection, classification, and location has been introduced in this chapter, individually. In Section 16.1.1, AI-based protection structure is discussed. An introduction of HVDC technology is proposed in Section 16.2. Section 16.3 describes the HVDC protection challenges and fault analysis. AI-based HVDC fault detection, classification, and location are provided in Sections 16.4, 16.5, and 16.6, respectively. Commutation fault identification is propounded in Section 16.7. A discussion of the advantages and disadvantages of the various investigated AI techniques in protection is presented in Section 16.8. Finally, Section 16.9 concludes this chapter.

16.1.1 Protection Tools Based on Artificial Intelligence

Since the main purpose of power system protection is to keep the power system stable by isolating the faulty section of line, fault identification in the minimum time is necessary. Emergence of smart numerical relay in the 1980s led to the introduction of AI and pattern recognition application to power system protection. The most important application of AI in protection is fault identification containing fault detection, classification, and location. The fault identification process contains three basic stages, shown in Figure 16.1.

16.1.1.1 Generation

The measured voltage and current are considered as inputs to AI-based protection. These sampled data are utilized in two forms. In some cases, raw data are directly

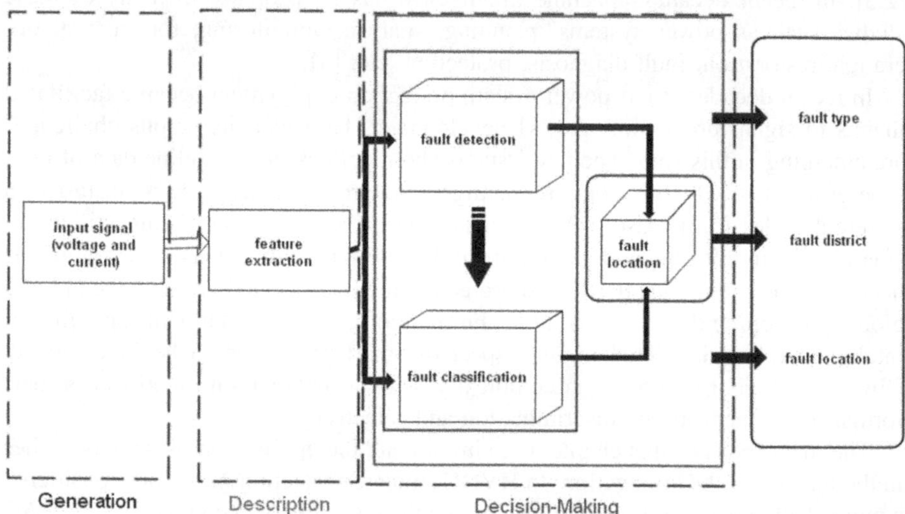

FIGURE 16.1 AI-based protection scheme.

applied as input features, without any preprocessing. Although this method has the privilege of simplicity, it suffers from large data volume and low generalization ability due to the presence of outlying vague data. In addition, in most of the cases, input data preprocessing is carried out to eliminate some unwanted attributes, such as noise effect using different types of filters. For instance, the Butterworth filter is utilized as a low-pass filter for strong distinction of fundamental components of input signals [10].

16.1.1.2 Description

In this stage, the feature extraction must be done using various processing tools. One of the most common approaches is employing signal transforms in three domains, which are frequency, time, and time-frequency. These signal transforms help to extract some helpful features of signals. The most popular transform in the frequency domain is Fourier transform (FT) and also its subroutines like discrete Fourier transform (DFT) and fast Fourier transform (FFT). Wavelet transforms (WT) and Stockwell transforms (ST) are members of the time-frequency family, and the third category denotes time domain transform e.g. empirical decomposition mode (EMD) [11]. After extracting different features of input faulty signals, dimensionality reduction is vital. Principle component analysis (PCA) is used for data mapping, including low- and high-dimensional subset to mitigate data dimensionality.

16.1.1.3 Decision Making

The main focus of this chapter is on the final stage of the AI-based protection scheme, which is called "decision making." This part of the training process contains fault detection, classification, and location using AI-based approaches. AI is one of the extensive fields of study consisting of several subsets that have many applications in various branches of science. Figure 16.2 shows a type of classification of the AI-based method according to the approaches. It is worth mentioning that in power system fault identification, some AI-based procedures are frequently applied for classification and prediction, such as artificial neural network (ANN), support vector machine (SVM), Fuzzy, adaptive neuro-fuzzy inference system (ANFIS), etc., which will be investigated in the following sections.

16.2 OVERVIEW OF HVDC TECHNOLOGY

In 1950, the first modern HVDC system was established in Sweden and the island Gotland[12]. Afterwards, the gradual development of valve technology and evolution of mercury valves led the experts to use HVDC systems increasingly, due to its attractive specifications compared with conventional high voltage alternative current (HVAC) grids such as [12]:

- Lower power loss and cost in long transmission distance
- Lesser number of conductors
- Asynchronous grid connection capability
- More controllability and flexibility of power flow

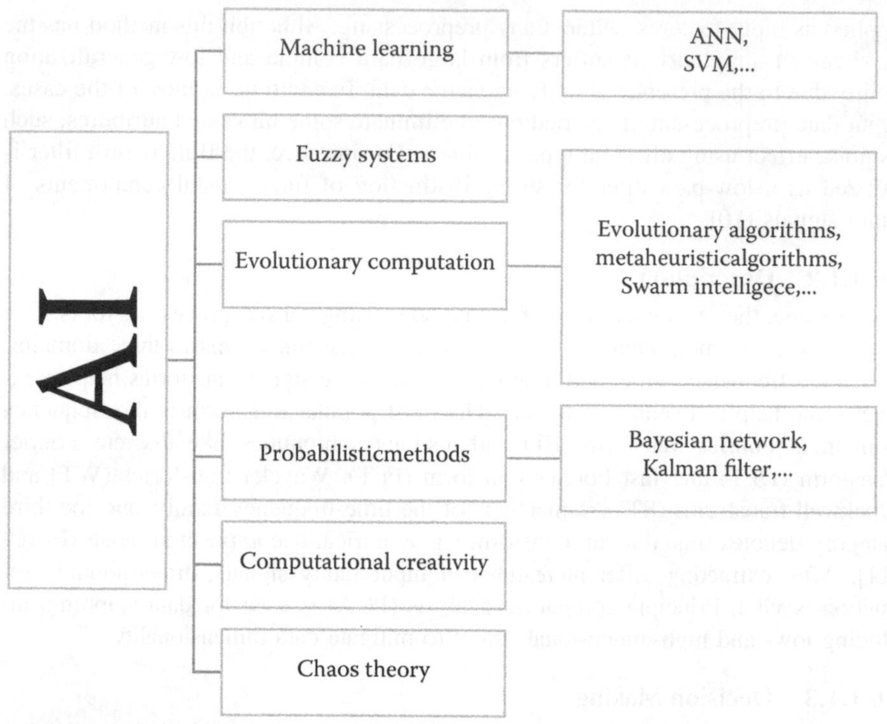

FIGURE 16.2 AI-based approaches.

- More sustainability for renewable energy source connection e.g. wind turbines
- No distance effect on stability
- No rise in reactive power demand

Early converter generation of HVDC was thyristor-based and called line commutated converters (LCC) technology with various advantages such as high power capability, overload withstand ability, etc. However, in recent years, voltage source converter (VSC) HVDC has emerged, which is a transistor-based (IGBT, GTO etc.) technology with some new attractive features e.g. high controllability (independent active and reactive control), modularity, etc. [12,13]. In the case of configuration, there are different topologies.

Most of the current HVDC systems have point-to-point topology. It is also envisaged that in the future, several offshore wind farms and existing HVDC lines will be interconnected to form a multi-terminal HVDC (MTDC) configuration, also termed "HVDC grids." Noted advantages of HVDC grids are optimal equipment utilization, flexibility, reliability, and cost reduction [14]. It will also pave the way for energy trading among regional countries [15]. The key milestones of HVDC technology are depicted in Figure 16.3.

1950: First modern HVDC system

1969: First HVDC system with solid state switch

1999: First voltage source converter in HVDC system (VSC-HVDC)

2009: First voltage source converter for offshore wind park

2011: First modular multilevel converter (MMC)

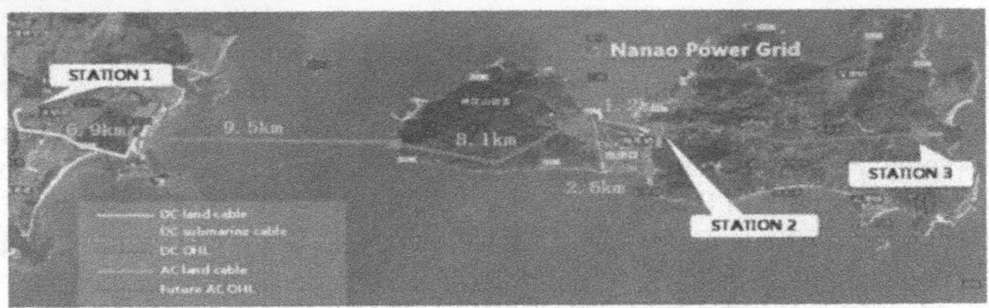

2013: First multi terminal VSC-HVDC

FIGURE 16.3 History of HVDC technology.

16.3 HVDC PROTECTION

Fault occurrence in a power system can be perilous not only for the system, but also for humans and components. Hence, it has to get cleared by the protection system in the minimal time. The major features of an efficient protection system include being fast, reliable, selective, independent, recoverable, and fail-safe [16]. Despite the numerous similarities between HVDC network protection system and HVAC, they have structural contrasts. The sensitivities and limitations of HVDC systems are more intense and stricter. This is because the fault current characteristics are different, and also that power electronic converters are highly sensitive to very high fault current in these systems [17]. Generally, in comparison with AC systems, DC fault current increases exponentially and just in some milliseconds (4–6 ms) reaches destructive levels, which is a result of low inductance in DC lines, while in HVAC systems, inductance and low rate of rise of fault current cause the protection system to function in 25–100 ms [18]. Figure 16.4 shows a comparative overview of the operation time of different parts of AC and DC networks. Another challenge is the lack of zero-crossing points in DC line protection systems, having been the most central issue in designing and developing HVDC circuit breakers in the last few decades [19,20]. The best time to clear the fault current is at zero-crossing points, and in AC systems there are at least 100 zero-crossing points per second. According to the above-mentioned explanations, designing and implementing ultra-fast protection systems seem to be essential for detecting and clearing faults.

In summary, the main challenges of protection in HVDC lines can be listed as below:

- Low impedance of DC line due to the lack of inductance (10 times resistive term in AC lines), leading to high rate of rise of fault current
- Lack of zero-crossing point to interrupt fault current
- High sensitivity of power electronic valves to sudden over voltage and current
- Complicated fault location due to low line impedance

HVDC protection system is divided into AC and DC sections; the AC section includes protection of converters transformers, AC filters, shunt capacitors and reactors, and busbars. And the DC section involves the protection of converters, smoothing reactors, and DC filters and lines [21]. Figure 16.5 depicts the most-used protection functions in HVDC systems.

Furthermore, Table 16.1 implies a statistical evaluation of DC fault variations and the possibility of their occurrence. It has to be mentioned that in this chapter DC protection link of HVDC lines will be focused on. Thus, DC fault phenomena will be analyzed to determine the contribution of different components in feeding fault current.

16.3.1 DC FAULT PHENOMENA

At the instant of fault occurrence in a DC line, the voltage of DC link declines drastically, and all sources of fault current start to contribute feeding short-circuit

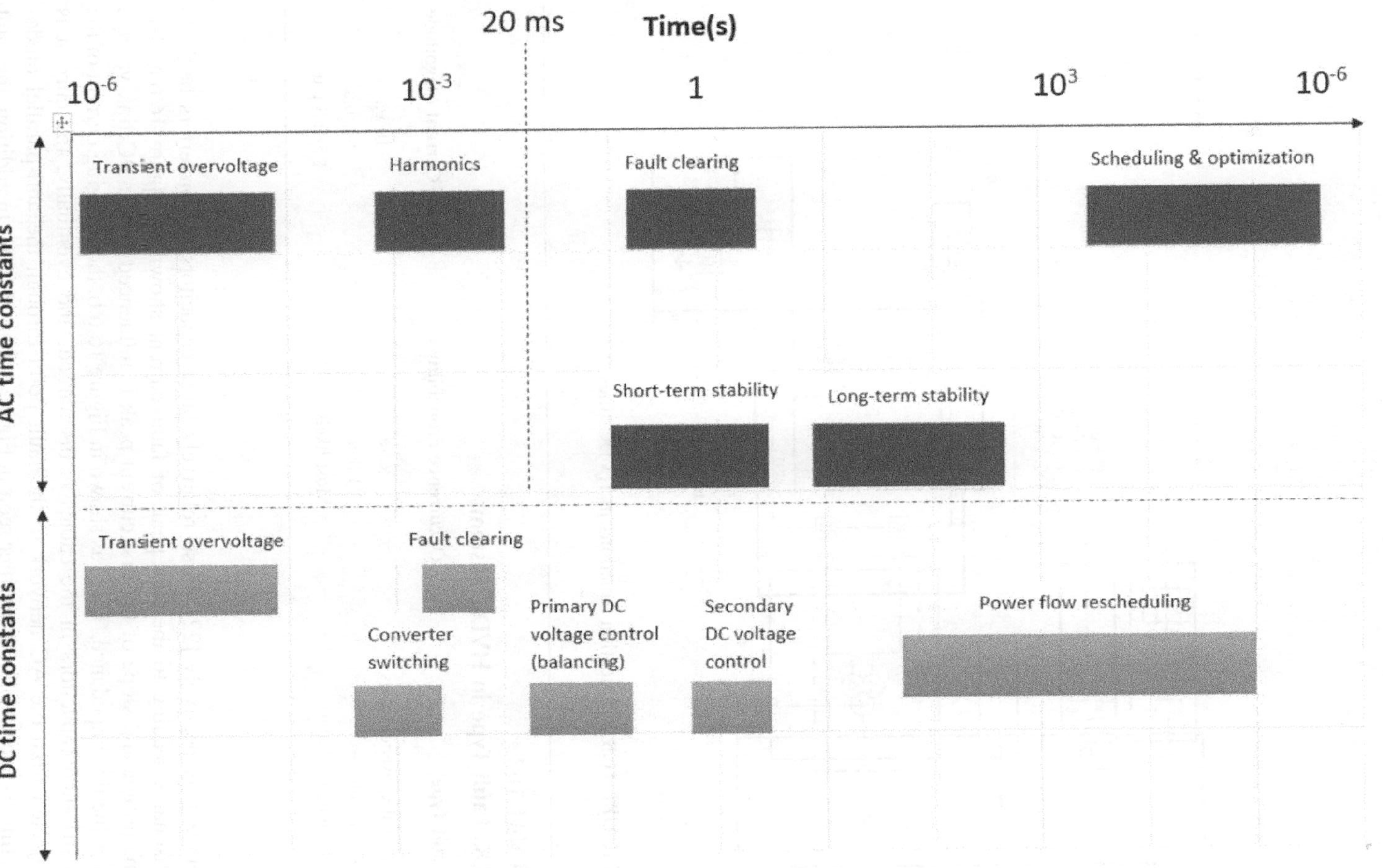

FIGURE 16.4 Cperation time of different parts of AC and DC grids.

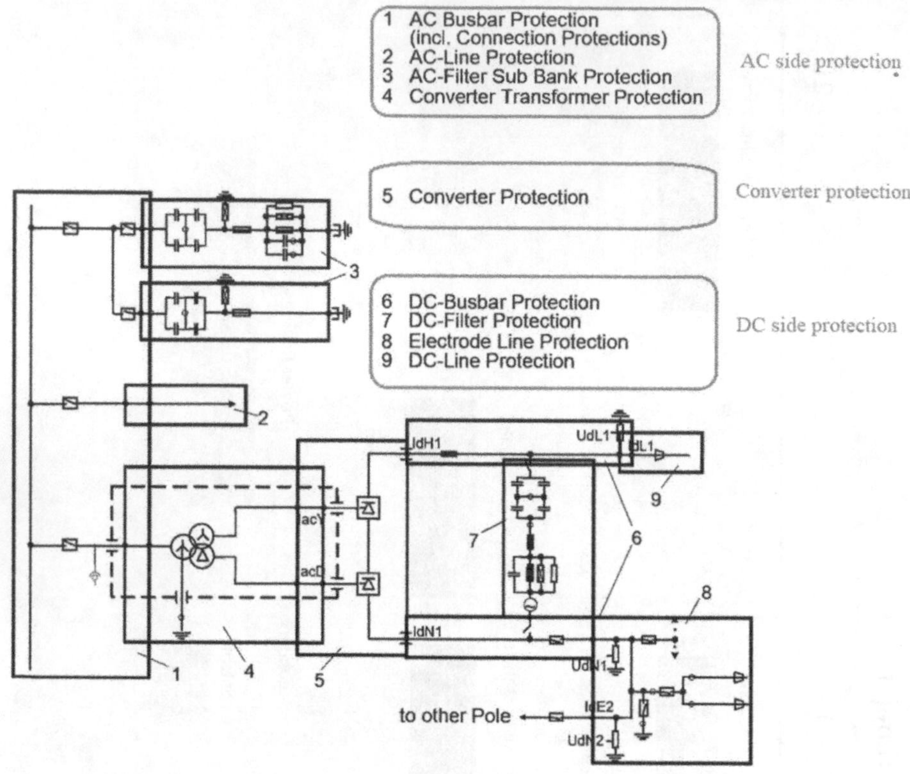

FIGURE 16.5 Protection functions in HVDC lines.

TABLE 16.1
DC Fault Type in HVDC Systems

Fault Type	Occurrence Probability	Fault Current Magnitude
Internal converter faults	Rare	10 p.u
DC line faults	Likely	2–3 p.u
Commutation failure	Most likely	1.5–2.5 p.u

current, respectively [22]. First, relatively large capacitors in converters begin to discharge leading to the first peak of fault current shown in Figure 16.6.a. The magnitude and slope of the first current peak is influenced by the DC link voltage and the line impedance. Then, as shown in Figure 16.6.b, the charging current of the distributed capacitors in the cables is discharged, and eventually, the current is injected from the AC network to the fault point through the anti-parallel diodes acting as uncontrollable rectifier seen in Figure 16.6.c [18,23]. In addition, the fault

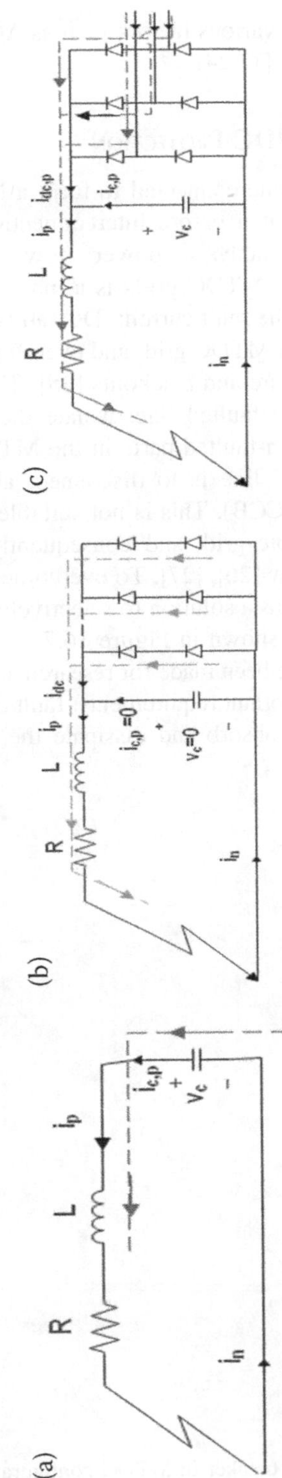

FIGURE 16.6 Equivalent circuits for fault current contribution (a) converter's capacitor discharge; (b) freewheel diode phase; and (c) AC grid feeding.

current amplitude is affected by various factors such as AC short-circuit ratio, cable impedance, and fault resistance [17,24].

16.3.2 Multi-Terminal HVDC Protection

Future HVDC scheme will be interconnected to form a MTDC grid. Four aspects need to be thoroughly investigated before interconnection of multiple DC lines: system integration, dynamic analysis, power flow control, and protection. Therefore, DC fault handling in MTDC grids is a major challenge due to its fast dynamics and the rapid rise of the fault current. DC faults propagate from the fault location to other healthy parts of MTDC grid, and even further to the tied AC grids, which may cause cascading failure and blackouts [25]. Thus, methods are required to identify and discriminate the faulted line, isolate the DC fault, and maintain power flow continuity in the non-faulted parts in the MTDC system. The existing protection of point-to-point DC line is to disconnect all tied DC systems with opening AC circuit breakers (ACCB). This is not suitable for DC grids because of the de-energization of the whole grid, and consequently, the considerable time needed to re-establish power flow [26], [27]. To overcome such an issue, DC circuit breakers (DCCB) are suggested as a solution to selectively isolate the faulty section during DC line short circuit, as shown in Figure 16.7.

Therefore, several efforts have been made for research and development of HVDC switch breakers to meet two important requirements: fault current interruption in few milliseconds, and the ability to absorb and dissipate the huge amounts of energy stored in DC systems [18].

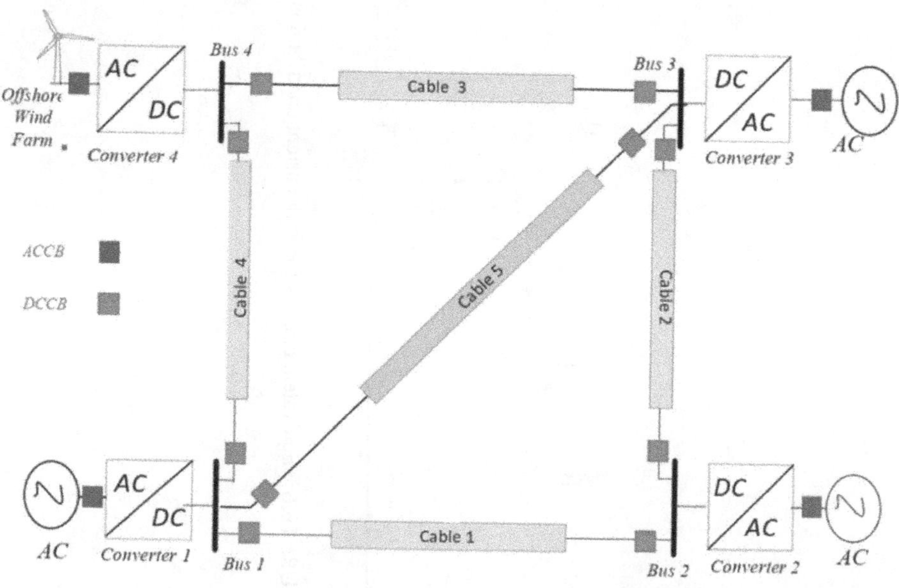

FIGURE 16.7 DC and AC circuit breaker in MTDC configuration.

16.4 AI-BASED FAULT DETECTION

As mentioned in the previous section, the fault characteristics of HVDC lines have a rapid rate of rise of fault current. This roots in the inherent structure of the DC line and lack of inductive property, causing damage to expensive equipment and converters. The fault characteristics of HVDC lines, due to the inherent structure and lack of inductive property in the line, have a rocketing rate of rise of fault current, which causes damage to expensive equipment and converters. Therefore, the first step in implementing a protection system proper to the requirements of the systems, and clearing faults from the lines as quickly as possible, is the ability to detect faults in the shortest time possible.

By and large, there are several approaches for fault detection in HVDC lines illustrated in Figure 16.8, which rely on methods such as traveling waves, differential protection, methods based on signal processing, hybrid methods, etc. However, in studies related to fault detection, other categories such as protections with or without telecommunication network methods have also been provided [28]. Since the speed of detection is quite determinative and effective, the traveling wave method is mostly used in current protection in HVDC lines. Also, application of protection studies using AI methods is more focused on fault classification and location. Later, some studies related to fault detection in HVDC lines using AI-based methods will be investigated.

ANN is a mathematical model inspired by biological neural networks, widely used in HVDC fault detection, classification, and location. This technique has outstanding characteristics such as generalization capability, noise immunity, robustness, and fault tolerance [29].

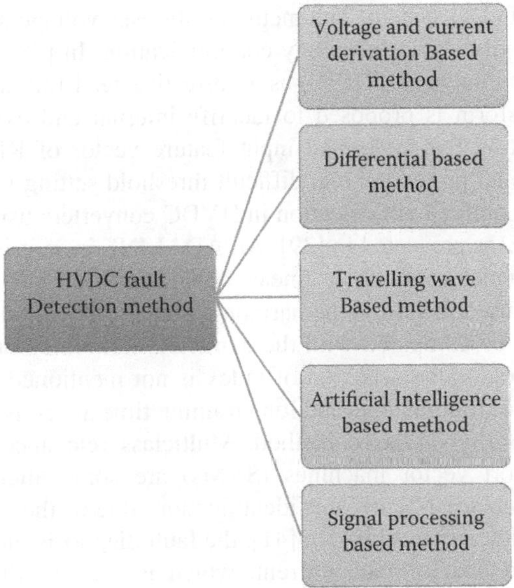

FIGURE 16.8 An overview of HVDC fault detection approaches.

"Networks" in "neural networks" refer to linkage between neurons from different layers. The higher the intricacy of problem is, the higher the requirement for number of hidden layers is. This leads to higher computational burden to the hardware. Thus, making a compromise between accuracy and complexity in AI-based method is obviously considered inevitable [30]. In addition, to develop relaying schemes, precise mathematical models of system in fault condition must be constructed. However, compared with conventional formal approaches, ANN can be trained to recognize non-linear relationships between input and output data without requiring knowledge of their internal processes [31].

Fault detection approaches containing ANN have been used to detect faults on HVDC systems. The main procedure of the method is training the system by creating numerous fault scenarios to reduce the error of location. They also use multilayer feed-forward networks and associated training, using error back-propagation (BP) [32,33]. Various studies have been conducted using neural networks.

In [34,35] a robust, ANN-based technique of fault detection and classification is proposed for a hybrid non-homogenous MTDC. The extracted features are the relationships between high and low frequency energy obtained via DFT. Also, the rate of change of fault current has been applied to ANN classifier for better discrimination of fault classification in [34]. An additional preprocessing stage before learning has been added to reduce the size of NN classifier and consequently to increase training process speed in [35].

In [36], a feature extraction technique is proposed, named independent component analysis (ICA), to overcome the noise effect on the ANN training process. In other words, ICA feature extraction is applied as an input of wavelet-based protection.

Merlin et al. in [37] applied 13 ANNs for the system to detect a fault condition in the two-terminal HVDC system. The method only uses voltage signal measured at the rectifier side and does not need any communication. In [38], a combination of radial basis function network (RBFN) as an effective feed-forward neural network (FNN) and S transform is proposed to identify internal and external fault in the system. By exploiting S transformed input feature vector of RBFN, some shortcomings of traditional protection e.g. difficult threshold setting has been solved. A novel approach for online fault detection in HVDC converters using adaptive linear neuron (ADALINE) is presented in [39]. An ADALINE is an n-input single output neural network, which outputs a linear combination of its inputs [39]. An ADALINE can be used to follow the harmonic content of a signal online. Then an index is defined to be compared with the extracted harmonic content for fault detection. The threshold value selection of index is not mentioned in the paper.

A large number of training data and long training time are considered as the main intrinsic drawbacks of NN-based method. Multiclass relevance vector machines (RVMs) and support vector machines (SVMs) are some alternative intelligent methods that can replace NN in fault identification, due to the fast training speed and regularized characteristics [40]. In [41], the fault diagnosis model is established with S transformed data of fault current, which is used as the input to SVM. Different fault scenarios such as commutation failure (CF), DC, and AC fault were simulated to investigate the validation of the method. A comprehensive study on

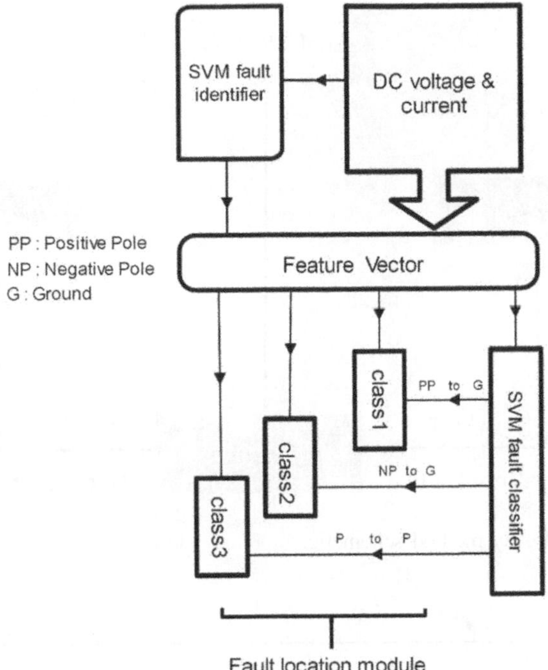

FIGURE 16.9 SVM-based fault identification method in bipolar HVDC system.

fault detection, classification, and location using the multi-class SVM is proposed in [42]. The fault detection criteria are according to continuous measurement of the time domain rectifier end AC rms voltage and DC voltage and current signals fed to the SVM-based fault detection module. A binary classification is carried out using SVM where the presence of a fault is indicated by "1" and the absence of fault is indicated by "0." The block diagram of the proposed method is shown in Figure 16.9.

A fuzzy logic–based method is constructed according to if-then rules. The main concept underlying fuzzy logic is that of a linguistic variable, which is a variable whose values are words rather than numbers. In summary, fuzzy logic is a rule-based approach where a set of rules are used to make the relevant decision. As shown in Figure 16.10, three basic stages of fuzzy inference system (FIS) are fuzzification, interference, and defuzzification [33]. In [43], fault detection and classification are proposed based on two separated FIS in the two-terminal HVDC. Regarding the above advantages and disadvantages of ANN and fuzzy logic, several studies have been carried out to combine fuzzy systems with neural networks to create a hybrid system named ANFIS. In [44], ANFIS-based fault identification method for converter faults of HVDC is proposed, being able to provide discrete indication of converter faults such as CF and arc through/fire through of a valve within the converter.

Table 16.2 illustrates the aforementioned papers, which have used AI-based application in HVDC fault detection.

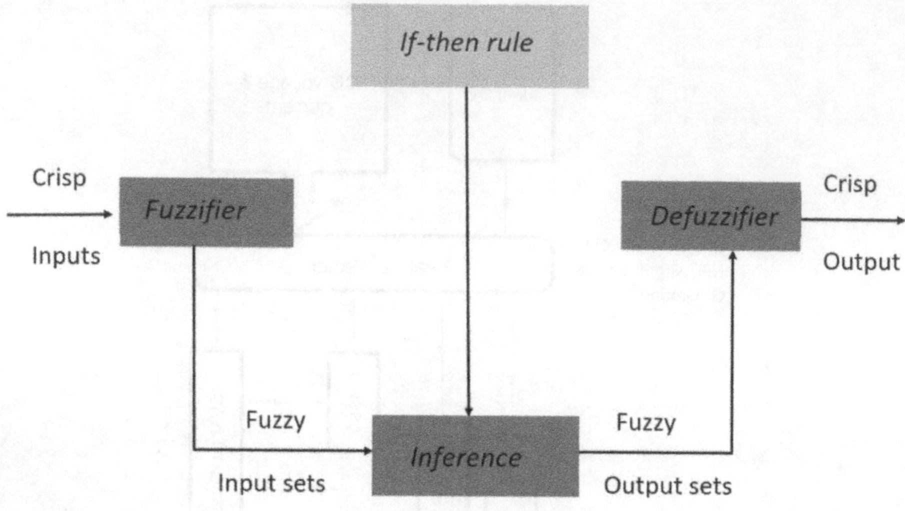

FIGURE 16.10 Fuzzy method schematic.

TABLE 16.2
Recent Publication in AI-based HVDC Fault Detection

Ref	Purpose	Features			Decision	Topology
		Input	DSP	Specification		
[34]	Detection and classification	Current's rate of change of DC current and DC current	DFT	• High and low frequency energy using DFT • Noise effect not discussed • 1 ms fault inception time	ANN	Hybrid MTDC
[35]	Detection and classification	DC current	DFT	• Preprocessing with Butterworth filter • 1 kHz sampling rate • 5 ms fault inception time	ANN	Two terminal HVDC
[36]	Detection	DC voltage	ICA-WT	• Noise effect suppression • 78% detection rate accuracy • Energy ration of wavelet band used as extracted features	ANN	Two terminal VSC HVDC

(Continued)

TABLE 16.2 (Continued)
Recent Publication in AI-based HVDC Fault Detection

Ref	Purpose	Features			Decision	Topology
		Input	DSP	Specification		
[37]	Detection, classification, and location	DC current and voltage		• A data window of 10 ms • Sampling rate of 2 kHz	ANN	Two terminal HVDC
[38]	Detection and classification	DC current	ST	• 20–200 kHz sampling rate • Satisfactory discrimination • 2 ms fault inception time	RBFN	Two terminal bipolar HVDC
[39]	Detection	DC voltage	–	• Harmonic content extraction • Lack of discrimination of different faults • 10 ms fault inception time	ADALINR	Two terminal LCC
[41]	Detection and classification	DC current and voltage	ST	• 4 class classification • Using RBF kernel function • 3.57% error rate in detection and classification	SVM	Two terminal LCC
[42]	Detection, classification, and location	DC/AC current and voltage	–	• Multiclass classifier • 1–4 kHz sampling rate • 100% detection rate accuracy	SVM	Two terminal LCC
[33]	Detection, classification	DC/AC voltage	–	• 2 separated FIS • 1 kHz sampling frequency • 100% detection rate accuracy • 15 ms maximum fault detection time	Fussy	Two terminal LCC
[44]	Detection	DC/AC current and voltage	–	• Converter fault identification e.g. commutation failure and arc through/fire through	ANFIS	LCC HVDC

16.5 AI-BASED FAULT CLASSIFICATION

Fault classification is a part of protection studies which can have different conditions based on the network structure. In AC networks, faults can be categorized according to three phases and neutral wire as follows:

- Single line-to-ground fault
- Line-to-line fault
- Double line-to-ground fault
- Three-phase fault
- Three-to-ground fault

In HVDC networks, faults can be categorized according to different topologies of HVDC networks as:

- Single-pole-to-ground fault (positive or negative pole)
- Pole-to-pole fault (particularly for bipolar grids)
- Pole-to-pole-to-ground fault (particularly for bipolar grids)

Methods relying on ANN are used for fault classification in various structures of HVDC grids, as suggested in [31,37,38,45–50].

The algorithm of fault classification via multilayer perceptron (MLP) and probabilistic neural network (PNN) is discussed in [46], in which the firing angle is estimated as the output of the neural network. Then, to complete the classification procedure, the firing angle is compared to its amount in a normal network condition.

In [45], PNN is used in order to detect and classify fault. This type is superior to the usual ANN methods due to having high speed, trusting convergence to the optimal answer, and solving local minimal issues. Ref [51] introduces a method based on fuzzy rules to detect and classify AC and DC network faults via measuring the three-phase AC current input to the rectifier. In this method, Park conversion is used to eliminate the phase mutual effect.

The use of fuzzy transform in [52] has been done for classification as well, which has a high detection speed in a half power cycle (10 ms). Ref [53] also classifies multi-terminal lines using fuzzy and neural methods for two-terminal and multi-terminal lines. In [49], classifying the fault of the HVDC line and distinguishing it from lightning strike fault was performed by convolutional neural network (CNN). On one hand, the CNN error rate in classification is lower but on the other hand, it requires far more training data. [37] Suggests an ANN-based method which leads to the detection of all various errors in a VSC HVDC system. Fault classification is done with the assistance of previous training data and is considered to be an offline method.

Also In [47] and [48], fault classification is presented using ANN and by means of the extracted WT feature. Ref [50] has used RBFN for fault classification in LCC HVDC. Fault detection and classification via k-nearest neighbors (KNN) method have been presented in [54], which is believed to be of the simplest machine learning methods. In ref [45], the fault is detected and classified through K-means algorithm which counts as one of the approaches of unsupervised clustering and has the simplest learning procedure and high speed.

Another method of fault classification which has been used in various studies is the SVM, as mentioned in [42,55,56]. In Ref [42] and [56], multi-class SVM was used to separate PG, NG, and PN fault in standard CIGRE benchmark with a sampling frequency of 1 kHz. In Ref [55] as well, the classification of strike and short-circuit fault has been performed by means of extracting wavelet entropy feature. In [57], feature extraction for fault classification is done by image processing.

A comparison of different fault classification techniques investigated in recent papers is listed on Table 16.3.

16.6 AL-BASED FAULT LOCATION

In power engineering, "fault" is attributed to any unusual condition. In power transmission lines, a short circuit occurs in unusual weather conditions such as lightning strike, snow, hail, and storm; as when lightning strikes a tower or lines. In all of the cases mentioned, the fault factors are coincidental. Therefore, faults can occur at any time or in any place. Hence, it is important to specify the fault location or to estimate it with acceptable accuracy. The faster this procedure gets done, the more quickly technicians are sent to the location of fault and the line returns to the circuit. Otherwise, the whole lines should be controlled. Therefore, facilitating power systems with fault locater brings the opportunity of saving money and time in repairing and returning the line to the grid.

Fault location in HDVC lines is of higher importance and some of its most significant reasons include the rate of rise of fault current, the variety of fault contingencies in HDVC systems, and also the wide use of submarine cables in offshore HVDC systems.

The methods of fault location are studied from different aspects. Some of the most important ones of which are mentioned below:

- **Time of using signals:** Most of the fault location methods utilize the recorded data while the fault is occurring. However, there are other approaches which use the data before the fault occurrence, being indeed less accurate and noticed [59].
- **Number of signals in use:** Fault location algorithms can be categorized into singleterminal, two-terminal, and multi-terminal methods [60]. Single-terminal approaches, which mostly rely on impedance, have the advantage of independence from telecommunication platiorms. However, most of the current systems of HDVC are equipped with the communication infrastructure of terminals.
- **Time synchronization:** Two-terminal fault location can be categorized based on whether synchronized or unsynchronized signals are being used [61].

By and large, fault location techniques can be divided into four main groups, shown in Figure 16.11.

TABLE 16.3
Recent Publication in AI-Based HVDC Fault Classification

Ref	Purpose	Features			Decision	Topology
		Input	DSP	Specifications		
[46]	Classification	DC current	–	• DC/AC fault classification • Firing angle estimation • 0.001% classification error accuracy	ANN	Two terminal LCC HVDC
[45]	Classification	DC current and voltage	–	• DC/AC fault classification • Commutation failure detection • Fast training process • Noise effect is not considered • No local minima issue	PNN	Two terminal LCC HVDC
[49]	Detection and classification	DC current and voltage	–	• 2–3 ms fault inception time • 96% classification accuracy • Lower error • Discrimination of fault and disturbance • Higher number of training data	CNN	MMC MTDC
[37]	Detection, classification and location	DC current and voltage	–	• Offline classification process • 2 kHz sampling rate • 49–77 ms classification time range	ANN	Two terminal VSC HVDC
[47]	Classification	DC current and voltage	DWT	• Converter fault classification such as commutation failure, misfire, backfire, fire-through, • 93.44% classification accuracy	ANN	Two terminal VSC HVDC
[48]	Classification	DC current and voltage	DWT	• 3 level decomposition • Low sampling frequency (1 kHz) • 100% classification accuracy	ANN	Two terminal bipolar CSC HVDC
[50]	Classification	AC/DC current and voltage	DWT	• Lower training time in comparison with BPNN	RBFNN	Two terminal bipolar CSC HVDC

(*Continued*)

TABLE 16.3 (Continued)
Recent Publication in AI-Based HVDC Fault Classification

Ref	Purpose	Features: Input	Features: DSP	Features: Specifications	Decision	Topology
[54]	Detection and classification	AC/DC current and voltage	–	• Simplicity • Low sampling rate (1 kHz) • 100% classification accuracy	KNN	Two terminal bipolar CSC HVDC
[45]	Detection, classification, and location	DC current and voltage	–	• 2 kHz sampling frequency • One end measurement • Robust and fast • Threshold values attain empirically • 0.001% classification error accuracy	K-means	Two terminal bipolar VSC HVDC
[42]	Classification	DC current and voltage	–	• 1–4 kHz sampling frequency • Multiclass SVM classifier • 100% classification rate accuracy	SVM	Two terminal bipolar VSC HVDC
[58]	Classification	DC current and voltage	–	• Various features like mean, median, standard deviation, weighted mean, skewness, and energy of the signal are extracted • 92% classification accuracy	SVM	Two terminal bipolar VSC HVDC
[55]	Classification	DC current and voltage	WT	• Wavelet entropy • DC fault and strike classification • 96.67% classification accuracy	SVM	Two terminal bipolar VSC HVDC
[57]	Classification	DC current and voltage	–	• AC/DC and converter classification	Image	
	processing	Two term-inal bipolar VSC HVDC				
[51]	Classification	AC current	–	• DC/AC fault classification • simplicity	Fuzzy	Two terminal VSC-HVDC

(Continued)

TABLE 16.3 (Continued)
Recent Publication in AI-Based HVDC Fault Classification

Ref	Purpose	Features			Decision	Topology
		Input	DSP	Specifications		
[52]	Classification	DC voltage and current	–	• DC/AC fault classification • 10 ms detection and classification	Fuzzy	Two terminal LCC-HVDC
[53]	Classification	DC/AC voltage and current	–	• DC/AC fault classification • 0.001% error accuracy	Fuzzy ANN	Two terminal and multi terminal HVDC

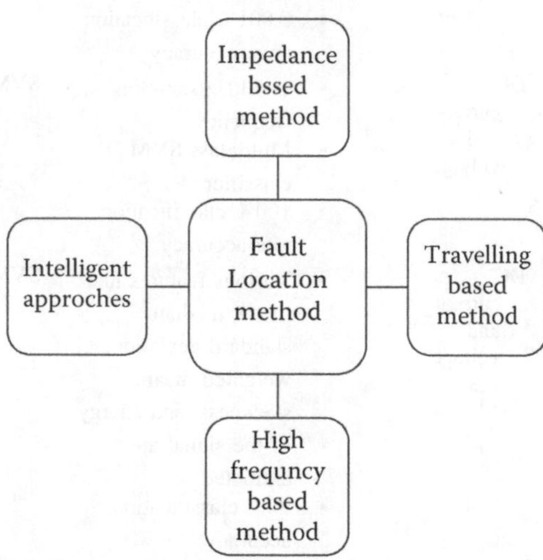

FIGURE 16.11 Fault location approaches.

- Techniques which work based on voltages and currents of the main frequency: These approaches are founded on the measurement of impedance. Notwithstanding the economic advantages of these approaches in comparison with the other methods, there are some limitations to point out, such as high dependence on network configuration including the fault loop impedance, electrical load, and source condition.
- Techniques doing the fault location based on traveling wave's phenomena: These techniques depend on the reflection of forward and backward traveling waves from the fault point to the two terminals of transmission line at the speed of light. By measuring the time length until these waves reach the terminals, the exact fault location is found. One of the advantages of this method is having high accuracy and not being influenced by network configuration in location.

However, the computational load due to quite high sampling rate and relatively high cost are of the disadvantages of these methods [61].
- Techniques which use the parameters of the high frequency being produced by fault: The basis of this method is the extraction of high frequency components generated at the instant of the fault, which has a remarkable accuracy. However, the need for special filters in measuring these components causes these methods to be costly and complex.
- Intelligent methods are based on training data.

In this section, fault location approaches using intelligent methods will be discussed.

In this section, different AI-based schemes for HVDC fault location have been investigated in terms of different aspects such as type of AI-based method, test system, accuracy, etc.

In the case of ANN-based methods for fault identification, there are two types of inputs. In the first method, the raw data of DC voltage and current samples are directly applied to ANN [28]. In the second method, measured samples are processed by various transforms e.g. DFT, WT [62,63], ST, etc.

NN functions can be examined from different aspects such as NN's type e.g. MLP, BP, RBF sampling frequency, number of inputs, speed of performance [30]. In [62,64,65], wavelet transform and neural network are applied for fault location.

Ref [65] introduces a method for two-terminal VSC-HVDC systems fault, which relies on ANN. In this reference, fault current data has been utilized for fault location. This data, after being analyzed by wavelet transform and obtaining partial coefficients, enters the ANN to form the fault pattern. In [64] as well, the signal of voltage and current of the AC network and DC has been used, with a sampling frequency of 20 kHz and a combination of wavelet transform and neural network for fault location in current source converter HVDC (CSC-HVDC).

In Ref [62], the wavelet's approximation and detailed coefficients in five levels are obtained using as input of ANN for accurate fault location in MTDC. But determination of proper decomposition level is considered a drawback of the method.

In Ref [66], a comparative evaluation concerning the process of fault location in MTDC networks has been carried out. This evaluation makes use of the DC current, measured directly from the terminals, and also the coefficients extracted from DFT and DWT as the input to ANN. Having said that, by using DWT coefficients, the accuracy of performance is higher.

In Ref [31], a comprehensive ANN-based study in fault detection, classification, and location was done, which used the frequency spectrum obtained from the Fourier transform of the signal right after the fault is presented. Unlike the traveling wave method, this method does not require wave front arrival time, which has been effective on system robustness and fault reduction. The system studied in this paper is a three-terminal MTDC network.

In Ref [28] a fault location scheme is proposed using DC voltage signal shape correlation during repetitive fault simulations with various impedance and locations. After extracting the training set, the exact place of fault is determined with ANN-based locator with rather low sampling frequency and high sensitivity

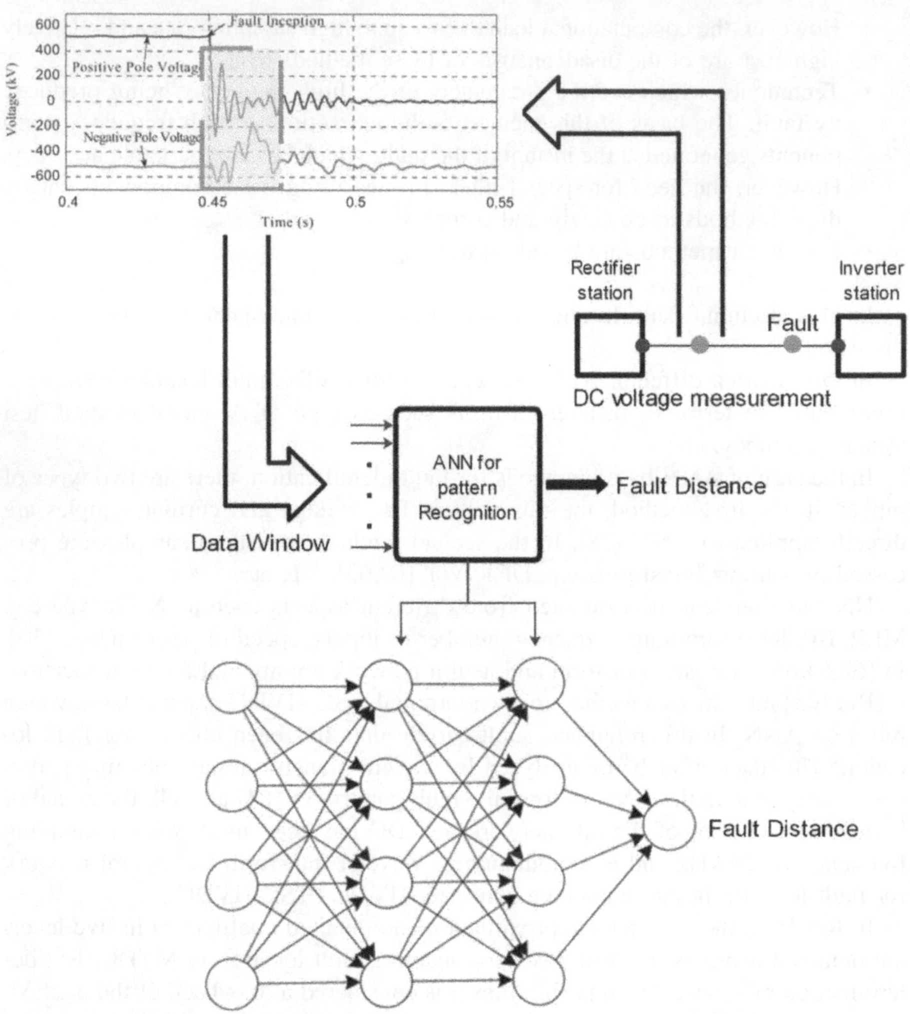

FIGURE 16.12 ANN-based fault location method.

compared with the AI-based method. The general ANN-based fault location scheme is depicted in Figure 16.12.

As previously mentioned, SVM as a supervised learning method for modeling and classification, has a significant role in different parts of a smart protection.

In the process of classification and fault location, the support vector method is used in Ref [42]. First, with the aid of binary SVM, occurrence or non-occurrence of the fault is identified. If it occurs, fault location is done with an accuracy of 0.3%, by making use of rms AC voltage and DC voltage on the rectifier side. Being a single-terminal location process and not requiring a communication medium are of the advantages of the proposed method.

Fault location in a two-terminal VSC HVDC network with support vector regression (SVR) algorithm has been done by extracting the time and frequency information feature, obtained through Hilbert Huang transform (HHT) in [67], in which by using the Bat algorithm (BA), optimization of the multiple SVR parameters with the minimal rate of error is achieved. The optimization algorithms, such as genetic algorithm in location, have been used in [68] to minimize location error mathematically, using the distributed line model.

In Refs [69] and [70], fault location is done with the help of training data obtained from WT and ST, which is used as the input to one SVR for the whole transmission line and also several SVRs; the results of multiple SVRs are more satisfying. Fault location estimation is provided in [71] by Gaussian process regression (GPR), which has a lower sampling frequency in comparison with ANN. In [72], the fault location estimation has been obtained with appropriate accuracy, using Pearson correlation coefficient calculation as the extracted feature. It should be noted that input data is obtained by utilizing optically-multiplexed DC current measurements.

Likewise, unsupervised learning methods such as KNN methods have been studied in fault location. In [63], Mohammad Farshad presents a training-based method for fault location in HVDC transmission lines, using the voltage signal measured at a line terminal to estimate the fault location. The method is based on the level of similarity between the current fault pattern and the pattern of database. In fact, the results of performed simulations for the fault occurrence at different points on the transmission line for different fault resistors and different pre-fault currents indicate that despite the difference in resistance and pre-fault current in a particular location, the shape of the voltage signal almost remains fixed, but then it changes as the fault location alters. Finally, the error location is estimated based on the calculation of the weighted average of the fault locations of these patterns (via the KNN parameters). In this paper, Pearson correlation coefficient is used to measure the similarity of signals.

In [52], fault detection, classification, and location are done using if-then fuzzy rules. The main advantage of this method is the simplicity of the implemented algorithm. Also in [73], the frequency spectrum is extracted by using DFT. Then, Random forest (RF) is applied to build fault-location estimator founded on a group of regression decision trees (DTs).

Another challenge of fault location in HVDC lines is the hybrid lines including cable and overhead line. The process of fault location by the traveling wave method is challenging. As shown in Figure 16.13, these traveling waves are propagated from the fault location at junction points of the line and cable, as a consequence of the refraction of traveling waves, and due to unequal impedance of the lines [74,75]. Thus, methods which are based on AI are used in limited references. In [76], to identify and locate the faulty part (cable or overhead line), an SVM is used by/through two classes, +1 and –1, for an overhead line and an underground cable. In this reference, which performs line measurements in a single-terminal way, the DWT energy feature is used for classification. However, with cable aging, the accuracy of the method decreases over time.

Recent publications in AI-based HVDC fault location with their specifications are listed in Table 16.4.

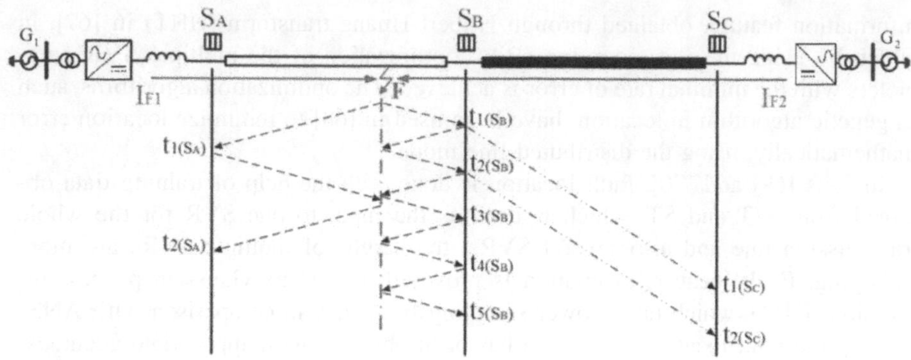

FIGURE 16.13 Bewley Lattice diagram incorporating OHL, UGC.

TABLE 16.4
Recent Publication in AI-Based HVDC Fault Location

Ref	Purpose	Features			Decision	Topology
		Input	DSP	Specification		
[66]	Fault location	current	DFT WT	• 2 kHz sampling rate • Fifth level decomposition • 0.96% average fault location error	ANN	Multi terminal
[65]	Fault location	current	WT	• Wavelet detailed coefficient as input data • 1.3% average fault location error	ANN	Bipolar two terminal
[64]	Fault location	AC/DC voltage and current	WT	• 1 kHz sampling rate • 0.0289% average fault location error	ANN	Two terminal CSC bipolar
[62]	Fault location	current	WT	• 10 kHz sampling rate. • 0.674% average fault location error	ANN	MTDC
[31]	Fault location	current	FFT	• 10 khz sampling rate • 1% average fault location error	ANN	MTDC
[28]	Fault location	voltage	–	• 20 kHz sampling rate • 2.9% average fault location error • 10 ms data window	ANN	Bipolar two terminal LCC HVDC
[42]	Fault location	AC/DC voltage	–	• 1–4 kHz sampling rate • 0.03% average fault location error • Single-ended scheme	SVM	Bipolar two terminal LCC HVDC
[67]	Fault location	DC current	HHT	• 1 MHz sampling rate • BA algorithm for SVR parameter optimization • 0.7% average fault location error	SVR	Two terminal VSC HVDC

(*Continued*)

TABLE 16.4 (Continued)
Recent Publication in AI-Based HVDC Fault Location

Ref	Purpose	Features			Decision	Topology
		Input	DSP	Specification		
[68]	Fault location	DC current	–	• 6.25 kHz sampling rate • Comparison with mathematical estimation • Weak performance in high impedance fault location • 1% average fault location error (low impedance fault)	GA	Two terminal VSC HVDC
[70]	Fault location	DC voltage and current	WT	• 500 kHz sampling rate • Less error comparing single SVR • 0.21% average fault location error	SVR	MTDC
[69]	Fault location	DC voltage and current	WT ST	• 500 kHz sampling rate • 0.0837% maximum fault location error	Multi ELM (Multi SVR)	MTDC
[71]	Fault location	AC/DC voltage and current	WT	• Double-ended measurement • 1 kHz sampling rate • 0.04% average fault location error	GPR	Bipolar two terminal LCC
[72]	Fault location	DC current	WT	• Utilizing optically multiplexed DC current measurements • 5 kHz sampling rate • 0.079% average fault location error	Pattern matching	MTDC
[63]	Fault location	DC voltage	–	• 80 kHz sampling rate • Pearson correlation coefficient • 0.077% average fault location error	KNN	Bipolar two terminal LCC
[73]	Fault location	DC voltage	DFT	• RF estimator • One end method • 16 kHz sampling rate • 0.244% average fault location error	DT	Bipolar two terminal LCC
[76]	Fault location	DC voltage and current	DWT	• Hybrid line fault location capability • 200 kHz sampling rate • 7% maximum fault location error	SVM	Hybrid LCC HVDC

16.7 AI-BASED COMMUTATION FAILURE (CF) IDENTIFICATION

Commutation failure (CF) is an unfavorable frequent problem, which occurs in thyristor-based HVDC systems like LCC. As the thyristor has only turn-on capability, a negative voltage must be applied to the valve to turn it off. When the time of applying negative voltage on the valve that has just been turned off is shorter than the time needed for it to restore the blocking capability, and the valve is conducted again when positive voltage is applied, commutation failure of the converter valve would happen [18]. This phenomenon will occur in inverter side of HVDC, when the extinction angle varies due to variation of system condition such as AC system faults. There are two types of CF: single CF which has self-clearing capability, and continual CF which could lead to blackout. Therefore CF identification is a necessity in HVDC systems.

Various studies have been carried out regarding CF identification using time or frequency domain analysis method like WT [77]. But in the case of AI approaches, Ref [47] applied combined DWT-ANN to detect and classify different types of converter fault like CF. The user's experience of setting is the shortcoming of this paper. To overcome this drawback, a CF identification technique is presented, applying singular value decomposition (SVD) as an extracted feature. Afterwards, a two-class SVM is applied for CF detection.

Table 16.5 represents AI-based CF identification studies carried out in recent years.

16.8 DISCUSSION

Regarding the investigation of various studies in AI-based HVDC protection, the strengths and weaknesses of the various techniques are identified and listed. It may give researchers some guidelines to appoint the best method according their requirements and purposes.

TABLE 16.5
Recent publication in AI-based HVDC CF identification

Ref	Purpose	Features			Decision	Topology
		Input	DSP	Specification		
[47]	CF identification	DC current and voltage	DWT	• High and low frequency energy using DFT • Noise effect not discussed	ANN	Hybrid MTDC
[78]	CF identification	DC current and voltage	DWT, SVD	• Extract feature using SVD • CF and AC side faults discrimination	SVM	Two terminal HVDC

- **Advantages:**
  - ANN-based approaches are widely used in different fields of study; thus, it counts as a mature machine learning method. In addition it has proper fault classification capability with simple implementation, noise immunity, adaptability, and fast response [29].
  - Fuzzy methods are easy to understand because of using natural language which leads to low computation process and simplicity. Also, the technique can tolerate vague training sets [43].
  - SVM technique is a strong tool in supervised learning and fault classification with low rate of error. Moreover, the misclassification error limits are not influenced by the training data. Also, the generalization capability of SVM classifier is suitable for small training spaces [79].
  - A decision tree has easy interpretation and the ability of setting rules.
  - ANFIS-based schemes are adaptable and fast convergent due to the hybrid learning process.

- **Disadvantages:**
  - The main drawbacks of ANN-based methods comprise high computational load and large training set. In some cases, there is no specific procedure for initial value of weight constraints assignment. Furthermore, low convergence and local minima trap would occur in the training process.
  - The FIS setting needs the experts to select best membership functions and fuzzy rules. Thus, the fuzzy methods cannot be categorized as a settings-free scheme.
  - The multiclass SVM classifier has a high training and testing dimension, which entails powerful hardware platform to implement protection scheme in smart numerical relays [80].
  - The shortcomings of DT are noise sensitivity, poor generalization, and overfitting [73].

16.9 CONCLUSION

In this chapter, a comprehensive review of AI-based schemes in HVDC application is investigated. It can be concluded that AI-based functions in protection are considered a powerful alternative for conventional present HVDC protection schemes like traveling wave technique. Before comparative evaluation of different methods, firstly, all steps of AI-based protection architecture have been illustrated comprising preprocessing input signals, feature extraction, dimensionality reduction, and decision making. Afterwards, widely used AI-based techniques in fault identification application have been reviewed, e.g. ANN, SVM, fuzzy, ANFIS from different aspect such as training set, input signal, sampling frequency, extracted features, etc. The literature review confirms that AI-based methods are efficient in DC/AC fault detection, classification, location, and internal converter faults, such as commutation failure identification in the HVDC systems. Ultimately, the advantages and disadvantages of different AI-based procedures are discussed.

REFERENCES

[1] S. J. Russell and P. Norvig, *Artificial intelligence–A modern approach*, 3rd International Edition. London, United Kingdom: Pearson Education, 2010.

[2] S. Russell and P. Norvig, *Artificial intelligence: A modern approach*. London, United Kingdom: Pearson education, 2002.

[3] N. J. Nilsson, *Principles of artificial intelligence*. Burlington, MA, USA: Morgan Kaufmann, 2014.

[4] M. Laughton, *Artificial Intelligence Techniques in Power Systems.*London, United Kingdom: IET, 1997.

[5] K. Chen, C. Huang, and J. He, "Fault detection, classification and location for transmission lines and distribution systems: a review on the methods," *High Voltage*, vol. 1, pp. 25–33, 2016.

[6] M. Kezunovic, "Smart fault location for smart grids," *IEEE Transactions on Smart Grid*, vol. 2, pp. 11–22, 2011.

[7] Y. Ouyang, J. He, J. Hu, and S. X. Wang, "A current sensor based on the giant magnetoresistance effect: Design and potential smart grid applications," *Sensors*, vol. 12, pp. 15520–15541, 2012.

[8] J. Han, J. Hu, Y. Yang, Z. Wang, S. X. Wang, and J. He, "A nonintrusive power supply design for self-powered sensor networks in the smart grid by scavenging energy from AC power line," *IEEE Transactions on Industrial Electronics*, vol. 62, pp. 4398–4407, 2014.

[9] B. Mitra, B. Chowdhury, and M. Manjrekar, "Fault analysis and hybrid protection scheme for multi-terminal HVDC using Wavelet transform," in 2016 North American Power Symposium (NAPS), 2016, pp. 1–6.

[10] Z. Moravej, M. Pazoki, and M. Khederzadeh, "New smart fault locator in compensated line with UPFC," *International Journal of Electrical Power & Energy Systems*, vol. 92, pp. 125–135, 2017.

[11] S. M. Kay, *Fundamentals of statistical signal processing*. Hoboken, NJ, USA: Prentice Hall PTR, 1993.

[12] O. Peake, "The history of high voltage direct current transmission," *Australian Journal of Multi-Disciplinary Engineering*, vol. 8, pp. 47–55, 2010.

[13] D. Larruskain, O. Abarrategui, I. Zamora, G. Buigues, V. Valverde, and A. Iturregi, "Requirements for fault protection in HVDC grids," in International Conference on Renewable Energies and Power Quality, E.T.S. of Computer and Telecommunication Engineering Source (Spain), 2015.

[14] D. Van Hertem, O. Gomis-Bellmunt, and J. Liang, *HVDC grids: For offshore and supergrid of the future*. Hoboken, NJ, USA: John Wiley & Sons, 2016.

[15] M. A. Ikhide, "DC line protection for multi-terminal high voltage DC (HVDC) transmission systems," Doctoral dissertation, Staffordshire University, 2017.

[16] M. J. Pérez-Molina, D. M. Larruskain, P. E. López, and G. Buigues, "Challenges for protection of future HVDC grids," *Frontiers in Energy Research*, vol. 8, pp. 33–39, 2020.

[17] G. Buigues, V. Valverde, I. Zamora, D. Larruskain, O. Abarrategui, and A. Iturregi, "DC fault detection in VSC-based HVDC grids used for the integration of renewable energies," in 2015 international conference on clean electrical power (ICCEP), 2015, pp. 666–673.

[18] K. Sharifabadi, L. Harnefors, H.-P. Nee, S. Norrga, and R. Teodorescu, *Design, control, and application of modular multilevel converters for HVDC transmission systems*. Hoboken, NJ, USA: John Wiley & Sons, 2016.

[19] M. K. Bucher and C. M. Franck, "Fault current interruption in multiterminal HVDC networks," *IEEE Transactions on Power Delivery*, vol. 31, pp. 87–95, 2015.

[20] B. Li, J. He, Y. Li, and B. Li, "A review of the protection for the multi-terminal VSC-HVDC grid," *Protection and Control of Modern Power Systems*, vol. 4, pp. 1–11, 2019.
[21] J. Liu, N. Tai, C. Fan, and W. Huang, "Protection scheme for high-voltage direct-current transmission lines based on transient AC current," *IET Generation, Transmission & Distribution*, vol. 9, pp. 2633–2643, 2015.
[22] L. Bin, H. Jiawei, T. Jie, F. Yadong, and D. Yunlong, "DC fault analysis for modular multilevel converter-based system," *Journal of Modern Power Systems and Clean Energy*, vol. 5, pp. 275–282, 2017.
[23] P. D. Judge, G. Chaffey, M. Wang, F. Z. Dejene, J. Beerten, T. C. Green, et al., "Power-system level classification of voltage-source HVDC converter stations based upon DC fault handling capabilities," *IET Renewable Power Generation*, vol. 13, pp. 2899–2912, 2019.
[24] X. Li, Q. Song, W. Liu, H. Rao, S. Xu, and L. Li, "Protection of nonpermanent faults on DC overhead lines in MMC-based HVDC systems," *IEEE Transactions on Power Delivery*, vol. 28, pp. 483–490, 2012.
[25] L. Sabug Jr., A. Musa, F. Costa, and A. Monti, "Real-time boundary wavelet transform-based DC fault protection system for MTDC grids," *International Journal of Electrical Power & Energy Systems*, vol. 115, Art. no. 105475, 2020.
[26] W. Leterme, "Communication-less protection algorithms for meshed VSC HVDC cable grids," 2016.
[27] M. Barnes, D. Van Hertem, S. P. Teeuwsen, and M. Callavik, "HVDC systems in smart grids," *Proceedings of the IEEE*, vol. 105, pp. 2082–2098, 2017.
[28] A. S. Silva, R. C. Santos, J. A. Torres, and D. V. Coury, "An accurate method for fault location in HVDC systems based on pattern recognition of DC voltage signals," *Electric Power Systems Research*, vol. 170, pp. 64–71, 2019.
[29] M. Pazoki, A. Yadav, and A. Y. Abdelaziz, "Pattern-recognition methods for decision-making in protection of transmission lines," in *Decision making applications in modern power systems*. Cambridge, MA, USA: Elsevier, 2020, pp. 441–472.
[30] K. Nagar and M. Shah, "A review on different ANN based fault detection techniques for HVDC systems," *International Journal of Innovative Research in Engineering & Management*, vol. 3, no. 6, pp. 477–487, 2016.
[31] Q. Yang, S. Le Blond, R. Aggarwal, Y. Wang, and J. Li, "New ANN method for multi-terminal HVDC protection relaying," *Electric Power Systems Research*, vol. 148, pp. 192–201, 2017.
[32] B. Mitra, B. Chowdhury, and M. Manjrekar, "HVDC transmission for access to off-shore renewable energy: a review of technology and fault detection techniques," *IET Renewable Power Generation*, vol. 12, pp. 1563–1571, 2018.
[33] G. Buigues, V. Valverde, D. Larruskain, P. Eguía, and E. Torres, "DC protection in modern HVDC networks: VSC-HVDC and MTDC systems," in the International Conference on Renewable Energies and Power Quality, Madrid, Spain, pp. 300–305, 2016.
[34] A. E. Abu-Elanien, "An artificial neural network based technique for protection of HVDC grids," in 2019 IEEE PES GTD Grand International Conference and Exposition Asia (GTD Asia), 2019, pp. 1004–1009.
[35] M. Sanaye-Pasand and H. Khorashadi-Zadeh, "Transmission line fault detection & phase selection using ANN," in International Conference on Power Systems Transients, 2003, pp. 1–6.
[36] Z. Li, C. Sheng, Y. Li, J. Xing, and B. Su, "Efficient fault feature extraction and fault isolation for high voltage DC transmissions," *Elektronika ir Elektrotechnika*, vol. 21, pp. 7–12, 2015.

[37] V. L. Merlin, R. C. dos Santos, S. Le Blond, and D. V. Coury, "Efficient and robust ANN-based method for an improved protection of VSC-HVDC systems," *IET Renewable Power Generation*, vol. 12, pp. 1555–1562, 2018.

[38] C. Ying, F. Songhai, W. Qiaomei, W. Tianbao, L. Lei, M. Xiaomin, et al., "A novel fault identification method for HVDC transmission line based on Stransform multi-scale area," in 2020 5th International Conference on Computational Intelligence and Applications (ICCIA), 2020, pp. 199–205.

[39] J. Moshtagh, M. Jannati, H. Baghaee, and E. Nasr, "A novel approach for online fault detection in HVDC converters," in 2008 12th International Middle-East Power System Conference, 2008, pp. 307–311.

[40] C. Wang, L. Zhou, and Z. Li, "Survey of switch fault diagnosis for modular multilevel converter," *IET Circuits, Devices & Systems*, vol. 13, pp. 117–124, 2018.

[41] X.-M. Liu, W.-Y. Wei, and F. Yu, "SVM theory and its application in fault diagnosis of HVDC system," in Third International Conference on Natural Computation (ICNC 2007), 2007, pp. 665–669.

[42] J. M. Johnson and A. Yadav, "Complete protection scheme for fault detection, classification and location estimation in HVDC transmission lines using support vector machines," *IET Science, Measurement & Technology*, vol. 11, pp. 279–287, 2016.

[43] S. Agarwal, A. Swetapadma, C. Panigrahi, and A. Dasgupta, "Fault analysis method of integrated high voltage direct current transmission lines for onshore wind farm," *Journal of Modern Power Systems and Clean Energy*, vol. 7, pp. 621–632, 2019.

[44] N. Bawane, A. G. Kothari, and D. P. Kothari, "ANFIS based HVDC control and fault identification of HVDC converter," *HAIT Journal of Science and Engineering B*, vol. 2, pp. 673–689, 2005.

[45] M. Farshad, "Detection and classification of internal faults in bipolar HVDC transmission lines based on K-means data description method," *International Journal of Electrical Power & Energy Systems*, vol. 104, pp. 615–625, 2019.

[46] P. Sanjeevikumar, B. Paily, M. Basu, and M. Conlon, "Classification of fault analysis of HVDC systems using artificial neural network," in 2014 49th International Universities Power Engineering Conference (UPEC), 2014, pp. 1–5.

[47] C. Venkatesh and P. V. Rao, "Wavelet-ANN based classification of HVDC converter faults," in 2016 IEEE 6th International Conference on Power Systems (ICPS), 2016, pp. 1–5.

[48] S. Ankar and A. Yadav, "ANN-based protection scheme for bipolar CSC-based HVDC transmission line," in 2019 Innovations in Power and Advanced Computing Technologies (i-PACT), 2019, pp. 1–5.

[49] J. Mei, R. Ge, Z. Liu, X. Zhan, G. Fan, P. Zhu, et al., "An auxiliary fault identification strategy of flexible HVDC grid based on convolutional neural network with branch structures," *IEEE Access*, vol. 8, pp. 115922–115931, 2020.

[50] K. Narendra, V. Sood, K. Khorasani, and R. Patel, "Application of a radial basis function (RBF) neural network for fault diagnosis in a HVDC system," *IEEE Transactions on Power Systems*, vol. 13, pp. 177–183, 1998.

[51] B. Paily, S. Kumaravel, M. Basu, and M. Conlon, "Fault analysis of VSC HVDC systems using fuzzy logic," in 2015 IEEE International Conference on Signal Processing, Informatics, Communication and Energy Systems (SPICES), 2015, pp. 1–5.

[52] S. Agarwal, A. Swetapadma, C. Panigrahi, and A. Dasgupta, "A method for fault section identification in high voltage direct current transmission lines using one end measurements," *Electric Power Systems Research*, vol. 172, pp. 140–151, 2019.

[53] B. Paily, "HVDC systems fault analysis using various signal processing techniques," 2015.

[54] J. M. Johnson and A. Yadav, "Fault detection and classification technique for HVDC transmission lines using KNN," in *Information and communication technology for sustainable development*. Springer, 2018, pp. 245–253.

[55] G. Luo, C. Yao, Y. Tan, and Y. Liu, "Transient signal identification of HVDC transmission lines based on wavelet entropy and SVM," *The Journal of Engineering*, vol. 2019, pp. 2414–2419, 2019.

[56] C. Xing, K. Tai, Y. Wang, and M. Liu, "Fault diagnosis for HVDC system based on wavelet entropy clustering and DS evidence fusion theory," in 2019 IEEE Innovative Smart Grid Technologies-Asia (ISGT Asia), 2019, pp. 344–348.

[57] R. Muzzammel, A. Raza, M. R. Hussain, G. Abbas, I. Ahmed, M. Qayyum, et al., "MT–HVdc systems fault classification and location methods based on traveling and non-traveling waves—A comprehensive review," *Applied Sciences*, vol. 9, Art. no. 4760, 2019.

[58] J. P. Keshri and H. Tiwari, "Fault classification in VSC-HVDC transmission system using machine learning approach," in 2019 8th International Conference on Power Systems (ICPS), 2019, pp. 1–6.

[59] M. M. Saha, J. J. Izykowski, and E. Rosolowski, *Fault location on power networks*. London, Dordrecht, Heidelberg, New York: Springer Science & Business Media, 2009.

[60] M. M. Saha, R. Das, P. Verho, and D. Novosel, "Review of fault location techniques for distribution systems," in Power Systems and Communications Infrastructures for the Future, Beijing. pp.1–6, 2002.

[61] Z. Yi-ning, L. Yong-hao, X. Min, and C. Ze-xiang, "A novel algorithm for HVDC line fault location based on variant travelling wave speed," in 2011 4th International Conference on Electric Utility Deregulation and Restructuring and Power Technologies (DRPT), 2011, pp. 1459–1463.

[62] J. O. A. Torres and R. C. dos Santos, "New method based on wavelet transform and ANN for multiterminal HVDC system protection," in 2019 IEEE Milan PowerTech, 2019, pp. 1–6.

[63] M. Farshad and J. Sadeh, "A novel fault-location method for HVDC transmission lines based on similarity measure of voltage signals," *IEEE Transactions on Power Delivery*, vol. 28, pp. 2483–2490, 2013.

[64] S. Ankar and A. Yadav, "Wavelet-ANN based fault location scheme for bipolar CSC-based HVDC transmission system," in 2020 First International Conference on Power, Control and Computing Technologies (ICPC2T), 2020, pp. 85–90.

[65] S. Vasanth, Y. M. Yeap, and A. Ukil, "Fault location estimation for VSC-HVDC system using artificial neural network," in 2016 IEEE Region 10 Conference (TENCON), 2016, pp. 501–504.

[66] J. A. Torres, R. C. dos Santos, and P. T. L. Asano, "A comparison of new methods based on ANNs for detecting and locating faults in MTDC systems," in 2018 International Conference on Smart Energy Systems and Technologies (SEST), 2018, pp. 1–6.

[67] Y. Hao, Q. Wang, Y. Li, and W. Song, "An intelligent algorithm for fault location on VSC-HVDC system," *International Journal of Electrical Power & Energy Systems*, vol. 94, pp. 116–123, 2018.

[68] Y. Li, S. Zhang, H. Li, Y. Zhai, W. Zhang, and Y. Nie, "A fault location method based on genetic algorithm for high-voltage direct current transmission line," *European Transactions on Electrical Power*, vol. 22, pp. 866–878, 2012.

[69] A. Hadaeghi, H. Samet, and T. Ghanbari, "Multi extreme learning machine approach for fault location in multi-terminal high-voltage direct current systems," *Computers & Electrical Engineering*, vol. 78, pp. 313–327, 2019.

[70] A. Hadaeghi, H. Samet, and T. Ghanbari, "Multi SVR approach for fault location in multi-terminal HVDC systems," *International Journal of Renewable Energy Research (IJRER)*, vol. 9, pp. 194–206, 2019.

[71] S. J. Ankar and A. Yadav, "A novel approach to estimate fault location in current source converter–based HVDC transmission line by Gaussian process regression," *International Transactions on Electrical Energy Systems*, vol. 30, Art. no. e12221, 2020.

[72] D. Tzelepis, A. Dyśko, G. Fusiek, P. Niewczas, S. Mirsaeidi, C. Booth, et al., "Advanced fault location in MTDC networks utilising optically-multiplexed current measurements and machine learning approach," *International Journal of Electrical Power & Energy Systems*, vol. 97, pp. 319–333, 2018.

[73] M. Farshad, "Locating short-circuit faults in HVDC systems using automatically selected frequency-domain features," *International Transactions on Electrical Energy Systems*, vol. 29, Art. no. e2765, 2019.

[74] D. Tzelepis, G. Fusiek, A. Dyśko, P. Niewczas, C. Booth, and X. Dong, "Novel fault location in MTDC grids with non-homogeneous transmission lines utilizing distributed current sensing technology," *IEEE Transactions on Smart Grid*, vol. 9, pp. 5432–5443, 2018.

[75] P. T. Lewis, B. M. Grainger, H. A. Al Hassan, A. Barchowsky, and G. F. Reed, "Fault section identification protection algorithm for modular multilevel converter-based high voltage DC with a hybrid transmission corridor," *IEEE Transactions on Industrial Electronics*, vol. 63, pp. 5652–5662, 2016.

[76] H. Livani and C. Y. Evrenosoglu, "A single-ended fault location method for segmented HVDC transmission line," *Electric Power Systems Research*, vol. 107, pp. 190–198, 2014.

[77] W. Yuhong, R. Zhen, and L. Qunzhan, "Wavelets selection for commutation failure detection in HVDC system," in TENCON 2006-2006 IEEE Region 10 Conference, 2006, pp. 1–4.

[78] C. Gao, Z. Liao, and S. Huang, "Fault diagnosis of commutation failures in the HVDC system based on wavelet singular value and support vector machine," in 2009 Asia-Pacific Power and Energy Engineering Conference, 2009, pp. 1–4.

[79] A. Raza, A. Benrabah, T. Alquthami, and M. Akmal, "A review of fault diagnosing methods in power transmission systems," *Applied Sciences*, vol. 10, Art. no. 1312, 2020.

[80] S. A. Kunsman, I. I. Jouny, and S. Kaprielian, "High impedance fault detection," 2006.

17 Intelligent Schemes for Fault Detection, Classification, and Location in HVDC Systems

Mohammad Farshad
Department of Electrical Engineering, Faculty of Basic Sciences and Engineering, Gonbad Kavous University, Iran

ABBREVIATIONS

AC	Alternating-current
CIGRE	International council on large electric systems (conseil international des grands réseaux électriques)
CSC	Current-source converter
CSC-HVDC	CSC-based HVDC
CWT	Continuous wavelet transform
DC	Direct-current
DCCB	DC circuit breaker
DFT	Discrete Fourier transform
DWT	Discrete wavelet transform
FIS	Fuzzy inference system
GRNN	Generalized regression neural network
HHT	Hilbert-Huang transform
HVAC	High-voltage alternating-current
HVDC	High-voltage direct-current
IGBT	Insulated-gate bipolar transistor
K-MDD	K-means data description
K-NN	K-nearest neighbors
LCC	Line-commutated converter
MLPNN	Multilayer perceptron neural network
MMC	Modular multilevel converter
MMC-HVDC	MMC-based HVDC
PA	Prony analysis

DOI: 10.1201/9780367552374-17

PCA	Principal component analysis
PWM	Pulse-width modulation
RF	Random forest
RMS	Root mean square
RReliefF	Regression relief feature selection
SAE	Stacked auto-encoder
SCC	Self-commutated converter
SVD	Singular value decomposition
SVM	Support vector machine
SVR	Support vector regression
VSC	Voltage-source converter
VSC-HVDC	VSC-based HVDC
WPD	Wavelet packet decomposition

17.1 INTRODUCTION

The complexity of power systems has increased dramatically over time due to the power industry restructuring, increasing loads, increasing penetration of renewable resources and distributed generations, and the developing power electronics technologies. In other words, in recent decades, the operation, control, management, planning, design, and protection of power systems have faced new challenges. Consequently, some of the traditional methods have lost their efficiency to be used in new situations. The power engineers have applied corrective modifications to the traditional methods or designed innovative algorithms to overcome the new complexities along with the changes in power systems. The pattern-recognition and machine-learning techniques are relatively new approaches that have been extensively utilized to solve the new power system problems in recent decades. These intelligent algorithms have high flexibility and, if implemented correctly, can perform well in the face of uncertainties and complexities. These algorithms are useful for solving various problems of the power system operation, control, management, planning, design, and protection (Chaturvedi, 2008; Eremia et al., 2016; Mielczarski, 1998; Ongsakul and Dieu, 2013; Warwick et al., 1997).

The use of high-voltage direct-current (HVDC) systems is efficiently expanding owing to their significant benefits over high-voltage alternating-current (HVAC) systems and the advancement of power electronics technologies. With the widening use of these systems, related issues such as their protection requirements have attracted much attention. Indeed, the various technologies employed in HVDC systems have their particular complexities and protection challenges. Therefore, it is essential to design appropriate fault detection, classification, and location schemes with specific consideration given to each technology. The pattern-recognition and machine-learning techniques can play a highly influential role in designing such schemes for HVDC systems and overcome the complexities and challenges by relying on their inherent capabilities.

In this chapter, after briefly stating the advantages and applications of HVDC systems, their different converter technologies are introduced, and their specific

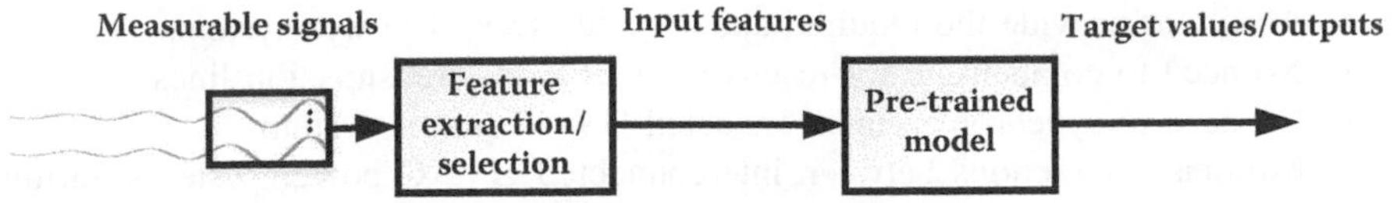

FIGURE 17.1 A general outline of intelligent protection and fault location schemes.

protection challenges are described. Next, the performed studies and achieved advances in fault detection, classification, and location in different HVDC systems are reviewed with specific focus given to the applied pattern-recognition and machine-learning procedures. Figure 17.1 presents a general outline of intelligent schemes. It should be noted that in recent studies, different input features have been extracted/selected using common tools (e.g. discrete Fourier transform (DFT), discrete wavelet transform (DWT), continuous wavelet transform (CWT), wavelet packet decomposition (WPD), singular value decomposition (SVD), Prony analysis (PA), Hilbert-Huang transform (HHT), principal component analysis (PCA), and regression relief feature selection (RReliefF) from measurable signals. Many machine-learning algorithms/models (e.g. multilayer perceptron neural network (MLPNN), support vector machine (SVM), support vector regression (SVR), fuzzy inference system (FIS), K-means data description (K-MDD), K-nearest neighbors (K-NN), random forest (RF), stacked auto-encoder (SAE), and generalized regression neural network (GRNN) have also been examined for fault detection, classification, and location in HVDC systems based on the extracted/selected input features. Hence, the intelligent schemes designed for different HVDC systems are discussed considering two main aspects: the extracted/selected input features and the employed learning algorithms/models, as indicated in Figure 17.1.

17.2 AN OVERVIEW OF HVDC SYSTEMS

Direct-current (DC) systems were used at the beginning of the history of electricity generation and supply. Then, as demand increased, these systems' limitations let alternating-current (AC) systems be selected as the primary platform for electrical energy generation, transmission, and distribution. In recent decades, with the technological advancement of power electronics, it has also become possible to take advantage of HVDC systems' unique benefits.

HVDC systems have been used for different aims, e.g. the long-distance transmission of bulk power, transmission of electrical energy generated from renewable resources, interconnecting nonsynchronous power systems, interconnecting power systems with different frequencies of 50 and 60 Hz, and safely interconnecting power systems for least interactions. Compared to HVAC systems, these systems have several advantages, some of which are (Eremia et al., 2016):

- Lower losses in transmission lines
- Less number of conductors in transmission lines
- A smaller cross-section of conductors in transmission lines
- Lower costs for long-distance power transmission

- Ability to provide the required ancillary services of power systems
- No need to compensate for reactive power along transmission lines
- No need to synchronize interconnected HVAC power systems
- Minimal interactions between interconnected HVAC power systems during disturbances and oscillations

Two general kinds of converters have been commonly practiced in HVDC systems, which are:

- Current-source converter (CSC) or line-commutated converter (LCC)
- Voltage-source converter (VSC) or self-commutated converter (SCC)

CSC technology is more mature than VSC technology. Indeed, the first commercial CSC-based HVDC (CSC-HVDC) system was commissioned in 1954. While, the first commercial VSC-based HVDC (VSC-HVDC) system was successfully commissioned after 45 years, with the development of controllable semiconductors (Eremia et al., 2016).

17.2.1 CSC-HVDC Systems

CSC technology, which is currently in use in conventional HVDC systems, is founded on thyristors. In a conventional CSC-HVDC system, the active power flow is controllable; but its direction is constant, from the rectifier station to the inverter station. However, this active power flow direction may be reversed by changing the voltage's polarity (Franck, 2011). In CSC-HVDC systems, the DC-side short-circuit fault currents have a relatively small rate-of-rise and can be easily controlled by the converter controllers (Faruque et al., 2006).

CSC-HVDC transmission systems may be installed/operated based on different configurations such as (Eremia et al., 2016):

- Monopolar configuration with the ground return
- Monopolar configuration with a metallic return
- Bipolar configuration

Figure 17.2 illustrates these configurations for two-terminal transmission systems. Despite the higher cost of the bipolar configuration compared to the monopolar configurations, its reliability is higher. In the case of a single-pole fault, to partially maintain the power transfer capacity, the bipolar system can be operated in the monopolar mode.

Converter stations of CSC-HVDC systems may have several components, some of which are indicated in Figure 17.3 based on the benchmark introduced by the international council on large electric systems (CIGRE) (Szechtman et al., 1994). Harmonics may be generated on the DC and AC-sides of the HVDC system due to the converters' operation. Therefore, AC and DC filters are installed on both sides to reduce the destructive effects of these harmonics. A capacitor bank is also installed on the AC-side to compensate for the converter's reactive power. Moreover,

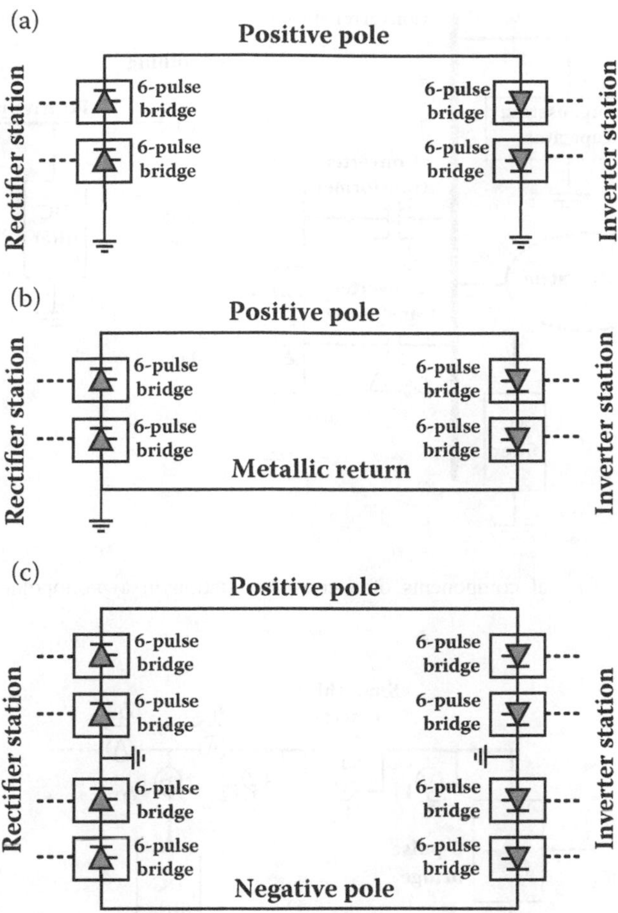

FIGURE 17.2 Typical configurations for two-terminal CSC-HVDC transmission systems: (a) monopolar with the ground return; (b) monopolar with a metallic return; (c) bipolar.

installing a smoothing reactor on the DC-side is necessary for smoothing the DC-side current and protection purposes.

One of the key points in designing protection and location schemes for CSC-HVDC systems is the exact place of the voltage and current measuring instruments on the DC-side. Figure 17.4 illustrates the possible places for the measuring instruments in a CSC-HVDC system, including before, after, and on the smoothing reactor for the voltage measuring instruments and before, after, and through the DC filter for the current measuring instruments. Figure 17.5 presents rectifier voltage and current signals for these possible places, considering a DC line fault at 1.2 s in the CIGRE 2-terminal benchmark (Szechtman et al., 1994) simulated via PSCAD/EMTDC (PSCAD User's Guide, 2018). As can be comprehended from this figure, voltage signals measured before,

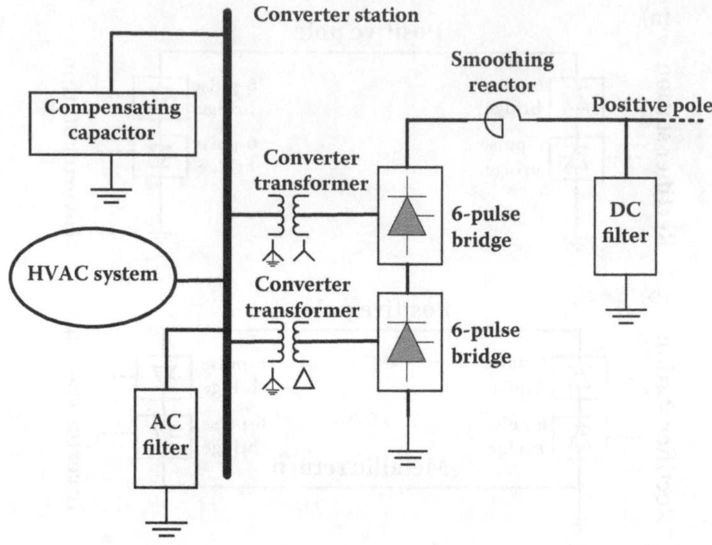

FIGURE 17.3 Typical components of a converter station in a monopolar CSC-HVDC system.

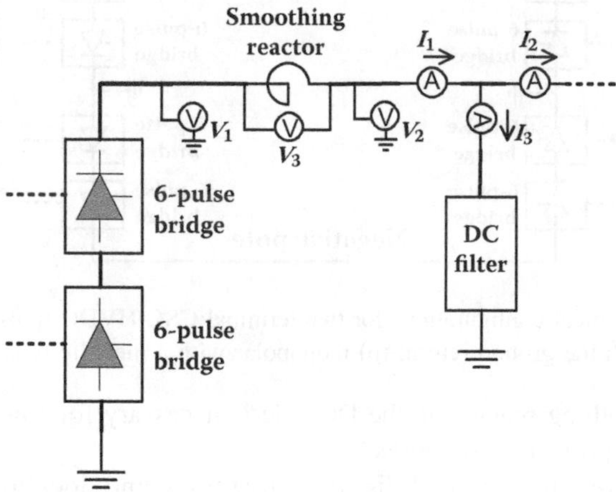

FIGURE 17.4 Possible places for the voltage and current measuring instruments on a CSC-HVDC system's DC side.

after, and on the smoothing reactor and current signals captured before, after, and through the DC filter have significant differences that may affect the performance of designed protection and location schemes. In other words, researchers should be aware of the importance of choosing the appropriate measurement place and its impact on the scheme's performance.

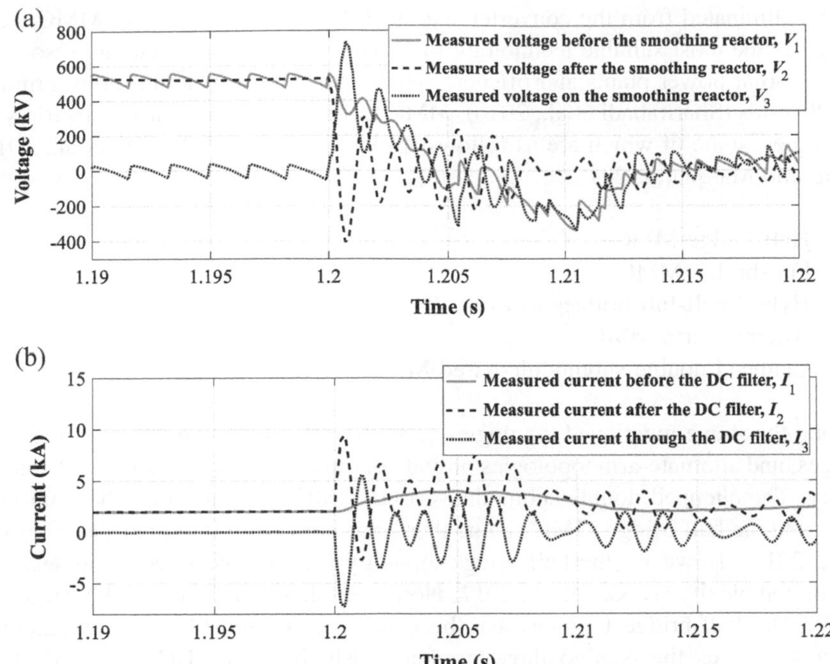

FIGURE 17.5 Measured signals for the measuring instruments' possible places, considering a DC line fault at 1.2 s in the CIGRE benchmark: (a) rectifier voltage before, after, and on the smoothing reactor; (b) rectifier current before, after, and through the DC filter.

17.2.2 VSC-HVDC Systems

VSC technology, as a relatively new trend for HVDC systems, is based on insulated-gate bipolar transistors (IGBTs). In VSC-HVDC systems, the active and reactive powers of converter stations are controllable. The power flow direction can also be easily changed by adjusting the pulse-width modulation (PWM) sequence (Franck, 2011). In these systems, short-circuit fault currents can be problematic due to their relatively high rate-of-rise (Elgeziry et al., 2017). However, there are some types of VSCs with controlled or uncontrolled fault blocking capability (Leterme et al., 2020). Indeed, different types of VSCs have been developed so far, some of which are as follows:

- Two-level VSC
- Three-level VSC
- Modular multilevel converter (MMC)

Among the above-mentioned types of VSCs, MMC technology has received more attention than others. This converter type has excellent flexibility in producing different voltage and power levels (Du et al., 2018). The output voltage waveform of this converter is of high quality, and consequently, AC and DC harmonic filters can be eliminated from the converter stations (Du et al., 2018). Also, MMC technology is the most suitable technology for connecting renewable energy resources such as solar power plants and offshore wind farms and expanding multi-terminal HVDC grids (Sharifabadi et al., 2016). MMC technology also comes in a variety of topologies, some of which are as follows (Dekka et al., 2017; Norrga et al., 2014; Wang and Marquardt, 2013):

- Half-bridge MMC
- Full-bridge MMC
- Hybrid half-full-bridges MMC
- Alternate-arm MMC
- Clamped double-submodule-based MMC

Among the above-mentioned topologies of MMCs, the full-bridge, hybrid half-full-bridges, and alternate-arm topologies provide the fault current limitation and control ability. The clamped double-submodule-based MMC can only limit the fault currents, and the half-bridge MMC can neither limit nor control fault currents (Norrga et al., 2014). However, the half-bridge topology has the lowest costs, losses, and design complexity (Dekka et al., 2017; Norrga et al., 2014; Wang and Marquardt, 2013). The half-bridge topology has been utilized in world-leading commercial projects such as the Nan'ao three-terminal MMC-based HVDC (MMC-HVDC) system (Rao, 2015), the Zhoushan 5-terminal MMC-HVDC system (Tang et al., 2015), and the Zhangbei four-terminal MMC-HVDC grid (Tong et al., 2019). The circuit diagram of a half-bridge MMC is depicted in Figure 17.6. As can be seen from this figure, each MMC arm contains multiple submodules, as the main reason for naming this type of converter.

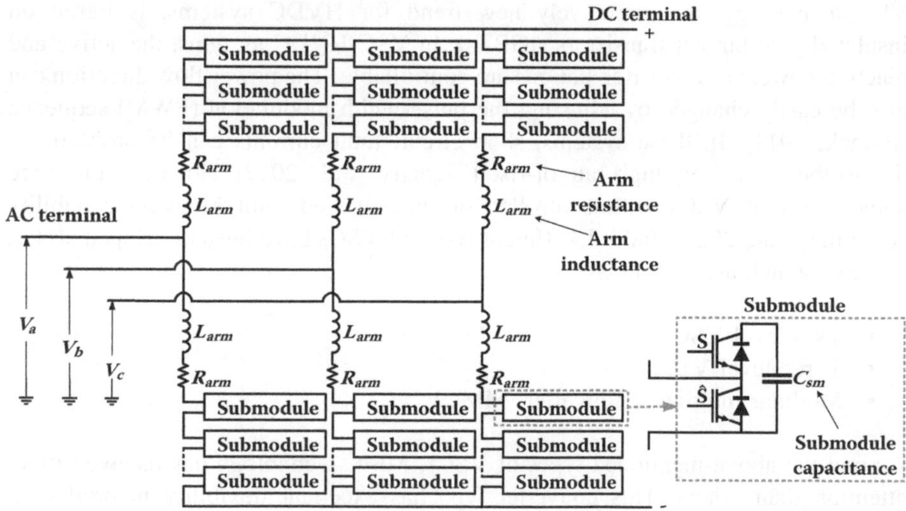

FIGURE 17.6 Circuit illustration of a half-bridge MMC.

VSC-HVDC transmission systems may be installed/operated based on different configurations such as (Sharifabadi et al., 2016):

- Symmetrical monopolar configuration
- Asymmetrical monopolar configuration with the ground return
- Asymmetrical monopolar configuration with a metallic return
- Bipolar configuration with the ground return
- Bipolar configuration with a metallic return

Figure 17.7 demonstrates these configurations for two-terminal VSC-HVDC transmission systems.

The VSC-HVDC system's grounding scheme is one of the critical factors that may influence the fault diagnosis schemes' performance due to its effect on the fault current path. According to the literature, the following grounding schemes are implementable in MMC-HVDC systems:

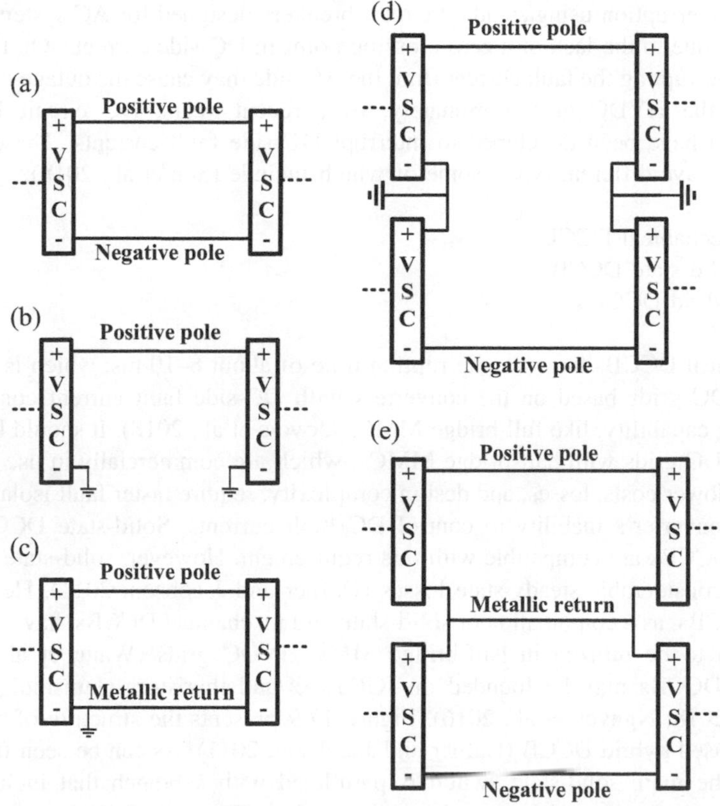

FIGURE 17.7 Typical configurations for 2-terminal VSC-HVDC transmission systems: (a) symmetrical monopolar; (b) asymmetrical monopolar with the ground return; (c) asymmetrical monopolar with a metallic return; (d) bipolar with the ground return; (e) bipolar with a metallic return.

- Grounding through a resistor at the neutral point of the star-connected transformer winding on the AC side (Hu et al., 2017; Li et al., 2017; Wang et al., 2014)
- Grounding through a three-phase star-connected reactor and a resistor at its neutral point on the AC side (Hu et al., 2017; Li et al., 2017)
- Grounding through a zig-zag transformer and a resistor at its neutral point on the AC side (Li et al., 2017; Wang et al., 2014)
- Grounding the midpoint of large parallel resistors on the DC side (Hu et al., 2017)
- Grounding the midpoint of small parallel capacitors on the DC side (Leterme et al., 2014)

The above-mentioned grounding schemes are illustrated for symmetrical monopolar MMC-HVDC systems in Figure 17.8.

One of the HVDC grid requirements is the timely isolation of short-circuit faults before severe equipment damage and bulky grid outage. However, the DC side fault current interruption using standard circuit breakers designed for AC systems is not practical due to the lack of a zero-crossing point in DC-side currents. On the other hand, interrupting the fault current from the AC side may cause the outage of a large part of the HVDC grid. Fortunately, over recent years, DC circuit breakers (DCCBs) have been developed to interrupt DC-side fault currents. These circuit breakers have different types, some of which include (Sen et al., 2018):

- Mechanical DCCB
- Solid-state DCCB
- Hybrid DCCB

Mechanical DCCBs have an interruption time of about 8–10 ms, which is suitable for HVDC grids based on the converters with DC-side fault current controlling/blocking capability, like full-bridge MMCs (Jovcic et al., 2018). It should be noted that HVDC grids with half-bridge MMCs, which are commercially in use because of their lower costs, losses, and design complexity, require faster fault isolation due to the converter's inability to control DC fault currents. Solid-state DCCBs and hybrid DCCBs are compatible with this requirement. However, solid-state DCCBs impose considerable steady-state losses (Häfner and Jacobson, 2011). Hence, hybrid DCCBs, as a combination of solid-state and mechanical DCCBs, have attracted attention to be utilized in half-bridge MMC-HVDC grids (Wang et al., 2018). Hybrid DCCBs may be founded on IGBTs or/and thyristors (Jamshidi Far and Jovcic, 2018; Nguyen et al., 2016). Figure 17.9 presents the structure of a typical IGBT-based hybrid DCCB (Häfner and Jacobson, 2011). As can be seen from this figure, the main solid-state switch is paralleled with a branch that includes the auxiliary solid-state and fast mechanical switches. The normal load current is bypassed via this parallel path. In the case of a trip command, the auxiliary solid-state switch commutates the current to the main solid-state switch, and the fast mechanical switch isolates the bypass path. Then, the main solid-state switch interrupts the current. The series reactor shown in Figure 17.9 is considered to limit the fault

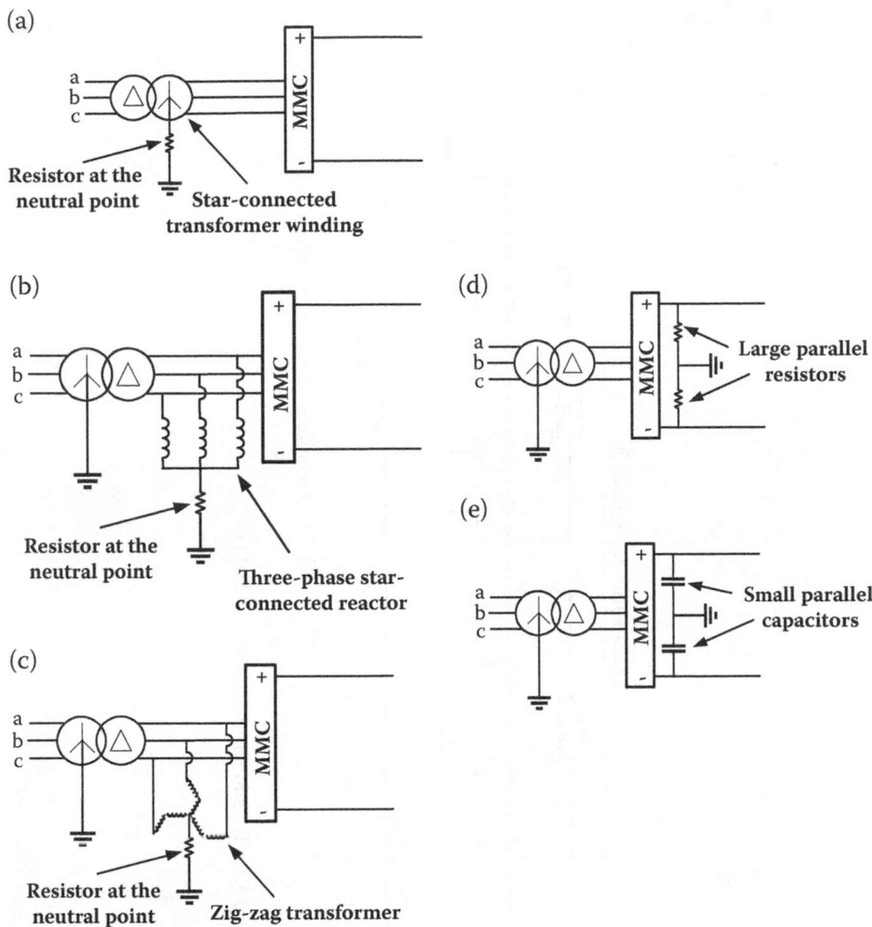

FIGURE 17.8 Different grounding schemes for symmetrical monopolar MMC-HVDC systems: (a) grounding through a resistor at the neutral point of the star-connected transformer winding; (b) grounding through a three-phase star-connected reactor and a resistor at its neutral point; (c) grounding through a zig-zag transformer and a resistor at its neutral point; (d) grounding the midpoint of large parallel resistors; (e) grounding the midpoint of small parallel capacitors.

current's rate-of-rise. This reactor helps hybrid DCCBs cut off the fault current before reaching the breaking capacity by prolonging the current rise time.

Since hybrid DCCBs are usually installed at the connection point of converter stations to transmission lines with their specific current limiting reactors, the voltage measurements before, after, or on these series reactors can affect the performance of designed protection and location schemes. In contrast, the current measurement place relative to these series components will not matter. Figure 17.10 depicts the single-line diagram of a four-terminal ±320 kV symmetrical monopolar half-bridge MMC-HVDC grid (Leterme et al., 2015) simulated via PSCAD/EMTDC (PSCAD

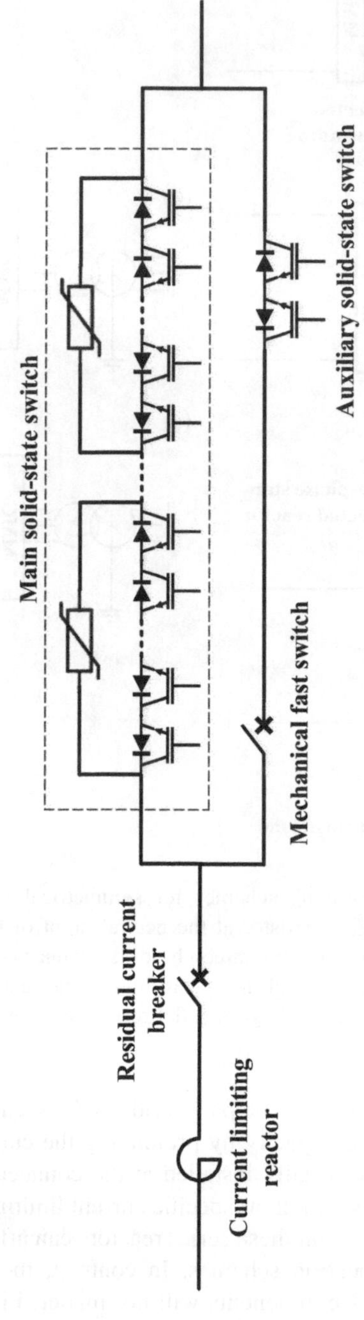

FIGURE 17.9 Structure of a typical IGBT-based hybrid DCCB (Häfner and Jacobson, 2011).

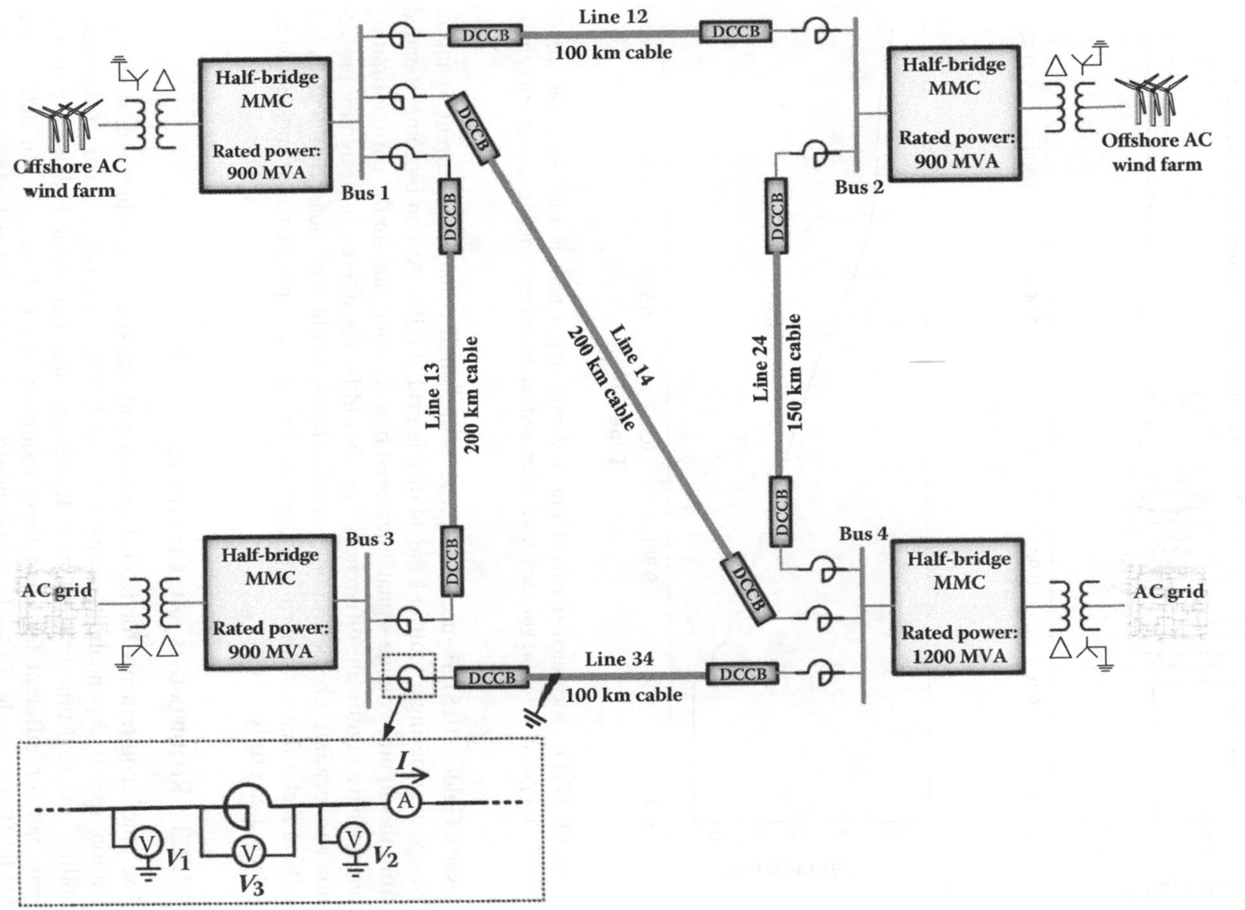

FIGURE 17.10 Illustration of a four-terminal MMC-HVDC grid (Leterme et al., 2015).

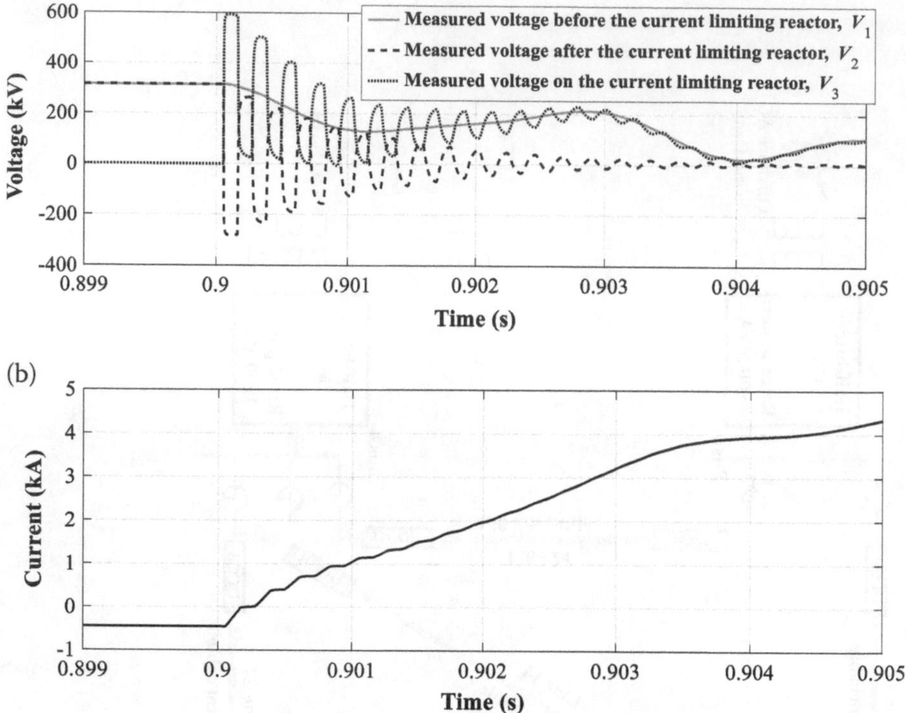

FIGURE 17.11 Measured signals at Bus 3, considering a fault in Line 34 of the MMC-HVDC test grid at 0.9 s: (a) voltage before, after, and on the current limiting reactor; (b) current.

User's Guide, 2018). Figure 17.11 presents measured voltage and current signals at Bus 3, considering a fault in Line 34 of this grid at 0.9 s. As can be comprehended from this figure, voltage signals measured before, after, and on the current limiting reactor have significant differences that should be considered in designing protection and location schemes. In other words, the exact place of measuring devices is a critical factor that can improve or degrade the scheme's performance and should be selected carefully.

17.2.3 Requirements and Challenges

The chief requirements for fault detection and classification schemes include operation speed, dependability, sensitivity, stability, selectivity, accuracy, practicability, and real-time executability. However, each particular HVDC technology may call for different levels of these requirements. For instance, in half-bridge MMC-HVDC grids, DC-side faults should be quickly detected within few milliseconds, e.g. in less than 3 ms for the Zhangbei four-terminal MMC-HVDC grid (Xiang et al., 2019; Zhang, Zou, et al., 2020), to prevent severe equipment damage and bulky grid outage. On the other hand, in two-terminal CSC-HVDC systems, a longer fault detection delay in the range of a few hundred milliseconds, even with an intentional 600-ms block time (Zheng et al., 2018), will not pose an acute problem thanks to the relatively small rate-of-rise of fault currents and the fault current control ability of their converters.

In an HVDC system, reliably detecting all solid and resistive internal faults in the designated protection zone with proper sensitivity and being completely stable against external faults or disturbances are of great importance for correctly isolating the faulty section and maintaining the continuity of power transmission. Moreover, the correct classification of faults is vital for a bipolar HVDC system since, in the event of single-pole faults, it is possible to partially maintain the power transfer capacity by operating the system in the monopolar mode. Correct identification and classification of faults are also critical for subsequent off-line steps such as the accurate fault location and maintenance operations. The use of easily measurable signals increases the practicability of the protection scheme. The computational burden and sampling frequency are also influential factors for real-time algorithms. The lower the computational burden and sampling frequency, the easier it will be to implement the real-time algorithm in reality.

As soon as a DC line fault is detected and classified, its location should be accurately estimated. Consequently, the maintenance crew should be dispatched to the estimated fault location for repairs. In the case of a permanent fault, these subsequent off-line steps help to restore the power transmission capacity as soon as possible. Furthermore, in the case of a temporary fault, these steps are necessary to inspect and prevent similar events in the future. While fault location schemes should be accurate and sensitive to the line faults' location, they should be implementable at the lowest costs, and their accuracy should accept the least impacts from other pre-fault and fault parameters.

It should be noted that fault location schemes should also be designed with specific consideration given to each HVDC technology and employed protection schemes. For example, according to ultra-high-speed protection schemes and hybrid DCCBs recommended for half-bridge MMC-HVDC grids, measurable post-fault data before the fault isolation stage may be extremely short, and fault location may be challenging in such systems. In contrast, it is not usually an issue in CSC-HVDC systems.

The pattern-recognition techniques and artificial intelligence have inherent capabilities that can be utilized to meet the abovementioned requirements and challenges. However, the extraction/selection of useful input characteristics and the utilization of appropriate learning algorithms/models are the critical steps towards contriving efficient schemes. It should be noted that fault detection and classification schemes require classifiers, while fault location schemes usually employ regression predictors.

17.3 FAULT DETECTION AND CLASSIFICATION IN CSC-HVDC SYSTEMS

In CSC-HVDC systems, reliably detecting internal faults in the protected zone with proper sensitivity and being stable against external faults or disturbances, as well as correctly identifying faulty poles, are vital for recovering all or part of the power transfer capacity and also critical for subsequent steps such as the fault location and repair operations. Some of the traditional protections used in CSC-HVDC systems include the voltage derivative, under-voltage, current differential, current derivative, and traveling-wave-based protections, each of which has its drawbacks (Kong et al., 2014; Suonan et al., 2013; Zheng et al., 2018). Researchers have attempted to improve these protections or present new schemes based on the single-end (Hao, Mirsaeidi, et al., 2018; Kong et al., 2016; Kong et al., 2014; Ma et al., 2018; Song et al., 2015; Suonan et al., 2013) and double-end (Li et al., 2019; Wang et al., 2017; Zhang, Li, et al., 2020; Zhang, Li, et al., 2019; Zheng et al., 2018; Zheng et al., 2020) measurements. These schemes have significant advantages over traditional protection schemes. However, they also have some disadvantages and have not entirely met all the protection requirements. Ideally, the correct operation of fault detection and classification schemes should not be affected by the fault type, fault resistance, fault location, pre-fault current and voltage, sampling frequency of signals, and the measurement/communication disturbances. In recent years, some researchers have tried to move towards this by relying on the pattern-recognition and machine-learning techniques' inherent capabilities (Agarwal et al., 2019; Farshad, 2019a; Johnson and Yadav, 2017).

17.3.1 Input Features

Johnson and Yadav (2017) have extracted the input features from the rectifier voltage and current signals sampled at the frequency of 1 kHz on the DC and AC sides for implementing an intelligent protection method for bipolar CSC-HVDC transmission systems. For detecting faults, they have directly used the samples of AC side's root mean square (RMS) voltage signal, DC side's voltage signals on both poles, and DC side's current signals on both poles as the input features. On the other hand, they have calculated the standard deviation of these signal samples from 10 ms before to 10 ms after the fault inception time for constructing the input feature vector to the fault type classifier.

Agarwal et al. (2019) have considered the RMS values of current signals as the input features for fulfilling an intelligent protection algorithm for bipolar CSC-HVDC transmission systems. They have measured the required current signals from both poles on the rectifier's DC-side with a 1-kHz sampling rate. It should be noted that the DC filters have not been considered in this study.

Farshad (2019a) has extracted four distinguishing features from the current and voltage data sampled at the frequency of 2 kHz on the inverter's DC side for fault detection and classification in bipolar CSC-HVDC transmission systems. He has considered the per-unit current and voltage samples' sum values on both DC poles over 10-ms time windows for constructing the input feature vector. In this study, the measuring instruments have been placed between the smoothing reactor and DC filter of each pole.

As is evident, one of the remarkable advantages of the abovementioned schemes is the comparatively low sampling frequencies used. Also, all of these methods have used only single-end measurements. In these schemes, it has been attempted to extract the input features with the least amount of calculations to be suitable for real-time execution. All these advantages have been made possible by relying on the intelligent techniques' inherent capabilities.

17.3.2 Learning Algorithms/Models

Johnson and Yadav (2017) have employed binary and multi-class SVM classifiers for fault detection and classification in bipolar CSC-HVDC transmission systems. Also, Agarwal et al. (2019) have examined FIS classifiers for this purpose. Besides these schemes, Farshad (2019a) has adapted one-class K-MDD classifiers for designing the protection scheme for bipolar CSC-HVDC transmission systems. Indeed, he has prepared three one-class K-MDD classifiers using the training patterns related to three types of faults in the protected zone.

All the previously mentioned learning algorithms/models with the suitably extracted input features have demonstrated promising results in protecting CSC-HVDC systems.

17.4 FAULT LOCATION IN CSC-HVDC SYSTEMS

Usually, high volumes of electric power are transmitted over long distances through CSC-HVDC transmission systems. Therefore, after detecting and classifying line faults in these systems, accurate fault location is of high importance to perform repair operations and restore the power transmission capacity to reduce the outage time and increase reliability. So far, several fault location schemes have been designed for CSC-HVDC transmission systems based on the single-end and double-end measurements, which can be categorized as follows:

- Single-ended traveling-wave-based schemes (Ando et al., 1985; He et al., 2014; Livani and Evrenosoglu, 2014)
- Double-ended traveling-wave-based schemes (Dewe et al., 1993; Nanayakkara et al., 2012a)
- Double-ended time-domain schemes (Li et al., 2012; Suonan et al., 2010; Yuansheng et al., 2015)
- Single-ended intelligent schemes (Farshad, 2018, 2019b; Farshad and Sadeh, 2013; Johnson and Yadav, 2017)

Generally, single-ended fault location schemes seem more attractive than double-ended ones due to their higher reliability in data availability and immunity against telecommunication link disturbances. However, single-ended traveling-wave-based schemes inherently face issues like the essential very high sampling frequencies and the challenge in detecting forward and backward waves, especially for high resistance faults (Suonan et al., 2010). Ideally, fault location schemes should accurately estimate the location of faults at any point along the line. Their accuracy should also not be affected by the fault type, fault resistance, pre-fault current and voltage, sampling frequency of signals, and the measurement/communication disturbances. The single-ended intelligent schemes can deal with these requirements thanks to the pattern-recognition and machine-learning approaches' competency in solving complex problems.

17.4.1 Input Features

Farshad and Sadeh (2013) have utilized the rare voltage samples on the faulty pole of CSC-HVDC systems from the sudden voltage drop instant to 10 ms thereafter for constructing the input feature vector to the fault locator. They have sampled the required voltage signal with the frequency of 80 kHz at the rectifier station (after the smoothing reactor). However, the analyses have confirmed that considering lower sampling frequencies, even as low as 2.5 kHz, will not significantly reduce the accuracy of the fault location scheme designed by Farshad and Sadeh (2013).

The complete CSC-HVDC protection scheme proposed by Johnson and Yadav (2017) also includes a fault location module. For this fault location module, Johnson and Yadav (2017) have used the same input features extracted for the fault classification module, i.e. the standard deviation of AC side's RMS voltage signal, DC side's voltage signals on both poles, and DC side's current signals on both poles from 10 ms before to 10 ms after the fault inception time. These required signals have also been captured at the rectifier station with a 1-kHz sampling rate. Farshad (2018) has examined a combination of SVD and PA for extracting useful fault location features from 5-ms frames of the after-fault voltage signals sampled on both poles with the frequency of 16 kHz at the rectifier station of CSC-HVDC systems (after the smoothing reactor). Indeed, he has constructed the input feature vector with eight features, including the normalized amplitude, phase angle, damping coefficient, and frequency values extracted for both poles.

In another work, Farshad (2019b) has utilized DFT for extraction of the harmonic spectrum from a 20-ms frame of the post-fault voltage data sampled on the faulted pole with a frequency of 16 kHz at the rectifier station (after the smoothing reactor). He has then constructed the input feature vector considering the most useful harmonic components pre-selected via a regression feature selection technique, i.e. RReliefF.

All the abovementioned schemes have used only single-end measurements and do not need to transmit/synchronize data measured at both line terminals. A relatively low sampling frequency requirement compared to the single-ended traveling-wave-based schemes is another advantage of these schemes. Since fault location algorithms usually do not require real-time execution, and the computational burden is usually not an issue in these algorithms, they have more freedom in examining various feature extraction/selection techniques and tools.

17.4.2 Learning Algorithms/Models

Farshad and Sadeh (2013) have employed a modified form of K-NN target predictor founded on the Pearson distance measure for locating faults in CSC-HVDC transmission lines. Johnson and Yadav (2017) have utilized an SVR target predictor for finding the relationship between the input features and the fault location in CSC-HVDC transmission lines. Farshad (2018) has found a GRNN target predictor more accurate to estimate the location of CSC-HVDC line faults based on the considered input features. Moreover, Farshad (2019b) has suggested an ensemble of regression decision trees, i.e. RF target predictor, for estimating the fault location in CSC-HVDC transmission lines using the selected frequency-domain input features.

It is noteworthy that the appropriate learning algorithm/model should be selected according to the input feature vector used. All the previously mentioned fault location schemes have demonstrated an acceptable estimation accuracy considering various pre-fault and fault conditions.

17.5 FAULT DETECTION AND CLASSIFICATION IN VSC-HVDC SYSTEMS

VSC-HVDC systems may have different protection requirements depending on their converter capability in controlling/blocking short-circuit fault currents. Most recent studies have focused on designing ultra-high-speed protection schemes for multi-terminal VSC-HVDC systems whose converters cannot control fault currents, like half-bridge MMC-HVDC grids with hybrid DCCBs embedded for very quickly isolation of faults. The ultra-high-speed fault detection and classification schemes designed for such systems should also be dependable, sensitive, stable, selective, accurate, efficiently and economically implementable, and executable in real time.

Some of the line protection schemes presented for VSC-HVDC systems require measured data from one line terminal and decide based on local calculations only at that terminal (Ikhide et al., 2018; Leterme et al., 2016; Sabug et al., 2020; Sneath and Rajapakse, 2016; Tang et al., 2019; Xiang et al., 2019; Yang et al., 2020; Zhang, Zou, et al., 2020). Some others decide based on exchanging local calculation results obtained at both line terminals using their local measurements (Huang et al., 2019; Li et al., 2018). Moreover, there are protection schemes that require measurements from both line terminals (Abu-Elanien et al., 2016; Hajian et al., 2015; Jovcic et al., 2018; Wang et al., 2019) or from sensors distributed along the protected line (Tzelepis et al., 2017) to perform calculations and make the decision. In addition to lines, VSC-HVDC systems' buses have also attracted researchers' attention to design differential (Azad and Hertem, 2017; Hajian et al., 2015; Tzelepis, Dyśko, Blair, et al., 2018; Zou et al., 2018) and non-differential (Elgeziry et al., 2019; Sneath and Rajapakse, 2016; Tzelepis et al., 2019) schemes tailored to their protection requirements. Of course, the problem has not been completely solved. Studies are still ongoing to design protection schemes with the least impact from the fault type, fault resistance, fault location, pre-fault current and voltage, sampling frequency of signals, and the measurement/communication disturbances. The system complexity and the critical protection requirements have motivated some researchers to benefit from the pattern-recognition and machine-learning procedures' inherent capabilities for designing efficient protection schemes for VSC-HVDC systems' lines (Bertho et al., 2018; Farshad, 2021; Xiang et al., 2020; Yang et al., 2017) and buses (Farshad 2020; Xiang et al., 2020).

17.5.1 Input Features

Yang et al. (2017) have employed DFT to extract the distinguishing features from 1-ms frames of the single-end current signals measured with a 10-kHz sampling rate on both poles to detect and classify line faults in multi-terminal two-level VSC-HVDC systems. They have constructed the input vector with 40 scaled features, considering

20 current high-frequency components in the range of 2800 Hz to 4700 Hz for each pole.

Bertho et al. (2018) have employed WPD and PCA to extract the distinguishing features from 3-ms frames of the single-end current data gathered on the DC-side with a 2-kHz sampling rate to protect lines in multi-terminal two-level VSC-HVDC systems. At first, they have calculated the normalized energy percentage of wavelet coefficients at the second level of WPD. They have then applied PCA to these values and constructed the input feature vector, including the scaled principal components.

Xiang et al. (2020) have designed a scheme to detect and classify bus and line faults in half-bridge MMC-HVDC grids. They have extracted the input features from 0.5-ms and 1.5-ms frames of the single-end voltage signals captured with a 10-kHz sampling rate before and after the current limiting reactor, respectively. They have constructed the input vector considering the following features:

- The energy of voltage detailed coefficients obtained by applying one level of DWT on a 0.5-ms data frame measured before the current limiting reactor
- The energy of voltage detailed coefficients obtained by applying one level of DWT on a 1.5-ms data frame measured after the current limiting reactor
- The integral of positive pole voltage changes over a 1.5-ms data frame measured after the current limiting reactor
- The integral of negative pole voltage changes over a 1.5-ms data frame measured after the current limiting reactor

Farshad (2020b) has designed a non-unit non-differential bus protection scheme for half-bridge MMC-HVDC grids. He has used 2-ms frames of the superimposed samples related to the voltage signal measured before the current limiting reactors and the current signals flowing through all lines connected to the protected bus. This bus protection scheme can work with a sampling frequency of 5 kHz and also can partially cover the AC-side of converter stations for severe faults that need to be swiftly detected and isolated as well as DC bus faults.

In another work, Farshad (2020a) has used 2-ms frames of the superimposed current and voltage signals measured after the current limiting reactors at both line terminals with the sampling frequency of 5 kHz for designing a pilot strategy for the line protection of half-bridge MMC-HVDC grids. This pilot scheme separately protects each line's pole and decides based on the exchange of fault detection results independently obtained at both line terminals using their local measurements.

Thanks to the inherent capabilities of pattern-recognition techniques, the required sampling frequencies of the schemes mentioned above are comparatively lower than most existing methods, making it easier to implement them in real-time applications (Farshad, 2021).

17.5.2 Learning Algorithms/Models

Yang et al. (2017) have employed separate MLPNN classifiers to detect and classify line faults in multi-terminal two-level VSC-HVDC systems. Bertho et al. (2018) have used a FIS classifier optimized by the genetic algorithm to detect line faults in multi-terminal two-level VSC-HVDC systems. Xiang et al. (2020) have also examined a multi-class MLPNN classifier to detect and classify bus and line faults in half-bridge MMC-HVDC grids. Farshad (2020b) has proposed a distance measure-based pattern scanning approach to design the non-unit, non-differential scheme and detect internal bus faults in the designated protection zone. In another work, Farshad (2020a) has separately implemented the distance measure-based pattern scanning approach at both line terminals and exchanged the binary fault detection results to decide on the occurrence of internal line faults in half-bridge MMC-HVDC grids. He has implemented the distance measure-based pattern scanning approach by comparing the cosine distance measure between real-time data frames and pre-recorded reference data vectors with pre-adjusted thresholds.

Thanks to the suitably utilized input features, all the previously mentioned learning algorithms/models have shown good achievement in suitably protecting VSC-HVDC systems' buses and lines.

17.6 FAULT LOCATION IN VSC-HVDC SYSTEMS

Accurate fault location is vital to quickly perform repair operations and reduce the outage time after detecting and clearing line faults in VSC-HVDC systems. Indeed, fault location in VSC-HVDC systems is a more challenging job due to a higher rate-of-rise of fault currents and fast protection and clearing strategies usually adopted according to the fault current control ability/inability of utilized converters in these systems (Leterme et al., 2019; Leterme and Van Hertem, 2015).

In general, input post-fault data used in fault location schemes can be captured from the measured data in three time-frames, including before the fault clearing stage, during the fault clearing stage, and after the fault clearing stage (as indicated in Figure 17.12). Naturally, in fault location schemes, it is common to use post-fault data before the fault clearing stage. However, in VSC-HVDC systems, especially in those outfitted with very-fast protection algorithms and hybrid DCCBs, available measured data before the fault clearing stage is very limited, making the fault location task difficult. Therefore, researchers may move to use data of the next time frames.

So far, some fault location schemes have been designed for two-terminal VSC-HVDC transmission systems using the single-end measurements during the fault existence (Ashouri et al., 2020; Song 2014; Zhang, Song, et al., 2019). Some fault location schemes have also been presented for multi-terminal VSC-HVDC systems based on the wide-area (Azizi, Afsharnia, et al., 2014; Azizi, Sanaye-Pasand, et al., 2014; Nanayakkara et al., 2012b), double-end (Li, Yang, Mu, Blond, and He 2018; Xu et al., 2020; Zhang et al., 2017), and along-line-distributed (Tzelepis, Fusiek, et al., 2018) measurements during the fault existence. Furthermore, there are fault location schemes founded on the pulse/signal injection from hybrid DCCBs of multi-terminal VSC-HVDC systems after the fault isolation stage (Song et al., 2019; Zhang, Zou, et al., 2019).

In addition to the abovementioned studies, some researchers have examined the pattern-recognition and learning procedures to design accurate fault location schemes for two-terminal (Hao, Wang, et al., 2018; Luo et al., 2018) and multi-terminal

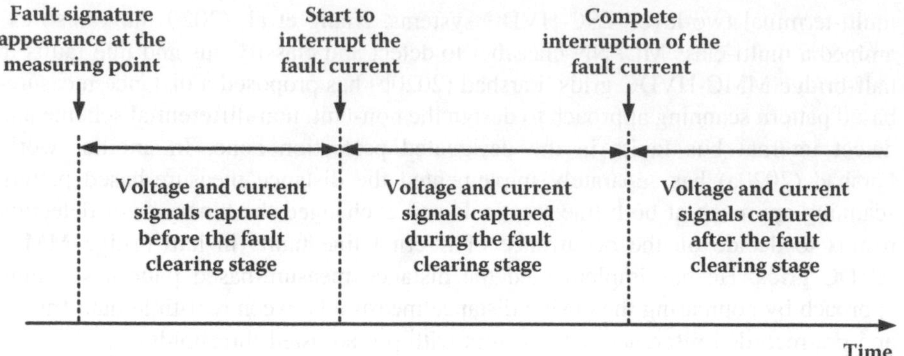

FIGURE 17.12 Time frames that can be considered for capturing input post-fault data for fault location schemes.

(Tzelepis, Dyśko, Fusiek, et al., 2018; Tzelepis et al., 2020; Yang et al., 2017) VSC-HVDC systems with the least impact from the pre-fault and fault parameters.

17.6.1 Input Features

Hao, Wang, et al. (2018) have decoupled the single-end current signal sampled at the frequency of 1 MHz on the DC side during the fault existence and employed HHT to extract the following scaled features for locating faults in two-terminal VSC-HVDC systems:

- The time delay between the modal components
- The characteristic frequency of mode 1
- The energy weakening coefficient of mode 0
- The energy weakening coefficient of mode 1
- The high-frequency energy of mode 0
- The high-frequency energy of mode 1

Luo et al. (2018) have directly used scaled samples of a 20-μs data frame from the single-end current signal sampled at the frequency of 5 MHz on the DC-side during the fault existence as the input features for locating faults in two-terminal two-level VSC-HVDC systems.

Yang et al. (2017) have employed DFT to extract the predictive features from 6-ms frames of the double-end current signals sampled at the frequency of 10 kHz on the faulty pole during the fault existence for locating faults in multi-terminal two-level VSC-HVDC systems. They have constructed the input vector with 98 scaled features, considering 49 current high-frequency components in the range of 3600 Hz to 4400 Hz for each terminal.

Tzelepis, Dyśko, Fusiek, et al. (2018) have utilized 5-ms frames of the current data measured by the along-line-distributed optical instruments with the sampling frequency of 5 kHz during the fault existence for fault location in half-bridge

MMC-HVDC grids. After the fault segment identification, they have constructed the input feature matrix by applying CWT and the double-ended traveling-wave-based fault location method considering all possible pairs of sensors placed at the left side and right side of the faulty segment. Indeed, they have considered the initial fault location results, with significant errors due to the low sampling frequency, as the input feature matrix elements.

Tzelepis et al. (2020) have used 6-ms data frames of the single-end current signals measured from three internal paths of hybrid DCCBs on both poles with the sampling frequency of 96 kHz during the fault clearing stage for locating faults in half-bridge MMC-HVDC grids.

As is evident in the previously mentioned methods, the pattern-recognition and feature extraction techniques have been chosen with more freedom to deal with the challenging requirements of VSC-HVDC systems since fault location is usually an offline task.

17.6.2 Learning Algorithms/Models

Hao, Wang, et al. (2018) have employed an SVR target predictor optimized with the bat algorithm for estimating the location of faults in 2-terminal VSC-HVDC systems. Also, Luo et al. (2018) have examined an SAE target predictor for finding relationships between the considered input features and the location of faults in two-terminal two-level VSC-HVDC systems.

Yang et al. (2017) have employed an MLPNN target predictor for predicting the location of faults in multi-terminal two-level VSC-HVDC systems. Tzelepis, Dyśko, Fusiek, et al. (2018) have attempted to increase the accuracy of fault location results for half-bridge MMC-HVDC grids using a K-NN idea founded on the Pearson distance measure. In another work, Tzelepis et al. (2020) have also estimated the location of faults in half-bridge MMC-HVDC grids, adapting a Pearson distance-based K-NN target predictor.

All of the previously mentioned learning algorithms/models have been selected with specific attention given to the extracted input features to achieve the best possible performance and accuracy.

17.7 CONSIDERATIONS FOR PRACTICAL IMPLEMENTATIONS

In addition to the general requirements and challenges mentioned in Section 17.2.3, specific challenges need to be considered in evaluating and practically implementing intelligent fault diagnosis schemes, some of which are as follow:

- Implementation costs
- Unseen new cases
- High-resistance faults
- Temporary arc faults
- Fault locations very close to line terminals
- Operation of adjacent circuit breakers
- Lightning disturbances

- Measurement noises/errors
- Inaccurate line parameters
- Communication delay, disturbance, and failure
- Time synchronization errors

These challenges are briefly discussed in this section.

17.7.1 Implementation Costs

The implementation cost of a fault diagnosis scheme should be as low as possible. If a scheme can be implemented with existing signal measurement/conditioning/transmitting infrastructure and with minimal modifications and additional equipment, its implementation cost will be satisfactorily reduced. Lower computational burden and required sampling frequency can also decrease the hardware cost and make it easier to implement real-time algorithms in reality.

17.7.2 Unseen New Cases

Intelligent fault diagnosis schemes should also perform well in the face of new pre-fault and fault conditions not seen in the training and preparation phase. The selected input features and learning model can affect the intelligent scheme's accuracy and generalizability in dealing with these new cases.

17.7.3 High-Resistance Faults

The fault resistance is one of the challenging factors for fault diagnosis schemes. No specific range is provided for the fault resistance. It can be 0 to several kΩ. The higher the fault resistance tolerable by the fault diagnosis scheme, the better its dependability and performance.

17.7.4 Temporary Arc Faults

Short-circuit faults can be permanent or temporary and can occur through linear or nonlinear arc resistances. Temporary faults with nonlinear arc resistances are among the most challenging cases for fault diagnosis schemes, specifically for fault location schemes. Location schemes should be able to deal with these faults as well. It is worth noting that locating temporary faults can be useful for preventive maintenance operations.

17.7.5 Fault Locations Very Close to Line Terminals

Measurement devices are usually placed at the line terminals. Faults very close to these measurement devices can generate higher frequency components in the measured signals. Therefore, they can be challenging for fault diagnosis schemes founded on the high-frequency components, depending on their sampling frequency. Single-ended fault detection schemes may also have difficulties distinguishing internal faults close to the remote terminal from the external ones, depending on their reach point.

17.7.6 Operation of Adjacent Circuit Breakers

As is clear, the fault detection scheme designed for HVDC grids should be stable for external faults outside of its protection zone. However, in an HVDC grid, the operation of protection relays and circuit breakers themselves can also generate considerable disturbances. The fault detection scheme should also remain stable during the disturbances caused by the adjacent circuit breakers' operation.

17.7.7 Lightning Disturbances

Overhead lines are exposed to lightning strikes. The lightning-generated disturbances can activate fault detection schemes and result in an undesirable outage. Therefore, it is necessary to design fault detection schemes in such a way as to withstand such disturbances.

17.7.8 Measurement Noises/Errors

Under real circumstances, measured and sampled signals may contain noise. Measuring equipment is also not ideal and may include a percentage of error in measurements. Therefore, it is vital to evaluate the designed scheme's tolerability to measurement noises/errors.

17.7.9 Inaccurate Line Parameters

Intelligent fault diagnosis schemes are usually trained and prepared using a data set generated through simulations in various conditions. Although the electromagnetic transient simulation tools have been significantly developed in accuracy and conformity with reality, a degree of discrepancy between real and simulated data is avoidable. The transmission line is one of the most affecting components that can exacerbate these differences between real and simulated data. For example, in reality, the height of conductors from the ground level and the ground resistivity may change when the transmission line passes from different geographical areas. Climatic conditions in different regions and seasons can also change the transmission line parameters. Therefore, the prepared fault diagnosis scheme should tolerate a reasonable level of changes in the line parameters relative to those used in the training phase.

17.7.10 Communication Delay, Disturbance, and Failure

The communication equipment's performance and reliability are important factors for double-ended fault detection, classification, and location schemes. For example, ultra-high-speed operation of pilot fault detection schemes can be undesirably affected by the communication delay, disturbance, and failure. The undesirable effects of communication delay and disturbance on the pilot fault detection schemes' performance can be even more challenging when required to transmit voltage and current signals sampled at high rates over long distances. Therefore, in pilot ultra-high-speed fault detection schemes, attempts have been made to perform part of the calculations locally and transmit only the calculations' result as a logical signal.

17.7.11 Time Synchronization Errors

Double-ended fault detection, classification, and location schemes, in addition to data transmission, may require time synchronization for two-terminal data. Although the fault diagnosis estimators may be trained based on ideally synchronized data, they should tolerate a reasonable time synchronization error level.

17.8 CONCLUSION

The present chapter has focused on applying pattern-recognition and machine-learning procedures in designing fault detection, classification, and location schemes for HVDC transmission systems with CSC and VSC technologies. Since these systems are relatively newer than HVAC systems, and their converter technologies are advancing rapidly, they have good potential for benefiting from the capabilities of pattern-recognition techniques and implementing artificial intelligence–based protection and location schemes. Since the intelligent schemes have high flexibility and, if implemented correctly, can perform well in the face of uncertainties and complexities, they can take the future of protection strategies for rapidly developing HVDC systems. However, although the intelligent schemes reviewed in this chapter have considerable advantages compared to the traditional ones, they need to be further developed, improved and matured. Additionally, the use of intelligent protection and fault location schemes in new HVDC systems, such as the full-bridge, hybrid half-full-bridges, alternate-arm, and clamped double-submodule-based MMC-HVDC systems, which has received less attention so far, can open a new field of study for researchers.

REFERENCES

[1] Abu-Elanien, A. E. B., A. A. Elserougi, A. S. Abdel-Khalik, A. M. Massoud, and S. Ahmed. 2016. A differential protection technique for multi-terminal HVDC. *Electric Power Systems Research* 130:78–88. https://doi.org/10.1016/j.epsr.2015.08.021

[2] Agarwal, S., A. Swetapadma, C. Panigrahi, and A. Dasgupta. 2019. A method for fault section identification in high voltage direct current transmission lines using one end measurements. *Electric Power Systems Research* 172:140–151. https://doi.org/10.1016/j.epsr.2019.03.008

[3] Ando, M., E. O. Schweitzer, and R. A. Baker. 1985. Development and field-data evaluation of single-end fault locator for two-terminal HVDV transmission lines-part 2: algorithm and evaluation. *IEEE Transactions on Power Apparatus and Systems* PAS-104 (12):3531–3537. https://doi.org/10.1109/TPAS.1985.318906

[4] Ashouri, M., F. F. D. Silva, and C. L. Bak. 2020. On the application of modal transient analysis for online fault localization in HVDC cable bundles. *IEEE Transactions on Power Delivery* 35 (3):1365–1378. https://doi.org/10.1109/TPWRD.2019.2942016

[5] Azad, S. P., and D. V. Hertem. 2017. A fast local bus current-based primary relaying algorithm for HVDC grids. *IEEE Transactions on Power Delivery* 32 (1):193–202. https://doi.org/10.1109/TPWRD.2016.2595323

[6] Azizi, S., S. Afsharnia, and M. Sanaye-Pasand. 2014. Fault location on multi-terminal DC systems using synchronized current measurements. *International Journal of Electrical Power & Energy Systems* 63:779–786. https://doi.org/10.1016/j.ijepes.2014.06.040

[7] Azizi, S., M. Sanaye-Pasand, M. Abedini, and A. Hasani. 2014. A traveling-wave-based methodology for wide-area fault location in multiterminal dc systems. *IEEE Transactions on Power Delivery* 29 (6):2552–2560. https://doi.org/10.1109/TPWRD.2014.2323356

[8] Bertho, R., V. A. Lacerda, R. M. Monaro, J. C. M. Vieira, and D. V. Coury. 2018. Selective nonunit protection technique for multiterminal VSC-HVDC grids. *IEEE Transactions on Power Delivery* 33 (5):2106–2114. https://doi.org/10.1109/TPWRD.2017.2756831

[9] Chaturvedi, D. K. 2008. *Soft Computing - Techniques and its Applications in Electrical Engineering*. Vol. 103, *Studies in Computational Intelligence*Janusz, Kacprzyk ed. Heidelberg, Germany: Springer-Verlag. https://doi.org/10.1007/978-3-540-77481-5

[10] Dekka, A., B. Wu, R. L. Fuentes, M. Perez, and N. R. Zargari. 2017. Evolution of topologies, modeling, control schemes, and applications of modular multilevel converters. *IEEE Journal of Emerging and Selected Topics in Power Electronics* 5 (4):1631–1656. https://doi.org/10.1109/JESTPE.2017.2742938

[11] Dewe, M. B., S. Sankar, and J. Arrillaga. 1993. The application of satellite time references to HVDC fault location. *IEEE Transactions on Power Delivery* 8 (3):1295–1302. https://doi.org/10.1109/61.252655

[12] Du, S., A. Dekka, B. Wu, and N. Zargari. 2018. *Modular Multilevel Converters: Analysis, Control, and Applications*. Hoboken, NJ, USA: Wiley-IEEE Press.

[13] Elgeziry, M. Z., M. A. Elsadd, N. I. Elkalashy, T. A. Kawady, and A. I. Taalab. 2017. AC spectrum analysis for detecting DC faults on HVDC systems. Paper read at Nineteenth International Middle East Power Systems Conference (MEPCON), 19-21 December, at Cairo, Egypt. https://doi.org/10.1109/MEPCON.2017.8301259

[14] Elgeziry, M., M. Elsadd, N. Elkalashy, T. Kawady, A.-M. Taalab, and M. A. Izzularab. 2019. Non-pilot protection scheme for multi-terminal VSC–HVDC transmission systems. *IET Renewable Power Generation* 13 (16):3033–3042. https://doi.org/10.1049/iet-rpg.2018.6265

[15] Eremia, M., C.-C. Liu, and A.-A. Edris, eds. 2016. *Advanced Solutions in Power Systems: HVDC, FACTS, and Artificial Intelligence*. Vol. 52, *IEEE Press Series on Power Engineering*. Hoboken, NJ, USA: Wiley-IEEE Press. https://doi.org/10.1002/9781119175391

[16] Farshad, M. 2018. Utilizing a combination of prony analysis and singular value decomposition for intelligent fault locating in bipolar high voltage direct current transmission lines. *Computational Intelligence in Electrical Engineering* 8 (4):31–44. https://doi.org/10.22108/isee.2018.105962.1063

[17] Farshad, M. 2019a. Detection and classification of internal faults in bipolar HVDC transmission lines based on K-means data description method. *International Journal of Electrical Power & Energy Systems* 104:615–625. https://doi.org/10.1016/j.ijepes.2018.07.044

[18] Farshad, M. 2019b. Locating short-circuit faults in HVDC systems using automatically selected frequency-domain features. *International Transactions on Electrical Energy Systems* 29 (3):e2765. https://doi.org/10.1002/etep.2765

[19] Farshad, M. 2020. Ultra-high-speed non-unit non-differential protection scheme for buses of MMC-HVDC grids. *IET Renewable Power Generation* 14 (9):1541–1549. https://doi.org/10.1049/iet-rpg.2019.1393

[20] Farshad, M. 2021. A pilot protection scheme for transmission lines of half-bridge MMC-HVDC grids using cosine distance criterion. *IEEE Transactions on Power Delivery* 36 (2): 1089–1096. https://doi.org/10.1109/TPWRD.2020.3001878

[21] Farshad, M., and J. Sadeh. 2013. A novel fault-location method for HVDC transmission lines based on similarity measure of voltage signals. *IEEE Transactions on Power Delivery* 28 (4):2483–2490. https://doi.org/10.1109/TPWRD.2013.2272436

[22] Faruque, M. O., Z. Yuyan, and V. Dinavahi. 2006. Detailed modeling of CIGRE HVDC benchmark system using PSCAD/EMTDC and PSB/SIMULINK. *IEEE Transactions on Power Delivery* 21 (1):378–387. https://doi.org/10.1109/TPWRD.2005.852376

[23] Franck, C. M. 2011. HVDC circuit breakers: A review identifying future research needs. *IEEE Transactions on Power Delivery* 26 (2):998–1007. https://doi.org/10.1109/TPWRD.2010.2095889

[24] Häfner, J., and B. Jacobson. 2011. Proactive hybrid HVDC breakers-A key innovation for reliable HVDC grids. Paper read at CIGRE Bologna Symposium - The Electric Power System of the Future: Integrating Supergrids and Microgrids, 13–15 September, at Bologna, Italy.

[25] Hajian, M., L. Zhang, and D. Jovcic. 2015. DC transmission grid with low-speed protection using mechanical DC circuit breakers. *IEEE Transactions on Power Delivery* 30 (3):1383–1391. https://doi.org/10.1109/TPWRD.2014.2371618

[26] Hao, Y., Q. Wang, Y. Li, and W. Song. 2018. An intelligent algorithm for fault location on VSC-HVDC system. *International Journal of Electrical Power & Energy Systems* 94:116–123. https://doi.org/10.1016/j.ijepes.2017.06.030

[27] Hao, W., S. Mirsaeidi, X. Kang, X. Dong, and D. Tzelepis. 2018. A novel traveling-wave-based protection scheme for LCC-HVDC systems using Teager Energy Operator. *International Journal of Electrical Power & Energy Systems* 99:474–480. https://doi.org/10.1016/j.ijepes.2018.01.048

[28] He, Z.-Y., K. Liao, X.-P. Li, S. Lin, J.-W. Yang, and R.-K. Mai. 2014. Natural frequency-based line fault location in HVDC lines. *IEEE Transactions on Power Delivery* 29 (2):851–859. https://doi.org/10.1109/TPWRD.2013.2269769

[29] Hu, J., K. Xu, L. Lin, and R. Zeng. 2017. Analysis and enhanced control of hybrid-MMC-based HVDC systems during asymmetrical dc voltage faults. *IEEE Transactions on Power Delivery* 32 (3):1394–1403. https://doi.org/10.1109/TPWRD.2016.2568240

[30] Huang, Q., G. Zou, S. Zhang, and H. Gao. 2019. A pilot protection scheme of DC lines for multi-terminal HVDC grid. *IEEE Transactions on Power Delivery* 34 (5):1957–1966. https://doi.org/10.1109/TPWRD.2019.2932188

[31] Ikhide, M., S. B. Tennakoon, A. L. Griffiths, H. Ha, S. Subramanian, and A. J. Adamczyk. 2018. A novel time domain protection technique for multi-terminal HVDC networks utilising travelling wave energy. *Sustainable Energy, Grids and Networks* 16:300–314. https://doi.org/10.1016/j.segan.2018.09.003

[32] Jamshidi Far, A., and D. Jovcic. 2018. Design, modeling and control of hybrid dc circuit breaker based on fast thyristors. *IEEE Transactions on Power Delivery* 33 (2):919–927. https://doi.org/10.1109/TPWRD.2017.2761022

[33] Johnson, J. M., and A. Yadav. 2017. Complete protection scheme for fault detection, classification and location estimation in HVDC transmission lines using

support vector machines. *IET Science, Measurement & Technology* 11 (3):279–287. https://doi.org/10.1049/iet-smt.2016.0244

[34] Jovcic, D., W. Lin, S. Nguefeu, and H. Saad. 2018. Low-energy protection system for DC grids based on full-bridge MMC converters. *IEEE Transactions on Power Delivery* 33 (4):1934–1943. https://doi.org/10.1109/TPWRD.2018.2791635

[35] Kong, F., Z. Hao, and B. Zhang. 2016. A novel traveling-wave-based main protection scheme for ±800 kV UHVDC bipolar transmission lines. *IEEE Transactions on Power Delivery* 31 (5):2159–2168. https://doi.org/10.1109/TPWRD.2016.2571438

[36] Kong, F., Z. Hao, S. Zhang, and B. Zhang. 2014. Development of a novel protection device for bipolar HVDC transmission lines. *IEEE Transactions on Power Delivery* 29 (5):2270–2278. https://doi.org/10.1109/TPWRD.2014.2305660

[37] Leterme, W., and D. Van Hertem. 2015. Classification of fault clearing strategies for HVDC grids. Paper read at CIGRE Lund Symposium, 27–28 May, at Lund, Sweden.

[38] Leterme, W., J. Beerten, and D. V. Hertem. 2016. Nonunit protection of HVDC grids with inductive DC Cable termination. *IEEE Transactions on Power Delivery* 31 (2):820–828. https://doi.org/10.1109/TPWRD.2015.2422145

[39] Leterme, W., P. Tielens, S. D. Boeck, and D. V. Hertem. 2014. Overview of grounding and configuration options for meshed HVDC grids. *IEEE Transactions on Power Delivery* 29 (6):2467–2475. https://doi.org/10.1109/TPWRD.2014.2331106

[40] Leterme, W., P. D. Judge, J. Wylie, and T. C. Green. 2020. Modeling of MMCs with controlled DC-side fault-blocking capability for DC protection studies. *IEEE Transactions on Power Electronics* 35 (6):5753–5769. https://doi.org/10.1109/TPEL.2019.2954743

[41] Leterme, W., I. Jahn, P. Ruffing, K. Sharifabadi, and D. V. Hertem. 2019. Designing for high-voltage dc grid protection: Fault clearing strategies and protection algorithms. *IEEE Power and Energy Magazine* 17 (3):73–81. https://doi.org/10.1109/MPE.2019.2897188

[42] Leterme, W., N. Ahmed, J. Beerten, L. Ängquist, D. V. Hertem, and S. Norrga. 2015. A new HVDC grid test system for HVDC grid dynamics and protection studies in EMT-type software. Paper read at 11th IET International Conference on AC and DC Power Transmission, 10–12 February, at Birmingham, UK. https://doi.org/10.1049/cp.2015.0068

[43] Li, Y., Y. Gong, and B. Jiang. 2018. A novel traveling-wave-based directional protection scheme for MTDC grid with inductive DC terminal. *Electric Power Systems Research* 157:83–92. https://doi.org/10.1016/j.epsr.2017.12.010

[44] Li, R., J. E. Fletcher, L. Xu, and B. W. Williams. 2017. Enhanced flat-topped modulation for MMC control in HVDC transmission systems. *IEEE Transactions on Power Delivery* 32 (1):152–161. https://doi.org/10.1109/TPWRD.2016.2561929

[45] Li, J., Q. Yang, H. Mu, S. Le Blond, and H. He. 2018. A new fault detection and fault location method for multi-terminal high voltage direct current of offshore wind farm. *Applied Energy* 220:13–20. https://doi.org/10.1016/j.apenergy.2018.03.044

[46] Li, Y., Y. Zhang, J. Song, L. Zeng, and J. Zhang. 2019. A novel pilot protection scheme for LCC-HVDC transmission lines based on smoothing-reactor voltage. *Electric Power Systems Research* 168:261–268. https://doi.org/10.1016/j.epsr.2018.12.012

[47] Li, Y., S. Zhang, H. Li, Y. Zhai, W. Zhang, and Y. Nie. 2012. A fault location method based on genetic algorithm for high-voltage direct current transmission line. *European Transactions on Electrical Power* 22 (6):866–878. https://doi.org/10.1002/etep.1659

[48] Livani, H., and C. Y. Evrenosoglu. 2014. A single-ended fault location method for segmented HVDC transmission line. *Electric Power Systems Research* 107: 190–198. https://doi.org/10.1016/j.epsr.2013.10.006

[49] Luo, G., C. Yao, Y. Liu, Y. Tan, J. He, and K. Wang. 2018. Stacked auto-encoder based fault location in VSC-HVDC. *IEEE Access* 6:33216–33224. https://doi.org/10.1109/ACCESS.2018.2848841

[50] Ma, Y., H. Li, G. Wang, and J. Wu. 2018. Fault analysis and traveling-wave-based protection scheme for double-circuit LCC-HVDC transmission lines with shared towers. *IEEE Transactions on Power Delivery* 33 (3):1479–1488. https://doi.org/10.1109/TPWRD.2018.2799323

[51] Mielczarski, W., ed. 1998. *Fuzzy Logic Techniques in Power Systems*. Vol. 11, *Studies in Fuzziness and Soft Computing*. Heidelberg, Germany: Physica-Verlag.

[52] Nanayakkara, O. M. K. K., A. D. Rajapakse, and R. Wachal. 2012a. Location of DC line faults in conventional HVDC systems with segments of cables and overhead lines using terminal measurements. *IEEE Transactions on Power Delivery* 27 (1):279–288. https://doi.org/10.1109/TPWRD.2011.2174067

[53] Nanayakkara, O. M. K. K., A. D. Rajapakse, and R. Wachal. 2012b. Traveling-wave-based line fault location in star-connected multiterminal HVDC systems. *IEEE Transactions on Power Delivery* 27 (4):2286–2294. https://doi.org/10.1109/TPWRD.2012.2202405

[54] Nguyen, A.-D., T.-T. Nguyen, and H.-M. Kim. 2016. A comparison of different hybrid direct current circuit breakers for application in HVDC system. *International Journal of Control and Automation* 9:381–394. https://doi.org/10.14257/ijca.2016.9.4.37

[55] Norrga, S., X. Li, and L. Ängquist. 2014. Converter topologies for HVDC grids. Paper read at IEEE International Energy Conference (ENERGYCON), 13–16 May, at Cavtat, Croatia. https://doi.org/10.1109/ENERGYCON.2014.6850630

[56] Ongsakul, W., and V. N. Dieu. 2013. *Artificial Intelligence in Power System Optimization*. Boca Raton, FL, USA: CRC Press.

[57] PSCAD User's Guide. 2018. Winnipeg, MB, Canada: Manitoba HVDC Research Ctr.https://www.pscad.com/knowledge-base/article/160

[58] Rao, H. 2015. Architecture of Nan'ao multi-terminal VSC-HVDC system and its multi-functional control. *CSEE Journal of Power and Energy Systems* 1 (1):9–18. https://doi.org/10.17775/CSEEJPES.2015.00002

[59] Sabug, L., A. Musa, F. Costa, and A. Monti. 2020. Real-time boundary wavelet transform-based DC fault protection system for MTDC grids. *International Journal of Electrical Power & Energy Systems* 115, Art. no. 105475. https://doi.org/10.1016/j.ijepes.2019.105475

[60] Sen, S., S. Mehraeen, and F. Ferdowsi. 2018. Improving dc circuit breaker performance through an alternate commutating circuit. Paper read at IEEE Energy Conversion Congress and Exposition (ECCE), 23–27 September, at Portland, OR, USA. https://doi.org/10.1109/ECCE.2018.8558468

[61] Sharifabadi, K., L. Harnefors, H.-P. Nee, S. Norrga, and R. Teodorescu. 2016. *Design, Control, and Application of Modular Multilevel Converters for HVDC Transmission Systems*. Chichester, West Sussex, UK: Wiley-IEEE Press.

[62] Sneath, J., and A. D. Rajapakse. 2016. Fault detection and interruption in an earthed HVDC grid using ROCOV and hybrid dc breakers. *IEEE Transactions on Power Delivery* 31 (3):973–981. https://doi.org/10.1109/TPWRD.2014.2364547

[63] Song, G., X. Chu, X. Cai, S. Gao, and M. Ran. 2014. A fault-location method for VSC-HVDC transmission lines based on natural frequency of current. *International Journal of Electrical Power & Energy Systems* 63:347–352. https://doi.org/10.1016/j.ijepes.2014.05.069

[64] Song, G., X. Chu, S. Gao, X. Kang, and Z. Jiao. 2015. A new whole-line quick-action protection principle for HVDC transmission lines using one-end current. *IEEE Transactions on Power Delivery* 30 (2):599–607. https://doi.org/10.1109/TPWRD.2014.2300183

[65] Song, G., T. Wang, and K. S. T. Hussain. 2019. DC line fault identification based on pulse injection from hybrid HVDC breaker. *IEEE Transactions on Power Delivery* 34 (1):271–280. https://doi.org/10.1109/TPWRD.2018.2865226

[66] Suonan, J., S. Gao, G. Song, Z. Jiao, and X. Kang. 2010. A novel fault-location method for HVDC transmission lines. *IEEE Transactions on Power Delivery* 25 (2):1203–1209. https://doi.org/10.1109/TPWRD.2009.2033078

[67] Suonan, J., J. Zhang, Z. Jiao, L. Yang, and G. Song. 2013. Distance protection for HVDC transmission lines considering frequency-dependent parameters. *IEEE Transactions on Power Delivery* 28 (2):723–732. https://doi.org/10.1109/TPWRD.2012.2232312

[68] Szechtman, M., T. Margaard, J. P. Bowles, C. V. Thio, D. Woodford, T. Wess, R. Joetten, G. Liss, M. Rashwan, P. C. Krishnayya, P. Pavlinec, V. Kovalev, K. Maier, J. Gleadow, J. L. Haddock, N. Kaul, R. Bunch, R. Johnson, G. Dellepiane, and N. Vovos 1994. The CIGRE HVDC benchmark model—A new proposal with revised parameters. *Electra* 157:61–65.

[69] Tang, L., X. Dong, S. Shi, and Y. Qiu. 2019. A high-speed protection scheme for the DC transmission line of a MMC-HVDC grid. *Electric Power Systems Research* 168:81–91. https://doi.org/10.1016/j.epsr.2018.11.008

[70] Tang, G., Z. He, H. Pang, X. Huang, and X. Zhang. 2015. Basic topology and key devices of the five-terminal DC grid. *CSEE Journal of Power and Energy Systems* 1 (2):22–35. https://doi.org/10.17775/CSEEJPES.2015.00016

[71] Tong, N., X. Lin, Y. Li, et al. 2019. Local measurement-based ultra-high-speed main protection for long distance VSC-MTDC. *IEEE Transactions on Power Delivery* 34 (1):353–364. https://doi.org/10.1109/TPWRD.2018.2868768

[72] Tzelepis, D., A. Dyśko, G. Fusiek, et al. 2017. Single-ended differential protection in MTDC networks using optical sensors. *IEEE Transactions on Power Delivery* 32 (3):1605–1615. https://doi.org/10.1109/TPWRD.2016.2645231

[73] Tzelepis, D., A. Dyśko, G. Fusiek, et al. 2018. Advanced fault location in MTDC networks utilising optically-multiplexed current measurements and machine learning approach. *International Journal of Electrical Power & Energy Systems* 97:319–333. https://doi.org/10.1016/j.ijepes.2017.10.040

[74] Tzelepis, D., A. Dyśko, S. M. Blair, et al. 2018. Centralised busbar differential and wavelet-based line protection system for multi-terminal direct current grids, with practical IEC-61869-compliant measurements. *IET Generation, Transmission & Distribution* 12 (14):3578–3586. https://doi.org/10.1049/iet-gtd.2017.1491

[75] Tzelepis, D., S. M. Blair, A. Dyśko, and C. Booth. 2019. DC busbar protection for HVDC substations incorporating power restoration control based on dyadic sub-band tree structures. *IEEE Access* 7:11464–11473. https://doi.org/10.1109/ACCESS.2019.2892202

[76] Tzelepis, D., G. Fusiek, A. Dyśko, P. Niewczas, C. Booth, and X. Dong. 2018. Novel fault location in MTDC grids with non-homogeneous transmission lines utilizing distributed current sensing technology *IEEE Transactions on Smart Grid* 9 (5):5432–5443. https://doi.org/10.1109/TSG.2017.2764025

[77] Tzelepis, D., S. Mirsaeidi, A. Dysko, Q. Hong, J. He, and C. Booth. 2020. Intelligent fault location in MTDC networks by recognising patterns in hybrid circuit breaker currents during fault clearance process. *IEEE Transactions on Industrial Informatics* 17(5): 3056–3068. https://doi.org/10.1109/TII.2020.3003476

[78] Wang, Y., and R. Marquardt. 2013. Future HVDC-grids employing modular multilevel converters and hybrid DC-breakers. Paper read at 15th European Conference on Power Electronics and Applications (EPE), 2–6 September, at Lille, France. https://doi.org/10.1109/EPE.2013.6631861

[79] Wang, H., G. Tang, Z. He, and J. Yang. 2014. Efficient grounding for modular multilevel HVDC converters (MMC) on the ac side. *IEEE Transactions on Power Delivery* 29 (3):1262–1272. https://doi.org/10.1109/TPWRD.2014.2311796

[80] Wang, D., H. L. Gao, S. B. Luo, and G. B. Zou. 2017. Travelling wave pilot protection for LCC-HVDC transmission lines based on electronic transformers' differential output characteristic. *International Journal of Electrical Power & Energy Systems* 93:283–290. https://doi.org/10.1016/j.ijepes.2017.06.004

[81] Wang, Y., Z. Hao, B. Zhang, and F. Kong. 2019. A pilot protection scheme for transmission lines in VSC-HVDC grid based on similarity measure of traveling waves. *IEEE Access* 7:7147–7158. https://doi.org/10.1109/ACCESS.2018.2889092

[82] Wang, Y., W. Wen, C. Zhang, Z. Chen, and C. Wang. 2018. Reactor sizing criterion for the continuous operation of meshed HB-MMC-based MTDC system under dc faults. *IEEE Transactions on Industry Applications* 54 (5):5408–5416. https://doi.org/10.1109/TIA.2018.2833819

[83] Warwick, K., A. Ekwue, and Raj Aggarwal, eds. 1997. *Artificial Intelligence Techniques in Power Systems*. Stevenage, UK: Institution of Engineering and Technology. https://doi.org/10.1049/PBPO022E

[84] Xiang, W., S. Yang, and J. Wen. 2020. ANN-based robust DC fault protection algorithm for MMC high-voltage direct current grids. *IET Renewable Power Generation* 14 (2):199–210. https://doi.org/10.1049/iet-rpg.2019.0733

[85] Xiang, W., S. Yang, L. Xu, J. Zhang, W. Lin, and J. Wen. 2019. A transient voltage-based DC fault line protection scheme for MMC-based DC grid embedding DC breakers. *IEEE Transactions on Power Delivery* 34 (1):334–345. https://doi.org/10.1109/TPWRD.2018.2874817

[86] Xu, J., Y. Lu, C. Zhao, and J. Liang. 2020. A model based dc fault location scheme for multi-terminal MMC-HVDC systems using a simplified transmission line representation. *IEEE Transactions on Power Delivery* 35 (1):386–395. https://doi.org/10.1109/TPWRD.2019.2932989

[87] Yang, Q., S. Le Blond, R. Aggarwal, Y. Wang, and J. Li. 2017. New ANN method for multi-terminal HVDC protection relaying. *Electric Power Systems Research* 148:192–201. https://doi.org/10.1016/j.epsr.2017.03.024

[88] Yang, S., W. Xiang, R. Li, X. Lu, W. Zuo, and J. Wen. 2020. An improved DC fault protection algorithm for MMC HVDC grids based on modal domain analysis. *IEEE Journal of Emerging and Selected Topics in Power Electronics* 8(8): 4086–4099. https://doi.org/10.1109/JESTPE.2019.2945200

[89] Yuansheng, L., W. Gang, and L. Haifeng. 2015. Time-domain fault-location method on HVDC transmission lines under unsynchronized two-end measurement and uncertain line parameters. *IEEE Transactions on Power Delivery* 30 (3):1031–1038. https://doi.org/10.1109/TPWRD.2014.2335748

[90] Zhang, X., N. Tai, Y. Wang, and J. Liu. 2017. EMTR-based fault location for DC line in VSC-MTDC system using high-frequency currents. *IET Generation, Transmission & Distribution* 11 (10):2499–2507. https://doi.org/10.1049/iet-gtd.2016.1215

[91] Zhang, C., G. Song, T. Wang, and L. Yang. 2019. Single-ended traveling wave fault location method in dc transmission line based on wave front information. *IEEE Transactions on Power Delivery* 34 (5):2028–2038. https://doi.org/10.1109/TPWRD.2019.2922654

[92] Zhang, S., G. Zou, B. Li, B. Xu, and J. Li. 2019. Fault property identification method and application for MTDC grids with hybrid DC circuit breaker. *International Journal of Electrical Power & Energy Systems* 110:136–143. https://doi.org/10.1016/j.ijepes.2019.02.048

[93] Zhang, Y., Y. Li, J. Song, B. Li, and X. Chen. 2019. A new protection scheme for HVDC transmission lines based on the specific frequency current of dc filter. *IEEE Transactions on Power Delivery* 34 (2):420–429. https://doi.org/10.1109/TPWRD.2018.2867737

[94] Zhang, S., G. Zou, C. Wang, J. Li, and B. Xu. 2020. A non-unit boundary protection of DC line for MMC-MTDC grids. *International Journal of Electrical Power & Energy Systems* 116, Art. no. 105538. https://doi.org/10.1016/j.ijepes.2019.105538

[95] Zhang, Y., Y. Li, J. Song, X. Chen, Y. Lu, and W. Wang. 2020. Pearson correlation coefficient of current derivatives based pilot protection scheme for long-distance LCC-HVDC transmission lines. *International Journal of Electrical Power & Energy Systems* 116, Art. no. 105526. https://doi.org/10.1016/j.ijepes.2019.105526

[96] Zheng, J., M. Wen, Y. Chen, and X. Shao. 2018. A novel differential protection scheme for HVDC transmission lines. *International Journal of Electrical Power & Energy Systems* 94:171–178. https://doi.org/10.1016/j.ijepes.2017.07.006

[97] Zheng, J., M. Wen, Y. Qin, X. Wang, and Y. Bai. 2020. A novel pilot directional backup protection scheme based on transient currents for HVDC lines. *International Journal of Electrical Power & Energy Systems* 115, Art. no. 105424. https://doi.org/10.1016/j.ijepes.2019.105424

[98] Zou, G., Q. Feng, Q. Huang, C. Sun, and H. Gao. 2018. A fast protection scheme for VSC based multi-terminal DC grid. *International Journal of Electrical Power & Energy Systems* 98:307–314. https://doi.org/10.1016/j.ijepes.2017.12.022

18 Fault Classification and Location in MT-HVDC Systems Based on Machine Learning

Raheel Muzzammel and Ali Raza

Department of Electrical Engineering, University of Lahore, Pakistan

18.1 INTRODUCTION

Integration of offshore renewable energy systems to conventional AC grids is made possible with the technology of voltage source converter based multi terminal high voltage direct current (MT-HVDC) grids [1,2]. Continuity of electric power supply is ensured in the increasing number of MT-HVDC grids with the availability of abrupt and trustworthy actions of protection circuitries. Minimum fault clearance time is the main objective behind deployment of protection system for MT-HVDC grids. This is because of the fact that DC faults are the main cause of developing DC fault currents of large magnitudes, resulting in high rate of rise of fault currents in very small time. This high rate of rise of fault current is vulnerable to the operation of VSCs and it creates difficulties in switching DC currents because of non-availability of natural; current zero. Discrimination and isolation of faulted line are the key challenges in development of MT-HVDC grid [3–5].

Researchers and engineers are very much interested in the development of protection of MT-HVDC systems to overcome the growing demands of reliable and green energy. Therefore, different methods are proposed and stated in literature.

A method based on current and voltage magnitude and direction known as handshaking is proposed to detect and to identify the faulted line. In addition to this, characteristics of fault current blocking and unblocking of VSCs are discussed which result in shut down of particular section of grid and complete shutdown of grid. Operation of different types of circuit breakers (AC and DC) with (DC) switches are analyzed [6]. Voltage and current magnitude and their derivatives based method of fault location is developed and tested for radial multi-terminal system [7]. In this method, capacitor discharge, diode-free wheel, and AC grid current fed stages are studied in the equivalent circuit of fault. Another method of fault location based on DC capacitor discharge stage is presented for meshed MT-HVDC grids [8].

DOI: 10.1201/9780367552374-18

Differential protection of DC grids based on current measurements made at two ends of DC transmission line is introduced [9,10]. A polarity-based protection method for DC transmission system is also developed. This technique is based on values of current measured at both ends of HVDC cables [11,12]. The major disadvantages associated with these methods are additional cost of synchronization and compromised reliability because of the fear of failure of communication links between two ends. In a single terminal measurement based method, difference between current and its average over a moving window are employed for protection of DC grids [13]. In this method, bus bar level communication is enough instead of system level. Effects of fault resistance on DC grid operation is not considered. Rate of change of DC voltage measured at the line side of series inductor is employed for fault detection and location [14]. This technique is tested on radial bipolar HVDC grid with a metallic return. It is found that this method is highly affected by fault loop resistance. This method is further improved by evaluating ratio of rate of change of peak voltage values at both ends of series inductor for the identification of fault [15]. Two different algorithms for DC fault protection are proposed based on rate of change of voltage and current and its magnitude [16]. In the first algorithm, product of rate of DC voltage and current is employed for fault protection. In the second algorithm, six different rules are developed based on rate of change of values of voltage and current and magnitude of current for interruption of fault currents in DC grids. Wavelet transform coefficients of voltage and current are evaluated for protection against DC fault currents [3]. This method is supported by magnitude of voltage and its rate of change in the case of fault discrimination in DC cable. Promising results are obtained in the case of solid faults but its accuracy is not tested for faults having resistance.

A non-unit protection technique is developed based on rate of change of voltage –voltage plane. This technique is supported with rate of change of current values [17]. Inductive termination of cables is used to state the open protection zones. Noise is the most influential factor in the methods based on measurement of rate of change of voltage and currents. In addition to this, transients of voltage generated during the opening of DC circuit breakers may create interruption in the correct operation of the protection system. The DC fault protection method is also developed based on the comparison of rate of change of series inductor voltage and predefined threshold values of voltage [18]. This method ensures better immunity against electromagnetic interference (EMI) and the structure of this method is equivalent to the second derivative of the DC current. The ratio of the mid-band frequencies transient voltage of the line side of series inductor to the mid-band frequencies transient voltage of DC terminal capacitor is employed for MT-HVDC protection. An additional voltage divider is added in these methods for the measurement of voltage across the series inductor and DC terminal capacitor [18,19]. High-frequency traveling waves transmitted from the faulted line are attenuated in case of a fault on the DC transmission lines. This attenuation mainly involves the series inductor. This concept is utilized for development of the protection system based on single-ended terminal-based measurements for creating a distinction between internal and external faults. Morphological gradients of voltage are compared with the predefined thresholds without the involvement of communication links to differentiate between internal and external faults [20–22].

In this chapter, a machine learning–based fault diagnostic technique is discussed and applied on a four-terminal HVDC test system. Based on DC voltage measurements, a support vector machine is applied to classify among normal, faulty and abnormal states of test system.

The rest of this chapter is organized as follows: Section 18.2 covers the introduction of machine learning–based fault diagnostic technique. Section 18.3 contains a brief description about DC faults in MT-HVDC systems. Section 18.4 covers a comprehensive review of advancement in VSC converters. Control of VSC converters and MT-HVDC systems are discussed in Sections 18.5 and 18.6, respectively. Simulation results are explained in Section 18.7. The conclusion is presented in Section 18.8.

18.2 MACHINE LEARNING–BASED FAULT DIAGNOSTIC TECHNIQUE

The setting of a threshold is the main problem in the fault diagnosis. Many false alarms will be generated in the case of setting a low threshold for diagnosis. This problem could make the system less reliable and effective. In a similar fashion, if the threshold is set at a very high value, it is possible that risks of non-detection of faults having characteristics closer to normal conditions will be higher. Because of rapid buildup of the DC fault current, such risks may not be tolerated; otherwise, serious security issues and concerns of operational safety of power electronic circuitries will be raised, resulting in failure of DC grids. Therefore, when the systems become complex and hard to define the analytical model, machine learning models are employed to cope with this challenge. Machine learning–based diagnosis has the ability to provide an accurate result in the presence of missing information, complex inputs, and mixed variables.

In the field of artificial intelligence, machine learning (ML) occupies an important space. It is an automated way of learning and making the basis of exposure and experience better. One of the interesting facts related to this tool is that learning and improvement do not require explicit programming. Hence, engineers and scientists are very much interested in developing such a computer program that can access information and can utilize information for learning by themselves [23].

Intelligent systems deal with machine learning as these systems can modify their behavior on the basis of input information. Moreover, these systems are capable of choosing a best-fitting function that can accomplish learning on the basis of input information [23]. Machine learning is in general, broadly classified into two categories that are supervised machine learning and unsupervised machine learning. In supervised machine learning, learning and training are accomplished by the labeled input information and know output information. In unsupervised machine learning, training and learning are accomplished by unlabeled input information and without any prior knowledge of output information. Unsupervised machine learning is beyond the scope of this chapter.

Classification is a subcategory of supervised machine learning, employed for the identification and detection of class. An algorithm employed for classification is known as a classifier. A successful classifier is dependent on learning. Data samples are divided into two subsamples that are training samples and testing samples. A training data sample is employed for training a classifier. Performance of the classifier is dependent upon the ratio of successful classification and total number of classifications. K-fold validation is also used to evaluate performance of classifier, applicable for small data sets. Accuracy of cross-validation is calculated as the average of success rate calculated for each of the k different testing data sets. In case of $k = N$, k-fold cross-validation is known as leave-one-out cross-validation [23].

Different types of classifiers are developed in literature and research [24,25]. k-nearest neighbor (kNN) classifier, Bayes classifier, logistic regression, Fisher's linear discriminant, decision tree, artificial neural networks (ANN), and support vector machines (SVMs) are popular among them. In this chapter, support vector machine (SVM) is discussed for fault diagnosis in MT-HVDC systems.

18.2.1 Support Vector Machines

Vapnik developed the support vector machines (SVMs) for binary classification initially [26]. A hyperplane is optimized so that data points could be discriminated among their classes. Moreover, non-linear classification is also carried out by SVM using the kernel trick. In this technique, inputs are mapped implicitly into higher-dimensional feature spaces. In this feature space, a linear decision surface is created. The high generalization ability of the learning machine is ensured with the special properties f the decision surface [27].

In SVM, a two-stage training data set $\{x_i, y_i\}_{i=1}^{N}$ contains N data points. ith real valued input vector is represented by x_i. Associated class of x_i is denoted by y_i. Its value is either +1 or –1. According to their different classifications, the hyperplane discriminating these points is given as:

$$w^T x_i + b = 0 \tag{18.1}$$

where w and b are the weight vector and the deviation parameter, respectively. Position of the segmentation is evaluated by the information of the weight vector and deviation parameter. The best values of w and b are determined by training the SVM. These best values maximize the separation between the categories. Separation margin (m) is given as [28]:

$$m = \frac{2}{\|w\|} \tag{18.2}$$

A better training model is obtained with the increase in the separation margin and with the decrease in the weight vector. Therefore, construction of SVM is maximized with $v(w)$ as:

$$v(w) = \frac{1}{2} w^T w \tag{18.3}$$

subject to:

$$y_i(w^T x_i + b) \geq 1 \tag{18.4}$$

In this study classification through SVM, is performed for fault diagnosis in MT-HVDC systems.

18.2.2 Feature Extraction and Selection

Dimensionality reduction stage of input data is divided into two sub-stages that are feature extraction and feature selection [23]. Pre-processed data is prepared by converting analog data to discrete data. The input signal for this stage is represented by $x[n]$, where $n = 1, 2, \ldots, 200$. This type of input signal is used to extract frequency-domain-based features. The lowest and highest peaks of frequency-based data is used as a feature for training the SVM. The lowest and highest peaks are obtained by Fourier transform, given as:

$$Y(k) = \sum_{j=1}^{N} X(j) W_n^{(j-1)(k-1)} \tag{18.5}$$

where $W_n = e^{-\frac{j2\pi i}{n}}$ is one of n roots of unity. Fast Fourier transform reduces the computational complexity by transforming the number of data points (400 values) to featured data of 26 values. These frequency-domain-based values are used as training data for SVM.

18.3 DC FAULTS IN MT-HVDC SYSTEMS

There are basically three classification of faults in MT-HVDC systems. These are i) pole to ground faults, ii) pole to pole faults, and iii) AC faults. The AC side of converter stations are equipped with conventional breakers to prevent the effects of AC faults from going into DC transmission link. Because of severity of DC faults, abrupt activation of interruption and differentiation must be features of DC grid protection [29,30]. This requirement has opened the doors of development of protection system which has the capability of providing isolation in a minimum possible time. In addition to this, it is found from research that in case of low impedance faults, low voltage surge originates from the fault and propagates towards terminal and reflects back as a high-voltage surge at the sensing terminal. In the overhead DC transmission system, pole-to-ground faults mainly occurred because of lightning [31]. Fault resistance, grounding type, impedance of grounding, and configuration of transformer of converter stations are the influencing factors of pole-to-ground faults. In the DC cables, most of the faults occurred due to failure of insulation. In the case of pole-to-pole faults, there are three stages of fault interval. In the first interval, discharging of capacitor takes place immediately with a sharp altitude. Blocking of insulated gate bipolar transistor

(IGBT) takes place in order to avoid harmful effects of overcurrent. In the same time, conduction takes place in the opposite direction via anti-parallel diodes. This helps against the effects of development of overvoltage at the valves. In the second interval, dropping of voltage takes place but feeding of series inductor takes place via freewheeling diodes even in the case of zero voltage. In the third interval, AC infeed is independent of the effects of DC faults and it feeds the faulty system via freewheeling diodes, particularly in the case of high bridge scheme of converters. It cannot be interrupted without the availability of protection scheme. As a result, overcurrent damages the converter stations, and other sensitive circuitries. Turning off IGBTs takes place for a very small instant during faults but flow of current towards AC side of converter station is not stopped because of freewheeling diodes. Therefore, only one cycle is available to prevent failure of converter station. As a result, reliable and immediate protection against DC faults is required [29].

18.4 VOLTAGE SOURCE CONVERTERS

Configuration of HVDC grid and its control are the influencing factors of the transient post-fault response sensed by the relay.

In an asymmetrical monopole, only one high voltage conductor is employed along with an earth return [32]. In this configuration, negative polarity is used because of less radio interference issues. In a symmetrical monopole, half voltage system is achieved with two poles connected to single converter. Bipolar configuration is a suitable candidate for MT-HVDC systems as this configuration offers redundancy and availability of 50% of power transfer in case of fault at one of the pole, thereby, working as an asymmetrical monopole. In this configuration, both poles are at full symmetrical voltage values. Each pole has its converter station [33]. In all the aforementioned arrangements, metallic return is added.

In a point-to-point HVDC link configurations, there are two classifications of converter technology i.e. line commutated converter (LCC) and voltage source converter (VSC). In LCC, thyristors are employed. Thyristors are line commutated because zero of the AC current is utilized for turning off of converter station and commutation the current to the next AC phase. External gate bias is the only way to control the firing of valves. LCC is also termed as current source converters (CSC) as it is taken as constant current source by an AC grid. Because of limited controlling ability, it is not a much favorable option for MT-HVDC systems. Reversal of current is not possible because of line commutation. Therefore, change in power flow demands is addressed only through reversal of voltage polarity which is not practical in MT-HVDC systems.

In voltage source converters (VSCs), reversal of current is possible for power transfer without changing the voltage polarity. Hence, the rectifier will become the inverter and an inverter will become a rectifier. Feature of self-commutation offers significant number of benefits in MT-HVDC systems. Active and reactive powers are controlled independently. VSC stations are less affected by disturbances originated at the AC grid and it has ability to establish connection with a weak AC grid and passive

AC network. External voltage bias can be applied to turn off and on semiconductor switches (insulated gate bipolar transistor (IGBT), gate turn off thyristors (GTOs), or integrated gate commutated thyristors (IGCT) of voltage source converters. It is not possible for a single switch to handle large voltages of HVDC transmission; therefore, many switched are engaged in series and switched simultaneously. This joint venture can be termed as one valve. Large valve halls are required for HVDC converter stations which is a challenging problem in offshore installation.

Two AC side voltage levels i.e. two-level converter and three voltage levels i.e. positive, negative, and neutral can be achieved for bipolar and symmetrical monopole HVDC system. Neutral point clamped (NPC) converter [34], active neutral point clamped (ANPC) converter [35], and flying capacitor (FC) converter [36] are common examples of three voltage level converters. AC side waveform is achieved by pulse width modulation (PWM). A saw-tooth carrier wave is compared with a sinusoid. Switching on of an appropriate valve is dependent upon exceeding the value of saw-tooth from sinusoid. Resulting waveforms can be made sinusoidal with appropriate filtering techniques [37]. A modular multi-level converter (MMC) is the latest VSC. In this MMC technology, sub-modules are composed of switches. Arbitrary number of voltage levels can be obtained for better approximation of AC sinusoidal waveform. In multi-level converters, a PWM with high frequency or nearest level modulation (NLM) [38] with lower frequency are employed for controlling. Actuation of valves improves the quality of waveform, resulting in no more requirements of filters. H-bridge modules based multi-level converters corrects the error in voltage in each cell with in the sub-cycles of a fundamental period. This initiates the concept of the small cell capacitor [39,40]. Hence, the need of more switches are compensated with this concept that results in small reduction in converter impressions. H-bridge based multi-level converters under various changes but these converters have the ability of not allowing the transfer of AC disturbances originated because of fault current, to DC side. Hence, the magnitude of the fault current and the speed required for interruption is reduced [41].

18.5 CONTROL SYSTEM OF VOLTAGE SOURCE CONVERTERS

There are a number of ways in which control of VSC can be designed. Vector-based control and direct power control are the most accepted methods. In the vector control method, threc phase currents and voltages are converted to direct quadrature domain. Stationary signals are obtained through a rotating coordinate system. Two-step transformations are involved in it. Firstly, AC currents and voltages are transformed into the $\alpha-\beta-\gamma$ domain via Clarke's transformation as:

$$i_{\alpha\beta\gamma}(t) = Transform(i_{abc}(t)) = \frac{2}{3}\begin{bmatrix} 1 & -\frac{1}{2} & -\frac{1}{2} \\ 0 & -\frac{\sqrt{3}}{2} & -\frac{\sqrt{3}}{2} \\ \frac{1}{2} & \frac{1}{2} & \frac{1}{2} \end{bmatrix}\begin{bmatrix} i_a(t) \\ i_b(t) \\ i_c(t) \end{bmatrix} \tag{18.6}$$

In case of a balanced system, α and β are sinusoidal 90° out of phase and γ is zero. Secondly, $\alpha - \beta - \gamma$ domain is then transformed into direct quadrature domain (dq0) through Park's transformation as:

$$i_{dq0}(t) = \begin{bmatrix} cos\theta & sin\theta & 0 \\ -sin\theta & cos\theta & 0 \\ 0 & 0 & 1 \end{bmatrix} \begin{bmatrix} i_\alpha(t) \\ i_\beta(t) \\ i_\gamma(t) \end{bmatrix} \tag{18.7}$$

Alignment of the d-axis of the rotating reference frame to AC voltage vector results in $V_q = 0$. Hence, the instantaneous active and reactive power absorbed or injected into the AC system is:

$$P = \frac{3}{2} v_d i_d \tag{18.8}$$

$$Q = -\frac{3}{2} v_d i_q \tag{18.9}$$

These equations depict the active and reactive power is controlled by voltage and current in the dq0 domain. There are basically two control loops involved for power controlling in VSC. An outer control loop is responsible for generating reference values i.e. $i_{d,ref}$ and $i_{q,ref}$. These reference values are used to evaluate reference values of power and voltage. In the inner control loop, the actual d and q voltages are regulated to find the desired currents. These d and q voltages are then transformed back to the a-b-c domain. These voltages are sent to the PWM or NLM controller for generating the waveforms by the direct control of firing of the valves [42]. Proportional integrator (PI) control is responsible for rapid action of control loops in response to a step change in system parameters originated because of faults. Therefore, a state of compromise is established between stability and responsiveness by designating the controller gain and integral time constant for PI loops. Settling time of faster inner current controller and voltage controller is of the order of 500 μs and 3 ms when the modulus optimum and symmetric optimum methods are used for tuning [43]. The inner current t control loop is more vital than the outer current control loop when fault transients are analyzed in the MT-HVDC system because of its rapid response but it does not mean that any one of them can be neglected. It is supposed that controller works on ideal analog input signals but measured system parameters are developed on sampled digital values. Therefore, it is required to model this sampling interval in order to avoid bearing on the upper limit of response time of controllers.

In direct power control, a switching table is made based on the instantaneous errors between the desired and measured values for P and Q [44]. A firing signal of the valve is received from switching table made from instantaneous active and reactive power control loops. An estimated flus vector is utilized instead of PWM and there is no inner current controller.

18.6 CONTROL OF MT-HVDC SYSTEM

In this research, droop control strategy is employed. The DC voltage droop control is employed in which more than one VSC station contributes towards controlling of DC link voltage. As a result, failure of one VSC station does not result in a black out of the whole MT-HVDC system [45,46]. In the test system, AC grid 1 and AC grid 2 operates at constant power; therefore, control is implemented for stabilizing DC link voltage. Variation in power is observed at wind farm 1 and wind farm 2; there P-VAC control is implemented to regulate the AC voltages of wind farms.

18.7 MT-HVDC TEST SYSTEM AND SIMULATION RESULTS

The test system of the MT-HVDC system is established in MATLAB/Simulink. This system consists of four HVDC transmission lines (L_1, L_2, L_3, L_4,) of lengths 300 km, 300 km, 200 km, and 200 km, respectively. This test system is made up of four converter stations. Two of them are rectifier stations and two of them are inverter stations, as shown in Figure 18.1.

Two offshore wind farms are connected to MT-HVDC system to show the favorable integration of renewable energy sources. Parameters of the four-terminal test circuit are given in Table 18.1.

18.7.1 DC Voltage Analysis

DC voltages are measured in a test system to analyze the behavior under normal and faulty conditions, as shown in Figure 18.2. Under normal conditions, DC voltage at four terminals are 1.0 per unit under steady state as shown in Figure 18.3. DC voltage attains a steady state in less than 0.05 s. It means that the normal condition

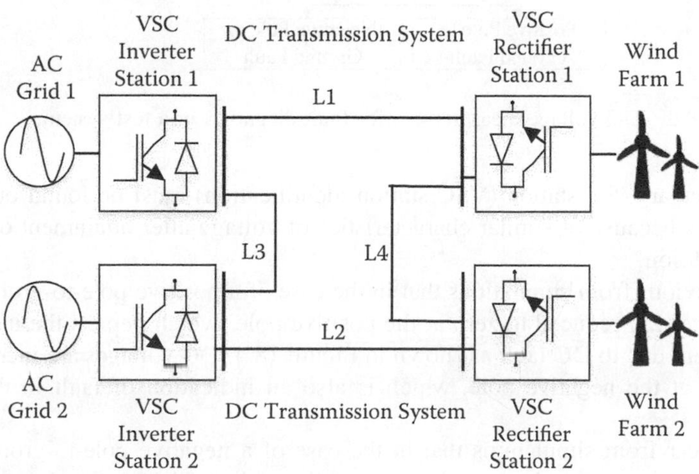

FIGURE 18.1 Four-terminal HVDC test system.

TABLE 18.1
Parameters of MT-HVDC Test System

Parameters	Inverter Station I (IS–I)	Inverter Station II ((IS –II)	Rectifier Station I (RS–I)	Rectifier Station II (RS –II)
Power (MVA)	200	200	200	200
Voltage (KV)	100	100	100	100
Phase Reactor (p.u.)	0.15	0.15	0.15	0.15
AC Filters (MVAR)	40	40	40	40
Three Phase Transformer (MVA)	200	200	200	200
DC Transmission Line (KV)	100	100	100	100
DC Capacitor (μF)	70	70	70	70
DC Filter (Ω, mH, μF)	0.147, 46.9, 12	0.147, 46.9, 12	0.147, 46.9, 12	0.147, 46.9,12
Smoothing Reactor (Ω, mH)	0.0251, 8	0.0251, 8	0.0251, 8	0.0251, 8

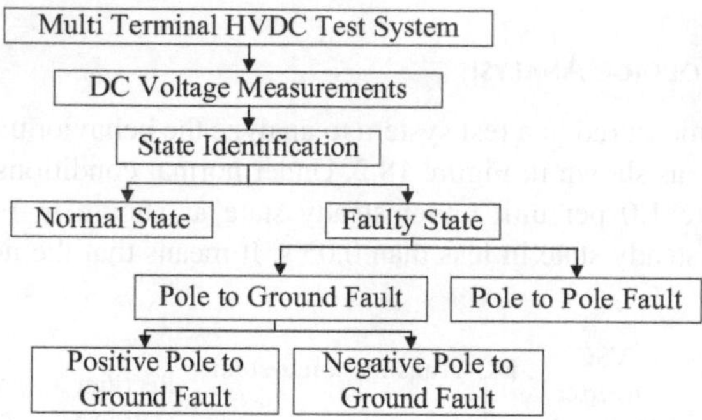

FIGURE 18.2 DC voltage measurement for fault diagnosis in a test system.

of a particular VSC station (VSC station identification) must be found out in less than 0.05 s because of similar characteristics of voltage after attainment of steady-state condition.

It is obvious from simulations that in the case of a positive pole-to-ground fault, DC voltages are reduced to zero at the positive pole, which depicts the increase in fault current due to DC fault as shown in Figure 18.4. DC voltages are increased by 1.6 times at the negative pole, which is also an indication of fault in the opposite pole.

It is clear from simulations that in the case of a negative pole-to-ground fault, DC voltages are reduced to zero at the negative pole, which shows the increase in fault current due to DC fault as shown in Figure 18.5. DC voltages are increased by

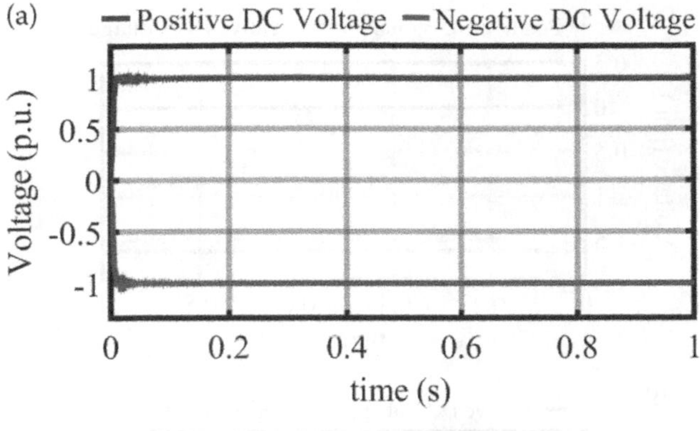

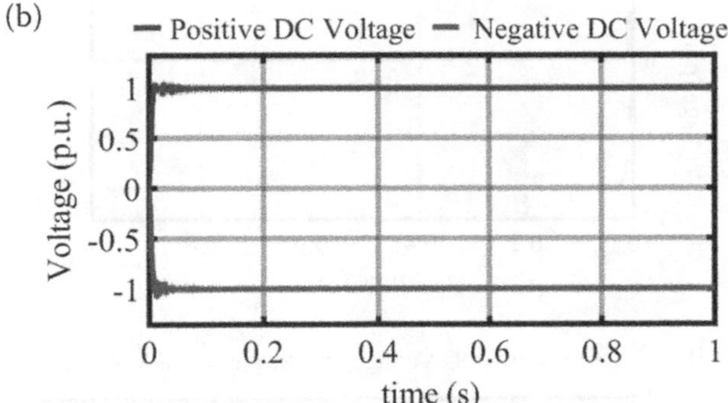

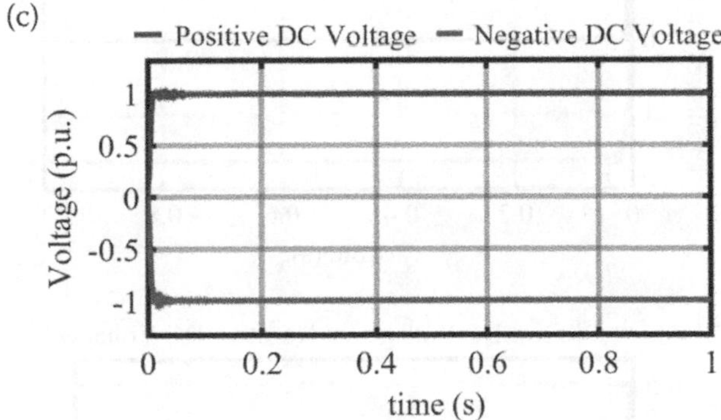

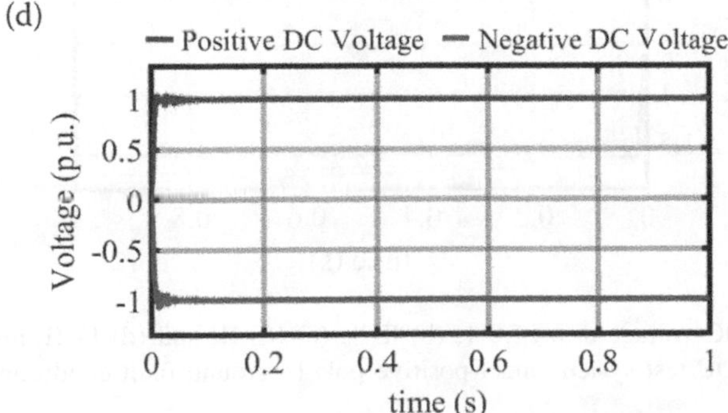

FIGURE 18.3 DC voltage at (a) RS-1; (b) IS-1; (c) RS-II; and (d) IS-II, respectively, in four-terminal HVDC test system under normal conditions.

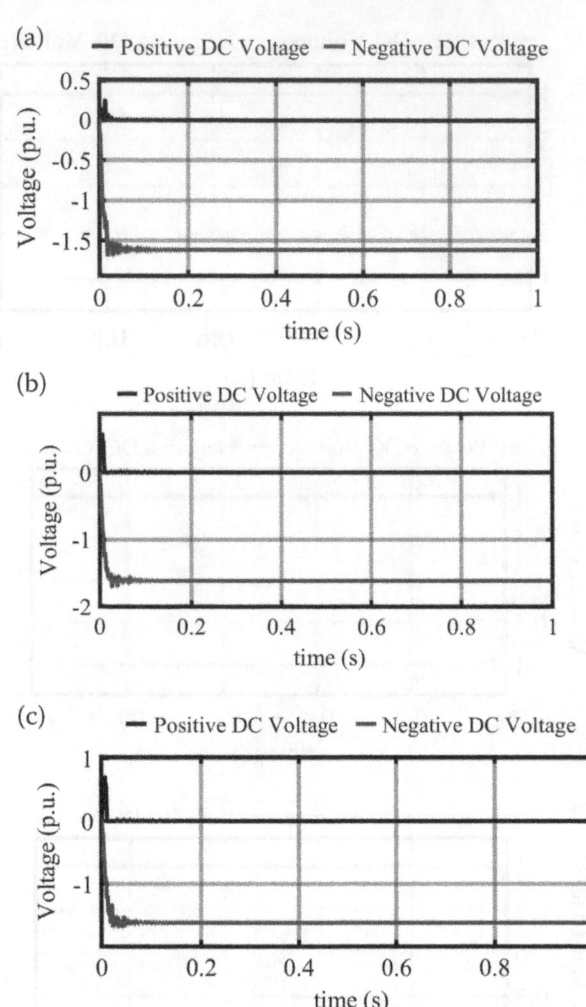

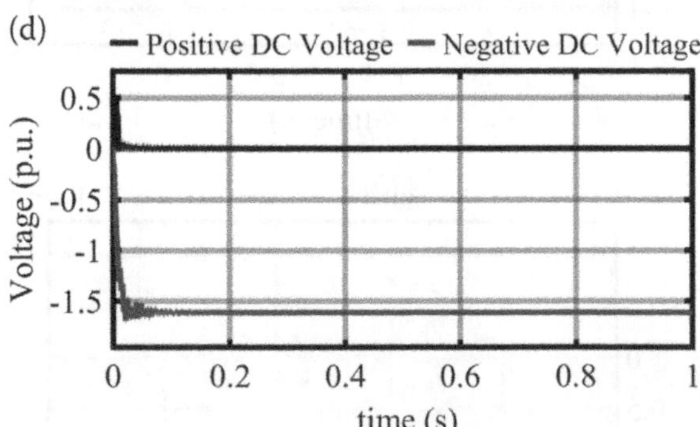

FIGURE 18.4 DC voltage at (a) RS-1; (b) IS-1; (c) RS-II; and (d) IS-II, respectively, in four-terminal HVDC test system under positive pole-to-ground fault conditions.

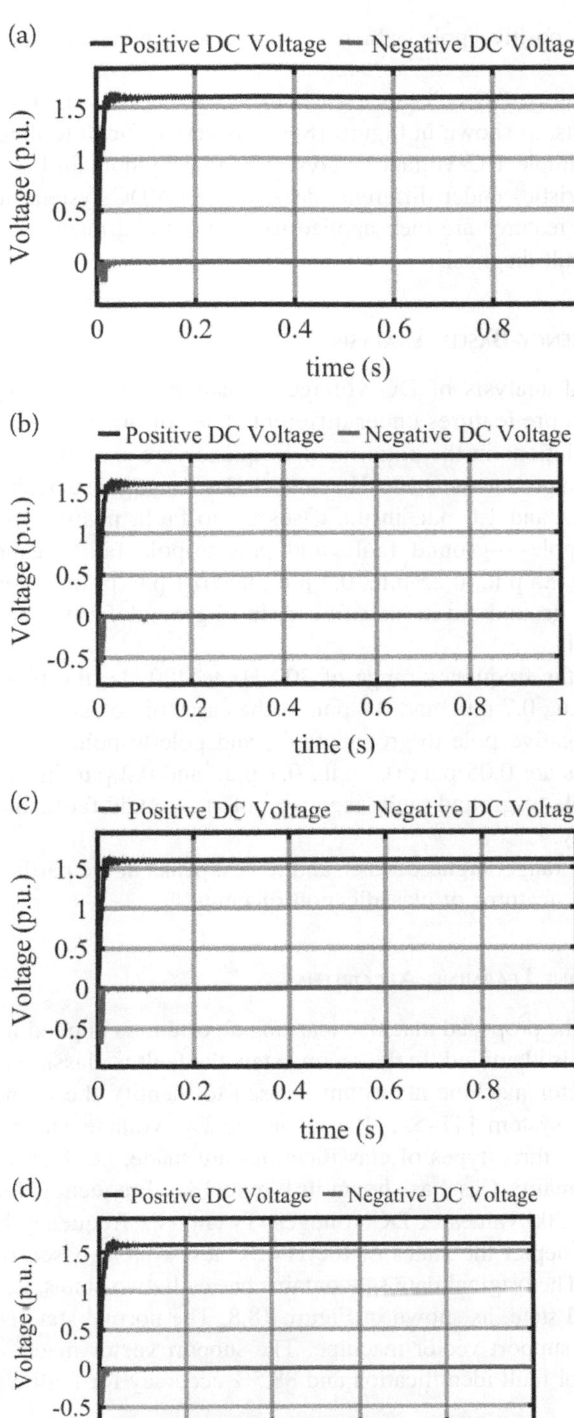

FIGURE 18.5 DC voltage at (a) RS-1; (b) IS-1; (c) RS-II; and (d) IS-II, respectively, in four-terminal HVDC test system under negative pole-to-ground fault conditions.

1.6 times at the positive pole, which is also an indication of fault in the opposite pole.

In the similar way, DC voltages exist in the range between 0.1 p.u. to 0.3 p.u. in pole-to-pole faults, as shown in Figure 18.6. It is very difficult to diagnose the type of faults with simple DC voltage analysis. Frequency-domain-based analysis of voltage characteristics under different states of the HVDC system are added. The frequency-based features are then applied to the proposed machine learning algorithm [40] for fault diagnosis.

18.7.2 Frequency-Based Analysis

Frequency-based analysis of DC voltage measurements at sensing terminal is done to extract more features under different states of the multi-terminal HVDC test system. Variations in the patterns of frequency are shown in Figure 18.7. In the frequency range between 100 Hz to 200 Hz, the highest peaks are 0.7 p.u., 0.5 p.u., 0.8 p.u., and 1.0 p.u. in the cases of no-fault, positive pole-to-ground fault, negative pole-to-ground fault, and pole-to-pole fault, respectively. The lowest peaks are 0.3 p.u., 0.25 p.u., 0.2 p.u., and 0.1 p.u. in the cases of no-fault, positive pole-to-ground fault, negative pole-to-ground fault, and pole-to-pole fault, respectively.

Similarly, in the frequency range of 300 Hz to 400 Hz, the highest peaks are 0.65 p.u., 0.55 p.u., 0.7 p.u., and 0.8 p.u. in the cases of no-fault, positive pole-to-ground fault, negative pole-to-ground fault, and pole-to-pole fault, respectively. The lowest peaks are 0.05 p.u., 0.3 p.u., 0.1 p.u., and 0.2 p.u. in the cases of no-fault, positive pole-to-ground fault, negative pole-to-ground fault, and pole-to-pole fault, respectively.

The frequency range, highest peaks, and lowest peaks in the particular frequency range are the key features of classification of faults.

18.7.3 Machine Learning Algorithm

In this research, the proposed machine learning algorithm is applied in two steps. In the first, the fault is identified. In the second step, the fault is classified into its types.

A support vector machine algorithm is used to identify the normal and faulty cases of the test system [47–52]. Based on the DC voltage values measured at converter stations, three types of classifications are made, i.e. Normal Case, Fault Case, and Abnormality Case, as shown in Figure 18.6. Frequency domain features are derived from 200 values of DC voltages. Twenty-six frequency-based features are employed to depict the states of the HVDC test system based on the support vector machine. The original data set contains basically two states, i.e. normal state and unrecognized state, as shown in Figure 18.8. The normal state is the deciding state used in the support vector machine. The support vector machine works with 100% accuracy for fault identification and 88.5% accuracy for fault classification in

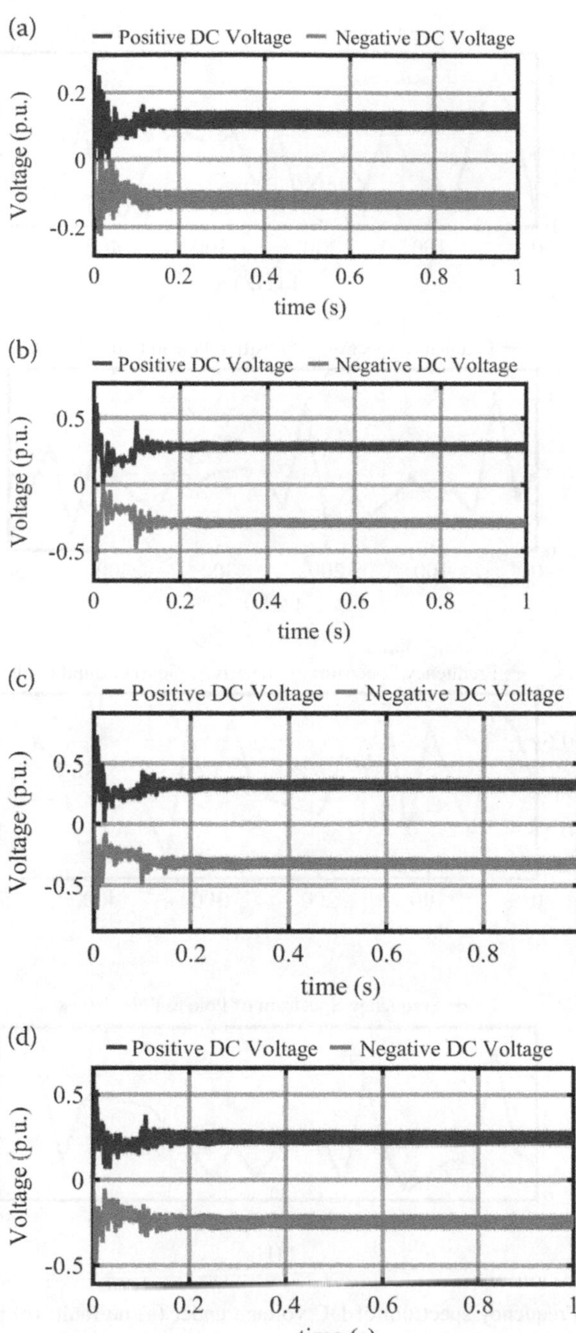

FIGURE 18.6 DC voltage at (a) RS-1; (b) IS-1; (c) RS-II; and (d) IS-II, respectively, in four-terminal HVDC test system under pole-to-pole fault conditions.

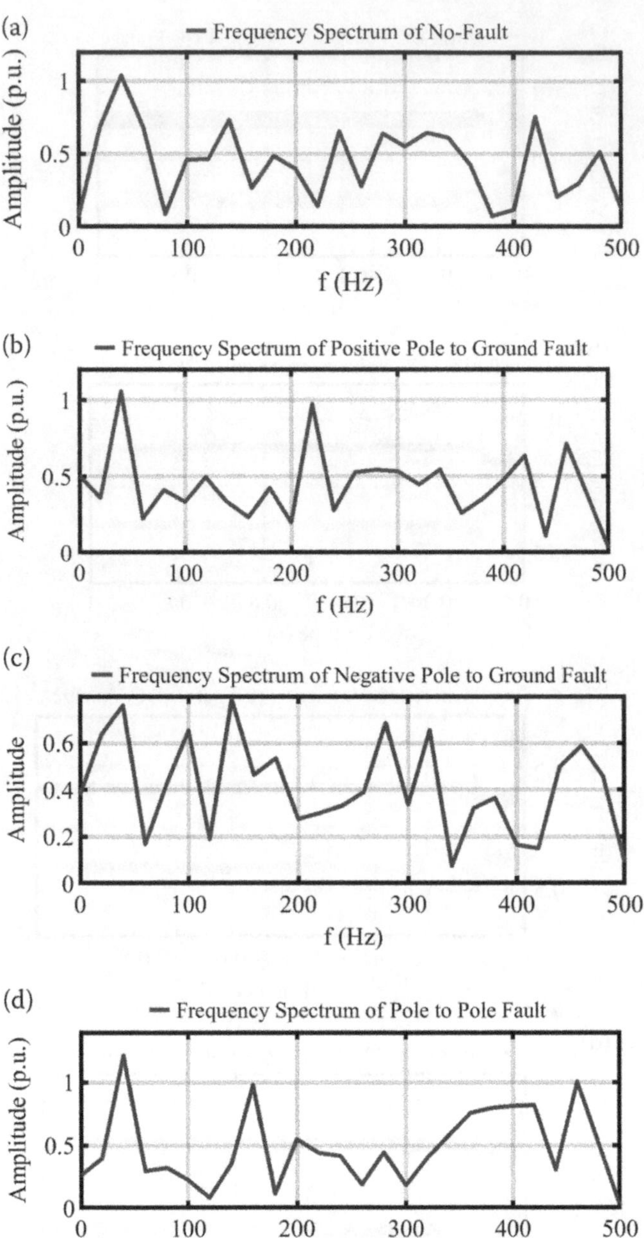

FIGURE 18.7 Frequency spectrum of DC voltage under (a) no fault; (b) positive pole-to-ground fault; (c) negative pole-to-ground fault; (d) pole-to-pole fault.

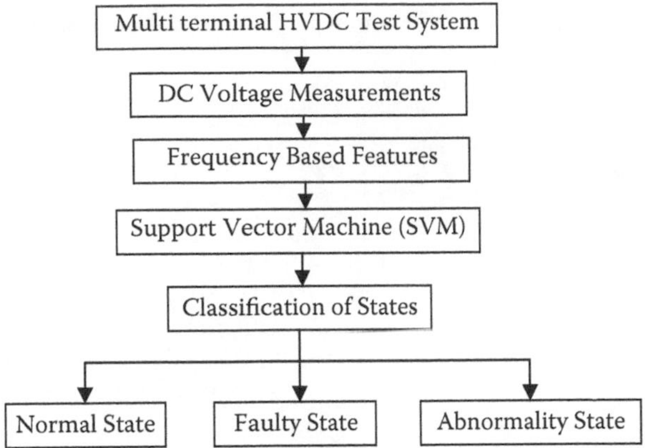

FIGURE 18.8 Flowchart of support vector machine for fault identification in test system.

this model. Data sets of no-fault state and unrecognized state obtained from the SVM are shown in Figure 18.9.

Performance of the proposed algorithm is evaluated by a confusion matrix. There are basically three confusion matrices obtained on the basis of number of observations, true positive and false negative rates, and positive predictive and fault discovery rates, as shown in Figures 18.10, 18.11, and 18.12. The accuracy of the machine learning algorithm would be enhanced so that the fault could be diagnosed in a better way.

Classification of faults is carried out in a similar fashion, as shown in Figure 18.13. The fault is identified by the decrease in the value of DC voltage. The decrease in the value of DC voltage happens in all types of faults. Therefore, frequency domain features are extracted from the 200 values of DC voltage so that computational time without compromise on accuracy could be reduced. Twenty-six frequency-based features are used to classify the faults into its types. The original data set and predicted data set for fault classification are shown in Figure 18.14.

Performance of the proposed algorithm is analyzed with the confusion matrix, as shown in Figure 18.15. Based on the number of observations, out of eight observations of negative pole-to-ground fault, seven are predicted accurately. In the case of no-fault, out of six observations, only one observation is predicted inaccurately. One hundred percent of observations are predicted accurately in the case of pole-to-pole fault. In the case of positive pole-to-ground fault, only one observation is wrong out of six observations.

On the basis of true positive and false negative rates, as shown in Figure 18.16, an accuracy of 88.5% of the proposed algorithm for fault classification is obtained. The prediction is done with 100% accuracy in the negative pole-to-ground fault and

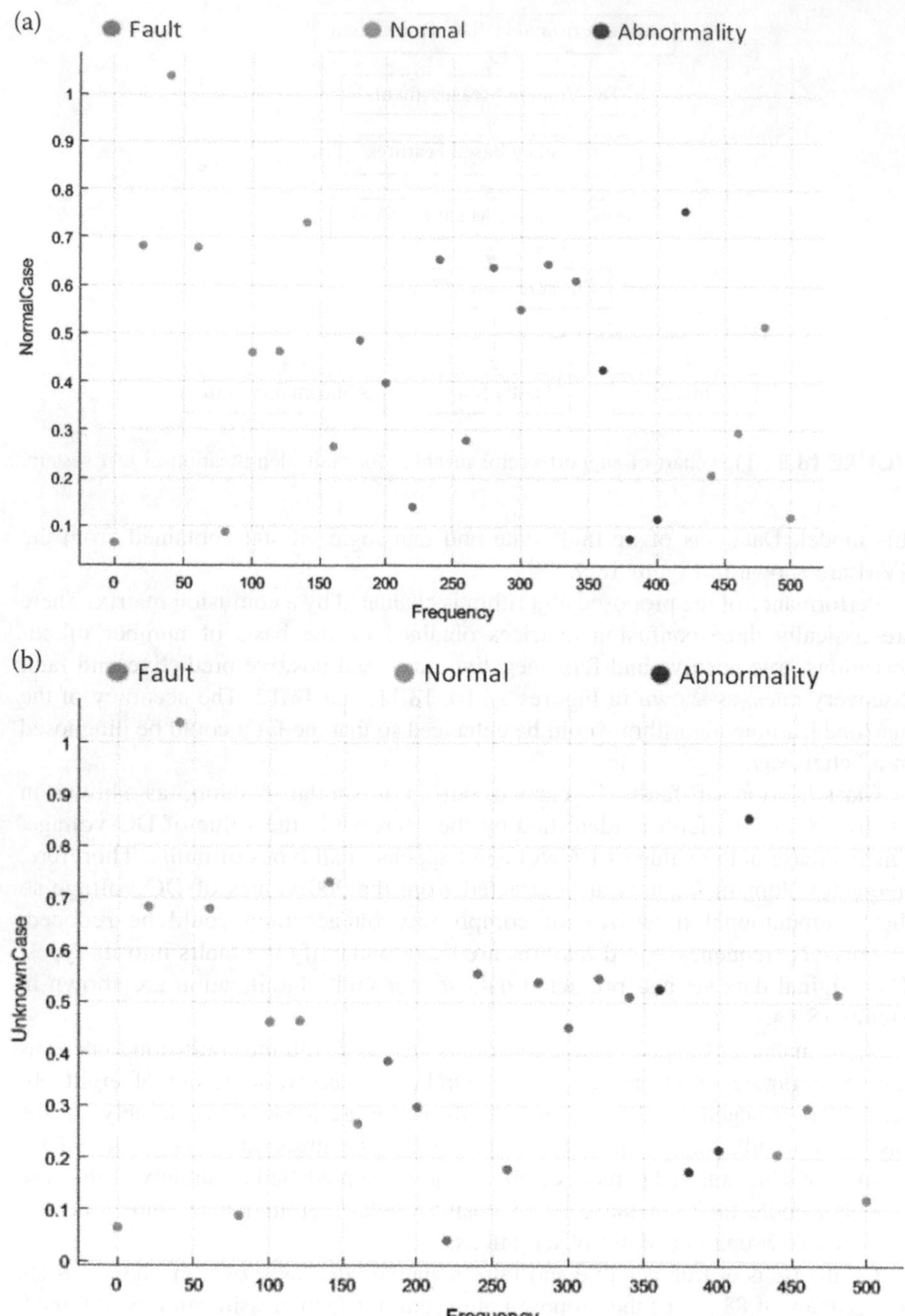

FIGURE 18.9 (a) Original and (b) predicted data set plotted for normal and unrecognized case.

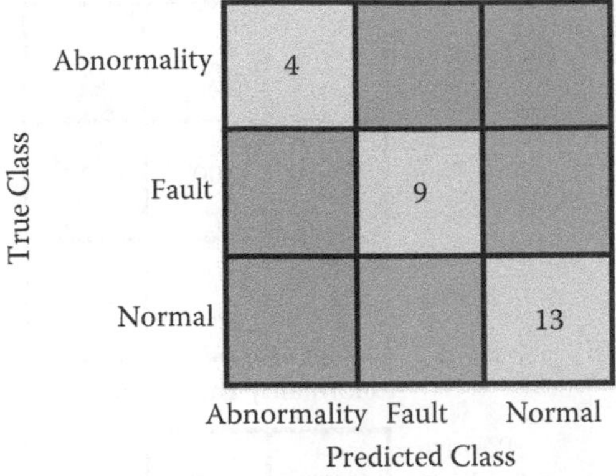

FIGURE 18.10 Confusion matrix of support vector machine based on number of observations.

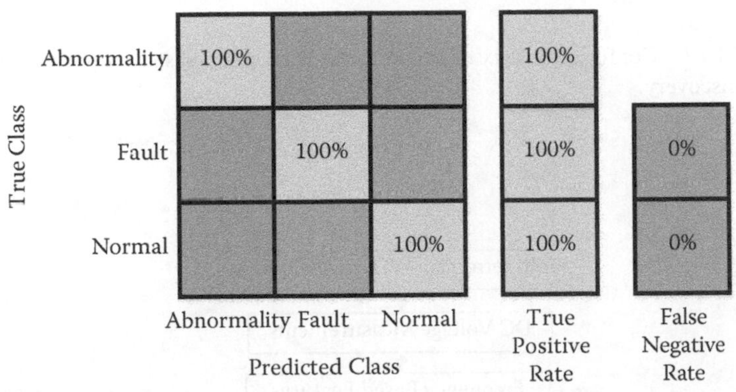

FIGURE 18.11 Confusion matrix of support vector machine based on true positive and false negative rates.

no-fault cases, 75% accuracy in pole-to-pole fault case, and 83% accuracy in positive pole-to-ground fault case, as shown in Figure 18.17.

These analyses of confusion matrices depict the accuracy of the proposed machine learning algorithm for fault classification.

The fault location is found by comparing the frequency domain features of DC voltage measurements, as shown in Figure 18.18. Faults are created at different locations (at the converter station, at 100 km from the converter station, at 200 km from the converter station, at 300 km from the converter station). Based on the

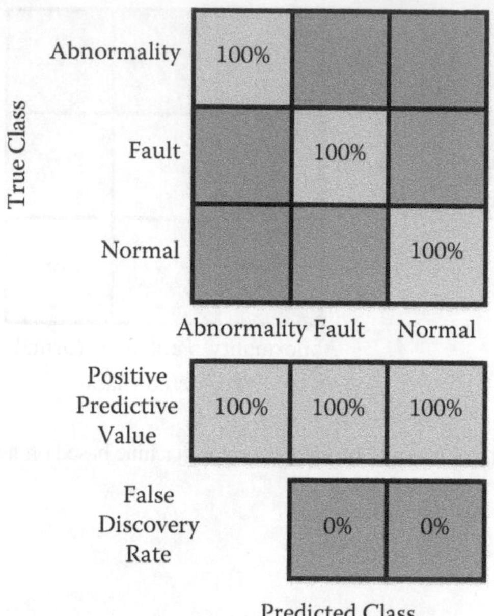

FIGURE 18.12 Confusion matrix of support vector machine based on positive predictive and false discovery.

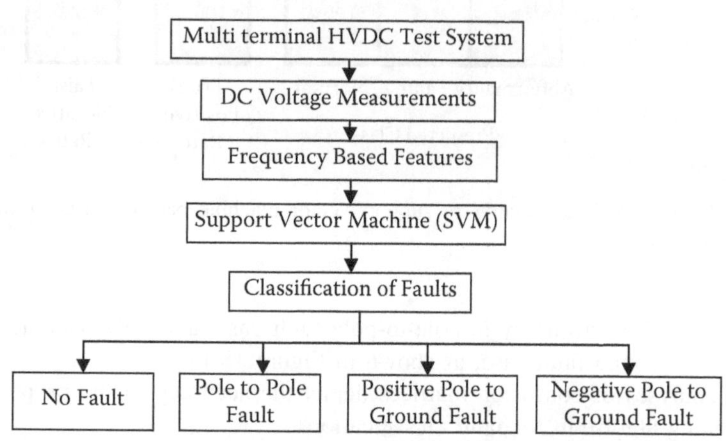

FIGURE 18.13 Flowchart of support vector machine for fault classification in test system.

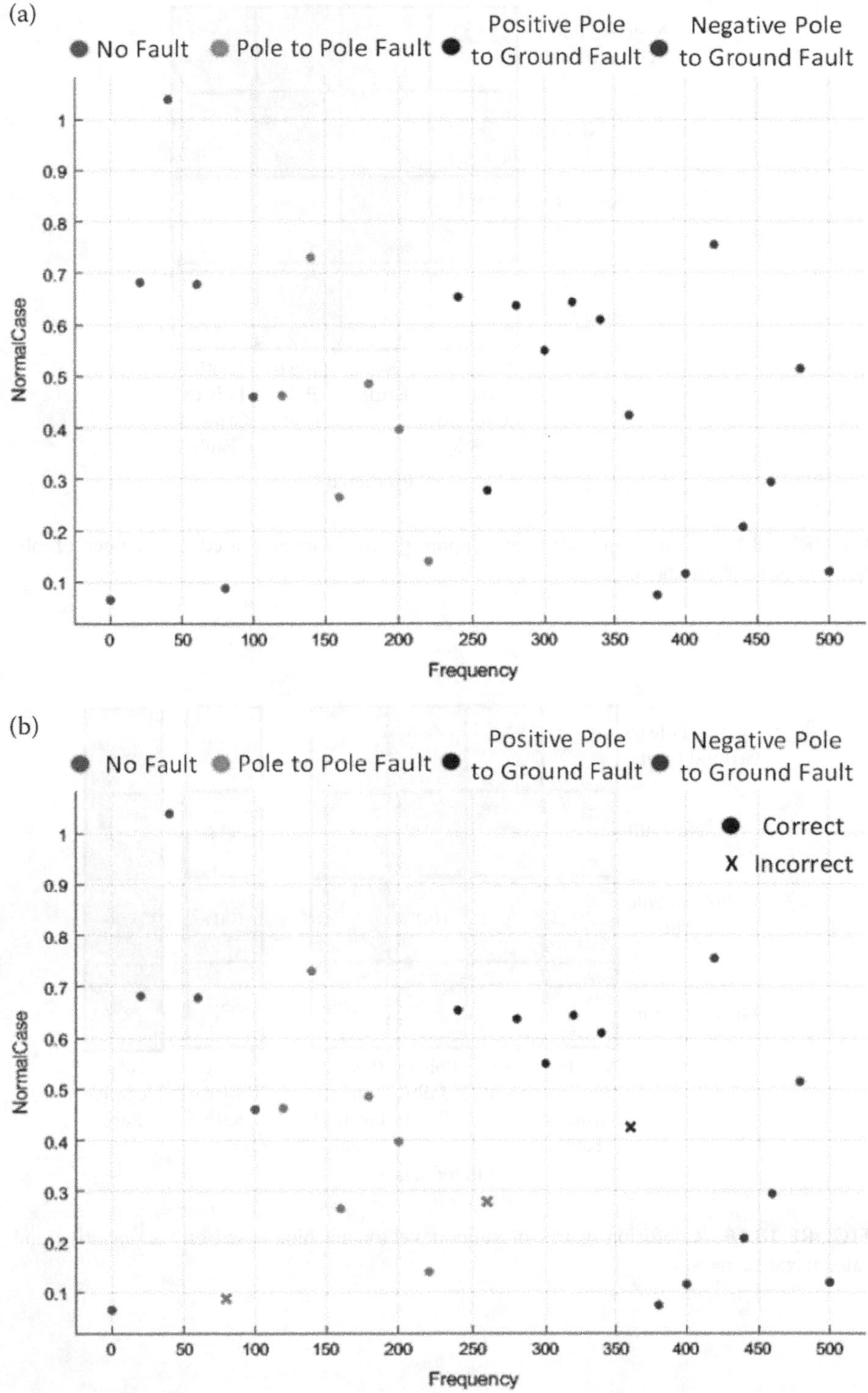

(a)

No Fault
Pole to Pole Fault
Positive Pole to Ground Fault
Negative Pole to Ground Fault

NormalCase
1
0.9
0.8
0.7
0.6
0.5
0.4
0.3
0.2
0.1
0 50 100 150 200 250 300 350 400 450 500
Frequency

(b)

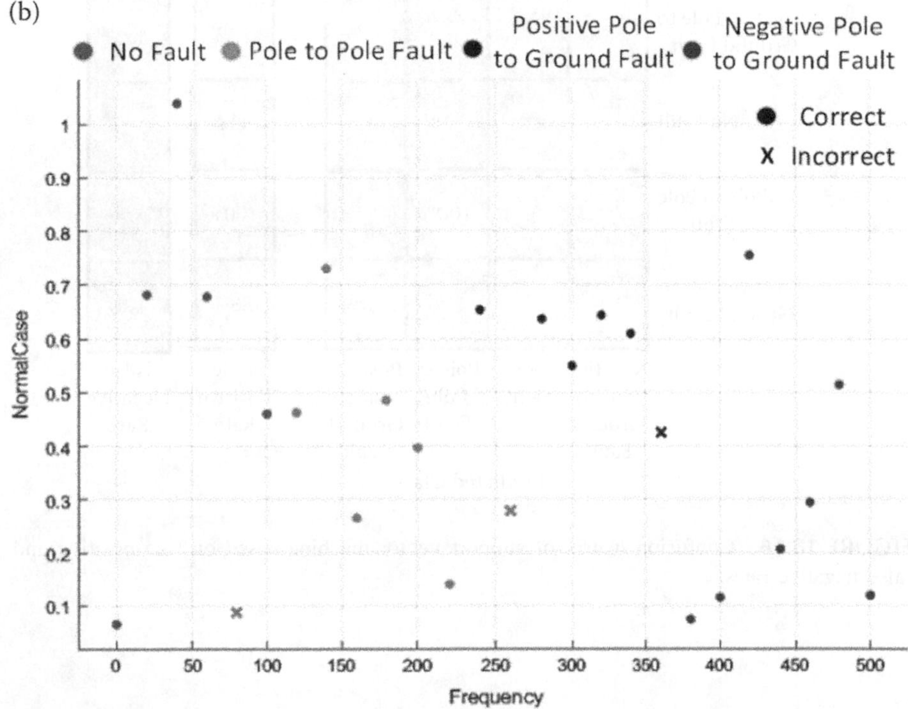

FIGURE 18.14 (a) Original and (b) predicted data set plotted for fault classification.

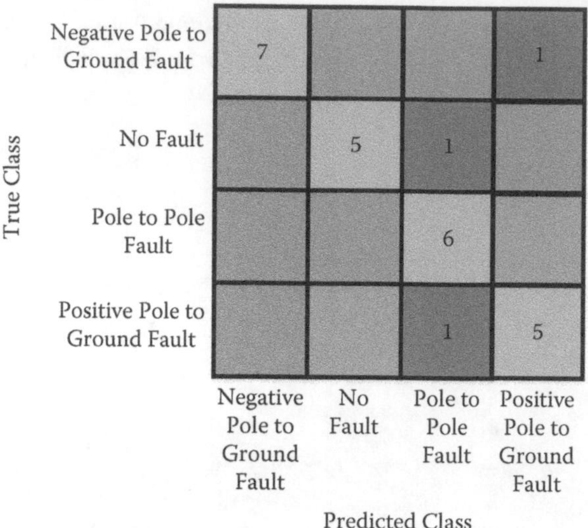

FIGURE 18.15 Confusion matrix of support vector machine based on number of observations for fault classification.

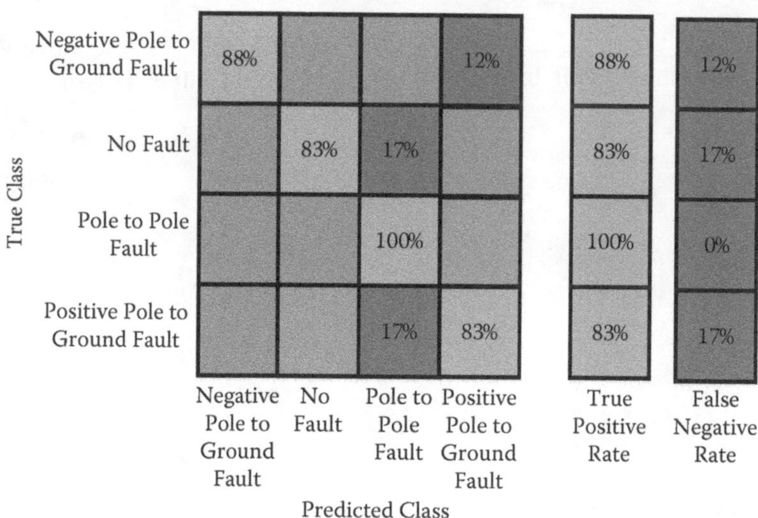

FIGURE 18.16 Confusion matrix of support vector machine based on true positive and false negative rates.

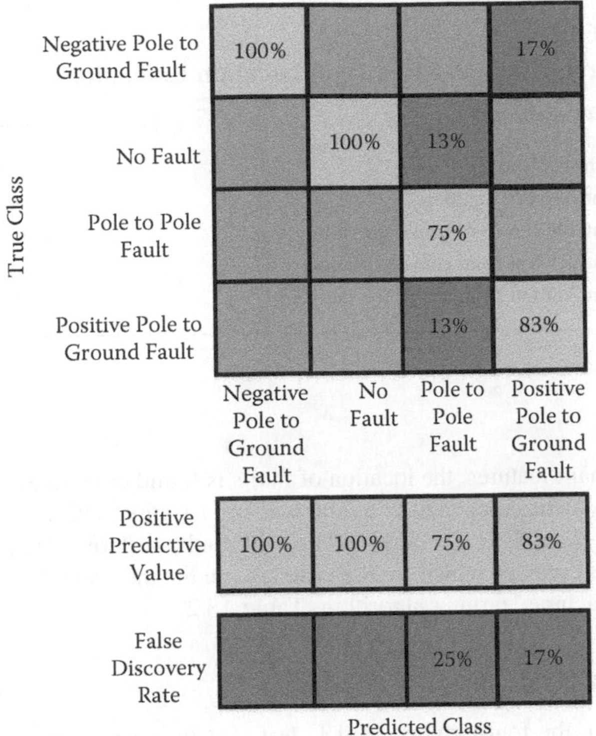

FIGURE 18.17 Confusion matrix of support vector machine based on positive predictive and false discovery.

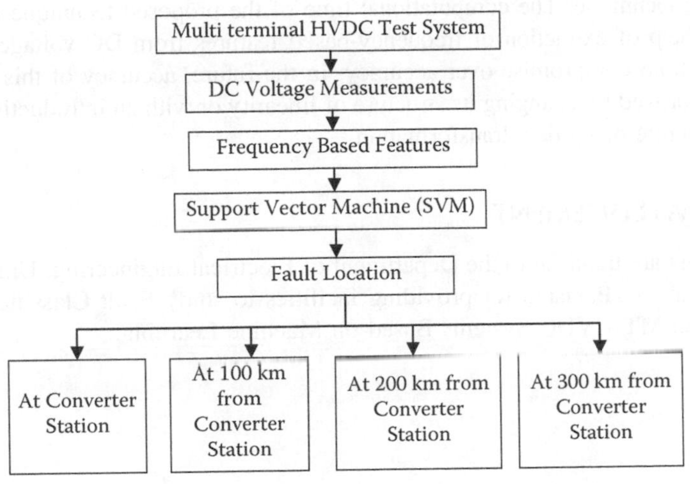

FIGURE 18.18 Flowchart of support vector machine for fault location in test system.

TABLE 18.2
Accuracy of SVM-Based Fault Location Estimation

Fault Location	Accuracy (%)
No fault	91.5
Fault at converter station	90.33
Fault at 100 km from converter station	90.33
Fault at 200 km from converter station	90.33
Fault at 300 km from converter station	90.33

frequency domain features, the location of faults is found out successfully. With the increase in the distance of fault from the converter station, DC voltage values will be decreased. Therefore, for any location of fault, different frequency domain features will be obtained which are then used to predict the accurate location of fault through the machine learning algorithm. Table 18.2

18.8 CONCLUSION

In this research, the four-terminal HVDC test system is analyzed and a fault diagnostic technique based on the support vector machine is developed and implemented. The fault diagnostic algorithm successfully carries out fault identification with 100% accuracy and fault classification with an accuracy of 88.5%. Faulty and normal states of the multi-terminal HVDC system are identified. Positive pole-to-ground faults, negative pole-to-ground faults, and pole-to-pole faults are classified with this fault diagnostic technique. The computational time of the proposed technique is reduced with the help of extraction of frequency-based features from DC voltage measurements with no compromise over accuracy. In the future, accuracy of this technique can be improved by changing its structure of linearity or with an introduction of non-linear structure or by data transformation.

ACKNOWLEDGEMENT

The authors are thankful to the Department of Electrical Engineering, University of Lahore, Lahore, Pakistan for providing facilities to study Fault Classification and Location in MT-HVDC Systems Based on Machine Learning.

REFERENCES

[1] D. Van Hertem, and M. Ghandhari, "Multi-terminal VSCHVDC for the European supergrid: Obstacles," *Renew. Sustain. Energy Rev.*, vol. 14, 3156–3163, 2010.

[2] O. Gomis-Bellmunt, J. Liang, J. Ekanayake, R. King R., and N. Jenkins, "Topologies of multi-terminal HVDC-VSC transmission for large offshore wind farms," *Electr. Power Syst. Res.*, vol. 81, pp. 271–281, 2011.

[3] K. De Kerf, K. Srivastava, M. Reza, D. Bekaert, S. Cole, D. Van Hertem, et al., "Wavelet based protection strategy for DC faults in multi-terminal VSC HVDC systems," *IET Gener. Transm. Distrib.*, vol. 5, no. 4, 496–503, 2011.

[4] M. K. Bucher, and C. M. Franck, "Contribution of fault current sources in multi-terminal HVDC cable networks," *IEEE Trans. Power Deliv.*, vol. 28, pp. 1796–1803, 2013.

[5] S. Liu, and M. Popov, "Development of HVDC system-level mechanical circuit breaker model," *Int. J. Elect. Power Energy Syst.*, vol. 103, pp. 159–167, 2018.

[6] L. Tang, and B. T. Ooi, "Locating and isolating DC faults in multi-terminal DC systems," *IEEE Trans. Power Deliv.*, vol. 22, pp. 1877–1884, 2007.

[7] J. Yang, J. E. Fletcher, and J. O'Reilly, "Multi-terminal DC wind farm collection grid internal fault analysis and protection design," *IEEE Trans. Power Deliv.*, vol. 25, pp. 2308–2318, 2010.

[8] J. Yang, J. E. Fletcher, and J. O'Reilly, "Short-circuit and ground fault analyses and location in VSC-based DC network cables," *IEEE Trans. Ind. Electron.*, vol. 59, pp. 3827–3837, 2012.

[9] J. Descloux, B. Raison, and J. B. Curis, "Protection strategy for undersea MTDC grids," in IEEE Grenoble Conf. PowerTech, Grenoble, 2013, pp. 1–6.

[10] S. Gao, Q. Liu, and G. Song, "Current differential protection principle of HVDC transmission system," *IET Gener., Transm. Distrib.* vol. 11, no. 5, pp. 1286–1292, 2017.

[11] G. Song, X. Chu, X. Cai, and S. Gao, "A novel pilot protection principle for VSC-HVDC cable lines based on fault component current," *Int. J. Electr. Power Energy Syst.*, vol. 53, pp. 426–433, 2013.

[12] P. Zhao, Q. Chen, and K. Sun, "A novel protection method for VSC-MTDC cable based on the transient DC current using the S transform," *Int. J. Electr. Power Energy Syst.*, vol. 97, pp. 299–308, 2018.

[13] S. P. Azad, and D. Van Hertem, "A fast local bus current-based primary relaying algorithm for HVDC grids," *IEEE Trans. Power Deliv.*, vol. 32, no. 1, pp. 193–202, 2017.

[14] J. Sneath, and A. D. Rajapakse, "Fault detection and interruption in an earthed HVDC grid using ROCOV and hybrid DC breakers," *IEEE Trans. Power Deliv.*, vol. 31, pp. 971–981, 2016.

[15] N. M. Haleem, and A. D. Rajapakse, "Local measurement based ultra-fast directional ROCOV scheme for protecting Bi-pole HVDC grids with a metallic return conductor," *Electr. Power Syst. Res.*, vol. 98, pp. 323–330, 2018.

[16] J. Wang, B. Berggren, K. Linden, J. Lan, and R. Nuqui, "Multi-terminal DC system line protection requirement and high speed protection solutions," in Proceedings of the 2015 CIGRE Symposium, Cape Town, South Africa, pp. 1–9, 2015.

[17] W. Leterme, J. Beerten, and D. Van Hertem, "Nonunit protection of HVDC grids with inductive DC cable termination," *IEEE Trans. Power Deliv.* vol. 31, pp. 820–828, 2016.

[18] J. Liu, N. Tai, and C. Fan, "Transient-voltage-based protection scheme for DC line faults in the multi-terminal VSC-HVDC system," *IEEE Trans. Power Deliv.*, vol. 32, pp. 1483–1494, 2017.

[19] R. Li, L. Xu, and L. Yao, "DC fault detection and location in meshed multi-terminal HVDC systems based on DC reactor voltage change rate," *IEEE Trans. Power Deliv.*, vol. 32, pp. 1516–1526, 2017.

[20] Q. H. Wu, Z. Lu, and T. Y. Ji, "Protective relaying of power systems using mathematical morphology," *Power Syst.*, vol. 45, pp. 1–208, 2009.

[21] R. Muzzammel, "Traveling waves-based method for fault estimation in HVDC transmission system," *Energies*, vol. 12, Art. no. 3614, 2019.

[22] R. Muzzammel, A. Raza, M. R. Hussain, G. Abbas, I. Ahmed, M. Qayyum, M. A. Rasool, and M. A. Khaleel, "MT–HVdc systems fault classification and location methods based on traveling and non-traveling waves—A comprehensive review," *Appl. Sci.*, vol. 9, Art. no. 4760, 2019.

[23] R. Ruiz-Gonzalez, J. Gomez-Gil, F. Gomez-Gil, and V. Martínez-Martínez, "An SVM-based classifier for estimating the state of various rotating components in agro-industrial machinery with a vibration signal acquired from a single point on the machine chassis," *Sensors*, vol. 14, no. 11, pp. 20713–20735, Nov. 2014.

[24] A. K. Jain, R. P. W. Duin, and J. Mao, "Statistical pattern recognition: A review," *IEEE Trans. Pattern Anal. Mach. Intell.*, vol. 22, pp. 4–37, 2000.

[25] S. B. Kotsiantis, "Supervised machine learning: A review of classification techniques," *Informatica*, vol. 31, pp. 249–268, 2007.

[26] V. Vapnik, *Statistical Learning Theory*. New York: Wiley, 1998.

[27] C. Cortes, V. Vapnik, "Support vector networks," *Mach. Learn.*, vol. 20, pp. 273–297, 1995.

[28] S. Haykin, "Neural networks, a comprehensive foundation," *International Journal of Neural Systems*, vol. 5, no. 4, pp. 363–364, 1994.

[29] S. Le Blond, R. Bertho, D. V. Coury, and J. C. M. Vieira, "Design of protection schemes for multi-terminal HVDC systems," *Renew. Sustain. Energy Rev.*, vol. 56, pp. 965–974, 2016.

[30] S. Tahir, J. Wang, M. H. Baloch, and G. S. Kaloi, "Digital control techniques based on voltage source inverters in renewable energy applications: A review," *Electronics*, vol. 7, pp. 1–18, 2018.

[31] A. J. Soares, M. A. O. Schroeder, and S. Visacro, "Transient voltages in transmission lines caused by direct lightning strikes," *IEEE Trans. Power Deliv.*, vol. 20, pp. 1447–1452, 2005.

[33] R. L. Sellick, and M. Akerberg, "Comparison of HVDC light (VSC) and HVDC classic (LCC) site aspects, for a 500 MW 400 kV HVDC transmission scheme," in Proceedings of the 10th IET International Conference on AC and DC Power Transmission, 2012, pp. 1–6.

[32] P. Kundur, N. Balu, *Power System Stability and Control*. New York, USA: McGraw-Hill Professional.

[34] A. Nabae, I. Takahashi, and H. Akagi, "A new neutral-point-clamped PWM inverter," *IEEE Trans. Ind. Appl.*, vol. IA-17, pp. 518–523, 1981.

[35] T. Bruckner, S. Bernet, and H. Guldner, "The active NPC converter and its loss balancing control," *IEEE Trans. Ind. Electron.*, vol. 52, pp. 855–868, 2005.

[36] L. Xu, and V. G. Agelidis, "VSC transmission system using flying capacitor multilevel converters and hybrid PWM control," *IEEE Trans. Power Deliv.*, vol. 22, pp. 693–702, 2007.

[37] B. T. Ooi, and X. Wang, "Boost-type PWM. HVDC transmission system," *IEEE Trans. Power Deliv.*, vol. 6, pp. 1557–1563, 1991.

[38] M. Guan, Z. Xu, and H. Chen, "Control and modulation strategies for modular multilevel converter based HVDC system," in Proceedings of the 37th Annual Conference of the IEEE Industrial Electronics Society, 2011, pp. 849–854.

[39] D. Soto-Sanchez, and T. C. Green, "Control of a modular multilevel converter-based HVDC transmission system," in Proceedings of the 14th European Conference on Power Electronics and Applications (EPE 2011), 2011, pp. 1–10.
[40] C. Chen, G. P. Adam, S. Finney, J. Fletcher, and B. Williams, "H-bridge modular multilevel converter: Control strategy for improved DC fault ride-through capability without converter blocking," *IET Power Electron.*, vol. 8, pp. 1996–2008, 2015.
[41] E. Kontos, R. T. Pinto, and P. Bauer, "Providing dc fault ride-through capability to H-bridge MMC-based HVDC networks," in Proceedings of the 9th International Conference on Power Electronics and ECCE Asia (ICPE-ECCE Asia), 2015, pp. 1542–1551.
[42] C. Bajracharya, and M. Molinas, "Control of VSC-HVDC for wind power [MSc dissertation]," Department of Electrical Power Engineering, Norwegian University of Science and Technology, Norway, 2008.
[43] C. Bajracharya, M. Molinas, J. A. Suul, and T. M. Undeland, "Understanding of tuning techniques of converter controllers for VSC-HVDC," Nordic Workshop on Power and Industrial Electronics, Helsinki University of Technology, Espoo, Finland, 2008, pp. 1–8.
[44] T. Noguchi, H. Tomiki, S. Kondo, and I. Takahashi, "Direct power control of PWM converter without power-source voltage sensors," *IEEE Trans. Ind. Appl.*, vol. 34, pp. 473–479, 1998.
[45] K. Rouzbehi, A. Miranian, A. Luna, and P. Rodriguez, "A generalized voltage droop strategy for control of multi-terminal DC grids," Proceedings of the 2013 IEEE Energy Conversion Congress and Exposition (ECCE), 2013, pp. 59–64.
[46] A.Raza, A. Shakeel, A. Altalbe, M. O. Alassafi, and A. R. Yasin, "Impacts of MT-HVDC systems on enhancing the power transmission capability," *Appl. Sci.*, vol. 10, no. 1, pp. 242, 2020.
[47] R. Muzzammel, "Machine learning based fault diagnosis in HVDC transmission lines," in *Intelligent Technologies and Applications, Communications in Computer and Information Science*, vol 932. Singapore: Springer, 2019.
[48] R. Muzzammel, "Restricted Boltzmann machines based fault estimation in multi terminal HVDC transmission," presented at the Intelligent Technologies and Applications, Bahawalpur, Pakistan, 2019, pp. 1–20.
[49] Q. Wang, Y. Yu, H. O. A. Ahmed, M. Darwish, and A. K. Nandi, "Fault Detection and classification in MMC-HVDC systems using learning methods," *Sensors*, vol. 20, no. 16, Art. no. 4438, Aug. 2020.
[50] R. Muzzammel, and A. Raza, "A support vector machine learning-based protection technique for MT-HVDC systems," *Energies*, vol. 13, Art. no. 6668, 2020.
[51] Y. Zhang, H. Hu, Z. Liu, M. Zhao, and L. Cheng, "Concurrent fault diagnosis of modular multilevel converter with Kalman filter and optimized support vector machine," *Syst. Sci. Control Eng.*, vol. 7, pp. 43–53, 2019.
[52] B. Zhu, H. Wang, S. Shi, and X. Dong, "Fault location in AC transmission lines with back-to-back MMC-HVDC using ConvNets," *J. Eng.*, vol. 2019, pp. 2430–2434, 2019.

Index

G

H

I

K

M

O

P

R

S

T

U

V

W